An Invitation to
Modern Number Theory

An Invitation to
Modern Number Theory

Steven J. Miller and
Ramin Takloo-Bighash

PRINCETON UNIVERSITY PRESS

PRINCETON AND OXFORD

In the United Kingdom: Princeton University Press, 3 Market Place, Woodstock, Oxfordshire OX20 1SY

Library of Congress Cataloging-in-Publication Data

Miller, Steven J.
    An invitation to modern number theory / Steven J. Miller and Ramin Takloo-Bighash.
       p. cm.
    Includes bibliographical references and index.
    ISBN-13: 978-0-691-12060-7 (cloth : alk. paper)
    ISBN-10: 0-691-12060-9 (cloth : alk. paper)
    1. Number theory. I. Takloo-Bighash, Ramin. II. Title.

QA241.M5344 2006
512.7—dc22   2005052165

British Library Cataloging-in-Publication Data is available

This book has been composed in Times Roman in LaTeX

The publisher would like to acknowledge the authors of this volume for providing the camera-ready copy from which this book was printed.

Printed on acid-free paper. ∞

pup.princeton.edu

Printed in the United States of America

10 9 8 7 6 5 4 3 2

To our wives, for their patience and constant encouragement, and to our students and colleagues, whose insights and exuberance made this project both possible and enjoyable.

Contents

Foreword

The beginning student of physics, chemistry, or computer science learns early on that in order to gain a proper understanding of the subject, one has to understand seemingly different topics and their relation to one another. While this is equally true in mathematics, this feature is not brought to the fore in most modern texts of pure mathematics. Different fields are usually presented as complete and isolated topics, and for the most part this is how it should be. However, modern mathematics, abstract as it might appear to the beginner, is driven by concrete and basic problems. In fact many of the different areas were developed in attempts, sometimes successful, to resolve such fundamental questions. Hence it should not be surprising that often the solution to a concrete long-standing problem involves combining different areas. This is especially true of modern number theory. The formulation of the problems is mostly elementary, and the expected truths were many times discovered by numerical experimentation.

In this lovely book the authors introduce number theory in terms of its connections to other fields of mathematics and its applications. While adhering to this theme, they also emphasize concrete problems (solved and unsolved). They develop the requisite mathematical background along the way to ensure proper and clear treatment of each of the many topics discussed. This allows the beginner to get an immediate taste of modern mathematics as well as of mathematical research. Naturally the treatments of various theories and theorems cannot be as complete as in books which are devoted to a single topic; however, through the indicated further reading, the many excellent exercises and the proposed research projects, the reader will get an excellent first understanding of the material.

Parts of this book have been used very successfully in undergraduate courses at the junior level at Princeton, New York University, Ohio State, and Brown. It will no doubt find similar success much more broadly, and it should appeal to both more- and less-advanced students. Covering most of the material in this book is a challenging task for both the student (or reader) and the instructor. My experience in co-teaching (with one of the authors) a version of part of the material is that this effort results in rich rewards for both the student and the instructor.

Peter Sarnak
Princeton, NJ
December 2005

Preface

This book on modern number theory grew out of undergraduate research seminars taught at Princeton University (2001–2003), and similar courses taught at New York University, Ohio State, Brown University and a summer Research Experience for Undergraduates at the American Institute of Mathematics. The purpose of these classes was to expose undergraduates to current research topics in mathematics. To supplement the standard lecture-homework classes, we wanted a course where students could work on outstanding conjectures and open problems and experience firsthand the kinds of problems mathematicians study. In the sciences and engineering, undergraduates are often exposed to state of the art problems in experimental laboratories. We want to bring a similar experience to students interested in mathematics. This book is the outcome of that effort, providing the novice with hints as to what we feel is a good path through the immense landscape of number theory, as well as the needed background material. We have tried to give students and their teachers a model which can be used to develop their own research program; to this end, throughout the book are detailed descriptions of accessible open problems and references to the literature. Though we encourage students and teachers to attempt some of the open problems, the book stands alone and may be used for a standard lecture course (especially for new subjects such as Random Matrix Theory where there are not many introductory works accessible to undergraduates). Our goal is to supplement the classic texts in the field by showing the connections between seemingly diverse topics, as well as making some of the subjects more accessible to beginning students and whetting their appetite for continuing in mathematics.

The book has five parts, though several themes run throughout the book.

- Part I deals with basic number theory (cryptography and basic group theory), elementary L-functions (including the connections between zeros of $\zeta(s)$ and primes), and solutions to Diophantine equations. The material in this part is fairly standard, and could serve as an introduction to number theory. In some sections a little group theory and first semester complex analysis is assumed for some advanced topics. Our purpose in the first chapter is not to write a treatise on cryptography, but to review some of the background necessary from basic number theory for later chapters. It is possible to *motivate* this material in the context of cryptography; though these applications are very important, this connection is meant only to interest the reader, as this is not a exposition on cryptography. Similarly, elliptic curves are a terrific example for some of the material in Chapter 4 (and later in the book); as such, we introduce just enough for these purposes. As there are numerous excellent books on both of these subjects, we have kept our treatments short and refer the interested reader to these for more details. One theme in these chapters is the search for efficient algorithms, which appears frequently in later parts as well.

- Part II has two connected themes: approximating numbers with rationals, and continued fractions. In the first, the basic properties of algebraic and transcendental numbers are discussed, and a proof of Roth's Theorem (on how well algebraic numbers can be approximated by rationals) is given in full detail. This is one of the great achievement of 20th century number theory. Roth's Theorem has now been greatly generalized, and there are a few different ways to prove it. Our formulation and proof follow Roth's original proof. The proof we present here, though long and technical, requires only knowledge of elementary calculus and linear algebra. The second part is an introduction to continued fractions (a subject of interest in its own right, but also of use in approximation theory) and culminates in several open problems; this chapter is independent of Roth's Theorem and may serve as a survey to the subject. Also, time and again (especially in Part III when we study digit bias and spacings between terms in certain sequences), we see that answers to many number theoretic questions depend on properties of the numbers in the problem; often the continued fraction expansion highlights these properties. There are references to open problems in continued fractions, many of which concern the distribution of digits (see Part III).

- Part III encompasses three themes. The first is the distribution of the first digit of several interesting sequences (for example, the Fibonacci numbers and iterates of the $3x + 1$ map). We use this problem as a motivation for hypothesis testing (whether or not numerical data supports or contradicts conjectured behavior). Hypothesis testing is an extremely important subject, especially as computers are used more and more frequently in mathematics. The second theme centers around the Gauss-Kuzmin Theorem for the distribution of digits of continued fractions. We then develop enough Fourier Analysis to prove various basic results, including a sketch of the proof of the Central Limit Theorem and Poisson Summation (one of the most used tools in number theory). We use these results to investigate the behavior of $n^k \alpha \bmod 1$ for fixed k, α (specifically, the spacings between these numbers in $[0, 1]$; for many k and α these spacings appear to be the same as the spacings between adjacent primes); we study other spacing problems in Part V; in fact, our results on the Fourier transform are needed in Chapter 18 when we investigate zeros of L-functions. Numerous open problems and references to the current literature are provided.

- Part IV is a brief introduction to the Circle Method, a powerful theory to study questions in additive number theory (such as writing a number as a sum of a fixed number of kth powers or primes). After developing the basics of the theory, we discuss in some detail why, using these methods, we cannot (yet?) show that any even number is the sum of two primes but we can show any large odd number is the sum of three primes. We use the Circle Method to predict how many Germain primes (p and $\frac{p-1}{2}$ both prime) are less than x. This example illustrates many of the key techniques of the theory, as well as the problems that arise in applications. Further, the density of these primes has recently been connected to fast primality testing algorithms. As usual we conclude with some open problems.

- Part V is an introduction to Random Matrix Theory and its interplay with number theory. What began as a model in the 1950s for physicists to study the energy levels of heavy nuclei has become a powerful tool after a chance encounter one day at tea in the 1970s (see [Ha] for an entertaining account of the meeting) for predicting the behavior of zeros of $\zeta(s)$ and other L-functions; knowledge of these zeros is intimately connected to properties of primes. The general result is that there is a striking similarity between the spacings between energy levels of heavy nuclei, eigenvalues of sets of matrices and zeros of L-functions. We take a classical approach to the subject. Results from linear algebra and occasionally first semester complex analysis are used (especially in the final chapter); a review of enough of the background material is provided for students to follow the key ideas in the proofs. There are numerous open problems requiring only elementary probability theory and linear algebra (at the level covered in this book); many have already been successfully investigated by our students.

There are several chapters throughout the book covering background material in basic number theory, algebra, Fourier analysis and probability theory, as well as two appendices on needed calculus, analysis and linear algebra results. Clearly our book is not meant to replace standard textbooks in these fields. We have two reasons for including these background chapters (in addition to the material being interesting in its own right). First, waiting for students to assemble such a background takes time, and the main purpose of our book is to show students in the early stages of their education what mathematicians do, and the interplay between the various parts of number theory and mathematics. Second, often very little of the background subjects is needed to understand the basic formulation and set-up of current work. Therefore a student who has not seen such material in a previous course can get a feel for these subjects by reading the review and background chapters, and then move on to the current research chapters. We have, however, written the chapters in such a way that there are often additional remarks or sections for students with stronger backgrounds. We have also included references throughout the book showing how the same methods and techniques are used for many different problems.

We have strived to keep the pre-requisites to a minimum: what is required is more a willingness to explore than a familiarity with the landscape. Several times we use results from later in the book in earlier investigations; our hope is that after seeing how these theorems are used and needed the reader will be motivated and interested enough to study the proofs. For most of the book one-variable calculus is the only requirement. We have also tried to emphasize common techniques in proofs (the reader is strongly encouraged to study the *techniques* entry in the index).

The book breaks naturally into five parts. Depending on the background of the students, and whether or not a class is going to explore open problems further, a typical semester class would cover material from one part of the book (as well as whatever background material is needed), though we recommend everyone at least skim Chapter 1 to ensure familiarity with the language and some of the motivating influences and themes of number theory. Many topics (such as applications to cryptography, algebraic structure of numbers and spacings between events) occur in various forms throughout the book. In a two semester course, one can cover two of the advanced parts and see these connections. We have also tried to give students the opportunity to discover the theory by themselves by giving many exercises. Mathematics is not meant to be a passive pursuit. Some of the problems are mere warm-ups; others are real problems that require time and effort. The reader should not be discouraged at being unable to work out all the problems. The value of an exercise is often in the time and energy spent on it, rather than the final solution. Many of the more difficult problems are standard theorems and can be seen proved in other textbooks. In this regard our manuscript is in the spirit of [Mu2]. **In Appendix C we have provided hints and further remarks to certain exercises; these problems are marked with either an (h) or (hr) in the text.**

We have assembled an extensive bibliography to aid the reader in further study. In addition to the excellent texts [AZ, Apo, BS, Da1, Da2, EE, Est2, Fe, HW, IR, IK, Kh, Kn, La2, Meh2, Na, NZM, ST, vdP6] on continued fractions, number theory and random matrix theory, we recommend the recent work of Narkiewicz [Nar] (where the reader will find proofs of many number theory results, as well as over 1800 references) as well as [Guy] (where there are extensive bibliographies for open problems). We conclude in Appendix D with some remarks on common themes running through this book and number theory.

The students in our courses used computers to assemble large amounts of data for some of the problems mentioned in the text, which then led us to appropriate conjectures and in some cases even gave us ideas on how to prove them. For links to previous student reports as well as some of the research papers mentioned in the bibliography, please visit

http://www.math.princeton.edu/mathlab/book/index.html

These include student programs (mostly in C++, Maple, Mathematica, MATLAB, or PARI) and detailed references for those interested in continuing these studies. Students should also consult MathSciNet [AMS], the arXiv [Cor1] and Project Euclid [Cor2] to find and download additional references.

It is a pleasure to thank the professors and teaching assistants who have helped run the class over the years (Alex Barnett, Vitaly Bergelson, João Boavida, Alexander Bufetov, Salman Butt, Brian Conrey, David Farmer, Harald Helfgott, Chris Hughes, James Mailhot, Atul Pokharel, Michael Rubinstein, Peter Sarnak, Lior Silberman, Yakov Sinai, Warren Sinnott, Florin Spinu and Andrew Wiles), as well as the students.

We would also like to thank several of our colleagues. In particular, we thank Eduardo Dueñez, Rob Gross and Amir Jafari for reviewing an early draft and providing numerous helpful suggestions to improve the presentation, and Timothy Abbot,

Mike Buchanan, Scott Craver, Kevin Dayaratna, Dean Eiger, Manfred Einsiedler, Dan File, Chris Hammond, Ted Hill, Alex Kontorovich, Josh Krantz, Matt Michelini, Jeff Miller, Liz Miller, Paria Mirmonsef, C. J. Mozzochi, Anna Pierrehumbert, Amitabha Roy, Zeév Rudnick, Eitan Sayag, Aaron Silberstein, Dan Stone, Howard Straubing, Yuri Tschinkel, Akshay Venkatesh and Bill Zaboski for discussions and comments on various chapters. The first author gave several lectures on the material to a summer research group and the Ross Program at Ohio State (summer 2004), and is indebted to the students for their comments and suggestions. We are grateful to Nicole, Michelle and Leo Beaupre, Andrew and David Norris, Joe Silverman and the staff at Princeton University Press for help with the illustrations, and to Stephen Kudla for mutually productive LaTeX discussions.

We are extremely grateful to Princeton University Press, especially to our editor Vickie Kearn, our production editor Lucy Day W. Hobor and our copyeditor Jennifer Slater, for all their help and aid, to Bob Gunning, for initiating contact between us and PUP and encouraging us to write the book, and to the National Science Foundation's VIGRE program, which helped fund many of the classes at Princeton, NYU and Ohio State.

The first author was partially supported by VIGRE post-doctoral fellowships at Princeton, New York University, The Ohio State University and Brown University, and enjoyed the hospitality of Boston University during the final stages of the project. The second author enjoyed the hospitality of the University of Maryland at College Park, Johns Hopkins University and The Ohio State University at various stages of working on the project. His work was partially supported by a Young Investigator's Award from the National Security Agency.

Steven J. Miller
Providence, RI
December 2005

Ramin Takloo-Bighash
Princeton, NJ
December 2005

Notation

$\mathbb{W}$: the set of whole numbers: $\{1, 2, 3, 4, \ldots\}$.

$\mathbb{N}$: the set of natural numbers: $\{0, 1, 2, 3, \ldots\}$.

$\mathbb{Z}$: the set of integers: $\{\ldots, -2, -1, 0, 1, 2, \ldots\}$.

$\mathbb{Q}$: the set of rational numbers: $\{x : x = \frac{p}{q}, p, q \in \mathbb{Z}, q \neq 0\}$.

$\mathbb{R}$: the set of real numbers.

$\mathbb{C}$: the set of complex numbers: $\{z : z = x + iy, \ x, y \in \mathbb{R}\}$.

$\Re z, \Im z$: the real and imaginary parts of $z \in \mathbb{C}$; if $z = x + iy$, $\Re z = x$ and $\Im z = y$.

$\mathbb{Z}/n\mathbb{Z}$: the additive group of integers mod n: $\{0, 1, \ldots, n-1\}$.

$(\mathbb{Z}/n\mathbb{Z})^*$: the multiplicative group of invertible elements mod n.

$\mathbb{F}_p$: the finite field with p elements: $\{0, 1, \ldots, p-1\}$.

$a|b$: a divides b.

$p^k||b$: p^k divides b and p^{k+1} does not divide b.

(a, b) : greatest common divisor (gcd) of a and b, also written $\gcd(a, b)$.

prime, composite : a positive integer a is prime if $a > 1$ and the only divisors of a are 1 and a; if $a > 1$ is not prime, we say a is composite.

coprime (relatively prime) : a and b are coprime (or relatively prime) if their greatest common divisor is 1.

$x \equiv y \mod n$: there exists an integer a such that $x = y + an$.

$\forall$: for all.

$\exists$: there exists.

Big-Oh notation : $A(x) = O(B(x))$, read "$A(x)$ is of order (or big-Oh) $B(x)$",
means $\exists C > 0$ and an x_0 such that $\forall x \geq x_0$, $|A(x)| \leq C\,B(x)$. This is also
written $A(x) \ll B(x)$ or $B(x) \gg A(x)$.

Little-Oh notation : $A(x) = o(B(x))$, read "$A(x)$ is little-Oh of $B(x)$", means
$\lim_{x \to \infty} A(x)/B(x) = 0$.

$|S|$ or $\#S$: number of elements in the set S.

p : usually a prime number.

i, j, k, m, n : usually an integer.

$[x]$ or $\lfloor x \rfloor$: the greatest integer less than or equal to x, read "the floor of x".

$\{x\}$: the fractional part of x; note $x = [x] + \{x\}$.

supremum : given a sequence $\{x_n\}_{n=1}^{\infty}$, the supremum of the set, denoted $\sup_n x_n$,
is the smallest number c (if one exists) such that $x_n \leq c$ for all n, and for any $\epsilon > 0$
there is some n_0 such that $x_{n_0} > c - \epsilon$. If the sequence has finitely many terms,
the supremum is the same as the maximum value.

infimum : notation as above, the infimum of a set, denoted $\inf_n x_n$, is the largest
number c (if one exists) such that $x_n \geq c$ for all n, and for any $\epsilon > 0$ there is some
n_0 such that $x_{n_0} < c + \epsilon$. If the sequence has finitely many terms, the infimum is
the same as the minimum value.

$\square$: indicates the end of a proof.

PART 1
Basic Number Theory

Chapter One

Mod p Arithmetic, Group Theory and Cryptography

In this chapter we review the basic number theory and group theory which we use throughout the book, culminating with a proof of quadratic reciprocity. Good introductions to group theory are [J, La3]; see [Da1, IR] for excellent expositions on congruences and quadratic reciprocity, and [Sil2] for a friendly introduction to much of the material below. We use cryptographic applications to motivate some basic background material in number theory; see [Ga] for a more detailed exposition on cryptography and [Lidl, vdP2] for connections with continued fractions. The guiding principle behind much of this chapter (indeed, much of this book and number theory) is the search for efficient algorithms. Just being able to write down an expression does not mean we can evaluate it in a reasonable amount of time. Thus, while it is often easy to prove a solution exists, doing the computations as written is sometimes impractical; see Chapter 6 of [BB] and [Wilf] for more on efficient algorithms.

1.1 CRYPTOGRAPHY

Cryptography is the science of encoding information so that only certain specified people can decode it. We describe some common systems. To prove many of the properties of these crypto-systems will lead us to some of the basic concepts and theorems of algebra and group theory.

Consider the following two password systems. In the first we choose two large distinct primes p and q; for example, let us say p and q have about 200 digits each. Let $N = pq$ and display the 400 digit number N for everyone to see. The password is any divisor of N greater than 1 and less than N. One very important property of the integers is unique factorization: any integer can be written uniquely as a product of prime powers. This implies that the only factorizations of N are $1 \cdot N$, $N \cdot 1$, $p \cdot q$ and $q \cdot p$. Thus there are two passwords, p and q. For the second system, we choose a 5000 digit number. We keep this number secret; to gain access the user must input this number.

Which method is more secure? While it is harder to correctly guess 5000 digits then 200, there is a danger in the second system: the computer needs to store the password. As there is no structure to the problem, the computer can only determine if you have entered the correct number by comparing your 5000 digit number to the one it was told is the password. Thus there is a code-book of sorts, and code-books can be stolen. In the first system there is no code-book to steal. The computer does not need to know p or q: it only needs to know N and how to divide, and it will know the password when it sees it!

There are so many primes that it is not practical to try all 200 digit prime numbers. The Prime Number Theorem (Theorem 2.3.7) states that there are approximately $\frac{x}{\log x}$ primes smaller than x; for $x = 10^{200}$, this leads to an impractically large number of numbers to check. What we have is a process which is easy in one direction (multiplying p and q), but hard in the reverse (knowing N, right now there is no "fast" algorithm to find p and q).

It is trivial to write an algorithm which is guaranteed to factor N: simply test N by all numbers (or all primes) at most $\sqrt{N}$. While this will surely work, this algorithm is so inefficient that it is useless for such large numbers. This is the first of many instances where we have an algorithm which will give a solution, but the algorithm is so slow as to be impractical for applications. Later in this chapter we shall encounter other situations where we have an initial algorithm that is too slow but where we can derive faster algorithms.

Exercise 1.1.1. *There are approximately 10^{80} elementary objects in the universe (photons, quarks, et cetera). Assume each such object is a powerful supercomputer capable of checking 10^{20} numbers a second. How many years would it take to check all numbers (or all primes) less than $\sqrt{10^{400}}$? What if each object in the universe was a universe in itself, with 10^{80} supercomputers: how many years would it take now?*

Exercise 1.1.2. *Why do we want p and q to be distinct primes in the first system?*

One of the most famous cryptography methods is RSA (see [RSA]). Two people, usually named Alice and Bob, want to communicate in secret. Instead of sending words they send numbers that represent words. Let us represent the letter a by 01, b by 02, all the way to representing z by 26 (and we can have numbers represent capital letters, spaces, punctuation marks, and so on). For example, we write 030120 for the word "cat." Thus it suffices to find a secure way for Alice to transmit numbers to Bob. Let us say a message is a number M of a fixed number of digits.

Bob chooses two large primes p and q and then two numbers d and e such that $(p-1)(q-1)$ divides $ed - 1$; we explain these choices in §1.5. Bob then makes publicly available the following information: $N = pq$ and e, but keeps secret p, q and d. It turns out that this allows Alice to send messages to Bob that only Bob can easily decipher. If Alice wants to send the message $M < N$ to Bob, Alice first calculates M^e, and then sends Bob the remainder after dividing by N; call this number X. Bob then calculates X^d, whose remainder upon dividing by N is the original message M! The proof of this uses modulo (or clock) arithmetic and basic group theory, which we describe below. Afterwards, we return and prove the claim.

Exercise 1.1.3. *Let $p = 101$, $q = 97$. Let $d = 2807$ and $e = 23$. Show that this method successfully sends "hi" (0809) to Bob. Note that $(0809)^{23}$ is a sixty-six digit number! See Remark 9.5.6 for one way to handle such large numbers.*

Exercise[hr] **1.1.4.** *Use a quadratic polynomial $ax^2 + bx + c$ to design a security system satisfying the following constraints:*

1. the password is the triple (a, b, c);

2. *each of* 10 *people is given some information such that any three of them can provide* (a, b, c), *but no two of them can.*

Generalize the construction: consider a polynomial of degree N such that some people "know more" than others (for example, one person can figure out the password with anyone else, another person just needs two people, and so on).

Remark 1.1.5. We shall see another important application of unique factorization in §3.1.1 when we introduce the Riemann zeta function. Originally defined as an infinite sum over the integers, by unique factorization we shall be able to express it as a product over primes; this interplay yields numerous results, among them a proof of the Prime Number Theorem.

1.2 EFFICIENT ALGORITHMS

For computational purposes, often having an algorithm to compute a quantity is not enough; we need an algorithm which will compute it *quickly*. We have seen an example of this when we tried to factor numbers; while we can factor any number, current algorithms are so slow that crypto-systems based on "large" primes are secure. For another example, recall Exercise 1.1.3 where we needed to compute a sixty-six digit number! Below we study three standard problems and show how to either rearrange the operations more efficiently or give a more efficient algorithm than the obvious candidate. See Chapter 6 of [BB] and [Wilf] for more on efficient algorithms.

1.2.1 Exponentiation

Consider x^n. The obvious way to calculate it involves $n - 1$ multiplications. By writing n in base two we can evaluate x^n in at most $2 \log_2 n$ steps, an enormous savings. One immediate application is to reduce the number of multiplications in cryptography (see Exercise 1.1.3). Another is in §1.2.33, where we derive a primality test based on exponentiation.

We are used to writing numbers in base 10, say

$$x = a_m 10^m + a_{m-1} 10^{m-1} + \cdots + a_1 10^1 + a_0, \quad a_i \in \{1, 2, 3, 4, 5, 6, 7, 8, 9\}. \tag{1.1}$$

Base two is similar, except each digit is now either 0 or 1. Let k be the largest integer such that $2^k \le x$. Then

$$x = b_k 2^k + b_{k-1} 2^{k-1} + \cdots + b_1 2 + b_0, \quad b_i \in \{0, 1\}. \tag{1.2}$$

It costs k multiplications to evaluate x^{2^i} for all $i \le k$. How? Consider $y_0 = x^{2^0}$, $y_1 = y_0 \cdot y_0 = x^{2^0} \cdot x^{2^0} = x^{2^1}$, $y_2 = y_1 \cdot y_1 = x^{2^2}, \ldots, y_k = y_{k-1} \cdot y_{k-1} = x^{2^k}$.

To evaluate x^n, note

$$
\begin{aligned}
x^n &= x^{b_k 2^k + b_{k-1} 2^{k-1} + \cdots + b_1 2 + b_0} \\
&= x^{b_k 2^k} \cdot x^{b_{k-1} 2^{k-1}} \cdots x^{b_1 2} \cdot x^{b_0} \\
&= \left(x^{2^k} \right)^{b_k} \cdot \left(x^{2^{k-1}} \right)^{b_{k-1}} \cdots \left(x^2 \right)^{b_1} \cdot \left(x^1 \right)^{b_0} \\
&= y_k^{b_k} \cdot y_{k-1}^{b_{k-1}} \cdots y_1^{b_1} \cdot y_0^{b_0}.
\end{aligned}
\tag{1.3}
$$

As each $b_i \in \{0,1\}$, we have at most $k + 1$ multiplications above (if $b_i = 1$ we have the term y_i in the product, if $b_i = 0$ we do not). It costs k multiplications to evaluate the x^{2^i} ($i \le k$), and at most another k multiplications to finish calculating x^n. As $k \le \log_2 n$, we see that x^n can be determined in at most $2\log_2 n$ steps. Note, however, that we do need more storage space for this method, as we need to store the values $y_i = x^{2^i}$, $i \le \log_2 n$. For n large, $2\log_2 n$ is much smaller than $n - 1$, meaning there is enormous savings in determining x^n this way. See also Exercise B.1.13.

Exercise 1.2.1. *Show that it is possible to calculate x^n storing only two numbers at any given time (and knowing the base two expansion of n).*

Exercise 1.2.2. *Instead of expanding n in base two, expand n in base three. How many calculations are needed to evaluate x^n this way? Why is it preferable to expand in base two rather than any other base?*

Exercise 1.2.3. *A better measure of computational complexity is not to treat all multiplications and additions equally, but rather to count the number of digit operations. For example, in 271×31 there are six multiplications. We then must add two three-digit numbers, which involves at most four additions (if we need to carry). How many digit operations are required to compute x^n?*

1.2.2 Polynomial Evaluation (Horner's Algorithm)

Let $f(x) = a_n x^n + a_{n-1} x^{n-1} + \cdots + a_1 x + a_0$. The obvious way to evaluate $f(x)$ is to calculate x^n and multiply by a_n (n multiplications), calculate x^{n-1} and multiply by a_{n-1} ($n - 1$ multiplications) and add, et cetera. There are n additions and $\sum_{k=0}^{n} k$ multiplications, for a total of $n + \frac{n(n+1)}{2}$ operations. Thus the standard method leads to about $\frac{n^2}{2}$ computations.

Exercise 1.2.4. *Prove by induction (see Appendix A.1) that $\sum_{k=0}^{n} k = \frac{n(n+1)}{2}$. In general, $\sum_{k=0}^{n} k^d = p_{d+1}(n)$, where $p_{d+1}(n)$ is a polynomial of degree $d + 1$ with leading term $\frac{n^{d+1}}{d+1}$; one can find the coefficients by evaluating the sums for $n = 0, 1, \ldots, d$ because specifying the values of a polynomial of degree d at $d + 1$ points uniquely determines the polynomial (see also Exercise 1.1.4). See [Mil4] for an alternate proof which does not use induction.*

Exercise 1.2.5. *Notation as in Exercise 1.2.4, use the integral test from calculus to show the leading term of $p_{d+1}(n)$ is $\frac{n^{d+1}}{d+1}$ and bound the size of the error.*

Exercise 1.2.6. *How many operations are required if we use our results on exponentiation?*

Consider the following grouping to evaluate $f(x)$, known as **Horner's algorithm**:

$$(\cdots((a_n x + a_{n-1})x + a_{n-2}) x + \cdots + a_1) x + a_0. \tag{1.4}$$

For example,

$$7x^4 + 4x^3 - 3x^2 - 11x + 2 = (((7x + 4)x - 3) x - 11) x + 2. \tag{1.5}$$

Evaluating term by term takes 14 steps; Horner's Algorithm takes 8 steps. One common application is in fractal geometry, where one needs to iterate polynomials (see also §1.2.4 and the references there). Another application is in determining decimal expansions of numbers (see §7.1).

Exercise 1.2.7. *Prove Horner's Algorithm takes at most $2n$ steps to evaluate $a_n x^n + \cdots + a_0$.*

1.2.3 Euclidean Algorithm

Definition 1.2.8 (Greatest Common Divisor). *Let $x, y \in \mathbb{N}$. The greatest common divisor of x and y, denoted by $\gcd(x, y)$ or (x, y), is the largest integer which divides both x and y.*

Definition 1.2.9 (Relatively Prime, Coprime). *If for integers x and y, $\gcd(x, y) = 1$, we say x and y are relatively prime (or coprime).*

The **Euclidean algorithm** is an efficient way to determine the greatest common divisor of x and y. Without loss of generality, assume $1 < x < y$. The obvious way to determine $\gcd(x, y)$ is to divide x and y by all positive integers up to x. This takes at most $2x$ steps; we show a more efficient way, taking at most about $2 \log_2 x$ steps.

Let $[z]$ denote the **greatest integer** less than or equal to z. We write

$$y = \left[\frac{y}{x}\right] \cdot x + r_1, \quad 0 \le r_1 < x. \tag{1.6}$$

Exercise 1.2.10. *Prove that $r_1 \in \{0, 1, \ldots, x - 1\}$.*

Exercise 1.2.11. *Prove $\gcd(x, y) = \gcd(r_1, x)$.*

We proceed in this manner until r_k equals zero or one. As each execution results in $r_i < r_{i-1}$, we proceed at most x times (although later we prove we need to apply these steps at most about $2 \log_2 x$ times).

$$x = \left\lfloor \frac{x}{r_1} \right\rfloor \cdot r_1 + r_2, \ 0 \le r_2 < r_1$$

$$r_1 = \left\lfloor \frac{r_1}{r_2} \right\rfloor \cdot r_2 + r_3, \ 0 \le r_3 < r_2$$

$$r_2 = \left\lfloor \frac{r_2}{r_3} \right\rfloor \cdot r_3 + r_4, \ 0 \le r_4 < r_3$$

$$\vdots$$

$$r_{k-2} = \left\lfloor \frac{r_{k-2}}{r_{k-1}} \right\rfloor \cdot r_{k-1} + r_k, \ 0 \le r_k < r_{k-1}. \tag{1.7}$$

Exercise 1.2.12. *Prove that if $r_k = 0$ then $\gcd(x, y) = r_{k-1}$, while if $r_k = 1$, then $\gcd(x, y) = 1$.*

We now analyze how large k can be. The key observation is the following:

Lemma 1.2.13. *Consider three adjacent remainders in the expansion: r_{i-1}, r_i and r_{i+1} (where $y = r_{-1}$ and $x = r_0$). Then $\gcd(r_i, r_{i-1}) = \gcd(r_{i+1}, r_i)$, and $r_{i+1} < \frac{r_{i-1}}{2}$.*

Proof. We have the following relation:

$$r_{i-1} = \left\lfloor \frac{r_{i-1}}{r_i} \right\rfloor \cdot r_i + r_{i+1}, \ 0 \le r_{i+1} < r_i. \tag{1.8}$$

If $r_i \le \frac{r_{i-1}}{2}$ then as $r_{i+1} < r_i$ we immediately conclude that $r_{i+1} < \frac{r_{i-1}}{2}$. If $r_i > \frac{r_{i-1}}{2}$, then we note that

$$r_{i+1} = r_{i-1} - \left\lfloor \frac{r_{i-1}}{r_i} \right\rfloor \cdot r_i. \tag{1.9}$$

Our assumptions on r_{i-1} and r_i imply that $\left\lfloor \frac{r_{i-1}}{r_i} \right\rfloor = 1$. Thus $r_{i+1} < \frac{r_{i-1}}{2}$. □

We count how often we apply these steps. Going from $(x, y) = (r_0, r_{-1})$ to (r_1, r_0) costs one application. Every two applications gives three pairs, say (r_{i-1}, r_{i-2}), (r_i, r_{i-1}) and (r_{i+1}, r_i), with r_{i+1} at most half of r_{i-1}. Thus if k is the largest integer such that $2^k \le x$, we see have at most $1 + 2k \le 1 + 2 \log_2 x$ pairs. Each pair requires one integer division, where the remainder is the input for the next step. We have proven

Lemma 1.2.14. *Euclid's algorithm requires at most $1 + 2 \log_2 x$ divisions to find the greatest common divisor of x and y.*

Euclid's algorithm provides more information than just the $\gcd(x, y)$. Let us assume that $r_i = \gcd(x, y)$. The last equation before Euclid's algorithm terminated was

$$r_{i-2} = \left\lfloor \frac{r_{i-2}}{r_{i-1}} \right\rfloor \cdot r_{i-1} + r_i, \ 0 \le r_i < r_{i-1}. \tag{1.10}$$

Therefore we can find integers a_{i-1} and b_{i-2} such that

$$r_i = a_{i-1}r_{i-1} + b_{i-2}r_{i-2}. \tag{1.11}$$

We have written r_i as a linear combination of r_{i-2} and r_{i-1}. Looking at the second to last application of Euclid's algorithm, we find that there are integers a'_{i-2} and b'_{i-3} such that

$$r_{i-1} = a'_{i-2}r_{i-2} + b'_{i-3}r_{i-3}. \tag{1.12}$$

Substituting for r_{i-1} in the expansion of r_i yields that there are integers a_{i-2} and b_{i-3} such that

$$r_i = a_{i-2}r_{i-2} + b_{i-3}r_{i-3}. \tag{1.13}$$

Continuing by induction and recalling $r_i = \gcd(x, y)$ yields

Lemma 1.2.15. *There exist integers a and b such that $\gcd(x, y) = ax + by$. Moreover, Euclid's algorithm gives a constructive procedure to find a and b.*

Thus, not only does Euclid's algorithm show that a and b exist, it gives an efficient way to find them.

Exercise 1.2.16. *Find a and b such that $a \cdot 244 + b \cdot 313 = \gcd(244, 313)$.*

Exercise 1.2.17. *Add the details to complete an alternate, non-constructive proof of the existence of a and b with $ax + by = \gcd(x, y)$:*

1. *Let d be the smallest positive value attained by $ax + by$ as we vary $a, b \in \mathbb{Z}$. Such a d exists. Say $d = \alpha x + \beta y$.*

2. *Show $\gcd(x, y)|d$.*

3. *Let $e = Ax + By > 0$. Then $d|e$. Therefore for any choice of $A, B \in \mathbb{Z}$, $d|(Ax + By)$.*

4. *Consider $(a, b) = (1, 0)$ or $(0, 1)$, yielding $d|x$ and $d|y$. Therefore $d \leq \gcd(x, y)$. As we have shown $\gcd(x, y)|d$, this completes the proof.*

Note this is a non-constructive proof. By minimizing $ax + by$ we obtain $\gcd(x, y)$, but we have no idea how many steps are required. Prove that a solution will be found either among pairs (a, b) with $a \in \{1, \dots, y - 1\}$ and $-b \in \{1, \dots, x - 1\}$, or $-a \in \{1, \dots, y-1\}$ and $b \in \{1, \dots, x-1\}$. Choosing an object that is minimal in some sense (here the minimality comes from being the smallest integer attained as we vary a and b in $ax + by$) is a common technique; often this number has the desired properties. See the proof of Lemma 6.4.3 for an additional example of this method.

Exercise 1.2.18. *How many steps are required to find the greatest common divisor of $x_1, \dots, x_N$?*

Remark 1.2.19. In bounding the number of computations in the Euclidean algorithm, we looked at three adjacent remainders and showed that a desirable relation held. This is a common technique, where it can often be shown that at least one of several consecutive terms in a sequence has some good property; see also Theorem 7.9.4 for an application to continued fractions and approximating numbers.

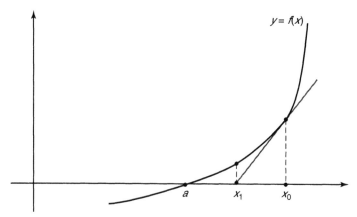

Figure 1.1 Newton's Method

1.2.4 Newton's Method and Combinatorics

We give some examples and exercises on efficient algorithms and efficient ways to arrange computations. The first assumes some familiarity with calculus, the second with basic combinatorics.

Newton's Method: Newton's Method is an algorithm to approximate solutions to $f(x) = 0$ for f a differentiable function on $\mathbb{R}$. It is much faster than the method of **Divide and Conquer** (see §A.2.1), which finds zeros by looking at sign changes of f, though this method is of enormous utility (see Remark 3.2.24 where Divide and Conquer is used to find zeros of the Riemann zeta function).

Start with x_0 such that $f(x_0)$ is small; we call x_0 the initial guess. Draw the tangent line to the graph of f at x_0, which is given by the equation

$$y - f(x_0) = f'(x_0) \cdot (x - x_0). \tag{1.14}$$

Let x_1 be the x-intercept of the tangent line; x_1 is the next guess for the root α. See Figure 1.1. Simple algebra gives

$$x_1 = x_0 - \frac{f(x_0)}{f'(x_0)}. \tag{1.15}$$

We now iterate and apply the above procedure to x_1, obtaining

$$x_2 = x_1 - \frac{f(x_1)}{f'(x_1)}. \tag{1.16}$$

If we let $g(x) = x - \frac{f(x)}{f'(x)}$, we notice we have the sequence

$$x_0, \; g(x_0), \; g(g(x_0)), \; \ldots \tag{1.17}$$

We hope this sequence will converge to the root, at least for x_0 close to the root and for f sufficiently nice. How close x_0 has to be is a delicate matter. If there are several roots to f, which root the sequence converges to depends crucially on the initial value x_0 and the function f. In fact its behavior is what is known technically as **chaotic**. Informally, we say that we have chaos when tiny changes in the initial value give us very palpable changes in the output. One common example is in iterates of polynomials, namely the limiting behavior of $f(x_0)$, $f(f(x_0))$, $f(f(f(x_0)))$ and so on; see [Dev, Edg, Fal, Man].

Exercise 1.2.20. *Let $f(x) = x^2 - a$ for some $a > 0$. Show Newton's Method converges to $\sqrt{a}$, and discuss the rate of convergence; i.e., if x_n is accurate to m digits, approximately how accurate is x_{n+1}? For example, look at $a = 3$ and $x_0 = 2$. Similarly, investigate $\sqrt[n]{a}$. Compare this with Divide and Conquer, where each iteration basically halves the error (so roughly every ten iterations yields three new decimal digits, because $\frac{1}{2^{10}} \approx \frac{1}{10^3}$).*

Remark 1.2.21. One big difference between Newton's Method and Divide and Conquer is that while both require us to evaluate the function, Newton's Method requires us to evaluate the derivative as well. Hence Newton's Method is not applicable to as wide of a class of functions as Divide and Conquer, but as it uses more information about f it is not surprising that it gives better results (i.e., converges faster to the answer).

Exercise 1.2.22. *Modify Newton's Method to find maxima and minima of functions. What must you assume about these functions to use Newton's method?*

Exercise 1.2.23. *Let $f(x)$ be a degree n polynomial with complex coefficients. By the Fundamental Theorem of Algebra, there are n (not necessarily distinct) roots. Assume there are m distinct roots. Assign m colors, one to each root. Given a point $x \in \mathbb{C}$, we color x with the color of the root that x approaches under Newton's Method (if it converges to a root). Write a computer program to color such sets for some simple polynomials, for example for $x^n - 1 = 0$ for $n = 2, 3$ or 4.*

Exercise 1.2.24. *Determine conditions on f, the root a and the starting guess x_0 such that Newton's Method will converge to the root. See page 212 of [BB] or page 118 of [Rud] for more details.*

Exercise[h] 1.2.25 (Fixed Points). *We say x_0 is a fixed point of a function h if $h(x_0) = x_0$. Let f be a continuously differentiable function. If we set $g(x) = x - \frac{f(x)}{f'(x)}$, show a fixed point of g corresponds to a solution to $f(x) = 0$.*
    Assume that $f : [a, b] \to [a, b]$ and there is a $C < 1$ such that $|f'(x)| < C$ for $x \in [a, b]$. Prove f has a fixed point in $[a, b]$. Is the result still true if we just assume $|f'(x)| < 1$? Fixed points have numerous applications, among them showing optimal strategies exist in n-player games. See [Fr] for more details.

Combinatorics: Below we describe a combinatorial problem which contains many common features of the subject. Assume we have 10 identical cookies and 5 distinct people. How many different ways can we divide the cookies among the people, such that all 10 cookies are distributed? Since the cookies are identical, we cannot tell which cookies a person receives; we can only tell how many. We could enumerate all possibilities: there are 5 ways to have one person receive 10 cookies, 20 ways to have one person receive 9 and another receive 1, and so on. While in principle we can solve the problem, in practice this computation becomes intractable, especially as the numbers of cookies and people increase.

We introduce common combinatorial functions. The first is the **factorial function**: for a positive integer n, set $n! = n \cdot (n-1) \cdots 2 \cdot 1$. The number of ways to choose r objects from n when order matters is $n \cdot (n-1) \cdots (n-(r-1)) = \frac{n!}{(n-r)!}$ (there are n ways to choose the first element, then $n-1$ ways to choose the second element, and so on). The **binomial coefficient** $\binom{n}{r} = \frac{n!}{r!(n-r)!}$ is the number of ways to choose r objects from n objects when order does not matter. The reason is that once we have chosen r objects there are $r!$ ways to order them. For convenience, we define $0! = 1$; thus $\binom{n}{0} = 1$, which may be interpreted as saying there is one way to choose zero elements from a set of n objects. For more on binomial coefficients, see §A.1.3.

We show the number of ways to divide 10 cookies among 5 people is $\binom{10+5-1}{5-1}$. In general, if there are C cookies and P people,

Lemma 1.2.26. *The number of distinct ways to divide C identical cookies among P different people is $\binom{C+P-1}{P-1}$.*

Proof. Consider $C + P - 1$ cookies in a line, and number them 1 to $C + P - 1$. Choose $P - 1$ cookies. There are $\binom{C+P-1}{P-1}$ ways to do this. This divides the cookies into P sets: all the cookies up to the first chosen (which gives the number of cookies the first person receives), all the cookies between the first chosen and the second chosen (which gives the number of cookies the second person receives), and so on. This divides C cookies among P people. Note different sets of $P - 1$ cookies correspond to different partitions of C cookies among P people, and every such partition can be associated to choosing $P - 1$ cookies as above.   □

Remark 1.2.27. In the above problem we do not care *which* cookies a person receives. We introduced the numbers for convenience: now cookies 1 through i_1 (say) are given to person 1, cookies $i_1 + 1$ through i_2 (say) are given to person 2, and so on.

For example, if we have 10 cookies and 5 people, say we choose cookies 3, 4, 7 and 13 of the $10 + 5 - 1$ cookies:

$$\odot \; \odot \; \otimes \; \otimes \; \odot \; \odot \; \otimes \; \odot \; \odot \; \odot \; \odot \; \odot \; \otimes \; \odot$$

This corresponds to person 1 receiving two cookies, person 2 receiving zero, person 3 receiving two, person 4 receiving five and person 5 receiving one cookie.

The above is an example of a partition problem: we are solving $x_1 + x_2 + x_3 + x_4 + x_5 = 10$, where x_i is the number of cookies person i receives. We may interpret Lemma 1.2.26 as the number of ways to divide an integer N into k non-negative integers is $\binom{N+k-1}{k-1}$.

Exercise 1.2.28. *Prove that*

$$\sum_{n=0}^{N} \binom{n+k-1}{k-1} = \binom{N+1+k-1}{k-1}. \tag{1.18}$$

We may interpret the above as dividing N cookies among k people, where we do not assume all cookies are distributed.

Exercise[h] **1.2.29.** *Let $\mathcal{M}$ be a set with $m > 0$ elements, $\mathcal{N}$ a set with $n > 0$ elements and $\mathcal{O}$ a set with $m + n$ elements. For $\ell \in \{0, \ldots, m + n\}$, prove*

$$\sum_{k=\max(0,\ell-n)}^{\min(m,\ell)} \binom{m}{k}\binom{n}{\ell - k} = \binom{m + n}{\ell}. \tag{1.19}$$

This may be interpreted as partitioning $\mathcal{O}$ into two sets, one of size ℓ.

In Chapter 13 we describe other partition problems, such as representing a number as a sum of primes or integer powers. For example, the famous Goldbach problem says any even number greater than 2 is the sum of two primes (known to be true for integers up to $6 \cdot 10^{16}$ [Ol]). While to date this problem has resisted solution, we have good heuristics which predict that, not only does a solution exist, but how many solutions there are. Computer searches have verified these predictions for large N of size 10^{10}.

Exercise 1.2.30 (Crude Prediction). *By the Prime Number Theorem, there are $\frac{N}{\log N}$ primes less than N. If we assume all numbers $n \leq N$ are prime with the same likelihood (a crude assumption), predict how many ways there are to write N as a sum of two primes.*

Exercise 1.2.31. *In partition problems, often there are requirements such as that everyone receives at least one cookie. How many ways are there to write N as a sum of k non-negative integers? How many solutions of $x_1 + x_2 + x_3 = 1701$ are there if each x_i is an integer and $x_1 \geq 2$, $x_2 \geq 4$, and $x_3 \geq 601$?*

Exercise 1.2.32. *In solving equations in integers, often slight changes in the coefficients can lead to wildly different behavior and very different sets of solutions. Determine the number of non-negative integer solutions to $x_1 + x_2 = 1996$, $2x_1 + 2x_2 = 1996$, $2x_1 + 2x_2 = 1997$, $2x_1 + 3x_2 = 1996$, $2x_1 + 2x_2 + 2x_3 + 2x_4 = 1996$ and $2x_1 + 2x_2 + 3x_3 + 3x_4 = 1996$. See Chapter 4 for more on finding integer solutions.*

Exercise[h] **1.2.33.** *Let f be a homogenous polynomial of degree d in n variables. This means*

$$f(x_1, \ldots, x_n) = \sum_{\substack{0 \leq k_1, \ldots, k_n \leq d \\ k_1 + \cdots + k_n = d}} a_{k_1, \ldots, k_n} x_1^{k_1} \cdots x_n^{k_n}, \quad a_{k_1, \ldots, k_n} x_1^{k_1} \in \mathbb{C}. \tag{1.20}$$

Prove for any $\lambda \in \mathbb{C}$ that

$$f(\lambda x_1, \ldots, \lambda x_n) = \lambda^d f(x_1, \ldots, x_n). \tag{1.21}$$

As a function of n and d, how many possible terms are there in f (each term is of the form $x_1^{k_1} \cdots x_n^{k_n}$)?

The above problems are a small set of interesting results in combinatorics; see also [Mil4] for other techniques to prove combinatorial identities. We give some additional problems which illustrate the subject; the Binomial Theorem (Theorem A.1.8) is useful for these and other investigations.

Exercise[h] **1.2.34.** *Let k be a positive integer and consider the sequence 1^k, 2^k, 3^k, ... (so $x_n = n^k$). Consider the new sequence obtained by subtracting adjacent terms: $2^k - 1^k$, $3^k - 2^k$, ... and so on. Continue forming new sequences by subtracting adjacent terms of the previous terms. Prove that each term of the kth sequence is $k!$.*

Exercise[hr] **1.2.35.** *Let k and d be positive integers. Prove*

$$k^d = \sum_{m=0}^{d-1} \sum_{\ell=0}^{k-1} \binom{d}{m} \ell^m. \tag{1.22}$$

1.3 CLOCK ARITHMETIC: ARITHMETIC MODULO n

Let $\mathbb{Z}$ denote the set of integers and for $n \in \mathbb{N}$ define $\mathbb{Z}/n\mathbb{Z} = \{0, 1, 2, \ldots, n-1\}$. We often read $\mathbb{Z}/n\mathbb{Z}$ as the **integers modulo** n.

Definition 1.3.1 (Congruence). *$x \equiv y \bmod n$ means $x - y$ is an integer multiple of n. Equivalently, x and y have the same remainder when divided by n.*

When there is no danger of confusion, we often drop the suffix mod n, writing instead $x \equiv y$.

Lemma 1.3.2 (Basic Properties of Congruences). *For a fixed $n \in \mathbb{N}$ and a, a', b, b' integers we have*

1. $a \equiv b \bmod n$ if and only if $b \equiv a \bmod n$.

2. $a \equiv b \bmod n$ and $b \equiv c \bmod n$ implies $a \equiv c \bmod n$.

3. $a \equiv a' \bmod n$ and $b \equiv b' \bmod n$, then $ab \equiv a'b' \bmod n$. In particular $a \equiv a' \bmod n$ implies $ab \equiv a'b \bmod n$ for all b.

Exercise 1.3.3. *Prove the above relations. If $ab \equiv cb \bmod m$, must $a \equiv c \bmod m$?*

For $x, y \in \mathbb{Z}/n\mathbb{Z}$, we define $x + y$ to be the unique number $z \in \mathbb{Z}/n\mathbb{Z}$ such that $n|(x + y - z)$. In other words, z is the unique number in $\mathbb{Z}/n\mathbb{Z}$ such that $x + y \equiv z \bmod n$. One can show that $\mathbb{Z}/n\mathbb{Z}$ is a finite group under addition; in fact, it is a finite ring. (See §1.4.1 for the definition of a group).

Exercise[h] **1.3.4** (Arithmetic Modulo n). *Define multiplication of $x, y \in \mathbb{Z}/n\mathbb{Z}$ by $x \cdot y$ is the unique $z \in \mathbb{Z}/n\mathbb{Z}$ such that $xy \equiv z \bmod n$. We often write xy for $x \cdot y$. Prove that this multiplication is well defined, and that an element x has a multiplicative inverse if and only if $(x, n) = 1$. Conclude that if every non-zero element of $\mathbb{Z}/n\mathbb{Z}$ has a multiplicative inverse, then n must be prime.*

Arithmetic modulo n is also called clock arithmetic. If $n = 12$ we have $\mathbb{Z}/12\mathbb{Z}$. If it is 10 o'clock now, in 5 hours it is 3 o'clock because $10 + 5 = 15 \equiv 3 \bmod 12$. See [Bob] for an analysis of the "randomness" of the inverse map in clock arithmetic.

Definition 1.3.5 (Least Common Multiple). *Let $m, n \in N$. The least common multiple of m and n, denoted by $\mathrm{lcm}(m, n)$, is the smallest positive integer divisible by both m and n.*

Exercise 1.3.6. *If $a \equiv b \bmod n$ and $a \equiv b \bmod m$, then $a \equiv b \bmod \mathrm{lcm}(m, n)$.*

Exercise 1.3.7. *Prove for all positive integers m, n that $\mathrm{lcm}(m, n) \cdot \gcd(m, n) = mn$.*

Are there integer solutions to the equation $2x + 1 = 2y$? The left hand side is always odd, the right hand side is always even. Thus there are no integer solutions. What we did is really arithmetic modulo 2 or arithmetic in $\mathbb{Z}/2\mathbb{Z}$, and indicates the power of congruence arguments.

Consider now $x^2 + y^2 + z^2 = 8n + 7$. This never has integer solutions. Let us study this equation modulo 8. The right hand side is 7 modulo 8. What are the squares modulo 8? They are $1^2 \equiv 1, 2^2 \equiv 4, 3^2 \equiv 1, 4^2 \equiv 0$, and then the pattern repeats (as modulo 8, k and $(8 - k)$ have the same square). We see there is no way to add three squares and get 7. Thus there are no solutions to $x^2 + y^2 + z^2 = 8n + 7$.

Remark 1.3.8 (Hasse Principle). In general, when searching for integer solutions one often tries to solve the equation modulo different primes. If there is no solution for some prime, then there are no integer solutions. Unfortunately, the converse is not true. For example, Selmer showed $3x^3 + 4y^3 + 5z^3 = 0$ is solvable modulo p for all p, but there are no rational solutions. We discuss this in more detail in Chapter 4.

Exercise 1.3.9 (Divisibility Rules). *Prove a number is divisible by 3 (or 9) if and only if the sum of its digits are divisible by 3 (or 9). Prove a number is divisible by 11 if and only if the alternating sum of its digits is divisible by 11 (for example, 341 yields 3-4+1). Find a rule for divisibility by 7.*

Exercise 1.3.10 (Chinese Remainder Theorem). *Let m_1, m_2 be relatively prime positive integers. Prove that for any $a_1, a_2 \in \mathbb{Z}$ there exists a unique $x \bmod m_1 m_2$ such that $x \equiv a_1 \bmod m_1$ and $x \equiv a_2 \bmod m_2$. Is this still true if m_1 and m_2 are not relatively prime? Generalize to $m_1, \ldots, m_k$ and $a_1, \ldots, a_k$.*

1.4 GROUP THEORY

We introduce enough group theory to prove our assertions about RSA. For more details, see [Art, J, La3].

1.4.1 Definition

Definition 1.4.1 (Group). *A set G equipped with a map $G \times G \to G$ (denoted by $(x, y) \mapsto xy$) is a group if*

1. *(Identity) $\exists e \in G$ such that $\forall x \in G$, $ex = xe = x$.*

2. *(Associativity)* $\forall x, y, z \in G$, $(xy)z = x(yz)$.

3. *(Inverse)* $\forall x \in G, \exists y \in G$ such that $xy = yx = e$.

4. *(Closure)* $\forall x, y \in G$, $xy \in G$.

We have written the group multiplicatively, $(x, y) \mapsto xy$; if we wrote $(x, y) \mapsto x + y$, we say the group is written additively. We call G a finite group if the set G is finite. If $\forall x, y \in G$, $xy = yx$, we say the group is **abelian** or **commutative**.

Exercise 1.4.2. *Show that under addition $\mathbb{Z}/n\mathbb{Z}$ is an abelian group.*

Exercise 1.4.3. *Consider the set of $N \times N$ matrices with real entries and non-zero determinant. Prove this is a group under matrix multiplication, and show this group is not commutative if $N > 1$. Is it a group under matrix addition?*

Exercise 1.4.4. *Let $(\mathbb{Z}/p\mathbb{Z})^* = \{1, 2, \ldots, p - 1\}$ where $a \cdot b$ is defined to be $ab \bmod p$. Prove this is a multiplicative group if p is prime. More generally, let $(\mathbb{Z}/m\mathbb{Z})^*$ be the subset of $\mathbb{Z}/m\mathbb{Z}$ of numbers relatively prime to m. Show $(\mathbb{Z}/m\mathbb{Z})^*$ is a multiplicative group.*

Exercise 1.4.5 (Euler's ϕ-function (or totient function)). *Let $\phi(n)$ denote the number of elements in $(\mathbb{Z}/n\mathbb{Z})^*$. Prove that for p prime, $\phi(p) = p - 1$ and $\phi(p^k) = p^k - p^{k-1}$. If p and q are distinct primes, prove $\phi(p^j q^k) = \phi(p^j)\phi(q^k)$. If n and m are relatively prime, prove that $\phi(nm) = \phi(n)\phi(m)$. Note $\phi(n)$ is the size of the group $(\mathbb{Z}/n\mathbb{Z})^*$.*

Definition 1.4.6 (Subgroup). *A subset H of G is a subgroup if H is also a group.*

Our definitions imply any group G has at least two subgroups, itself and the empty set.

Exercise 1.4.7. *Prove the following equivalent definition: A subset H of a group G is a subgroup if for all $x, y \in H$, $xy^{-1} \in H$.*

Exercise 1.4.8. *Let G be an additive subgroup of $\mathbb{Z}$. Prove that there exists an $n \in \mathbb{N}$ such that every element of G is an integral multiple of n.*

Exercise 1.4.9. *Let $GL_n(\mathbb{R})$ be the multiplicative group of $n \times n$ invertible matrices with real entries. Let $SL_n(\mathbb{Z})$ be the subset with integer entries and determinant 1. Prove $SL_n(\mathbb{Z})$ is a subgroup. This is a very important subgroup in number theory; when $n = 2$ it is called the **modular group**. See §7.7 for an application to continued fractions.*

1.4.2 Lagrange's Theorem

We prove some basic properties of **finite groups** (groups with finitely many elements).

Definition 1.4.10 (Order). *If G is a finite group, the number of elements of G is the order of G and is denoted by $|G|$. If $x \in G$, the order of x in G, $\mathrm{ord}(x)$, is the least positive power m such that $x^m = e$, where $e \in G$ is the identity of the group.*

Exercise[(h)] **1.4.11.** *Prove all elements in a finite group have finite order.*

Theorem 1.4.12 (Lagrange). *Let H be a subgroup of a finite group G. Then $|H|$ divides $|G|$. In particular, taking H to be the subgroup generated by $x \in G$, $\mathrm{ord}(x)|\mathrm{ord}(G)$.*

We first prove two useful lemmas.

Lemma 1.4.13. *Let H be a subgroup of G, and let $h \in H$. Then $hH = H$.*

Proof. It suffices to show $hH \subset H$ and $H \subset hH$. By closure, $hH \subset H$. For the other direction, let $h' \in H$. Then $hh^{-1}h' = h'$; as $h^{-1}h' \in H$, every $h' \in H$ is also in hH. $\qquad\square$

Lemma 1.4.14. *Let H be a subgroup of a group G. Then for all $g_i, g_j \in G$ either $g_iH = g_jH$ or the two sets are disjoint.*

Proof. Assume $g_iH \cap g_jH$ is non-empty; we must show they are equal. Let $x = g_ih_1 = g_jh_2$ be in the intersection. Multiplying on the right by $h_1^{-1} \in H$ (which exists because H is a subgroup) gives $g_i = g_jh_2h_1^{-1}$. So $g_iH = g_jh_2h_1^{-1}H$. As $h_2h_1^{-1}H = H$, we obtain $g_iH = g_jH$. $\qquad\square$

Definition 1.4.15 (Coset). *We call a subset gH of G a coset (actually, a left coset) of H. In general the set of all gH for a fixed H is not a subgroup.*

Exercise[(h)] **1.4.16.** *Show not every set of cosets is a subgroup.*

We now prove Lagrange's Theorem.

Proof of Lagrange's theorem. We claim

$$G = \bigcup_{g \in G} gH. \qquad (1.23)$$

Why is there equality? As $g \in G$ and $H \subset G$, each $gH \subset G$, hence their union is contained in G. Further, as $e \in H$, given $g \in G$, $g \in gH$. Thus, G is a subset of the right side, proving equality.

By Lemma 1.4.13, two cosets are either identical or disjoint. By choosing a subset of the cosets, we show the union in (1.23) equals a union of disjoint cosets. There are only finitely many elements in G. As we go through all g in G, if the coset gH equals one of the cosets already chosen, we do not include it; if it is new, we do. Continuing this process, we obtain

$$G = \bigcup_{i=1}^{k} g_iH \qquad (1.24)$$

for some finite k, and the k cosets are disjoint. If $H = \{e\}$, k is the number of elements of G; in general, however, k will be smaller. Each set g_iH has $|H|$ elements, and no two cosets share an element. Thus $|G| = k|H|$, proving $|H|$ divides $|G|$. $\qquad\square$

Exercise 1.4.17. *Let $G = (\mathbb{Z}/15\mathbb{Z})^*$. Find all subgroups of G and write G as the union of cosets for some proper subgroup H (H is a* **proper subgroup** *of G if H is neither $\{1\}$ nor G).*

Exercise 1.4.18. *Let $G = (\mathbb{Z}/p_1p_2\mathbb{Z})^*$ for two distinct primes p_1 and p_2. What are the possible orders of subgroups of G? Prove that there is either a subgroup of order p_1 or a subgroup of order p_2 (in fact, there are subgroups of both orders).*

1.4.3 Fermat's Little Theorem

We deduce some consequences of Lagrange's Theorem which will be useful in our cryptography investigations.

Corollary 1.4.19 (Fermat's Little Theorem). *For any prime p, if $\gcd(a, p) = 1$ then $a^{p-1} \equiv 1 \bmod p$.*

Proof. As $|(\mathbb{Z}/p\mathbb{Z})^*| = p - 1$, the result follows from Lagrange's Theorem. $\qquad\square$

Exercise[h] **1.4.20.** *One can reformulate Fermat's Little Theorem as the statement that if p is prime, for all a we have $p|a^p - a$. Give a proof for this formulation without using group theory. Does $n|a^n - a$ for all n?*

Exercise 1.4.21. *Prove that if for some a, $a^{n-1} \not\equiv 1 \bmod n$ then n is composite.*

Thus Fermat's Little Theorem is a fast way to show certain numbers are composite (remember exponentiation is fast: see §1.2.1); we shall also encounter Fermat's Little Theorem in §4.4.3 when we count the number of integer solutions to certain equations. Unfortunately, it is not the case that $a^{n-1} \equiv 1 \bmod n$ implies n is prime. There are composite n such that for all positive integers a, $a^{n-1} \equiv 1 \bmod n$. Such composite numbers are called Carmichael numbers (the first few are 561, 1105 and 1729). More generally, one has

Theorem 1.4.22 (Euler). *If $\gcd(a, n) = 1$, then $a^{\phi(n)} \equiv 1 \bmod n$.*

Proof. Let $(a, n) = 1$. By definition, $\phi(n) = |(\mathbb{Z}/n\mathbb{Z})^*|$. By Lagrange's Theorem the order of $a \in (\mathbb{Z}/n\mathbb{Z})^*$ divides $\phi(n)$, or $a^{\phi(n)} \equiv 1 \bmod n$. $\qquad\square$

Remark 1.4.23. For our applications to RSA, we only need the case when n is the product of two primes. In this case, consider the set $\{1, \dots, pq\}$. There are pq numbers, q numbers are multiples of p, p numbers are multiples of q, and one is a multiple of both p and q. Thus, the number of numbers in $\{1, \dots, pq\}$ relatively prime to pq is $pq - p - q + 1$ (why?). Note this equals $\phi(p)\phi(q) = (p-1)(q-1)$. This type of argument is known as **Inclusion - Exclusion**. See also Exercise 2.3.18.

Exercise 1.4.24. *Korselt [Kor] proved that a composite number n is a Carmichael number if and only if n is square-free and if a prime $p|n$, then $(p-1)|(n-1)$. Prove that if these two conditions are met then n is a Carmichael number.*

Research Project 1.4.25 (Carmichael Numbers). It is known (see [AGP]) that there are infinitely many Carmichael numbers. One can investigate the spacings between adjacent Carmichael numbers. For example, choose a large X and look at all Carmichael numbers in $[X, 2X]$, say $c_1, \ldots, c_{n+1}$. The average spacing between these numbers is about $\frac{2X - X}{n}$ (they are spread out over an interval of size X, and there are n differences: $c_2 - c_1, \ldots, c_{n+1} - c_n$. How are these differences distributed? Often, it is more natural to rescale differences and spacings so that the average spacing is 1. The advantage of such a renormalization is the results are often scale invariant (i.e., unitless quantities). For more on investigating such spacings, see Chapter 12.

Exercise[h] **1.4.26.** *Prove an integer is divisible by 3 (resp., 9) if and only if the sum of its digits is divisible by 3 (resp., 9).*

Exercise[h] **1.4.27.** *Show an integer is divisible by 11 if and only if the alternating sum of its digits is divisible by 11; for example, 924 is divisible by 11 because $11|(9 - 2 + 4)$. Use Fermat's Little Theorem to find a rule for divisibility by 7 (or more generally, for any prime).*

Exercise[h] **1.4.28.** *Show that if x is a positive integer then there exists a positive integer y such that the product xy has only zeros and ones for digits.*

1.4.4 Structure of $(\mathbb{Z}/p\mathbb{Z})^*$

The multiplicative group $(\mathbb{Z}/p\mathbb{Z})^*$ for p prime has a rich structure which will simplify many investigations later.

Theorem 1.4.29. *For p prime, $(\mathbb{Z}/p\mathbb{Z})^*$ is cyclic of order $p - 1$. This means there is an element $g \in (\mathbb{Z}/p\mathbb{Z})^*$ such that*

$$(\mathbb{Z}/p\mathbb{Z})^* = \{1, 2, \ldots, p - 2, p - 1\} = \{g^1, g^2, \ldots, g^{p-2}, g^{p-1}\}. \qquad (1.25)$$

We say g is a **generator** of the group. For each x there is a unique integer $k \in \{1, \ldots, p - 1\}$ such that $x \equiv g^k \bmod p$. We say k is the **index** of x relative to g. For each $x \in (\mathbb{Z}/p\mathbb{Z})^*$, the **order** of x is the smallest positive integer n such that $x^n \equiv 1 \bmod p$. For example, if $p = 7$ we have

$$\{1, 2, 3, 4, 5, 6\} = \{3^6, 3^2, 3^1, 3^4, 3^5, 3^3\}, \qquad (1.26)$$

which implies 3 is a generator (and the index of 4 relative to 3 is 4, because $4 \equiv 3^4 \bmod 7$). Note 5 is also a generator of this group, so the generator need not be unique.

Sketch of the proof. We will use the fact that $(\mathbb{Z}/p\mathbb{Z})^*$ is a commutative group: $xy = yx$. Let $x, y \in (\mathbb{Z}/p\mathbb{Z})^*$ with orders m and n for the exercises below. The proof comes from the following:

Exercise 1.4.30. *Assume $m = m_1 m_2$, with m_1, m_2 relatively prime. Show x^{m_1} has order m_2.*

Exercise[h] **1.4.31.** *Let ℓ be the least common multiple of m and n (the smallest number divisible by both m and n). Prove that there is an element z of order ℓ.*

Exercise 1.4.32. *By Lagrange's Theorem, the order of any x divides $p-1$ (the size of the group). From this fact and the previous exercises, show there is some d such that the order of every element divides $d \le p-1$, and there is an element of order d and no elements of larger order.*

The proof is completed by showing $d = p - 1$. The previous exercises imply that every element satisfies the equation $x^d - 1 \equiv 0 \bmod p$. As every element in the group satisfies this, and there are $p-1$ elements in the group, we have a degree d polynomial with $p-1$ roots. We claim this can only occur if $d = p - 1$.

Exercise[h] **1.4.33.** *Prove the above claim.*

Therefore $d = p-1$ and there is some element g of order $p-1$; thus, g's powers generate the group.   □

Exercise 1.4.34. *For $p > 2$, $k > 1$, what is the structure of $(\mathbb{Z}/p^k\mathbb{Z})^*$? If all the prime divisors of m are greater than 2, what is the structure of $(\mathbb{Z}/m\mathbb{Z})^*$? For more on the structure of these groups, see any undergraduate algebra textbook (for example, [Art, J, La3]).*

1.5 RSA REVISITED

We have developed sufficient machinery to prove why RSA works. Remember Bob chose two primes p and q, and numbers d (for decrypt) and e (for encrypt) such that $de \equiv 1 \bmod \phi(pq)$. He made public $N = pq$ and e and kept secret the two primes and d. Alice wants to send Bob a number M (smaller than N). She encrypts the message by sending $X \equiv M^e \bmod N$. Bob then decrypts the message by calculating $X^d \bmod N$, which we claimed equals M.

As $X \equiv M^e \bmod N$, there is an integer n such that $X = M^e + nN$. Thus $X^d = (M^e + nN)^d$, and the last term is clearly of the form $(M^e)^d + n'N$ for some n'. We need only show $(M^e)^d \equiv M \bmod N$. As $ed \equiv 1 \bmod \phi(N)$, there is an m such that $ed = 1 + m\phi(N)$. Therefore

$$(M^e)^d = M^{ed} = M^{1+m\phi(N)} = M \cdot M^{m\phi(N)} = M \cdot (M^{\phi(N)})^m. \quad (1.27)$$

If M is relatively prime to N then By Euler's Theorem (Theorem 1.4.22), $M^{\phi(N)} \equiv 1 \bmod N$, which completes the proof. Thus we can only send messages relatively prime to N. In practice this is not a problem, as it is very unlikely to stumble upon a message that shares a factor with N; of course, if we did find such a message we could quickly find the factors of N. If our initial message has a factor in common with N, we need only tweak our message (add another letter or spell a word incorrectly).

Why is RSA secure? Assume a third person (say Charlie) intercepts the encrypted message X. He knows X, N and e, and wants to recover M. Knowing d such that $de \equiv 1 \bmod \phi(N)$ makes decrypting the message trivial: one need only compute $X^d \bmod N$. Thus Charlie is trying to solve the equation $ed \equiv 1 \bmod \phi(N)$; fortunately for Alice and Bob this equation has two unknowns, d and $\phi(N)$! Right now, there is no known fast way to determine $\phi(N)$ from N. Charlie can of course factor N; once he has the factors, he knows $\phi(N)$ and can find d; however, the fastest factorization algorithms make 400 digit numbers unaccessible for now.

This should be compared to primality testing, which was only recently shown to be fast ([AgKaSa]). Previous deterministic algorithms to test whether or not a number is prime were known to be fast only if certain well believed conjectures are true. It was an immense achievement showing that there is a deterministic, efficient algorithm. The paper is very accessible, and worth the read.

Remark 1.5.1. Our simple example involved computing a sixty-six digit number, and this was for a small N ($N = 9797$). Using binary expansions to exponentiate, as we need only transmit our message modulo N, we never need to compute anything larger than the product of four digit numbers.

Remark 1.5.2. See [Bon] for a summary of attempts to break RSA. Certain products of two primes are denoted RSA challenge numbers, and the public is invited to factor them. With the advent of parallel processing, many numbers have succumbed to factorization. See http://www.rsasecurity.com/rsalabs/node.asp?id=2092 for more details.

Exercise 1.5.3. *If $M < N$ is not relatively prime to N, show how to quickly find the prime factorization of N.*

Exercise 1.5.4 (Security Concerns). *In the system described, there is no way for Bob to verify that the message came from Alice! Design a system where Alice makes some information public (and keeps some secret) so that Bob can verify that Alice sent the message.*

Exercise 1.5.5. *Determining $\phi(N)$ is equivalent to factoring N; there is no computational shortcut to factoring. Clearly, if one knows the factors of $N = pq$, one knows $\phi(N)$. If one knows $\phi(N)$ and N, one can recover the primes p and q. Show that if $K = N + 1 - \phi(N)$, then the two prime factors of N are $(K \pm \sqrt{K^2 - 4N})/2$, and these numbers are in fact integers.*

Exercise[hr] **1.5.6** (Important). *If e and $(p-1)(q-1)$ are given, show how one may efficiently find a d such that $ed - 1$ divides $(p-1)(q-1)$.*

1.6 EISENSTEIN'S PROOF OF QUADRATIC RECIPROCITY

We conclude this introduction to basic number theory and group theory by giving a proof of quadratic reciprocity (we follow the beautiful exposition in [LP] of Eisenstein's proof; see the excellent treatments in [IR, NZM] for alternate proofs). In §1.2.4, we described Newton's Method to find square roots of real numbers. Now we turn our attention to a finite group analogue: for a prime p and an $a \not\equiv 0 \bmod p$, when is $x^2 \equiv a \bmod p$ solvable? For example, if $p = 5$ then $(\mathbb{Z}/p\mathbb{Z})^* = \{1, 2, 3, 4\}$. Squaring these numbers gives $\{1, 4, 4, 1\} = \{1, 4\}$. Thus there are two solutions if $a \in \{1, 4\}$ and no solutions if $a \in \{2, 3\}$. The problem of whether or not a given number is a square is solvable: we can simply enumerate the group $(\mathbb{Z}/p\mathbb{Z})^*$, square each element, and see if a is a square. This takes about p steps; quadratic reciprocity will take about $\log p$ steps. For applications, see §4.4.

1.6.1 Legendre Symbol

We introduce notation. From now on, p and q will always be distinct odd primes.

Definition 1.6.1 (Legendre Symbol $\left(\frac{\cdot}{p}\right)$). *The Legendre Symbol $\left(\frac{a}{p}\right)$ is*

$$\left(\frac{a}{p}\right) = \begin{cases} 1 & \text{if } a \text{ is a non-zero square modulo } p \\ 0 & \text{if } a \equiv 0 \text{ modulo } p \\ -1 & \text{if } a \text{ is a not a square modulo } p. \end{cases} \quad (1.28)$$

The Legendre symbol is a function on $\mathbb{F}_p = \mathbb{Z}/p\mathbb{Z}$. We extend the Legendre symbol to all integers by $\left(\frac{a}{p}\right) = \left(\frac{a \bmod p}{p}\right)$.

Note a is a square modulo p if there exists an $x \in \{0, 1, \ldots, p-1\}$ such that $a \equiv x^2 \bmod p$.

Definition 1.6.2 (Quadratic Residue, Non-Residue). *For $a \not\equiv 0 \bmod p$, if $x^2 \equiv a \bmod p$ is solvable (resp., not solvable) we say a is a quadratic residue (resp., non-residue) modulo p. When p is clear from context, we just say residue and non-residue.*

Exercise 1.6.3. *Show the Legendre symbol is multiplicative: $\left(\frac{ab}{p}\right) = \left(\frac{a}{p}\right)\left(\frac{b}{p}\right)$.*

Exercise[h] 1.6.4 (Euler's Criterion). *For odd p, $\left(\frac{a}{p}\right) \equiv a^{\frac{p-1}{2}} \bmod p$.*

Exercise 1.6.5. *Show $\left(\frac{-1}{p}\right) = (-1)^{\frac{p-1}{2}}$ and $\left(\frac{2}{p}\right) = (-1)^{\frac{p^2-1}{8}}$.*

Lemma 1.6.6. *For p an odd prime, half of the non-zero numbers in $(\mathbb{Z}/p\mathbb{Z})^*$ are quadratic residues and half are quadratic non-residues.*

Proof. As p is odd, $\frac{p-1}{2} \in \mathbb{N}$. Consider the numbers $1^2, 2^2, \ldots, (\frac{p-1}{2})^2$. Assume two numbers a and b are equivalent mod p. Then $a^2 \equiv b^2 \bmod p$, so $(a-b)(a+b) \equiv 0 \bmod p$. Thus either $a \equiv b \bmod p$ or $a \equiv -b \bmod p$; in other words, $a \equiv p - b$. For $1 \leq a, b \leq \frac{p-1}{2}$ we cannot have $a \equiv p - b \bmod p$, implying the $\frac{p-1}{2}$ values above are distinct. As $(p-r)^2 \equiv r^2 \bmod p$, the above list is all of the non-zero squares modulo p. Thus half the non-zero numbers are non-zero squares, half are non-squares. □

Remark 1.6.7. By Theorem 1.4.29, $(\mathbb{Z}/p\mathbb{Z})^*$ is a cyclic group with generator g. Using the group structure we can prove the above lemma directly: once we show there is at least one non-residue, the g^{2k} are the quadratic residues and the g^{2k+1} are the non-residues.

Exercise 1.6.8. *Show for any $a \not\equiv 0 \bmod p$ that*

$$\sum_{t=0}^{p-1} \left(\frac{t}{p}\right) = \sum_{t=0}^{p-1} \left(\frac{at+b}{p}\right) = 0. \quad (1.29)$$

Exercise 1.6.9. *For $x \in \{0, \ldots, p-1\}$, let $F_p(x) = \sum_{a \leq x} \left(\frac{n}{p}\right)$; note $F_p(0) = F_p(p-1) = 0$. If $\left(\frac{-1}{p}\right) = 1$, show $F_p\left(\frac{p-1}{2}\right) = 0$. Do you think $F(x)$ is more likely to be positive or negative? Investigate its values for various x and p.*

Initially the Legendre symbol is defined only when the bottom is prime. We now extend the definition. Let $n = p_1 \cdot p_2 \cdots p_t$ be the product of t distinct odd primes. Then $\left(\frac{a}{n}\right) = \left(\frac{a}{p_1}\right)\left(\frac{a}{p_2}\right) \cdots \left(\frac{a}{p_t}\right)$; this is the **Jacobi symbol**, and has many of the same properties as the Legendre symbol. We will study only the Legendre symbol (see [IR] for more on the Jacobi symbol). Note the Jacobi symbol does *not* say that if a is a square (a quadratic residue) mod n, then a is a square mod p_i for each prime divisor.

The main result (which allows us to calculate the Legendre symbol quickly and efficiently) is the celebrated

Theorem 1.6.10 (The Generalized Law of Quadratic Reciprocity). *For m, n odd and relatively prime,*

$$\left(\frac{m}{n}\right)\left(\frac{n}{m}\right) = (-1)^{\frac{m-1}{2}\frac{n-1}{2}}. \tag{1.30}$$

Gauss gave eight proofs of this deep result when m and n are prime. If either p or q are equivalent to 1 mod 4 then we have $\left(\frac{q}{p}\right) = \left(\frac{p}{q}\right)$, i.e., p has a square root modulo q if and only if q has a square root modulo p. We content ourselves with proving the case with m, n prime.

Exercise 1.6.11. *Using the Generalized Law of Quadratic Reciprocity, Exercise 1.6.5 and the Euclidean algorithm, show one can determine if $a < m$ is a square modulo m in logarithmic time (i.e., the number of steps is at most a fixed constant multiple of $\log m$). This incredible efficiency is just one of many important applications of the Legendre and Jacobi symbols.*

1.6.2 The Proof of Quadratic Reciprocity

Our goal is to prove

Theorem 1.6.12 (Quadratic Reciprocity). *Let p and q be distinct odd primes. Then*

$$\left(\frac{q}{p}\right)\left(\frac{p}{q}\right) = (-1)^{\frac{p-1}{2}\frac{q-1}{2}}. \tag{1.31}$$

As p and q are distinct, odd primes, both $\left(\frac{q}{p}\right)$ and $\left(\frac{p}{q}\right)$ are ± 1. The difficulty is figuring out which signs are correct, and how the two signs are related. We use Euler's Criterion (Exercise 1.6.4).

The idea behind Eisenstein's proof is as follows: $\left(\frac{q}{p}\right)\left(\frac{p}{q}\right)$ is -1 to a power. Further, we only need to determine the power modulo 2. Eisenstein shows many expressions are equivalent modulo 2 to this power, and eventually we arrive at an expression which is trivial to calculate modulo 2. We repeatedly use the fact that as p and q are distinct primes, the Euclidean algorithm implies that q is invertible modulo p and p is invertible modulo q.

We choose to present this proof as it showcases many common techniques in mathematics. In addition to using the Euclidean algorithm and modular arithmetic, the proof shows that quadratic reciprocity is equivalent to a theorem about the number of integer solutions of some inequalities, specifically the number of pairs of integers strictly inside a rectangle. This is just one of many applications of counting solutions; we discuss this topic in greater detail in Chapter 4.

1.6.3 Preliminaries

Consider all multiples of q by an even $a \leq p-1$: $\{2q, 4q, 6q, \ldots, (p-1)q\}$. Denote a generic multiple by aq. Recall $[x]$ is the greatest integer less than or equal to x. By the Euclidean algorithm,

$$aq = \left[\frac{aq}{p}\right] p + r_a, \quad 0 < r_a < p - 1. \tag{1.32}$$

Thus r_a is the least non-negative number equivalent to aq mod p. The numbers $(-1)^{r_a} r_a$ are equivalent to even numbers in $\{0, \ldots, p-1\}$. If r_a is even this is clear; if r_a is odd, then $(-1)^{r_a} r_a \equiv p - r_a$ mod p, and as p and r_a are odd, this is even. Finally note $r_a \neq 0$; if $r_a = 0$ then $p | aq$. As p and q are relatively prime, this implies $p | a$; however, p is prime and $a \leq p - 1$. Therefore p cannot divide a and thus $r_a \neq 0$.

Lemma 1.6.13. *If* $(-1)^{r_a} r_a \equiv (-1)^{r_b} r_b$ *then* $a = b$.

Proof. We quickly get $\pm r_a \equiv r_b$ mod p. If the plus sign holds, then $r_a \equiv r_b$ mod p implies $aq \equiv bq$ mod p. As q is invertible modulo p, we get $a \equiv b$ mod p, which yields $a = b$ (as a and b are even integers between 2 and $p - 1$).

If the minus sign holds, then $r_a + r_b \equiv 0$ mod p, or $aq + bq \equiv 0$ mod p. Multiplying by q^{-1} mod p now gives $a + b \equiv 0$ mod p. As a and b are even integers between 2 and $p - 1$, $4 < a + b \leq 2p - 2$. The only integer strictly between 4 and $2p - 2$ which is equivalent to 0 mod p is p; however, p is odd and $a + b$ is even. Thus the minus sign cannot hold, and the elements are all distinct. $\square$

Remark 1.6.14. The previous argument is very common in mathematics. We will see a useful variant in Chapter 5, where we show certain numbers are irrational by proving that if they were not then there would have to be an integer strictly between 0 and 1.

Lemma 1.6.15. *We have*

$$\left(\frac{q}{p}\right) = (-1)^{\sum_{a \text{ even}, a \neq 0} r_a}, \tag{1.33}$$

where a *even*, $a \neq 0$ *means* $a \in \{2, 4, \ldots, p - 3, p - 1\}$.

Proof. For each even $a \in \{2, \ldots, p-1\}$, $aq \equiv r_a \bmod p$. Thus modulo p

$$\prod_{\substack{a \text{ even} \\ a \neq 0}} aq \equiv \prod_{\substack{a \text{ even} \\ a \neq 0}} r_a$$

$$q^{\frac{p-1}{2}} \prod_{\substack{a \text{ even} \\ a \neq 0}} a \equiv \prod_{\substack{a \text{ even} \\ a \neq 0}} r_a$$

$$\left(\frac{q}{p}\right) \prod_{\substack{a \text{ even} \\ a \neq 0}} a \equiv \prod_{\substack{a \text{ even} \\ a \neq 0}} r_a, \tag{1.34}$$

where the above follows from the fact that we have $\frac{p-1}{2}$ choices for an even a (giving the factor $q^{\frac{p-1}{2}}$) and Euler's Criterion (Exercise 1.6.4). As a ranges over all even numbers from 2 to $p-1$, so too do the distinct numbers $(-1)^{r_a} r_a \bmod p$. Note how important it was that we showed $r_a \neq 0$ in (1.32), as otherwise we would just have $0 = 0$ in (1.34). Thus modulo p,

$$\prod_{\substack{a \text{ even} \\ a \neq 0}} a \equiv \prod_{\substack{a \text{ even} \\ a \neq 0}} (-1)^{r_a} r_a$$

$$\prod_{\substack{a \text{ even} \\ a \neq 0}} a \equiv (-1)^{\sum_{a \text{ even}, a \neq 0} r_a} \prod_{\substack{a \text{ even} \\ a \neq 0}} r_a. \tag{1.35}$$

Combining gives

$$\left(\frac{q}{p}\right)(-1)^{\sum_{a \text{ even}, a \neq 0} r_a} \prod_{\substack{a \text{ even} \\ a \neq 0}} r_a \equiv \prod_{\substack{a \text{ even} \\ a \neq 0}} r_a \bmod p. \tag{1.36}$$

As each r_a is invertible modulo p, so is the product. Thus

$$\left(\frac{q}{p}\right)(-1)^{\sum_{a \text{ even}, a \neq 0} r_a} \equiv 1 \bmod p. \tag{1.37}$$

As $\left(\frac{q}{p}\right) = \pm 1$, the lemma follows by multiplying both sides by $\left(\frac{q}{p}\right)$. □

Therefore it suffices to determine $\sum_{a \text{ even}, a \neq 0} r_a \bmod 2$. We make one last simplification. By the first step in the Euclidean algorithm (1.32), we have $aq = \left[\frac{aq}{p}\right] p + r_a$ for some $r_a \in \{2, \ldots, p-1\}$. Hence

$$\sum_{\substack{a \text{ even} \\ a \neq 0}} aq = \sum_{\substack{a \text{ even} \\ a \neq 0}} \left(\left[\frac{aq}{p}\right] p + r_a\right) = \sum_{\substack{a \text{ even} \\ a \neq 0}} \left[\frac{aq}{p}\right] p + \sum_{\substack{a \text{ even} \\ a \neq 0}} r_a. \tag{1.38}$$

As we are summing over even a, the left hand side above is even. Thus the right hand side is even, so

$$\sum_{\substack{a \text{ even} \\ a \neq 0}} \left[\frac{aq}{p}\right] p \equiv \sum_{\substack{a \text{ even} \\ a \neq 0}} r_a \bmod 2$$

$$\sum_{\substack{a \text{ even} \\ a \neq 0}} \left[\frac{aq}{p}\right] \equiv \sum_{\substack{a \text{ even} \\ a \neq 0}} r_a \bmod 2, \tag{1.39}$$

where the last line follows from the fact that p is odd, so modulo 2 dropping the factor of p from the left hand side does not change the parity. We have reduced the proof of quadratic reciprocity to calculating $\sum_{a \text{ even}, a \neq 0} \left[\frac{aq}{p} \right]$. We summarize our results below.

Lemma 1.6.16. *Define*

$$\mu = \sum_{\substack{a \text{ even} \\ a \neq 0}} \left[\frac{aq}{p} \right]$$

$$\nu = \sum_{\substack{a \text{ even} \\ a \neq 0}} \left[\frac{ap}{q} \right]. \tag{1.40}$$

Then

$$\left(\frac{q}{p} \right) = (-1)^{\mu}$$

$$\left(\frac{p}{q} \right) = (-1)^{\nu}. \tag{1.41}$$

Proof. By (1.37) we have

$$\left(\frac{q}{p} \right) = (-1)^{\sum_{a \text{ even}, a \neq 0} r_a}. \tag{1.42}$$

By (1.39) we have

$$\sum_{\substack{a \text{ even} \\ a \neq 0}} \left[\frac{aq}{p} \right] \equiv \sum_{\substack{a \text{ even} \\ a \neq 0}} r_a \bmod 2, \tag{1.43}$$

and the proof for $\left(\frac{q}{p} \right)$ is completed by recalling the definition of μ; the proof for the case $\left(\frac{p}{q} \right)$ proceeds similarly. $\qquad \square$

1.6.4 Counting Lattice Points

As our sums are not over all even $a \in \{0, 2, \ldots, p-1\}$ but rather just over even $a \in \{2, \ldots, p-1\}$, this slightly complicates our notation and forces us to be careful with our book-keeping. We urge the reader not to be too concerned about this slight complication and instead focus on the fact that we are able to show quadratic reciprocity is equivalent to counting the number of pairs of integers satisfying certain relations.

Consider the rectangle with vertices at $A = (0,0)$, $B = (p,0)$, $C = (p,q)$ and $D = (0,q)$. The upward sloping diagonal is given by the equation $y = \frac{q}{p}x$. As p and q are distinct odd primes, there are no pairs of integers (x,y) on the line AC. See Figure 1.2.

We add some non-integer points: $E = (\frac{p}{2}, 0)$, $F = (\frac{p}{2}, \frac{q}{2})$, $G = (0, \frac{q}{2})$ and $H = (\frac{p}{2}, q)$. Let $\#ABC_{\text{even}}$ denote the number of integer pairs **strictly inside** the triangle ABC with even x-coordinate, and $\#AEF$ denote the number of integer

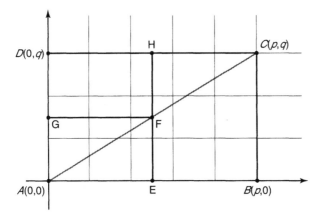

Figure 1.2 Lattice for the proof of Quadratic Reciprocity. Points $E(\frac{p}{2},0)$, $F(\frac{p}{2},\frac{q}{2})$, $G(0,\frac{q}{2})$, $H(\frac{p}{2},q)$

pairs **strictly inside** the triangle AEF; thus, we do not count any integer pairs on the lines AB, BC, CD or DA.

We now interpret $\sum_{a \text{ even}, a\neq 0} \left[\frac{aq}{p}\right]$. Consider the vertical line with x-coordinate a. Then $\left[\frac{aq}{p}\right]$ gives the number of pairs (x,y) with x-coordinate equal to a and y-coordinate a positive integer at most $\left[\frac{aq}{p}\right]$. To see this, consider the line AC (which is given by the equation $y = \frac{q}{p}x$). For definiteness, let us take $p = 5$, $q = 7$ and $a = 4$. Then $\left[\frac{aq}{p}\right] = \left[\frac{28}{5}\right] = 5$, and there are exactly five integer pairs with x-coordinate equal to 4 and positive y-coordinate at most $\left[\frac{28}{5}\right]$: $(4,1)$, $(4,2)$, $(4,3)$, $(4,4)$ and $(4,5)$. The general proof proceeds similarly.

Thus $\sum_{a \text{ even}, a\neq 0} \left[\frac{aq}{p}\right]$ is the number of integer pairs **strictly inside** the rectangle $ABCD$ with even x-coordinate that are below the line AC, which we denote $\#ABC_{\text{even}}$. We prove

Lemma 1.6.17. *The number of integer pairs under the line AC strictly inside the rectangle with even x-coordinate is congruent modulo 2 to the number of integer pairs under the line AF strictly inside the rectangle. Thus $\#ABC_{\text{even}} = \#AEF$.*

Proof. First observe that if $0 < a < \frac{p}{2}$ is even then the points under AC with x-coordinate equal to a are exactly those under the line AF with x-coordinate equal to a. We are reduced to showing that the number of points under FC strictly inside the rectangle with even x-coordinate is congruent modulo 2 to the number of points under the line AF strictly inside the rectangle with odd x-coordinate. Therefore let us consider an even a with $\frac{p}{2} < a < p-1$.

The integer pairs on the line $x = a$ strictly inside the rectangle are $(a,1)$, $(a,2), \ldots, (a,q-1)$. There are $q-1$ pairs. As q is odd, there are an even number of integer pairs on the line $x = a$ strictly inside the rectangle. As there are no integer pairs on the line AC, for a fixed $a > \frac{p}{2}$, modulo 2 there are the same number of integer pairs *above* AC as there are *below* AC. The number of integer pairs *above* AC on the line $x = a$ is equivalent modulo 2 to the number of integer pairs below AF on the line $x = p - a$. To see this, consider the map which takes (x, y) to $(p - x, q - y)$. As $a > \frac{p}{2}$ and is even, $p - a < \frac{p}{2}$ and is odd. Further, every odd $a < \frac{p}{2}$ is hit (given $a_{\text{odd}} < \frac{p}{2}$, start with the even number $p - a_{\text{odd}} > \frac{p}{2}$). A similar proof holds for $a < \frac{p}{2}$.   □

Exercise 1.6.18. *Why are there no integer pairs on the line AC?*

We have thus shown that

$$\sum_{\substack{a \text{ even} \\ a \neq 0}} \left\lceil \frac{aq}{p} \right\rceil \equiv \#AEF \text{ mod } 2; \tag{1.44}$$

remember that $\#AEF$ is the number of integer pairs strictly inside the triangle AEF. From Lemma 1.6.16 we know the left hand side is μ and $\left(\frac{q}{p}\right) = (-1)^{\mu}$. Therefore

$$\left(\frac{q}{p}\right) = (-1)^{\mu} = (-1)^{\#AEF}. \tag{1.45}$$

Reversing the rolls of p and q, we see that

$$\left(\frac{p}{q}\right) = (-1)^{\nu} = (-1)^{\#AGF}, \tag{1.46}$$

where $\nu \equiv \#AGF \text{ mod } 2$, with $\#AGF$ equal to the number of integer pairs strictly inside the triangle AGF.

Exercise 1.6.19. *Prove 1.46.*

Combining our expressions for μ and ν yields

$$\mu + \nu = \#AEF + \#AGF \text{ mod } 2, \tag{1.47}$$

which is the number of integer pairs strictly inside the rectangle $AEFG$. There are $\frac{p-1}{2}$ choices for x ($x \in \{1, 2, \ldots, \frac{p-1}{2}\}$) and $\frac{q-1}{2}$ choices for y ($y \in \{1, 2, \ldots, \frac{q-1}{2}\}$), giving $\frac{p-1}{2} \frac{q-1}{2}$ pairs of integers strictly inside the rectangle $AEFG$. Thus,

$$\left(\frac{q}{p}\right)\left(\frac{p}{q}\right) = (-1)^{\mu+\nu}$$
$$= (-1)^{\#AEF + \#AGF}$$
$$= (-1)^{\frac{p-1}{2} \frac{q-1}{2}}, \tag{1.48}$$

which completes the proof of Quadratic Reciprocity.

Exercise 1.6.20 (Advanced). *Let p be an odd prime. Are there infinitely many primes q such that q is a square mod p? The reader should return to this problem after Dirichlet's Theorem (Theorem 2.3.4).*

Chapter Two

Arithmetic Functions

We introduce many of the common functions of number theory, and the techniques to analyze them. We determine how rapidly many of these functions grow; for example, how many numbers at most x are prime. As seen in §1.1, this has immediate applications to cryptography (ensuring that there are sufficiently many primes to make certain codes reasonably secure), though such applications are not the driving force of our investigations. Primes are the building blocks of numbers; it is natural to study how many primes there are, and how they are distributed. Many of the other functions introduced here will reappear throughout the book (in Chapter 3 when we investigate primes in arithmetic progression, and Chapters 13 and 14 when we investigate writing integers as the sum of primes); for more on basic number theory see [Apo, HW, IR, NZM] as well as the first chapter of [IK]. In studying the number of primes we introduce numerous standard techniques of number theory, such as partial summation, telescoping series and estimating integrals.

2.1 ARITHMETIC FUNCTIONS

Definition 2.1.1. *An arithmetic function is a function from the natural numbers* $\mathbb{N} = \{0, 1, 2, \dots\}$ *to the complex numbers* $\mathbb{C}$.

Some of the most important examples of arithmetic functions are the following:

1. Set $e(n)$ equal to one if $n = 1$ and zero otherwise.

2. For s a complex number, set $\sigma_s(0) = 0$ and for positive integers n,

$$\sigma_s(n) = \sum_{d|n} d^s. \tag{2.1}$$

   $\sigma_0(n)$ is the number of divisors of n. The usual notation for the **divisor function** σ_0 is $d(n)$ or $\tau(n)$.

3. Set $\phi(n)$ to be the number of natural numbers $a \in \{1, \dots, n\}$ with the property that $(a, n) = 1$ (see Exercise 1.4.5); ϕ is called the **totient function**.

4. The **Möbius function** μ is defined by $\mu(1) = 1$ and

$$\mu(n) = \begin{cases} (-1)^s & \text{if } n = p_1 p_2 \cdots p_s, \ p_i \text{ distinct primes} \\ 0 & \text{otherwise.} \end{cases} \tag{2.2}$$

5. The **von Magnoldt function** Λ is defined by

$$\Lambda(n) = \begin{cases} \log p & \text{if } n = p^k, p \text{ prime} \\ 0 & \text{otherwise.} \end{cases} \tag{2.3}$$

6. Set $\pi(n) = \#\{p : p \text{ is a prime}, p \le n\}$, the number of primes at most x.

Note the Möbius function is zero if n is not square-free, and the von Magnoldt function is zero if n is not a power of a prime.

Definition 2.1.2 (Multiplicative Function). *An arithmetic function f is said to be **multiplicative** if for all $m, n \in \mathbb{N}$ with $(m, n) = 1$, we have*

$$f(mn) = f(m)f(n). \tag{2.4}$$

It is called completely multiplicative *if $f(mn) = f(m)f(n)$ for all $m, n \in \mathbb{N}$.*

It is not hard to see that $e(n)$, $\mu(n)$ and $\sigma_s(n)$ are multiplicative. The function $\phi(n)$ is also multiplicative, though it is slightly harder to prove. The von Magnoldt function is not multiplicative.

Exercise[h] **2.1.3.** *Prove the above statements.*

Exercise[hr] **2.1.4.** *In this exercise we collect some basic properties of $\phi(n)$. Verify the following statements:*

1. *$\phi(p^\alpha) = p^\alpha - p^{\alpha-1}$, p prime.*

2. *$\phi(n) = n \prod_{p|n} \left(1 - \frac{1}{p}\right)$.*

3. *$\sum_{d|n} \phi(d) = n$.*

4. *If $(a, n) = 1$, we have $a^{\phi(n)} \equiv 1 \bmod n$.*

Exercise 2.1.5. *Here we list some problems related to the divisor function $d = \sigma_0$.*

1. *Let $n = p_1^{\alpha_1} \cdots p_k^{\alpha_k}$. Prove that*

$$d(n) = \prod_{i=1}^{k} (\alpha_i + 1). \tag{2.5}$$

2. *For any fixed $A, B > 0$ one can find an n such that $d(n) \ge A \log^B n$; however, given any $\epsilon > 0$ there exists a C_ϵ such that for all n, $d(n) \le C_\epsilon n^\epsilon$.*

3. *Prove that a number n is a perfect square if an only if $d(n)$ is odd.*

4. *Suppose we have 10000 doors numbered $1, 2, \ldots, 10000$. Suppose at first all of them are locked. First we go through all the doors and unlock all of them. Then we pick all even numbered doors, and change the status of their lock. Then we do the same with all the doors whose number is a multiple of three, then with four and so on, all the way to 10000. How many doors will be left unlocked?*

If f, g are two arithmetic functions, we define their **convolution** $f * g$ by

$$f * g(n) = \sum_{d|n} f(d) g\left(\frac{n}{d}\right). \tag{2.6}$$

The convolution is also know as the **Dirichlet product**.

Lemma 2.1.6. *The convolution $*$ has the following properties: for all arithmetic functions f, g, h we have*

1. $f * e = f$,

2. $f * g = g * f$,

3. $f * (g * h) = (f * g) * h$,

4. $f * (g + h) = f * g + f * h$.

Exercise 2.1.7. *Prove Lemma 2.1.6.*

Remark 2.1.8 (For Those Knowing Abstract Algebra). Lemma 2.1.6 shows that the collection of all arithmetic functions under pointwise addition and $*$ multiplication forms a commutative ring. Note an arbitrary arithmetic function need not have a $*$ inverse.

Lemma 2.1.9. *If f, g are multiplicative functions, then so is $f * g$.*

Exercise 2.1.10. *Prove Lemma 2.1.9.*

We often let 1 denote the constant function that takes the value 1 at all natural numbers: $1(n) = 1$. We have

Lemma 2.1.11. $\mu * 1 = e$.

Proof. We need to show that

$$\sum_{d|n} \mu(d) = 0 \tag{2.7}$$

whenever $n \neq 1$; the case $n = 1$ is straightforward. Let $n = p_1^{\alpha_1} \cdots p_r^{\alpha_r}$. Any divisor d of n can be written as $d = p_1^{\beta_1} \cdots p_r^{\beta_r}$, with $\beta_i \leq \alpha_i$. By definition $\mu(d) = 0$ unless $\beta_i \in \{0, 1\}$. The number of divisors d with exactly k of the $\beta_i = 1$ is equal to $\binom{r}{k}$. For such divisors we have $\mu(d) = (-1)^r$. Consequently

$$\sum_{d|n} \mu(d) = \sum_{k=0}^{r} (-1)^k \binom{n}{k}, \tag{2.8}$$

which by the Binomial Theorem (Theorem A.1.8) equals $(1 - 1)^k = 0$. $\square$

Exercise 2.1.12. *Prove*

$$\sum_{d^k|n} \mu(d) = \begin{cases} 1 & \text{if } n \text{ is } k^{\text{th}}\text{-power-free,} \\ 0 & \text{otherwise.} \end{cases} \tag{2.9}$$

Theorem 2.1.13 (Möbius Inversion Formula). *Let f, g be two arithmetic functions. If $f = g * 1$, then $g = f * \mu$, and vice versa.*

Proof. We only prove one direction, the other one being similar. We have $g = g * e = g * (1 * \mu) = (g * 1) * \mu = f * \mu$. □

We say an arithmetic function f is **invertible** if there is a unique arithmetic function g such that $f * g = g * f = e$; we commonly denote g by f^{-1}.

Exercise 2.1.14. *Let f be a multiplicative function. Prove that f is invertible if and only if $f(1) \neq 0$. In fact, f^{-1} is given by the recursive formula*

$$f^{-1}(1) = \frac{1}{f(1)} \tag{2.10}$$

and

$$f^{-1}(n) = \frac{-1}{f(1)} \sum_{\substack{d|n \\ d<n}} f\left(\frac{n}{d}\right) f^{-1}(d). \tag{2.11}$$

Exercise[(h)] **2.1.15** (Continuation of the previous exercise). *Show that the inverse of a multiplicative function is multiplicative. If f is completely multiplicative, show its inverse f^{-1} is given by the formula*

$$f^{-1}(n) = \mu(n)f(n). \tag{2.12}$$

Use the last statement to compute the inverse of Euler's ϕ-function.

For more basic properties of arithmetic functions see [Ap, Na].

2.2 AVERAGE ORDER

Often it is important to know how large the values of an arithmetic function can be, or how the values of the function behave for large values of the argument. It is often hard to obtain precise results in this direction, as the values of arithmetic functions can behave in odd ways. For example, the divisor function $d(n)$ equals 2 infinitely often, but $d(n)$ is usually much larger than 2; we determine its average value in Exercise 2.2.26. For many applications, however, it is sufficient to know how a function behaves on average. Given an arithmetic function $f(n)$, we define its average by

$$\overline{f}(n) = \frac{1}{n} \sum_{k=1}^{n} f(k). \tag{2.13}$$

For most arithmetically interesting functions the average is a much better behaved function. We will see below that sometimes it is easier to study the sum $\sum_{k \leq x} f(k)$ as a function of the real variable x rather than the discrete variable n.

Averaging techniques are particularly important in applications to non-vanishing of values of arithmetically interesting *non-negative* functions. The basic idea is that if we can show that the average of the values of a non-negative arithmetic function f

on the interval $[1, x]$ when $x \to \infty$ is non-zero, then infinitely many values of f are non-zero; for an example, see Exercise 2.2.27. In this book, however, we will not be concerned too much with such applications. We do give one application of passing from a good estimate for the average value to non-vanishing in Theorem 18.2.7, where we study the values of Dirichlet L-functions. For a more thorough study of such techniques see [MM]. See also [CL1, CL2] for interesting applications of averaging techniques to studying the orders of certain groups.

One problem we will study, however, is how many numbers have a certain property. For example, if $P(k)$ is 1 if k is a prime and 0 otherwise, then $\overline{P}(n)$ is the percentage of numbers at most n that are prime. We show later that $\lim_{n \to \infty} \overline{P}(n) = 0$; thus, in the limit most numbers are not prime. Similarly we could consider $S(k) = 1$ if k is a perfect square and 0 otherwise. This yields $\overline{S}(n)$, the percentage of numbers at most n that are perfect squares. While $\lim_{n \to \infty} \overline{S}(n) = 0$ as well, $\overline{S}(n)$ tends to zero faster than $\overline{P}(n)$, which implies (in some sense) that there are more primes than perfect squares. As another application, often sums of arithmetic functions arise in investigations (see §4.3.2), and an understanding of the average size suffices.

To complete these and other investigations, we prove some common summation formulas. The most important is partial summation, the discrete analogue of integration by parts. Its utility is that it often allows us to pass from a sum we want to know to a sum which is easier to evaluate. For example, often the sum we want to study is not what naturally arises, but rather a weighted analogue. In §3.2.2 we study $\pi(x) = \sum_{p \text{ prime}} 1$, the number of primes at most x; however the sum that naturally occurs and is easier to evaluate is $\sum_{p \text{ prime}} \log p$. As the logarithm is a slowly varying function, it is not surprising that we can pass from this weighted sum to the unweighted one. Another spectacular application will be in §14.7 when we investigate Germain primes (p and $\frac{p-1}{2}$ both prime), though the same technique works for twin primes and Goldbach's problem (writing even numbers as the sum of two primes).

Exercise 2.2.1. *For what integers a does the divisor function equal a infinitely often? Does $\phi(n)$ take on all positive integer values? How large can $d(n)$ or $\phi(n)$ be relative to n?*

Exercise 2.2.2. *Let p be an odd prime and set $f(n) = 1$ if n is a non-zero square modulo p and 0 otherwise. What is the average order of f?*

2.2.1 Notation

We often want to compare the average value of arithmetic functions to standard functions whose growth properties are well known. To do so we use the following notation:

Definition 2.2.3 (Big-Oh, Little-Oh). *If F and G are two real functions with $G(x) > 0$ for x large, we write*

$$F(x) = O(G(x)) \tag{2.14}$$

if there exist $M, x_0 > 0$ *such that* $|F(x)| \leq MG(x)$ *for all* $x > x_0$. *If*

$$\lim_{x \to \infty} \frac{F(x)}{G(x)} = 0, \qquad (2.15)$$

we write $F(x) = o(G(x))$ *and say* F *is little-oh of* G.

An alternative notation for $F(x) = O(G(x))$ is $F(x) \ll G(x)$. If the constant depends on parameters α, β but not on parameters a, b, we sometimes write $F(x) \ll_{\alpha,\beta} G(x)$. We write $F(x) \sim G(x)$ if

$$F(x) = O(G(x)) \quad \text{and} \quad G(x) = O(F(x)), \qquad (2.16)$$

and say F is of order G. We write $H(x) = F(x) + O(G(x))$ if $H(x) - F(x) = O(G(x))$, and similarly for little oh.

Exercise 2.2.4. *Prove for any* $r, \epsilon > 0$, *as* $x \to \infty$ *we have* $x^r = O(e^x)$ *and* $\log x = O(x^\epsilon)$.

2.2.2 Summation Formulas

We state some summation formulas that are often useful in the study of averages of arithmetic functions. The first is a discrete version of integration by parts: $\int_a^b u\, dv = u(x)v(x)|_a^b - \int_a^b v\, du$. These formulas are extremely important, and allow us to pass from knowledge of one sum to another, related sum.

Theorem 2.2.5 (Partial Summation: Discrete Version). *Let* $A_N = \sum_{n=1}^{N} a_n$. *Then*

$$\sum_{n=M}^{N} a_n b_n = A_N b_N - A_{M-1} b_M + \sum_{n=M}^{N-1} A_n (b_n - b_{n+1}). \qquad (2.17)$$

Proof. Since $A_n - A_{n-1} = a_n$,

$$\sum_{n=M}^{N} a_n b_n = \sum_{n=M}^{N} (A_n - A_{n-1}) b_n$$
$$= (A_N - A_{N-1}) b_N + (A_{N-1} - A_{N-2}) b_{N-1} + \cdots + (A_M - A_{M-1}) b_M$$
$$= A_N b_N + (-A_{N-1} b_N + A_{N-1} b_{N-1}) + \cdots$$
$$\cdots + (-A_M b_{M+1} + A_M b_M) - a_{M-1} b_M$$
$$= A_N b_N - a_{M_1} b_M + \sum_{n=M}^{N-1} A_n (b_n - b_{n+1}). \qquad (2.18)$$

$\square$

We now state an integral version of the above identity:

Theorem 2.2.6 (Partial Summation: Integral Version). *Let* $h(x)$ *be a continuously differentiable function. Let* $A(x) = \sum_{n \leq x} a_n$. *Then*

$$\sum_{n \leq x} a_n h(n) = A(x) h(x) - \int_1^x A(u) h'(u)\, du. \qquad (2.19)$$

Exercise 2.2.7. *Prove Theorem 2.2.6 (see, for example, [Rud], page 70).*

Exercise 2.2.8. *Prove the following:*

$$\sum_{n\leq x} \log n \ = \ x\log x - x + O(\log x) \tag{2.20}$$

and

$$\sum_{n\leq x} \frac{1}{n} \ = \ \log x + O(1). \tag{2.21}$$

Another useful, more powerful summation formula is the following:

Lemma 2.2.9 (Euler's Summation Formula). *If f is continuously differentiable on the interval $[x, y]$, where $0 < x < y$, then*

$$\sum_{x<n\leq y} f(n) \ = \ \int_x^y f(t)\,dt + \int_x^y (t-[t])f'(t)\,dt + f(y)\,([y]-y) - f(x)\,([x]-x).$$

$$\tag{2.22}$$

Exercise 2.2.10. *Prove Lemma 2.2.9 (see for example [Apo], page 54).*

Example 2.2.11. *In this example we show that*

$$\sum_{n\leq x} \frac{1}{n} \ = \ \log x + \gamma + O\left(\frac{1}{x}\right), \tag{2.23}$$

where γ is Euler's constant, defined by

$$\gamma \ = \ \lim_{n\to\infty} \left(\sum_{n\leq x} \frac{1}{n} - \log x\right) \ = \ .5772156649\ldots. \tag{2.24}$$

This is clearly much better than the corresponding estimate from partial summation in (2.21). Take $f(t) = \frac{1}{t}$ in Euler's summation formula to obtain

$$\sum_{n\leq x} \frac{1}{n} \ = \ \int_1^x \frac{dt}{t} - \int_1^x \frac{t-[t]}{t^2}\,dt + 1 - \frac{x-[x]}{x}$$

$$= \ \log x - \int_1^x \frac{t-[t]}{t^2}\,dt + 1 + O\left(\frac{1}{x}\right). \tag{2.25}$$

To finish the proof, note

$$\gamma \ = \ -\lim_{x\to\infty} \int_1^x \frac{t-[t]}{t^2}\,dt + 1. \tag{2.26}$$

Exercise 2.2.12. *Prove (2.26).*

Exercise 2.2.13. *Prove for $x \geq 1$,*

1. $\sum_{n\leq x} \frac{1}{n^s} = \frac{x^{1-s}}{1-s} + O(x^{-s})$ if $s > 0, s \neq 1$.

2. $\sum_{n>x} \frac{1}{n^s} = O(x^{1-s})$ if $s > 1$.

3. $\sum_{n\leq x} n^\alpha = \frac{x^{\alpha+1}}{\alpha+1} + O(x^\alpha)$ if $\alpha \geq 0$.

Exercise 2.2.14. *Find a good estimate for $n!$ (when n large) by using $\log n! = \sum_{k=1}^n \log k$. Estimate the error. Better approximations are given by Stirling's Formula (Lemma 8.4.5), which states $n! \sim n^n e^{-n}\sqrt{2\pi n}$.*

2.2.3 Sums and Averages of Arithmetic Functions

Definition 2.2.15. *Suppose f is an arithmetic function and g is an increasing function of the real variable x. We say that g is the average order of f if as $x \to \infty$*

$$\sum_{n \leq x} f(n) = xg(x) + o(xg(x)). \tag{2.27}$$

The usual technique in calculating the average of an arithmetic function is to represent the function as the convolution product of two other functions, and then use summation techniques to analyze the resulting expression. When one of the functions is trivial, we have the following proposition (which allows us to pass from sums of one function to sums of another).

Proposition 2.2.16. *If $f = g * 1$, then*

$$\sum_{n \leq x} f(n) = \sum_{d \leq x} g(d) \left[\frac{x}{d}\right]. \tag{2.28}$$

Proof. We have

$$\sum_{n \leq x} f(n) = \sum_{n \leq x} \sum_{d|n} g(d)$$

$$= \sum_{d \leq x} g(d) \# \{n \leq x \mid d|n\}$$

$$= \sum_{d \leq x} g(d) \left[\frac{x}{d}\right]. \tag{2.29}$$

We have used the fact that the number of natural numbers $n \leq x$ which are divisible by a given natural number d is equal to $\left[\frac{x}{d}\right]$. $\square$

Exercise 2.2.17. *Generalize the above proposition to the Dirichlet product of arbitrary functions f and g.*

Example 2.2.18. *The following result is needed for §4.3.2:*

$$\sum_{n \leq x} \phi(n) = \frac{3}{\pi^2} x^2 + O(x \log x), \tag{2.30}$$

which implies that the average order of $\phi(n)$ is $\frac{3}{\pi^2} n$. To see this, note

$$\phi(n) = \sum_{d|n} \frac{n}{d} \mu(d). \tag{2.31}$$

We then obtain

$$\sum_{n \leq x} \phi(n) = \sum_{n \leq x} \sum_{d|n} \frac{n}{d} \mu(d)$$

$$= \sum_{d \leq x} \sum_{m \leq \frac{x}{d}} m\mu(d)$$

$$= \sum_{d \leq x} \frac{\mu(d)}{2} \left[\frac{x}{d}\right] \left(\left[\frac{x}{d}\right] + 1\right). \tag{2.32}$$

We now use the fact that $[y]([y]+1) = y^2 + O(y)$ *to obtain*

$$\sum_{n\leq x} \phi(n) = \frac{1}{2}x^2 \sum_{d\leq x} \frac{\mu(d)}{d^2} + O\left(x\sum_{d\leq x}\frac{1}{d}\right). \tag{2.33}$$

By Exercise 2.2.8

$$\sum_{d\leq x}\frac{1}{d} = \log x + O(1), \tag{2.34}$$

and

$$\sum_{d\leq x}\frac{\mu(d)}{d^2} = \sum_{d=1}^{\infty}\frac{\mu(d)}{d^2} + O\left(\sum_{d>x}\frac{1}{d^2}\right)$$
$$= \sum_{d=1}^{\infty}\frac{\mu(d)}{d^2} + O\left(\frac{1}{x}\right) \tag{2.35}$$

by Exercise 2.2.13. The result now follows from these observations and the identity

$$\sum_{d=1}^{\infty}\frac{\mu(d)}{d^2} = \frac{6}{\pi^2}, \tag{2.36}$$

whose proof we leave as an exercise (see §11.3.4).

Exercise 2.2.19. *Prove* (2.31).

Exercise 2.2.20. *Prove that the average of the arithmetic function $\sigma_1(n)$ is Cx for some constant C. We will see in §11.3.4 that the constant is $\frac{\pi^2}{12}$.*

Exercise 2.2.21. *Suppose we have an infinite array $[a_{ij}]_{i,j\in\mathbb{N}}$ consisting entirely of natural numbers. Also suppose that for each i,j we have $a_{ij} \leq ij$. Prove that for each natural number k there is a number that is repeated at least k times among the entries of the array.*

Exercise 2.2.22. *Use the inclusion-exclusion principle (see Exercise 2.3.18 and its generalization Exercise 2.3.19) to compute the probability of two randomly chosen numbers to be coprime.*

Exercise 2.2.23. *What is the probability of a randomly chosen number to be square-free? To be k^{th}-power-free?*

Exercise 2.2.24. *Show that*

$$\sum_{\substack{n\leq x\\(n,k)=1}}\frac{1}{n} = \frac{\phi(k)}{k}\log x + O(1). \tag{2.37}$$

Can you generalize this identity?

See [Mu2] for more examples.

Exercise 2.2.25 (Dirichlet's Hyperbola Method). *Let f, g be two arithmetic functions. Set*

$$F(x) = \sum_{n \leq x} f(n) \tag{2.38}$$

and

$$G(x) = \sum_{n \leq x} g(n). \tag{2.39}$$

Then for $1 \leq y \leq x$, prove

$$\sum_{n \leq x} f * g(n) = \sum_{n \leq y} g(n) F\left(\frac{x}{n}\right) + \sum_{m \leq \frac{x}{y}} f(m) G\left(\frac{x}{m}\right) - F\left(\frac{x}{y}\right) G(y). \tag{2.40}$$

(We invite the reader to consult [Te], Chapter I.3 for various interesting application of the hyperbola method.)

Exercise 2.2.26 (Average Order of the Divisor Function). *Apply the previous exercise to $f(n) = g(n) = 1$ with $y = \sqrt{x}$ to obtain*

$$\sum_{n \leq x} d(n) = x(\log x + 2\gamma - 1) + O(\sqrt{x}). \tag{2.41}$$

Deduce the average order of $d(x)$.

Exercise[h] **2.2.27.** *Let $I_p(s)$ be average of $L(s, \chi)$ over all non-trivial characters* $\mod p$. *Show that*

$$I_p\left(\frac{1}{2}\right) = 1 + O\left(\frac{1}{\sqrt{p}}\right). \tag{2.42}$$

Conclude that as $p \to \infty$, almost all $L(s, \chi)$ are non-vanishing at $s = \frac{1}{2}$.

Remark 2.2.28 (Advanced). Another very powerful method in obtaining average orders for arithmetic functions is the method of formal Dirichlet series combined with Tauberian theorems. In short, the idea is to associate to the arithmetic function $f(n)$ the Dirichlet series

$$\sum_{n=1}^{\infty} \frac{f(n)}{n^s} \tag{2.43}$$

as a formal expression. Sometimes it is possible to say something about the properties of this formal expression as a genuine complex function of the variable s; Tauberian theorems then relate such properties, such as location of poles and their orders, to asymptotic statements about the arithmetic function $f(n)$. For more details see for example [Te], Chapter 17.

See Chapter 4 of [IK] for more on summation formulas.

2.3 COUNTING THE NUMBER OF PRIMES

While we have known for over 2000 years that there are infinitely many primes, it was only in the 1800s that the correct order of magnitude estimates for the number of primes at most x were proved. We prove the order of magnitude estimates, and sketch the proofs of the more exact formulas. See [Da2, Ed, EE] for more details.

2.3.1 Euclid's Theorem

The following proof is one of the oldest mathematical arguments still in use:

Theorem 2.3.1 (Euclid). *There are infinitely many primes.*

Proof. Assume there are only finitely many primes, say $p_1, p_2, \ldots, p_n$. Consider

$$x = p_1 p_2 \cdots p_n + 1; \tag{2.44}$$

x cannot be prime, as we are assuming p_1 through p_n is a complete list of primes. Thus x is composite, and divisible by a prime. However, p_i cannot divide x, as it gives a remainder of 1. Therefore x must be divisible by some prime not in our list, contradicting the assumption that p_1 through p_n is a complete enumeration of the primes. □

For more proofs of the infinitude of primes, see [AZ, Nar].

Exercise 2.3.2. *Use Euler's totient function to prove there are infinitely many primes: let $D = p_1 \cdots p_n$ be the product of all the primes, and consider $\phi(D)$.*

Exercise 2.3.3. *Using Euclid's argument, find an explicit lower bound (growing with x) to*

$$\pi(x) = \#\{p : p \text{ is prime and } p \le x\} = \sum_{p \le x} 1. \tag{2.45}$$

Does Euclid's argument prove there are more primes at most x than perfect squares at most x?

See Exercise 2.3.15 for some results on the size of gaps between adjacent primes.

2.3.2 Dirichlet's Theorem

Theorem 2.3.4 (Dirichlet's Theorem on Primes in Arithmetic Progressions). *Let a and m be relatively prime integers. Then there are infinitely many primes in the progression $nm + a$. Further, for a fixed m to first order all relatively prime a give progressions having the same number of primes. This means that if $\pi_{m,b}(x)$ denotes the number of primes at most x congruent to b modulo m then $\lim_{x \to \infty} \frac{\pi_{m,a}(x)}{\pi_{m,b}(x)} = 1$ for all a, b relatively prime to m. As there are $\phi(m)$ numbers relatively prime to m and $\pi(x)$ primes at most x, we have that as $x \to \infty$*

$$\pi_{m,a}(x) = \frac{\pi(x)}{\phi(m)} + \text{lower order terms.} \tag{2.46}$$

Notice that the condition $(m, a) = 1$ is necessary. If $\gcd(m, a) > 1$, $nm + a$ can be prime for at most one choice of n. Dirichlet's remarkable result is that this condition is also sufficient. We describe this problem in detail in §3.3, culminating in a sketch of the proof in §3.3.3.

Exercise 2.3.5. *Dirichlet's Theorem is not easy to prove in general, although some special cases can be handled with a generalization of Euclid's argument. Prove the case when $m = 4$, $a = 3$. Proving there are infinitely many primes of the form $4n+1$ is harder. For which a, m can Euclid's argument be generalized? See [Mul] and Chapter 3.3.3 for more details.*

Exercise[h] 2.3.6. *Reconsider Exercise 1.6.20.*

2.3.3 Chebyshev's Theorem

Theorem 2.3.7 (Prime Number Theorem (PNT)). *As $x \to \infty$*

$$\pi(x) \sim \frac{x}{\log x}. \tag{2.47}$$

By (2.47) we mean that $\pi(x) - \frac{x}{\log x} = o(\frac{x}{\log x})$. The Prime Number Theorem, which was suggested by Gauss, was proved in 1896 by Jacques Hadamard and Charles Jean Gustave Nicolas Baron de la Vallée Poussin. Looking at tables of primes, in 1792 Gauss proposed that $\pi(x) \sim \frac{x}{\log x}$; he later wrote a better estimate is $\mathrm{Li}(x) = \int_2^x \frac{dt}{\log t}$. It is natural to consider such an integral: tables of primes showed the density of primes around t was approximately $\frac{1}{\log t}$, and integrating the density should give the number of primes. See Theorem 13.2.10 for bounds of the error term.

Exercise 2.3.8. *Prove for any fixed positive integer k that $\mathrm{Li}(x)$ has the expansion*

$$\mathrm{Li}(x) = \int_2^x \frac{dt}{\log t} = \frac{x}{\log x} + \frac{1!x}{\log^2 x} + \cdots + \frac{(k-1)!x}{\log^k x} + O\left(\frac{x}{\log^{k+1} x}\right). \tag{2.48}$$

A weaker version of the Prime Number Theorem was proved by Pafnuty Chebyshev around 1850:

Theorem 2.3.9 (Chebyshev's Theorem). *There exist explicit positive constants A and B such that, for $x > 30$:*

$$\frac{Ax}{\log x} \le \pi(x) \le \frac{Bx}{\log x}. \tag{2.49}$$

Chebyshev showed one can take $A = \log\left(\frac{2^{\frac{1}{2}} 3^{\frac{1}{3}} 4^{\frac{1}{4}}}{30^{\frac{1}{30}}}\right) \approx .921$ and $B = \frac{6A}{5} \approx 1.105$, which are very close to 1. To highlight the method, we use cruder arguments and prove the theorem for a smaller A and a larger B. We follow the presentation in [Da2].

Chebyshev's argument uses an identity of the von Mangoldt Lambda $\Lambda(n)$-function (recall $\Lambda(n) = \log p$ if $n = p^k$ for some prime p, and 0 otherwise). Define

$$T(x) = \sum_{1 \le n \le x} \Lambda(n) \left[\frac{x}{n}\right] = \sum_{n \ge 1} \Lambda(n) \left[\frac{x}{n}\right]. \tag{2.50}$$

Exercise 2.3.10. *Show that $T(x) = \sum_{n \le x} \log n$.*

Exercise[h] 2.3.11. *Show $0 \le \left[\frac{x}{n}\right] - 2\left[\frac{x}{2n}\right] \le 1$.*

Proof of Chebyshev's Theorem. Using Exercise 2.3.10 and comparing the upper and lower sums gives

$$T(x) = \sum_{n \le x} \log n = \int_1^x \log t \, dt + O(\log x) = x \log x - x + O(\log x), \tag{2.51}$$

a good approximation to $T(x)$. As we will have errors of size x, there is no gain from improving the $O(\log x)$ error above. The trick is to look at

$$T(x) - 2T\left(\frac{x}{2}\right) = \sum_n \Lambda(n)\left(\left[\frac{x}{n}\right] - 2\left[\frac{x}{2n}\right]\right). \tag{2.52}$$

By (2.51), the left hand side of (2.52) is

$$T(x) - 2T\left(\frac{x}{2}\right) = x\log 2 + O(\log x). \tag{2.53}$$

For the right hand side, by Exercise 2.3.11 the difference in (2.52) is at most 1. The numbers $p, p^2, \ldots, p^k$ contribute at most $\log p$ each, provided $p^k \leq x$. There are at most $\left[\frac{\log x}{\log p}\right] \leq \frac{\log x}{\log p}$ such prime powers. Therefore the right hand side satisfies

$$\sum_n \Lambda(n)\left(\left[\frac{x}{n}\right] - 2\left[\frac{x}{2n}\right]\right) \leq \sum_{p \leq x}(\log p)\frac{\log x}{\log p} = \pi(x)\log x. \tag{2.54}$$

Hence we immediately obtain the lower bound (which is better than what Euclid's argument gives)

$$\pi(x) \geq \frac{x\log 2}{\log x} + O(\log x). \tag{2.55}$$

To obtain an upper bound for $\pi(x)$, we notice that since $[2\alpha] \geq 2[\alpha]$, the sum in (2.52) has only positive terms. By dropping terms we obtain a lower bound.

$$
\begin{aligned}
T(x) - 2T\left(\frac{x}{2}\right) &\geq \sum_{x/2 < n \leq x} \Lambda(n)\left(\left[\frac{x}{n}\right] - 2\left[\frac{x}{2n}\right]\right) \\
&\geq \sum_{x/2 < p \leq x} \log p \\
&\geq \log\left(\frac{x}{2}\right) \sum_{x/2 < p \leq x} 1 \\
&= \log\left(\frac{x}{2}\right)\left(\pi(x) - \pi\left(\frac{x}{2}\right)\right). \tag{2.56}
\end{aligned}
$$

Remark 2.3.12 (Important: Approximating Technique). This is a very common technique: for $p \in [x/2, x]$, $\log p$ is approximately constant, and there is very little error in replacing the slowly varying function with $\log(x/2)$. Further, for $\frac{x}{2} < n \leq x$, $\left[\frac{x}{n}\right] - 2\left[\frac{x}{2n}\right] = 1$, which means that for such n, there is no error in replacing this difference with 1. In estimating sums, one should always explore how much information is "lost" by any approximation.

We have found an upper bound for the number of primes between $\frac{x}{2}$ and x:

$$\pi(x) - \pi(x/2) \leq \frac{x\log 2}{\log(\frac{x}{2})} + O(1). \tag{2.57}$$

Now if we write inequality (2.57) for x, $\frac{x}{2}$, $\frac{x}{2^2}$, ... and use $\log 2 < 1$ then we obtain

$$\pi(x) - \pi(x/2) \leq 2\frac{x/2}{\log(x/2)} + O(1),$$

$$\pi(x/2) - \pi(x/2^2) \leq 2\frac{x/2^2}{\log(x/2^2)} + O(1),$$

$$\vdots$$

$$\pi(x/2^{k-1}) - \pi(x/2^k) \leq 2\frac{x/2^k}{\log(x/2^k)} + O(1), \tag{2.58}$$

as long as $\frac{x}{2^k} > 1$, i.e., $k < [\log_2 x] = k_0$. Summing the above inequalities we get on the left hand side a *telescoping sum*; this is a common technique known as a *dyadic decomposition* as we have a union of intervals of the form $[m, 2m]$. All the terms cancel except for the leading term $\pi(x)$ and $\pi(x/2^{k_0})$, which is either 0 or 1 as $1 < x/2^{k_0} \leq 2$.

It simplifies later arguments if we choose k_0 so that $3 \leq x/2^{k_0} \leq 6$. The reason for this is we see above that we are dividing by $\log(x/2^k)$; if $x/2^k$ is close to 1 then this logarithm is close to zero. The above restriction ensures that this logarithm is always at least 1; as there are only three primes in $[1, 6]$, our bound will miss at most three primes. By adding three to the sum of the terms in (2.58) we obtain an upper bound for $\pi(x)$. Thus

$$\pi(x) \leq 3 + 2\sum_{k=1}^{k_0} \frac{x/2^k}{\log(x/2^k)} + O(k_0). \tag{2.59}$$

As $k_0 = [\log_2 x]$, the error term is of size $O(\log x)$. To evaluate the sum in (2.59), we split it into two parts, k small and k large. The problem is the competing influence of k: when k is large, $x/2^k$ is small but we are dividing by the logarithm of a small number; when k is small $x/2^k$ is large but we are dividing by the logarithm of a large number. This is a common technique and allows us to exploit $\log(x/2^k)$ is large for k small, and $x/2^k$ is small for k large. More precisely, let $n_0 = \log_2(x^{1/10})$ so that $2^{n_0} = x^{1/10}$. For $k > n_0$ we have $\log(x/2^k) > 1$ and note that

$$2\sum_{k>n_0} \frac{x/2^k}{\log(x/2^k)} \leq 2\sum_{k>n_0} \frac{x}{2^k} \leq \frac{2x}{2^{n_0}} = \frac{2x}{x^{1/10}} = 2x^{9/10}. \tag{2.60}$$

Hence the contribution from k large is very small compared to what we expect (i.e., order of magnitude $\frac{x}{\log x}$), or we can say that the main term comes from the sum over k small.

We now evaluate the contribution from small k.

$$2\sum_{k=1}^{n_0} \frac{x}{2^k}\frac{1}{\log(x/2^k)} \leq \frac{2x}{\log(x/2^{n_0})}\sum_{k=1}^{n_0}\frac{1}{2^k} \leq \frac{2x}{\log(x^{9/10})} = \frac{20}{9}\frac{x}{\log x}. \tag{2.61}$$

Therefore the right hand side of the (2.59) is made up of three parts: a main term of size $\frac{Bx}{\log x}$ coming from (2.61), a lower order term coming from (2.60) and $3 + O(\log x)$. For x sufficiently large,

$$\pi(x) \leq \frac{Bx}{\log x}, \tag{2.62}$$

where B can be any constant strictly bigger than $\frac{20}{9}$. To obtain Chebyshev's better constant we would have to work a little harder along these lines, but it is the same method.

Gathering equations (2.55) and (2.62), we see we have proven that for x large

$$\frac{Ax}{\log x} \leq \pi(x) \leq \frac{Bx}{\log x}. \tag{2.63}$$

While this is not an asymptotic for $\pi(x)$, it does give the right order of magnitude for $\pi(x)$, namely $\frac{x}{\log x}$. □

For many applications, the full strength of the Prime Number Theorem is not needed. We give two such applications:

Exercise 2.3.13. *Using Chebyshev's Theorem, prove Bertrand's Postulate: for any integer $n \geq 1$, there is always a prime number between n and $2n$.*

Exercise 2.3.14. *Using Chebyshev's Theorem, prove for any integer M there exists an even integer $2k$ such that there are at least M primes p with $p + 2k$ also prime. Unfortunately $2k$ will depend on M. If it did not, we would have solved the Twin Prime Conjecture, namely, there are infinitely many primes p such that $p + 2$ is also prime; see also §14.1.*

Exercise 2.3.15 (Prime Deserts). *Let p_n denote the nth prime. The Prime Number Theorem or Chebyshev's Theorem imply that $p_{n+1} - p_n$ is typically about $\log p_n$. A fascinating question is how large or small this difference can be. Prove there are arbitrarily large gaps between adjacent primes. Namely, show for any N there is an m such that $p_{m+1} - p_m \geq N$. We know there is a prime in every interval $[x, 2x]$; therefore N must be very small relative to p_m. Let $G(x)$ be the largest gap between adjacent primes that are at most x. Find a lower bound for $G(x)$ as a function of x. See Chapter 10 of [IK] (pages 264–267) for more on primes in small intervals and differences between adjacent primes.*

Exercise[h] **2.3.16.** *Prove the sum of the reciprocals of the primes diverges; as the sum of the reciprocals of perfect squares converges, this provides another measure that there are more primes than perfect squares (see also §2.2).*

We know that $\sum_{n \leq x} \frac{1}{n} = \log x + O(1)$; how large is $\sum_{p \leq x} \frac{1}{p}$? For sums like these which diverge very slowly, it is easy to be mislead into thinking they converge by evaluating them for modest sized x. Mertens showed

$$\sum_{p \leq x} \frac{1}{p} = \log \log x + B_1 + o(1), \quad B_1 \approx .261496; \tag{2.64}$$

see [Da2] for a proof, and Project 4.5.1 and Remark 7.10.2 for interesting examples where data can be misleading.

Exercise 2.3.17. *Let $\pi_2(x)$ denote the number of twin primes at most x; p is a twin prime if p and $p + 2$ are both prime. Brun (see [Na]) proved that $\pi_2(x) = O\left(\frac{x(\log \log x)^2}{\log^2 x}\right)$. Prove that if the sum of the reciprocals of the twin primes diverges then there are infinitely many twin primes. Using the methods of Exercise*

2.3.16, *show this bound on* $\pi_2(x)$ *implies the sum of the reciprocals of twin primes converges. Unfortunately, it has been shown that*

$$\sum_{\substack{p \\ p,p+2 \; prime}} \left(\frac{1}{p} + \frac{1}{p+2} \right) \approx 1.90216058; \tag{2.65}$$

this is called Brun's constant. Interestingly, the Pentium bug was discovered by Nicely [Ni1, Ni2] in his investigations of this sum!

Exercise 2.3.18. *The following series counts the number of primes at most N. Let $\lfloor x \rfloor$ denote the greatest integer at most x. Any composite number that is at most N is divisible by a prime that is at most $\sqrt{N}$. There are $N - 1$ integers in $\{2, 3, 4, \ldots, N\}$: these are the candidate prime numbers. The number of these divisible by p is $\lfloor \frac{N}{p} \rfloor$; we remove all but one of these from our candidate list (as $1 \cdot p$ is prime and the other multiples are composite). Our first approximation to the number of primes at most N is*

$$N - 1 - \sum_{\substack{p \leq \sqrt{N} \\ p \; prime}} \left(\left\lfloor \frac{N}{p} \right\rfloor - 1 \right). \tag{2.66}$$

The above formula is slightly off: numbers divisible by two distinct primes were discarded twice. We must add back numbers that are multiples of two primes:

$$N - 1 - \sum_{\substack{p \leq \sqrt{N} \\ p \; prime}} \left(\left\lfloor \frac{N}{p} \right\rfloor - 1 \right) + \sum_{\substack{p_1,p_2 \leq \sqrt{N} \\ p_1 \neq p_2, p_i \; prime}} \left\lfloor \frac{N}{p_1 p_2} \right\rfloor. \tag{2.67}$$

Of course, numbers divisible by three distinct primes have been kept in the list, and must be removed. Finish the argument, and write down a formula for $\pi(N)$. This process is known as Inclusion – Exclusion (see also Remark 1.4.23); while we have an explicit formula for $\pi(N)$, this formula is computationally difficult to use.

Exercise 2.3.19 (Inclusion-Exclusion). *One can easily generalize the result of the previous exercise. Suppose $A_1, A_2, \ldots, A_n$ is a collection of sets. Then the Inclusion-Exclusion Principle asserts that*

$$\left| \bigcup_{i=1}^{n} A_i \right| = \sum_i |A_i| - \sum_{i,j} |A_i \cap A_j| + \sum_{i,j,k} |A_i \cap A_j \cap A_k| - \cdots . \tag{2.68}$$

Use the Inclusion-Exclusion Principle to determine how many square-free numbers there are less than N (as $N \to \infty$). See also Exercise 3.1.8.

2.3.4 Weighted Sums

It is often easier to count weighted sums (and then remove the weights by partial summation) because it is the weighted sum which naturally arises; we give one such example below. Another surfaces in our investigations of Germain primes (Chapter 14). Partial Summation is one of the main tools in a number theorist's arsenal: the point of the exercise below is to show that "nice" weights can readily be removed (see §14.7 for more on weighted sums).

Remark 2.3.20 (Advanced). In Chapter 3 we study the Riemann zeta function

$$\zeta(s) = \sum_{n=1}^{\infty} \frac{1}{n^s} = \prod_{p \text{ prime}} \left(1 - \frac{1}{p^s}\right)^{-1}, \tag{2.69}$$

which converges for $\Re(s) > 1$. From complex analysis, contour integrals of the logarithmic derivative provide information about the zeros and poles. The **logarithmic derivative** is

$$\frac{d \log \zeta(s)}{ds} = \frac{\zeta'(s)}{\zeta(s)} = -\sum_{n=1}^{\infty} \frac{\Lambda(n)}{n^s}. \tag{2.70}$$

Conversely, knowledge of the distribution of the zeros and poles can be used to obtain information about $\sum_{n \leq x} \Lambda(n)n^{-s}$, which by partial summation and trivial estimation yields information about $\sum_{p \leq x} 1$. This shows why it is "natural" to weight the prime p by $\log p$ in counting arguments. See §3.2.2 for complete details.

Suppose we know that

$$\sum_{p \leq x} \log p = x + O(x^{\frac{1}{2}+\epsilon}). \tag{2.71}$$

We use this to glean information about $\pi(x) = \sum_{p \leq x} 1$. Define

$$h(n) = \frac{1}{\log n} \quad \text{and} \quad a_n = \begin{cases} \log n & \text{if } n \text{ is prime} \\ 0 & \text{otherwise.} \end{cases} \tag{2.72}$$

Applying partial summation to $\sum_{p \leq x} a_n h(n)$ yields knowledge about $\sum_{p \leq x} 1$. Note as long as $h(n) = \frac{1}{\log n}$ for n prime, it does not matter how we define $h(n)$ elsewhere; however, to use the integral version of Partial Summation (Theorem 2.2.6), we need h to be a differentiable function. Thus

$$\sum_{p \leq x} 1 = \sum_{p \leq x} a_n h(n)$$

$$= \left(x + O(x^{\frac{1}{2}+\epsilon})\right) \frac{1}{\log x} - \int_2^x \left(u + O(u^{\frac{1}{2}+\epsilon})\right) h'(u) du. \tag{2.73}$$

The main term $A(x)h(x)$ equals $\frac{x}{\log x}$ plus a significantly smaller error. We now calculate the integral, noting $h'(u) = -\frac{1}{u \log^2 u}$. The error piece in the integral gives a constant multiple of

$$\int_2^x \frac{u^{\frac{1}{2}+\epsilon}}{u \log^2 u} du. \tag{2.74}$$

As $\frac{1}{\log^2 u} \leq \frac{1}{\log^2 2}$ for $2 \leq u \leq x$, the integral is bounded by

$$\frac{1}{\log^2 2} \int_2^x u^{-\frac{1}{2}+\epsilon} < \frac{1}{\log^2 2} \cdot \frac{x^{\frac{1}{2}+\epsilon}}{\frac{1}{2}+\epsilon}, \tag{2.75}$$

which is significantly less than $A(x)h(x) = \frac{x}{\log x}$. We now need to handle the other integral:

$$\int_2^x \frac{u}{u \log^2 u} du = \int_2^x \frac{1}{\log^2 u} du. \tag{2.76}$$

The obvious approximation to try is $\frac{1}{\log^2 u} \leq \frac{1}{\log^2 2}$. Unfortunately, inserting this bounds the integral by $\frac{x}{\log^2 2}$. This is larger than the expected main term, $A(x)h(x)$! The reason trivial bounds failed is the length of the integral is too large (of size x) and we lost too much decay by approximating $\frac{1}{\log^2 u}$ with $\frac{1}{\log^2 2}$.

As a rule of thumb, in trying to bound some quantity try the simplest, most trivial bounds first. Only if they fail should one attempt to be clever. Here we need to be clever as the obvious approach bounds the integral by something larger than the observed terms. We split the integral into two pieces:

$$\int_2^x = \int_2^{\sqrt{x}} + \int_{\sqrt{x}}^x. \qquad (2.77)$$

For the first piece, we use the trivial bound for $\frac{1}{\log^2 u}$. Note the interval has length $\sqrt{x} - 2 < \sqrt{x}$. Thus, the first piece contributes at most $\frac{\sqrt{x}}{\log^2 2}$, significantly less than $A(x)h(x)$. The advantage of splitting the integral in two is that in the second piece, even though most of the length of the original interval is here (it is of length $x - \sqrt{x} \approx x$), the function $\frac{1}{\log^2 u}$ is small here. Instead of bounding this function by a constant everywhere, we now bound it by a constant only on this interval. Thus the contribution from this integral is at most $\frac{x - \sqrt{x}}{\log^2 \sqrt{x}} < \frac{4x}{\log^2 x}$. Note that this is significantly less than the main term $A(x)h(x) = \frac{x}{\log x}$.

Splitting the integral as above is another standard technique. The advantage is that, similar to the proof of Chebyshev's Theorem in §2.3.3, we can use different bounding functions in the different regions.

Exercise 2.3.21. *Determine a good order of magnitude estimate as $x \to \infty$ for*

$$\int_1^x \frac{1}{t \log(t+2)} \, dt. \qquad (2.78)$$

Exercise 2.3.22. *There are of course analogues to splitting the integral for sums. Consider*

$$\sum_{n \leq \frac{M}{2}} \frac{1}{n^2 \log \frac{M}{n}}. \qquad (2.79)$$

Show the sum is $O\left(\frac{1}{\log M}\right)$. This problem arises in the following context. Let G be a finite group. By Lemma 1.4.13, G can be embedded in S_n, the group of permutations on n letters. Let $d(G)$ be the smallest n such that G can be embedded in S_n, $|G|$ the number of elements in G and $\alpha(G) := \frac{d(G)}{|G|}$. Therefore $0 < \alpha(G) \leq 1$. See [EST] for how the above sum arises in estimating the average order of $\alpha(G)$.

Chapter Three

Zeta and L-Functions

In §2.3 we gave two proofs that there are infinitely many primes. Euclid's proof led to a weak lower bound for $\pi(x)$, the number of primes at most x. Chebyshev proved that there exist constants A and B such that $\frac{Ax}{\log x} \leq \pi(x) \leq \frac{Bx}{\log x}$. Gauss conjectured that the correct order of magnitude of $\pi(x)$ is $\frac{x}{\log x}$. Hadamard and de la Vallée Poussin proved this and more in the 1890s: they were able to obtain bounds for $|\pi(x) - \frac{x}{\log x}|$. Their proofs involved powerful techniques from complex analysis, which might seem surprising as there are no complex numbers in sight. This is because the auxiliary functions (called L-functions) which are introduced to help study the primes are functions of complex variables. It was not until almost 50 years later that "elementary" (this does not mean easy: it means using the standard arithmetic functions of Chapter 2) proofs were found by Erdös and Selberg. See [Gol2] for the history and [Ed, EE] for an exposition of the elementary proof.

In this chapter we explain the connection of complex analysis to the study of primes. The introductory sections require only basic arithmetic and calculus; except for §3.2.2, the later sections assume some familiarity with complex analysis. While we give a quick review of some of the tools and techniques (both in this chapter and later in the book), this is meant only to give a flavor of the subject. The interested reader should read the statements of the theorems and then consult the references for more details. We shall use much of this material in Chapters 15 to 18 when we investigate the distribution of zeros of L-functions.

3.1 THE RIEMANN ZETA FUNCTION

A series of the form

$$\sum_{n=1}^{\infty} \frac{a(n)}{n^s} \tag{3.1}$$

with $s \in \mathbb{C}$ and $a(n)$ a sequence of complex numbers is called a **Dirichlet series**. It is common to write $s = \sigma + it$, with $\sigma, t \in \mathbb{R}$. Thus the real part of s is $\Re s = \sigma$ and the imaginary part is $\Im s = t$.

The most exciting Dirichlet series is the simplest one of the them all. The **Riemann zeta function** is defined by

$$\zeta(s) = \sum_{n=1}^{\infty} \frac{1}{n^s}. \tag{3.2}$$

Exercise[h] **3.1.1.** *Prove $\zeta(s)$ converges for $\Re s > 1$. If $s \in [0, 1]$, show the series for $\zeta(s)$ diverges.*

While $\zeta(s)$ as defined only makes sense for $\Re s > 1$, we will show that we can "extend" this to a new function defined for all $s \in \mathbb{C}$. In §3.1.1 we show how $\zeta(s)$ is built from the primes, and it is this connection that will prove fruitful below. Explicitly, tools from elementary analysis and complex analysis allow us to pass from knowledge about $\zeta(s)$ to knowledge about the primes. See the survey article [Wei2] for the origins of $\zeta(s)$, and [Roc] for a popular account of its history and connections to various parts of mathematics. Riemann's classic paper [Ri], as well as an expanded version, are available in [Ed]. This short paper is Riemann's only published paper on $\zeta(s)$. It is one of the most influential papers in mathematics, and we strongly encourage everyone to read it.

3.1.1 Euler Product

The following property shows the connection between $\zeta(s)$ and the primes. The proof depends crucially on the **unique factorization** property of integers, namely any positive integer n can be written uniquely as a product of powers of primes (if $n = p_1^{r_1} \cdots p_k^{r_k} = q_1^{s_1} \cdots q_m^{s_m}$, then $k = m$ and after possibly reordering, $p_i = q_i$).

Exercise 3.1.2. *Prove unique factorization.*

Theorem 3.1.3 (Euler Product of $\zeta(s)$). *We have*

$$\zeta(s) = \sum_n \frac{1}{n^s} = \prod_p \left(1 - \frac{1}{p^s}\right)^{-1}, \quad \Re s > 1. \tag{3.3}$$

Sketch of Proof. The product expansion follows from the absolute convergence of the series for $\Re s > 1$, and the unique factorization property of the integers. Namely, by the geometric series formula we have

$$\frac{1}{1 - r} = 1 + r + r^2 + r^3 + \cdots = \sum_{k=0}^{\infty} r^k \tag{3.4}$$

for $|r| < 1$. Applying this to $(1 - p^{-s})^{-1}$ for $\Re s > 1$ we obtain

$$\left(1 - \frac{1}{p^s}\right)^{-1} = 1 + \frac{1}{p^s} + \left(\frac{1}{p^s}\right)^2 + \cdots; \tag{3.5}$$

multiplying these expressions and using the unique factorization of natural numbers into products of prime powers completes the proof. □

Exercise[hr] **3.1.4.** *Prove the infinite geometric series formula (3.4), as well as the finite geometric series formula*

$$\sum_{k=m}^{n} r^k = \frac{r^m - r^{n+1}}{1 - r}. \tag{3.6}$$

Exercise[h] **3.1.5.** *Rigorously prove Theorem 3.1.3.*

The product expansion

$$\zeta(s) \;=\; \prod_p \left(1 - \frac{1}{p^s}\right)^{-1} \tag{3.7}$$

is called the **Euler product** after Euler, who initiated such investigations. This product expansion is one of the most important properties of $\zeta(s)$. Initially $\zeta(s)$ is defined as a sum over integers — the Euler product connects $\zeta(s)$ to the primes. As the integers are well understood, this allows us to pass from knowledge of integers to knowledge of primes, as the next two exercises illustrate.

Exercise[hr] 3.1.6 (Important: $\zeta(1)$). *Show $\lim_{x\to\infty} \sum_{n\le x} \frac{1}{n}$ diverges by comparing it with $\log x$ (use the Integral test). Use the Euler product to show that there are infinitely many prime numbers by investigating $\lim_{s\to 1} \zeta(s)$.*

Exercise 3.1.7 (Important: $\zeta(2)$). *Assume that π^2 is irrational; thus $\pi^2 \ne \frac{a}{b}$ for any $a, b \in \mathbb{N}$. We sketch a proof of the irrationality of π^2 in Exercise 5.4.17. In §11.3.4 we show that*

$$\zeta(2) \;=\; \sum_{n=1}^{\infty} \frac{1}{n^2} \;=\; \frac{\pi^2}{6}. \tag{3.8}$$

Using the Euler product, show there must be infinitely many primes. See Theorem 205 of [HW] for a proof of the transcendence of π (which implies π^2 is irrational), and Chapter 11 of [BB] or [Sc] for a history of calculations of π (as well as many entertaining exercises). This is just one of many applications of Fourier analysis (Chapter 11) to Number Theory.

Exercise 3.1.8. *Write $\frac{1}{\zeta(2)}$ as a sum over primes and prime powers, and interpret this number as the probability that as $N \to \infty$ a number less than N is square-free. See also Exercise 2.3.19.*

The method of proof in Exercise 3.1.7 is known as a **special value proof**, and is the first example of a very important phenomenon: Dirichlet series evaluated at certain "natural" points encode interesting algebraic and arithmetic information (see also Exercise 3.3.28). Unlike Euclid and Chebyshev's Theorems, $\zeta(2)$ irrational provides *no* information on how many primes there are at most x; by Partial Summation one can obtain bounds on $\pi(x)$ from $\lim_{s\to 1} \zeta(s)$. See [BP] for connections between $\zeta(k)$ and continued fractions. For some results on the distribution of digits of values of $\zeta(s)$ and Benford's Law, see Remark 9.4.5.

Exercise 3.1.9. *Use the product expansion to prove $\zeta(s) \ne 0$ for $\Re s > 1$; this important property is not at all obvious from the series expansion. While it is clear from the series expansion that $\zeta(s) \ne 0$ for real $s > 1$, what happens for complex s is not apparent.*

Exercise 3.1.10. *Suppose a_n is a sequence of positive rational numbers such that $P = \prod_{n=1}^{\infty} a_n$ is convergent; this means that $\lim_{N\to\infty} \prod_{n=1}^{N} a_n$ converges to a finite non-zero number. Must P be rational? If so, this would provide a very easy proof of the irrationality of π! See also Exercises 3.3.8 and 5.6.9.*

Exercise[h] **3.1.11.** *Prove (see Exercise 3.3.8) that $\prod_{n=2}^{\infty} \zeta(n)$ converges and is finite, and bound $\left| \prod_{n=2}^{\infty} \zeta(n) - \prod_{n=2}^{N} \zeta(n) \right|$. See [CL1, CL2] for applications of this number to average orders of groups.*

3.1.2 Functional Equation and Analytic Continuation

The Euler product is the first of many important properties of $\zeta(s)$. The next is that $\zeta(s)$ has a meromorphic continuation to the entire plane. We first give a definition and an example of meromorphic continuation, and then return to our study of $\zeta(s)$.

Definition 3.1.12 (Zero, Pole, Order, Residue). *Assume f has a convergent Taylor series expansion (see §A.2.3) about z_0:*

$$f(z) = a_n(z - z_0)^n + a_{n+1}(z - z_0)^{n+1} + \cdots = \sum_{m=n}^{\infty} a_m(z - z_0)^m, \quad (3.9)$$

with $a_n \neq 0$. Thus n is the location of the first non-zero coefficient. If $n > 0$ we say f has a zero of order n at z_0; if $n < 0$ we say f has a pole of order $-n$ at z_0. If $n = -1$ we say f has a simple pole with residue a_{-1}. We denote the order of the f at z_0 by $\mathrm{ord}_f(z_0) = n$.

Definition 3.1.13 (Meromorphic Function). *We say f is meromorphic at z_0 if the Taylor expansion about z_0 converges for all z close to z_0. In particular, there is a disk about z_0 of radius r and an integer n_0 such that for all z with $|z - z_0| < r$,*

$$f(z) = \sum_{n \geq n_0} a_n(z - z_0)^n. \quad (3.10)$$

*If f is meromorphic at each point in a disk, we say f is meromorphic in the disk. If $n_0 \geq 0$, we say f is **analytic**.*

We have already seen an example of meromorphic continuation (to be defined below). Consider the geometric series $G(r) = \sum_{n=1}^{\infty} r^n$. This series converges for all r with $|r| < 1$, and for such r we have

$$\sum_{n=1}^{\infty} r^n = \frac{1}{1 - r}. \quad (3.11)$$

Let us denote the right hand side of (3.11) by $H(r)$. By Exercise 3.1.4, $G(r) = H(r)$ for $|r| < 1$; however, $H(r)$ is well defined for *all* r except for $r = 1$, where $H(r)$ is undefined (it has a simple pole with residue -1). As $H(r)$ agrees with $G(r)$ wherever $G(r)$ is defined and is defined for additional r, we say H is a continuation of G. Since H has a pole, we say H is a **meromorphic continuation**; if H had no poles, we would have an **analytic continuation**. If a function defined for all $z \in \mathbb{C}$ has a convergent Taylor expansion at each point (which implies it has no poles), the function is said to be **entire**.

To study $\zeta(s)$, we need to recall the definition and basic properties of the Gamma function $\Gamma(s)$. For $\Re s > 0$ set

$$\Gamma(s) = \int_0^{\infty} e^{-t} t^{s-1} \, dt. \quad (3.12)$$

Exercise 3.1.14. *Prove for $\Re s > 0$ that*

$$\Gamma(s+1) \;=\; s\Gamma(s). \tag{3.13}$$

One can then use (3.13) to extend the definition of $\Gamma(s)$ to all values of s. It is then seen that the value of $\Gamma(s)$ is always finite unless $s = 0, -1, -2, \ldots$.

Exercise 3.1.15. *Prove the above claims. For n a positive integer, show $\Gamma(n) = (n-1)!$ (remember $0!$ is defined to be 1). Thus $\Gamma(s)$ is a generalization of the factorial function $n!$.*

The following theorem collects some of the most important properties of the Gamma function. We refer the reader to [WW], Chapters 12 and 13 for proofs.

Theorem 3.1.16. *The Γ-function has the following properties:*

1. *$\Gamma(s)$ has a meromorphic continuation to the entire complex plane with simple poles at $s = 0, -1, -2, \ldots$, and the residue at $s = -k$ is $\frac{(-1)^k}{k!}$. The meromorphically continued function is never zero.*

2. *For all s we have*

$$\Gamma(s)\Gamma(1-s) \;=\; \frac{\pi}{\sin \pi s} \tag{3.14}$$

and

$$\Gamma(s)\Gamma\left(s+\frac{1}{2}\right) \;=\; 2^{1-2s}\pi^{\frac{1}{2}}\Gamma(2s). \tag{3.15}$$

3. *For any fixed $\delta > 0$, as $|s| \to \infty$ in $-\pi + \delta < \arg s < \pi - \delta$,*

$$\log \Gamma(s) \;=\; \left(s - \frac{1}{2}\right)\log s - s + \frac{1}{2}\log 2\pi + O(|s|^{-1}). \tag{3.16}$$

Remark 3.1.17. Theorem 3.1.16(3) is called Stirling's Formula. It can be used to give an asymptotic formula for $n!$. See also Lemma 8.4.5.

Exercise[(h)] **3.1.18.** *Show $\Gamma(s)$ has a simple pole with residue 1 at $s = 0$. We will need this in Chapter 18 when we derive the explicit formula (which relates sums over zeros of $\zeta(s)$ to sums over primes; this formula is the starting point for studying properties of the zeros of $\zeta(s)$). More generally, for each non-negative integer k show that $\Gamma(s)$ has a pole at $s = -k$ with residue $\frac{(-1)^k}{k!}$. Finally, show $\Gamma(s)$ is never zero.*

The following theorem is one of the most important theorems in mathematics:

Theorem 3.1.19 (Analytic Continuation of the Completed Zeta Function). *Define the completed zeta function by*

$$\xi(s) \;=\; \frac{1}{2}s(s-1)\Gamma\left(\frac{s}{2}\right)\pi^{-\frac{s}{2}}\zeta(s); \tag{3.17}$$

$\xi(s)$, originally defined for $\Re s > 1$, has an analytic continuation to an entire function and satisfies the functional equation $\xi(s) = \xi(1-s)$.

Proof. For the functional equation we follow Riemann's original argument as described in [Da2]. For $\Re s > 0$, by definition of the Gamma function and change of variables we have

$$\int_0^\infty x^{\frac{1}{2}s-1} e^{-n^2\pi x}\, dx = \frac{\Gamma\left(\frac{s}{2}\right)}{n^s \pi^{\frac{s}{2}}}. \tag{3.18}$$

Summing over $n \in \mathbb{N}$, with $\Re s > 1$ to guarantee convergence, we obtain

$$\pi^{-\frac{1}{2}s}\Gamma\left(\frac{s}{2}\right)\zeta(s) = \int_1^\infty x^{\frac{1}{2}s-1}\left(\sum_{n=1}^\infty e^{-n^2\pi x}\right)dx$$

$$= \int_0^\infty x^{\frac{1}{2}s-1}\omega(x)\, dx$$

$$= \int_1^\infty x^{\frac{1}{2}s-1}\omega(x)\, dx + \int_1^\infty x^{-\frac{1}{2}s-1}\omega\left(\frac{1}{x}\right)dx, \tag{3.19}$$

with $\omega(x) = \sum_{n=1}^{+\infty} e^{-n^2\pi x}$. Note we divided the integral into two pieces and changed variables by $x \mapsto x^{-1}$ in the second integral; this leads to rapidly converging integrals. The absolute convergence of the sum-integral justifies the rearrangement of the terms in (3.19) (see Theorem A.2.8, but see Exercise 11.4.12 for an example where the orders cannot be interchanged). We will show in a moment that the function ω satisfies the functional equation

$$\omega\left(\frac{1}{x}\right) = -\frac{1}{2} + \frac{1}{2}x^{\frac{1}{2}} + x^{\frac{1}{2}}\omega(x) \tag{3.20}$$

for $x > 0$. Simple algebra then shows that for $\Re s > 1$ we have

$$\pi^{-\frac{1}{2}s}\Gamma\left(\frac{s}{2}\right)\zeta(s) = \frac{1}{s(s-1)} + \int_1^\infty (x^{\frac{1}{2}s-1} + x^{-\frac{1}{2}s-\frac{1}{2}})\omega(x)\, dx. \tag{3.21}$$

Because of the rapid decay of $\omega(x)$, the integral on the right converges absolutely for *any* s and represents an entire function of s. The remaining assertions of the theorem are easy consequences of the location of poles of $1/s(s-1)$ and the invariance of the right hand side of (3.21) under $s \mapsto 1-s$. It remains to verify the functional equation of ω. For this we write

$$\omega(x) = \frac{\theta(x) - 1}{2} \tag{3.22}$$

with

$$\theta(x) = \sum_{n=-\infty}^{+\infty} e^{-\pi n^2 x}. \tag{3.23}$$

Note this series is rapidly converging for $x > 0$. The desired functional equation for ω easily follows from

$$\theta(x^{-1}) = x^{\frac{1}{2}}\theta(x), \quad x > 0, \tag{3.24}$$

which we now prove. Without worrying about convergence, we have

$$\theta(x^{-1}) = \sum_{n=-\infty}^\infty e^{-\pi n^2 x^{-1}}$$

$$= \sum_{\nu=-\infty}^\infty \int_{-\infty}^\infty e^{-\pi t^2 x^{-1} + 2\pi i \nu t}\, dt = x^{\frac{1}{2}}\theta(x) \tag{3.25}$$

by the Poisson summation formula (see §11.4.2) and Exercise 3.1.22. □

Remark 3.1.20. Theorem 3.1.19 furnishes us with a meromorphic continuation of $\zeta(s)$ to all $s \in \mathbb{C}$, with its only pole at $s = 1$. We denote the meromorphic continuation by $\zeta(s)$; note, however, that $\zeta(s)$ is only given by the series and product expansions if $\Re s > 1$.

Remark 3.1.21 (Technique: Splitting the Integral). A common method in number theory for obtaining analytic properties of functions defined by integral transforms (here a multiple of $\zeta(s)$ equals $\int_0^\infty x^{\frac{1}{2}s-1}w(x)dx$) is to write the integration as $\int_0^1 + \int_1^\infty$ and then use functional relations of the integrand to relate values at $\frac{1}{x}$ to those at x.

Exercise 3.1.22 (Advanced). *Justify the use of the Poisson summation formula in the proof of Theorem 3.1.19. In particular, if $f(t) = e^{-\pi t^2}$ then calculate $\hat{f}(u) = \int_{\mathbb{R}} f(t)e^{2\pi itu}dt$. See §11.4 for the definition of the Fourier transform.*

Weil [Wei2] calls the Euler product expansion the pre-history of the Riemann zeta function. It seems that the functional equation and the meromorphic continuation were first obtained by Riemann in the monumental paper [Ri]. Because the functional equation connects values of $\zeta(s)$ with those of $\zeta(1 - s)$, it suffices to investigate $\zeta(s)$ for $\Re s \geq \frac{1}{2}$.

Exercise[h] 3.1.23. *The following gives an elementary proof that $\zeta(s)$ can be continued to a meromorphic function for $\Re s > 0$. Show*

$$\zeta(s) = \frac{s}{s-1} - s \int_1^\infty \{x\}x^{-s-1}dx, \qquad (3.26)$$

where $\{x\}$ is the fractional part of x.

3.1.3 Values of the Zeta Function at Integer Points

In Exercises 3.1.6 and 3.1.7 we showed how knowledge of $\zeta(s)$ at special s led to information about primes. We needed a result, to be proved in §11.3.4, that

$$\zeta(2) = \frac{\pi^2}{6}. \qquad (3.27)$$

In fact, for positive integers n, $\zeta(2n)$ can be explicitly computed. It turns out that

$$\zeta(2n) = r(n)\pi^{2n}, \qquad (3.28)$$

where for each n, $r(n)$ is a rational number. For example,

$$\zeta(4) = \frac{\pi^4}{90}. \qquad (3.29)$$

The rational number $r(n)$ is expressible in terms of Bernoulli numbers (see [Ed, La6, Se] for a proof, and [CG] or Exercise 3.1.24 for properties of Bernoulli numbers). It is then clear that $\zeta(2n)$ is always irrational, indeed transcendental (see definitions 5.0.3 and 5.0.4 for definitions of algebraic and transcendental numbers). The question of whether the values of $\zeta(s)$ are irrational at odd positive numbers not equal to one is a much more complex problem. The irrationality of $\zeta(3)$ was established by Apéry [Ap] in 1979 (see also the simple proof [Mill]). It is not known whether $\zeta(3)$ is transcendental. It is a very recent theorem that infinitely many of $\zeta(2n+1)$ are irrational [BR]. We invite the reader to consult the above references for details and further remarks.

Exercise 3.1.24. *The Bernoulli numbers B_n are defined by*

$$\frac{z}{e^z - 1} = \sum_{n=0}^{\infty} \frac{B_n}{n!} z^n. \tag{3.30}$$

Calculate the first few Bernoulli numbers, especially B_{2m+1}.

Exercise[h] 3.1.25 (Advanced)**.** *Show*

$$\zeta(2n) = \frac{(-1)^{n-1} B_{2n}}{2(2n)!} (2\pi)^{2n}. \tag{3.31}$$

Exercise[h] 3.1.26. *Verify that $\zeta(s)$ has a simple pole at $s = 1$, with reside equal to 1. Prove that $\zeta(0) = -\frac{1}{2}, \zeta(-1) = -\frac{1}{12}$. Compute the values of zeta at all negative integers.*

Remark 3.1.27. Note that the value of $\zeta(-1)$ comes from the continuation of $\zeta(s)$, and not from substituting $s = 1$ in the series expansion: $1+2+3+4+\cdots \neq -\frac{1}{12}$.

Exercise 3.1.28. *Here is another way of computing $\zeta(2)$ (due to G. Simmons). Show that*

$$\zeta(2) = \int_0^1 \int_0^1 \frac{dx\,dy}{1 - xy}. \tag{3.32}$$

Now compute the integral using the change of variable

$$x = (u - v)\sqrt{2}/2,$$
$$y = (u + v)\sqrt{2}/2. \tag{3.33}$$

Exercise 3.1.29. *Show the number of square-free numbers at most x is $\frac{x}{\zeta(2)} + o(x)$. Generalize to k^{th}-power-free numbers.*

For more on $\zeta(2)$, including complete details of the above exercise, see the entry "Riemann Zeta Function Zeta(2)" in [We].

3.2 ZEROS OF THE RIEMANN ZETA FUNCTION

For this section we assume the reader has more than a passing familiarity with complex analysis, except for §3.2.1 and §3.2.2 where we give an explanation of the interplay between zeros of $\zeta(s)$ and the distribution of primes. In §3.2.1 we introduce the terminology, and in §3.2.2 highlight some of the key results from complex analysis (with sketches of proofs) and describes how these results connect the zeros of $\zeta(s)$ to properties of the primes.

3.2.1 Definitions

From the functional equation of $\zeta(s)$, it is clear that $\zeta(s)$ cannot have any zeros for $\Re s < 0$ except for those forced upon it by the poles of the Γ-function, namely at $-2m$ for $m \in \mathbb{N}$; these are called the **trivial zeros**. In the **critical strip** $0 \le \Re s \le 1$, again a simple application of the functional equation implies that the zeros must lie symmetrically around the critical line $\Re s = \frac{1}{2}$, called the **critical line**. The **Riemann Hypothesis** asserts that all the zeros of the zeta function in the critical strip lie on the critical line. These are called the **non-trivial zeros**. The Riemann Hypothesis is an extremely difficult unsolved problem, and it has been the subject of much research since its announcement in 1860. The Riemann Hypothesis (often abbreviated RH) has very important consequences in number theory, especially in problems related to the study of the distribution of prime numbers. It is justly one of the central themes of research in modern mathematics. It was proved by Hardy in 1914 that infinitely many of the zeros lie on the critical line, and by Selberg in 1942 that a positive proposition of all the zeros lie on the line. While we cannot prove the Riemann Hypothesis in its full strength, one can still obtain some interesting results regarding the distribution of the zeros of the zeta function. For example, the following proposition was stated by Riemann in his 1860 paper and was proved by von Magnoldt in 1905:

Proposition 3.2.1. *The number $N(T)$ of zeros of $\zeta(s)$ in the critical strip with $0 < t \le T$ satisfies*

$$N(T) = \frac{T}{2\pi} \log \frac{T}{2\pi} - \frac{T}{2\pi} + O(\log T). \tag{3.34}$$

For a proof see [Da2], §15. This proposition will be used in Chapter 18 as a guide for scaling the zeros of the zeta function.

Remark 3.2.2 (Important). Using techniques from complex analysis, Riemann [Ri] calculated (but did not mention in his paper!) that the first few zeros of $\zeta(s)$ lie on the critical line. This is one of the most powerful examples of numerical evidence leading to an important conjecture. An English translation of Riemann's paper is available in [Ed]. The first trillion zeros have all been shown to lie on the critical line; for an introduction to such researches see [Od1, Od2] and Zeta-Grid [Wed], where one can download a screensaver to join a world-wide, parallel process investigation of the zeros where more than a billion zeros are checked each day. See [Con2] for a survey of some of the approaches towards proving RH.

Exercise 3.2.3. *Prove that the only zeros of $\zeta(s)$ for $\Re s < 0$ are at $s = -2m$ for $m \in \mathbb{N}$. In the critical strip, prove that the zeros lie symmetrically around the critical line $\Re s = \frac{1}{2}$ (if $\rho = \sigma + it$ is a zero, so is $1 - \sigma + it$).*

3.2.2 Zeros of $\zeta(s)$ and Primes

The connection between the zeros of the Riemann zeta function and the distribution of primes was mentioned earlier. The connection comes through the Euler product expansion (the Euler factors play the role of the polynomial coefficients in the heuristic below).

Why does knowledge about the zeros of $\zeta(s)$ yield information about prime numbers? A simple heuristic is the following. Let $P(x)$ be a polynomial with zeros $r_1, \dots, r_N$ and leading term Ax^n. Then

$$P(x) = A \cdot (x - r_1)(x - r_2) \cdots (x - r_n)$$

$$= A\left(x^n + a_{n-1}(r_1, \dots, r_n)x^{n-1} + \cdots + a_0(r_1, \dots, r_n)\right), \quad (3.35)$$

where

$$a_{n-1}(r_1, \dots, r_n) = -(r_1 + \cdots + r_n)$$

$$\vdots$$

$$a_0(r_1, \dots, r_n) = r_1 r_2 \cdots r_n. \quad (3.36)$$

Hence knowledge of the zeros gives knowledge about the coefficients of the polynomial, and vice versa. See [J] for more on relations between roots and coefficients, as well as other related materials (i.e., Newton's identities and symmetric polynomials). For $\zeta(s)$ the coefficients are related to the primes.

Exercise 3.2.4. *Are there nice formulas for the roots r_i in terms of the coefficients a_i? Investigate for various choices of n.*

A better explanation of the connection between primes and zeros of $\zeta(s)$ is through complex analysis and contour integration, which we now quickly review and then apply. Two functions play a central role in what follows. The first one (see §2.1) is the von Magnoldt function Λ, defined by

$$\Lambda(n) = \begin{cases} \log p & \text{if } n = p^k, p \text{ prime} \\ 0 & \text{otherwise.} \end{cases} \quad (3.37)$$

We also define a function ψ on the set of positive real numbers by

$$\psi(x) = \sum_{n \le x} \Lambda(x). \quad (3.38)$$

Exercise 3.2.5. *Use Chebyshev's Theorem (Theorem 2.3.9) to prove*

$$x \ll \psi(x) \ll x. \quad (3.39)$$

Exercise 3.2.6. *Prove most of the contribution to $\psi(x)$ comes from primes and not prime powers. Explicitly, show the contribution from p^k ($k \ge 2$) is $O(x^{1/2} \log x)$. Note $\psi(x)$ is a weighted counting function of primes, and the techniques of §2.3.4 allow us to remove these weights easily. See also §3.2.2 and §14.7.*

We assume basic knowledge of complex numbers, though knowledge of Green's Theorem from multivariable calculus (see Theorem A.2.9) is needed to complete the proofs of some claims. We provide a heuristic as to how properties of the zeros of $\zeta(s)$ give information about the primes. Any complex number z can be written either as $x + iy$ or as $re^{i\theta}$, with $r \in [0, \infty)$ and $\theta \in [0, 2\pi)$; note by De Moivre's Theorem that $e^{i\theta} = \cos(\theta) + i\sin(\theta)$. If $z \ne 0$ then there is a unique representation of z as $re^{i\theta}$, with r and θ restricted as above.

Exercise 3.2.7. *Write down the change of variables from* $(x, y) \to (r, \theta)$ *and from* $(r, \theta) \to (x, y)$.

Theorem 3.2.8. *Let* $r > 0$ *and* $n \in \mathbb{Z}$. *Then*

$$\frac{1}{2\pi i} \int_{\theta=0}^{2\pi} (re^{i\theta})^n ire^{i\theta} d\theta = \begin{cases} 1 & \text{if } n = -1 \\ 0 & \text{otherwise.} \end{cases} \tag{3.40}$$

Sketch of the proof. In (3.40) we have $e^{i(n+1)\theta} = \cos((n+1)\theta) + i\sin((n+1)\theta)$ (de Moivre's Theorem). If $n = -1$ the integrand is 1 and the result follows; if $n \neq -1$ the integral is zero because we have an integral number of periods of sin and cos (see the following exercise). $\square$

Exercise 3.2.9. *For* $m > 0$, *show that* $\sin(mx)$ *and* $\cos(mx)$ *are periodic functions of period* $\frac{2\pi}{m}$ *(i.e.,* $f(x + \frac{2\pi}{m}) = f(x)$*). If* m *is a positive integer, these functions have an integral number of periods in* $[0, 2\pi]$.

If we consider all $z \in \mathbb{C}$ with $|z| = r > 0$, this set is a circle of radius r about the origin. As only θ varies, we have $dz = izd\theta$, and the integral in (3.40) becomes

$$\frac{1}{2\pi i} \oint_{|z|=r} z^n dz = \begin{cases} 1 & \text{if } n = -1 \\ 0 & \text{otherwise.} \end{cases} \tag{3.41}$$

The symbol $\oint$ denotes that we are integrating about a closed curve, and $|z| = r$ states which curve. Note the answer is independent of the radius.

Exercise 3.2.10. *Prove*

$$\frac{1}{2\pi i} \oint_{|z-z_0|=r} (z - z_0)^n dz = \begin{cases} 1 & \text{if } n = -1 \\ 0 & \text{otherwise.} \end{cases} \tag{3.42}$$

We state, but do not prove, the following theorems on **contour integration** (see, for example, [Al, La6, SS2]):

Theorem 3.2.11. *Let* $f(z)$ *be a meromorphic function (see Definition 3.1.13) in a disk of radius* r *about* z_0, *say*

$$f(z) = \sum_{n \geq n_0} a_n(z - z_0)^n. \tag{3.43}$$

If $f(z)$ *is not identically zero, for* r *sufficiently small* f *has no poles in this disk except possibly at* z_0; *it has a pole if* $n_0 < 0$. *Then*

$$\frac{1}{2\pi i} \oint_{|z-z_0|=r} f(z)dz = a_{-1}. \tag{3.44}$$

Sketch of the proof. If we could interchange summation and integration, we would have

$$\frac{1}{2\pi i} \oint f(z)dz = \frac{1}{2\pi i} \oint \sum_{n \geq n_0} a_n(z - z_0)^n dz = \sum_{n \geq n_0} a_n \frac{1}{2\pi i} \oint (z - z_0)^n dz. \tag{3.45}$$

The only non-zero integral is when $n = -1$, which gives 1. The difficulty is in justifying the interchange. $\square$

Theorem 3.2.12. *Let $f(z)$ be a meromorphic function in a disk of radius r about the origin, with finitely many poles (at $z_1, z_2, \ldots, z_N$). Then*

$$\frac{1}{2\pi i} \oint_{|z|=r} f(z)dz = \sum_{n=1}^{N} a_{-1}(z_n), \qquad (3.46)$$

*where $a_{-1}(z_n)$ is the **residue** (the -1 coefficient) of the Taylor series expansion of $f(z)$ at z_n.*

The proof is similar to the standard proof of Green's Theorem (see Theorem A.2.9). In fact, the result holds not just for integrating over circular regions, but over any "nice" region. *All that matters are whether there are any poles of f in the region, and if so, what their residues are.*

Exercise 3.2.13 (Logarithmic Derivative). *Let $f(z) = a_n(z - z_0)^n$, $n \neq 0$. Show*

$$\frac{f'(z)}{f(z)} = \frac{na_n}{z - z_0}. \qquad (3.47)$$

Note $\frac{f'(z)}{f(z)} = \frac{d \log f(z)}{dz}$. In particular, this implies

$$\frac{1}{2\pi i} \oint_{|z-b|=r} \frac{f'(z)}{f(z)} dz = \begin{cases} na_n & \text{if } z_0 \text{ is in the disk of radius } r \text{ about } b \\ 0 & \text{otherwise.} \end{cases} \qquad (3.48)$$

In the above exercise, note it does not matter if $n > 0$ (z_0 is a zero) or $n < 0$ (z_0 is a pole): $\frac{f'(z)}{f(z)}$ has a simple pole at z_0 with residue na_n. The above result generalizes to $f(z) = \sum_{m \geq n} a_m(z - z_0)^m$. One can show

$$\frac{f'(z)}{f(z)} = \frac{n}{z - z_0} + h(z), \qquad (3.49)$$

where

$$h(z) = \sum_{m=0}^{\infty} b_m(z - z_0)^m, \qquad (3.50)$$

and if $f(z)$ has at most one zero or pole in the disk of radius r about b (namely, z_0), then

$$\frac{1}{2\pi i} \oint_{|z-b|=r} \frac{f'(z)}{f(z)} dz = \begin{cases} n & \text{if } z_0 \text{ is in the disk of radius } r \text{ about } b \\ 0 & \text{otherwise.} \end{cases} \qquad (3.51)$$

Theorem 3.2.14. *Let $f(z)$ be a meromorphic function on a disk of radius r about z_0. Assume the only zeros or poles of $f(z)$ inside a disk of radius r about b are $z_1, \ldots, z_N$. Then*

$$\frac{1}{2\pi i} \oint_{|z-b|=r} \frac{f'(z)}{f(z)} dz = \sum_{n=1}^{N} \text{ord}_f(z_n). \qquad (3.52)$$

(See Definition 3.1.12 for the definition of $\text{ord}_f(z_n)$.)

This result is also true if we consider "nice" regions instead of a disk. Note $\frac{f'(z)}{f(z)} = \frac{d \log f(z)}{dz}$ is the logarithmic derivative of $f(z)$; these results state that the logarithmic derivative encodes information about the zeros and poles of f. We now use these results to sketch the link between primes and zeros of $\zeta(s)$.

Exercise[h] **3.2.15** (Important). *For $\Re s > 1$ show that*

$$\frac{\zeta'(s)}{\zeta(s)} = -\sum_n \frac{\Lambda(n)}{n^s}. \tag{3.53}$$

A common method to deal with products is to take the logarithm (see Theorem 10.2.4 for another example) because this converts a product to a sum, and we have many methods to understand sums.

The idea is as follows: we integrate each side of (3.53) over the perimeter of a box (the location is chosen to ensure convergence of all integrals). For convenience, we first multiply each side by $\frac{x^s}{s}$, where x is an arbitrary parameter that we send to infinity. This will yield estimates for $\pi(x)$. We then have

$$\oint_{\text{perimeter}} \frac{\zeta'(s)}{\zeta(s)} \frac{x^s}{s} ds = -\oint_{\text{perimeter}} \sum_n \frac{\Lambda(n)}{n^s} \frac{x^s}{s} ds. \tag{3.54}$$

Using results from complex analysis, one shows that $\oint \frac{(x/n)^s}{s} ds$ is basically 1 if $n < x$ and 0 otherwise. For the left hand side, we get contributions from the zeros and poles of $\zeta(s)$ in the region. $\zeta(s)$ has a pole at $s = 1$ of residue 1. Let ρ range over the zeros of $\zeta(s)$. One needs to do some work to calculate the residues of $\frac{\zeta'(s)}{\zeta(s)} \frac{x^s}{s}$ at the zeros and poles of $\zeta(s)$; the answer is basically $\frac{x^\rho}{\rho}$. Combining the pieces and multiplying by -1 yields

$$x - \sum_{\rho:\zeta(\rho)=0} \frac{x^\rho}{\rho} = \sum_{n \leq x} \Lambda(n). \tag{3.55}$$

The $x = x^1$ is from the pole of $\zeta(s)$ at $s = 1$. The above is known as an Explicit Formula (to be discussed in detail in §3.2.5). Note $|x^\rho| = x^{\Re \rho}$. If we knew all the zeros had $\Re \rho = \frac{1}{2}$, then the left hand side of (3.55) is approximately $x + O(x^{1/2})$ (we are ignoring numerous convergence issues in an attempt to describe the general features). One can pass from knowledge of $\sum_{n \leq x} \Lambda(x)$ to $\pi(x) = \sum_{n \leq x} 1$: first note the contributions from the prime powers in the sum is at most $x^{1/2} \log x$, and then use Partial Summation (see §2.3.4 for details).

We begin to see now why the Euler Product is such an important property of $\zeta(s)$. In Exercise 3.2.15 we considered the logarithmic derivative of $\zeta(s)$; because of the Euler Product we have a very tractable expression in terms of sums over the primes. Contour integration then relates sums over zeros to sums over primes. If we did not have an Euler Product then we would not have a nice expression on the prime side.

Exercise 3.2.16. *Using Partial Summation, show if $\sum_{n \leq x} \Lambda(x) = x + O(x^{\frac{1}{2}+\epsilon})$ then $\pi(x) = \frac{x}{\log x}$ plus a smaller error term. How small can one take the error term to be?*

Thus, using complex analysis, information on the location of the zeros of $\zeta(s)$ translates to information on the number of primes. To date, it is unknown that there is some number $\sigma_0 \in (\frac{1}{2}, 1)$ such that all zeros have real part at most σ_0. The best known results are that there are no zeros on the line $\Re s = 1$, and there are no zeros "slightly" to the left of this line (the amount we can move left decreases with how high we are on the line). In fact, $\zeta(1 + it) \neq 0$ is equivalent to the Prime Number Theorem! See [Da2, EE] for more details.

Remark 3.2.17. Another way to see the connection between zeros of $\zeta(s)$ and primes is through the product expansion of $\zeta(s)$ in terms of its zeros (see §3.2.3).

Remark 3.2.18. As we saw in Remark 2.3.20, because the logarithmic derivative of $\zeta(s)$ involves the von Magnoldt $\Lambda(n)$-function, we see it is often easier to weight the primes by slowly varying logarithmic factors. This is present throughout modern number theory, namely it is a common technique to study weighted prime sums because they occur more naturally and then remove the weights by partial summation.

Exercise 3.2.19. *The Prime Number Theorem is essentially equivalent to the statement that if $\zeta(\rho) = 0$ then $\Re(\rho) < 1$ (the technical difficulty is exchanging the summation on ρ with the limit as $x \to \infty$ in (3.55): see [Ed] for details). Mertens gave an elegant proof that $\Re(\rho) < 1$ by a clever application of a trigonometric identity. We sketch his argument. Prove the following statements:*

1. $3 + 4\cos\theta + \cos 2\theta \geq 0$ *(Hint: Consider $(\cos\theta - 1)^2$);*

2. *For $s = \sigma + it$, $\log\zeta(s) = \sum_p \sum_{k=1}^{\infty} \frac{p^{-k\sigma}}{k} e^{-itk\log p}$;*

3. $\Re\log\zeta(s) = \sum_p \sum_{k=1}^{\infty} \frac{p^{-k\sigma}}{k} \cos\left(t\log p^k\right)$;

4. $3\log\zeta(\sigma) + 4\Re\log\zeta(\sigma + it) + \Re\log\zeta(\sigma + 2it) \geq 0$;

5. $\zeta(\sigma)^3 |\zeta(\sigma + it)^4 \zeta(\sigma + 2it)| \geq 1$;

6. *If $\zeta(1 + it) = 0$, then as σ decreases to 1 from above, $|\zeta(\sigma + it)| < A(\sigma - 1)$ for some A;*

7. *As $\zeta(\sigma) \sim (\sigma - 1)^{-1}$ ($\zeta(s)$ has a simple pole of residue 1 at $s = 1$) and $\zeta(\sigma + 2it)$ is bounded as $\sigma \to 1$ (the only pole of $\zeta(s)$ is at $s = 1$), the above implies that if $\zeta(1 + it) = 0$ then as $\sigma \to 1$, $\zeta(\sigma)^3 |\zeta(\sigma + it)^4 \zeta(\sigma + 2it)| \to 0$. As the product must be at least 1, this proves $\zeta(1 + it) \neq 0$.*

The key to Mertens' proof is the positivity of a certain trigonometric expression.

3.2.3 Product Expansion

Below we assume the reader is familiar with complex analysis [Al, La6, SS2]. Another interesting property, first noted in Riemann's original paper, is the following:

Proposition 3.2.20. *The function $\xi(s)$ has the following product representation*

$$\xi(s) \;=\; e^{A+Bs} \prod_{\rho} \left(1 - \frac{s}{\rho}\right) e^{\frac{s}{\rho}}, \tag{3.56}$$

with A, B are constants and ρ runs through the non-trivial zeros of the zeta function.

Sketch of proof. A non-constant entire function $f(z)$ is said to be of **finite order** if there exists a number $\alpha > 0$ such that

$$|f(z)| \;=\; O(e^{|z|^{\alpha}}) \quad \text{as } |z| \to \infty. \tag{3.57}$$

If $f(z)$ is of finite order, the infimum of the numbers α is called the order of $f(z)$. If $r_1, r_2, \ldots$ are the absolute values of the zeros of $f(z)$, then it follows from Jensen's formula that for $\beta >$ order of $f(z)$, the series $\sum_1^{\infty} r_n^{-\beta}$ converges, provided of course $f(0) \neq 0$. It is then a theorem of Weierstrass that if $f(z)$ is an entire function of order 1 with zeros $z_1, z_2, \ldots$ (none of which are zero), then there are constants A, B with

$$f(z) \;=\; e^{A+Bz} \prod_{n} \left(1 - \frac{z}{z_n}\right) e^{\frac{z}{z_n}}. \tag{3.58}$$

To prove the proposition we need to show that for any $\alpha > 1$ we have

$$|\xi(s)| \;=\; O(e^{|s|^{\alpha}}) \tag{3.59}$$

as $|s| \to \infty$. In fact, one shows

$$|\xi(s)| \;<\; \exp(C|s| \log|s|) \tag{3.60}$$

as $|s| \to \infty$. It is clear that there is a constant C_1 with

$$\left| \frac{1}{2} s(s-1) \pi^{-\frac{1}{2}s} \right| \;<\; \exp(C_1|s|). \tag{3.61}$$

Also Stirling's formula (Theorem 3.1.16(3)) shows the existence of C_2 such that

$$\left| \Gamma\left(\frac{s}{2}\right) \right| \;<\; \exp(C_2|s| \log|s|). \tag{3.62}$$

We need to control $\zeta(s)$. For this we have the following:

$$\begin{aligned}
\zeta(s) &= \sum_{n=1}^{\infty} n \left[\frac{1}{n^s} - \frac{1}{(n+1)^s} \right] \\
&= s \sum_{n=1}^{\infty} n \int_{n}^{n+1} t^{-s-1} dt \\
&= s \sum_{n=1}^{\infty} \int_{n}^{n+1} [t] t^{-s-1} dt \\
&= s \int_{1}^{\infty} [t] t^{-s-1} dt \\
&= s \int_{1}^{\infty} t^{-s} dt - s \int_{1}^{\infty} \{t\} t^{-s-1} dt \\
&= \frac{s}{s-1} + O(s). \tag{3.63}
\end{aligned}$$

This shows that $|\zeta(s)| < C_3|s|$ for $|s|$ large. Multiplying the inequalities gives (3.60). $\qquad \square$

Remark 3.2.21. Fine tuning the same argument shows that $\zeta(s)$ has an infinite number of zeros in the critical strip; see Chapter 12 of [Da2] for more details. Comparing formula (3.58) with the Euler product hints at a mysterious relationship between the zeros of the zeta function and prime numbers!

Remark 3.2.22. Note the above argument gives a meromorphic continuation of $\zeta(s)$ to $\Re s > 0$ with a simple pole of residue 1 at $s = 1$.

3.2.4 Riemann-Siegel Zeta Function and Hardy's Theorem

The Riemann-Siegel zeta function $\Xi(t)$ is defined by

$$\Xi(t) \;=\; -\frac{1}{2}\left(t^2 + \frac{1}{4}\right)\pi^{-\frac{1}{4} - \frac{it}{2}}\,\Gamma\left(\frac{1}{4} + \frac{it}{2}\right)\zeta\left(\frac{1}{2} + it\right). \tag{3.64}$$

Exercise[h] 3.2.23. *Prove that $\Xi(t)$ is an entire function. It is real for real t, and is an even function: $\Xi(t) = \Xi(-t)$.*

A zero of $\zeta(s)$ on $\Re s = \frac{1}{2}$ corresponds to a real zero of $\Xi(t)$. As $\Xi(t)$ is continuous, in order to detect a real zero we simply need to search for places where the sign of $\Xi(t)$ changes (see also Exercise 5.3.21). This is a very useful observation, and in fact is one of the most important methods to locate zeros of $\zeta(s)$.

Remark 3.2.24 (Divide and Conquer). In practice one chooses a small step size h and evaluates $\Xi(t_0 + nh)$ for $n \in \{0, \dots, N\}$. Every sign change corresponds to a zero of $\zeta(s)$, and repeating the calculation near the sign change with a smaller step size yields better approximations to the zero; see Exercise 5.3.21. If there is a zero of even order, however, there will not be a sign change to detect while for zeros of odd order there is only one sign change and only one zero is detected. Thus the method cannot detect multiplicities of zeros. Using contour integration (which gives the number of zeros in a region), we can ensure that we have found all the zeros. Moreover, once we have seen there are M zeros with imaginary part at most T, and if we have shown there are M such zeros on the critical line, we will have experimentally verified RH in this region. See [Od1, Od2] for more on these algorithms, and Chapter 15 for a connection between zeros of $\zeta(s)$ and eigenvalues of matrices!

In the following series of exercises we show that infinitely many of the zeros of the Riemann zeta function lie on the critical line. This was originally proved by Hardy in 1914; in 1942 Selberg proved that in fact a positive proportion of the zeros lie on the line. The proof of this fact, though within the scope of this manuscript, is very technical. See [Ed, EE, Ti] for details.

Theorem 3.2.25 (Hardy, 1914). *Infinitely many of the zeros of the Riemann zeta function lie on the line $\Re s = \frac{1}{2}$.*

There are various proofs of this fact; a few are recorded in [Ti]. We sketch the argument on page 258 of [Ti].

Sketch of proof. We show that the function $\Xi(t)$ has infinitely many sign changes. For this we use the following theorem of Fejér:

Theorem 3.2.26 (Fejér). *Let n be a positive integer. Then the number of sign changes in the interval $(0, a)$ of a continuous function $f(x)$ is not less than the number of sign changes of the sequence*

$$f(0), \quad \int_0^a f(t)\, dt, \quad \ldots, \quad \int_0^a f(t)t^n\, dt. \tag{3.65}$$

We prove Fejér's theorem after proving Hardy's theorem. To apply Fejér's theorem, one proves

$$\lim_{\alpha \to \frac{\pi}{4}} \int_0^\infty \frac{\Xi(t)}{t^2 + \frac{1}{4}} t^{2n} \cosh \alpha t\, dt = \frac{(-1)^n \pi \cos \frac{\pi}{8}}{2^{2n}}. \tag{3.66}$$

as sketched in the following exercise:

Exercise 3.2.27. *Use (3.21) to prove that*

$$\int_0^\infty \frac{\Xi(t)}{t^2 + \frac{1}{4}} \cos xt\, dt = \frac{1}{2}\pi \left[e^{\frac{1}{2}x} - 2e^{-\frac{1}{2}x}\omega(e^{-2x}) \right]. \tag{3.67}$$

Inserting $x = -i\alpha$ gives

$$\frac{2}{\pi} \int_0^\infty \frac{\Xi(t)}{t^2 + \frac{1}{4}} \cosh \alpha t\, dt = e^{-\frac{1}{2}i\alpha} - 2e^{\frac{1}{2}i\alpha}\omega(e^{2i\alpha}). \tag{3.68}$$

1. *Show that the above can be differentiated with respect to α any number of times if $\alpha < \frac{1}{4}\pi$.*

2. *Use the first part to show*

$$\frac{2}{\pi} \int_0^\infty \frac{\Xi(t)}{t^2 + \frac{1}{4}} t^{2n} \cosh \alpha t\, dt$$

$$= \frac{(-1)^n \cos \frac{1}{2}\alpha}{2^{2n-1}} - 2 \left(\frac{d}{d\alpha} \right)^{2n} e^{\frac{1}{2}i\alpha} \left[\frac{1}{2} + \omega(e^{2i\alpha}) \right]. \tag{3.69}$$

3. *Show that $\omega(x + i) = 2\omega(4x) - \omega(x)$. Use the functional equation of ω to conclude*

$$\omega(x + i) = \frac{1}{\sqrt{x}}\omega\left(\frac{1}{4x} \right) - \frac{1}{\sqrt{x}}\omega\left(\frac{1}{x} \right) - \frac{1}{2}. \tag{3.70}$$

4. *Use (3.70) to show that $\frac{1}{2} + \omega(x)$ and all its derivatives tend to zero as $x \to i$ in the angle $|\arg(x - i)| < \frac{1}{2}\pi$. Show that this observation combined with (3.69) gives (3.66).*

Note (3.66) implies that given any N there is an $a = a(N)$ large and an $\alpha = \alpha(N)$ very close to $\frac{\pi}{4}$ such that the expression

$$\int_0^a \frac{\Xi(t)}{t^2 + \frac{1}{4}} t^{2n} \cosh \alpha t\, dt \tag{3.71}$$

has the same sign as $(-1)^n$ for $n \in \{0, \ldots, N\}$. Now Fejér's theorem implies that $\Xi(t)$ has at least N sign changes in $a(N)$. As N is arbitrary the theorem follows. $\square$

Proof of Fejér's theorem. We first need a lemma:

Lemma 3.2.28 (Fekete). *The number of changes in sign in the interval $(0, a)$ of a continuous function $f(x)$ is not less than the number of changes in sign in the sequence*

$$f_0(a), \; f_1(a), \; \ldots, \; f_n(a), \tag{3.72}$$

where $f_0(x) = f(x)$ and

$$f_\nu(x) \; = \; \int_0^x f_{\nu-1}(t) \, dt \quad (\nu = 1, 2, \ldots, n). \tag{3.73}$$

Exercise[h] **3.2.29.** *Use induction and interchanging the order of integration to show that*

$$f_\nu(x) \; = \; \frac{1}{(\nu - 1)!} \int_0^x (x - t)^{\nu-1} f(t) \, dt, \tag{3.74}$$

and prove that Fekete's Lemma implies Fejér's Theorem.

By the above exercise, we just need to prove Fekete's Lemma. We proceed by induction. If $n = 1$, the lemma is obvious. Now assume the lemma for $n - 1$. Suppose there are at least k changes of sign in the sequence $f_1(a), \ldots, f_n(a)$. Then $f_1(x)$ has at least k changes of sign. The following exercise completes the proof:

Exercise 3.2.30. *In the above setting,*

1. *if $f(a)$ and $f_1(a)$ have the same sign, $f(x)$ has at least k changes of sign;*

2. *if $f(a)$ and $f_1(a)$ have different signs, $f(x)$ has at least $k + 1$ changes of sign.*

This completes the proof of Fejér's theorem.   □

3.2.5 Explicit Formula

The following proposition, first proved by von Magnoldt in 1895, highlights the relationship between the prime numbers and the zeros of the Riemann zeta function that we alluded to in (3.55). For more on Explicit Formulas, see Chapter 18. Recall

$$\psi(x) \; = \; \sum_{n \le x} \Lambda(n). \tag{3.75}$$

Theorem 3.2.31 (Explicit Formula for $\zeta(s)$). *We have*

$$\psi(x) \; = \; x - \sum_\rho \frac{x^\rho}{\rho} - \frac{\zeta'(0)}{\zeta(0)} - \frac{1}{2} \log(1 - x^2), \tag{3.76}$$

where ρ ranges over all the non-trivial zeros of the Riemann zeta function (i.e., $\mathbb{R}\rho \in [0, 1]$).

Exercise 3.2.32. *This exercise is the starting point of the proof of the above theorem. Set*

$$I(x, R) = \frac{1}{2\pi i} \int_{c-iR}^{c+iR} \frac{x^s}{s} \, ds, \tag{3.77}$$

and

$$\delta(x) = \begin{cases} 0 & 0 < x < 1 \\ \frac{1}{2} & x = 1 \\ 1 & x > 1. \end{cases} \tag{3.78}$$

Prove that

$$|I(x, R) - \delta(x)| < \begin{cases} x^c \min(1, R^{-1}|\log x|^{-1}) & \text{if } x \neq 1 \\ \frac{c}{R} & \text{if } x = 1. \end{cases} \tag{3.79}$$

Exercise 3.2.33. *Prove that*

$$\psi(x) = \sum_{n=1}^{\infty} \Lambda(n)\delta\left(\frac{n}{x}\right) \tag{3.80}$$

and

$$-\frac{\zeta'(s)}{\zeta(s)} = \sum_{n=1}^{\infty} \frac{\Lambda(n)}{n^s}. \tag{3.81}$$

The proof of the proposition then follows from a careful analysis of the difference

$$\left| \psi(x) - \frac{1}{2\pi i} \int_{c-iR}^{c+iR} \left(-\frac{\zeta'(s)}{\zeta(s)} \right) \frac{x^s}{s} \, ds \right|, \tag{3.82}$$

using the previous exercise with appropriate choices of c and $R \to \infty$. One then shows that with such choices the difference approaches 0, and the proposition then follows from the computation of the integral appearing in the expression. For this one uses Cauchy's theorem, noting the function $\frac{\zeta'(s)}{\zeta(s)}$ has its poles at the zeros of $\zeta(s)$ with an additional pole coming from $s = 0$ which is responsible for the term $-\frac{\zeta'(0)}{\zeta(0)}$. The non-trivial zeros of the zeta contribute $-\sum_\rho x^\rho/\rho$ and the trivial ones contribute $-\frac{1}{2}\log(1 - x^2)$. We refer the reader to §17 of [Da2] for the details; see also [Sch].

Exercise 3.2.34. *Assuming the Riemann hypothesis, prove that*

$$\psi(x) = x + O(x^{\frac{1}{2}} \log^2 x), \tag{3.83}$$

and vice versa.

3.2.6 Li's constants and the Riemann Hypothesis

We state an interesting criterion for the validity of the Riemann hypothesis. This section is independent of the rest of the book, and the interested reader should see [BoLa, Vo] for more details. Our exposition of this topic closely follows [BoLa].

As usual, denote a typical complex zero of $\zeta(s)$ by ρ. For each positive integer n we define Li's constants λ_n by

$$\lambda_n = \sum_\rho \left[1 - \left(1 - \frac{1}{\rho} \right)^n \right], \tag{3.84}$$

where the sum over ρ is taken as

$$\sum_\rho \left[1 - \left(1 - \frac{1}{\rho} \right)^n \right] = \lim_{T \to \infty} \sum_{|\Im\rho| \le T} \left[1 - \left(1 - \frac{1}{\rho} \right)^n \right]. \tag{3.85}$$

Theorem 3.2.35 (Li's criterion). *The Riemann hypothesis is equivalent to $\lambda_n > 0$ for $n = 1, 2, 3, \ldots$.*

Another interpretation of the sequence λ_n is the following. As before we set

$$\xi(s) = \frac{1}{2} s(s-1) \pi^{-\frac{s}{2}} \Gamma\left(\frac{s}{2}\right) \zeta(s). \tag{3.86}$$

Then

$$\lambda_n = \frac{1}{(n-1)!} \frac{d^n}{ds^n} \left[s^{n-1} \log \xi(s) \right] |_{s=1}. \tag{3.87}$$

Exercise 3.2.36. *Prove the above statement.*

We sketch the proof in [BoLa] of Li's criterion. This simple proof shows that Li's criterion is combinatorial in nature, and holds for much more general settings. We begin with some needed terminology.

Definition 3.2.37. *A **multiset** is a set whose elements have positive integral multiplicities assigned to them.*

Definition 3.2.38. *Let $\mathcal{R}$ be a multiset and $(a_\rho)_{\rho \in \mathcal{R}}$ a sequence indexed by the elements of $\mathcal{R}$. Then if the limit*

$$\lim_{T \to \infty} \sum_{|\Im\rho| \le T} a_\rho \tag{3.88}$$

exists we say the sum is $*$-convergent, and we denote this by $\sum_{\rho \in \mathcal{R}} a_\rho$.

As we will ultimately be interested in multisets defined by zeros of $\zeta(s)$, we will assume that our multisets appearing below do not contain 0 or 1.

Lemma 3.2.39. *Let $\mathcal{R}$ be a multiset of complex numbers ρ with*

$$\sum_{\rho \in \mathcal{R}} \frac{1 + |\Re\rho|}{(1 + |\rho|)^2} < \infty. \tag{3.89}$$

Then for all integers n the sum

$$\sum_{\rho \in \mathcal{R}} \Re\left[1 - \left(1 - \frac{1}{\rho} \right)^n \right] \tag{3.90}$$

converges absolutely. Moreover, if $\sum_{\rho \in \mathcal{R}} \frac{1}{\rho}$ is $*$-convergent, then

$$\lambda_n = \sum_{\rho \in \mathcal{R}} \left[1 - \left(1 - \frac{1}{\rho} \right)^n \right] \tag{3.91}$$

is also $*$-convergent.

Exercise[h] **3.2.40.** *Prove Lemma 3.2.39.*

The main result is the following:

Theorem 3.2.41. *Let $\mathcal{R}$ be a multiset that satisfies the conditions of Lemma 3.2.39. Then the following conditions are equivalent:*

1. $\Re\rho \le \frac{1}{2}$;

2. $\sum_{\rho \in \mathcal{R}} \Re\left[1 - \left(1 - \frac{1}{\rho}\right)^{-n}\right] \ge 0$ for $n = 1, 2, 3, \ldots$;

3. *for every $\epsilon > 0$ there is a constant $c(\epsilon)$ such that for $n = 1, 2, 3, \ldots$*

$$\sum_{\rho \in \mathcal{R}} \Re\left[1 - \left(1 - \frac{1}{\rho}\right)^{-n}\right] \ge -c(\epsilon)e^{\epsilon n}. \tag{3.92}$$

The following exercise shows how Li's criterion for $\zeta(s)$ (Theorem 3.2.35) follows from Theorem 3.2.41 (in particular, how we pass from having an exponent of $-n$ in Theorem 3.2.41 to an exponent of n in Theorem 3.2.35).

Exercise[h] **3.2.42.** *Prove the following:*

1. *Let $\mathcal{R}_\zeta$ be the multiset consisting of the non-trivial zeros of $\zeta(s)$. Prove that $\mathcal{R}_\zeta$ satisfies the conditions of Lemma 3.2.39. Also prove that if $\rho \in \mathcal{R}_\zeta$, then $\bar{\rho}, 1 - \rho, 1 - \bar{\rho} \in \mathcal{R}_\zeta$.*

2. *Apply Theorem 3.2.41 to $\mathcal{R}_\zeta$ and $1 - \mathcal{R}_\zeta = \{1 - \rho; \rho \in \mathcal{R}_\zeta\}$ to deduce Theorem 3.2.35.*

Proof of Theorem 3.2.41. For $\rho \ne 1$ we have

$$\left|1 - \frac{1}{\rho}\right|^{-2} = 1 + \frac{2\Re\rho - 1}{|1 - \rho|^2}; \tag{3.93}$$

giving $(1) \Rightarrow (2)$. Note $(2) \Rightarrow (3)$ is obvious. To see $(3) \Rightarrow (1)$, we proceed as follows. Suppose (1) does not hold. Then for at least one $\rho \in \mathcal{R}$ we have $\Re\rho > \frac{1}{2}$. On the other hand, as

$$\sum_{\rho \in \mathcal{R}} \frac{1}{(1 + |\rho|)^2} < +\infty \tag{3.94}$$

we have $|\rho| \to \infty$, which then implies that $\rho^{-1} \to 0$ and consequently

$$\left|1 - \frac{1}{\rho}\right| \to 1. \tag{3.95}$$

This combined with the elementary inequality (3.93) from above gives

$$\frac{2\Re\rho - 1}{|1 - \rho|^2} \to 0, \tag{3.96}$$

hence the maximum over ρ of this quantity is attained and there are finitely many elements $\rho_k \in \mathcal{R}$, $k = 1, 2, \ldots, K$, such that

$$\left| 1 - \frac{1}{\rho} \right|^{-1} = 1 + t \tag{3.97}$$

is a maximum. We have $t > 0$ as $\Re\rho > \frac{1}{2}$ for at least one ρ. For the other ρ we have

$$\left| 1 - \frac{1}{\rho} \right|^{-1} < 1 + t - \delta \tag{3.98}$$

for a fixed small positive δ. Let ϕ_k be the argument of $1 - 1/\rho_k$. Then by definition

$$1 - \left(1 - \frac{1}{\rho_k} \right)^{-n} = 1 - (t + 1)^n e^{-in\phi_k}. \tag{3.99}$$

For $\rho \neq \rho_k$ we have

$$\left| 1 - \frac{1}{\rho} \right|^{-1} < (1 + t - \delta)^n, \tag{3.100}$$

and also if $|\rho| > n$ then

$$\Re \left[1 - \left(1 - \frac{1}{\rho_k} \right)^{-n} \right] = O\left(\frac{n|\Re\rho| + n^2}{|\rho|^2} \right). \tag{3.101}$$

Exercise[h] **3.2.43.** *Verify the last inequality.*

The last inequality implies that the sum over $|\rho| > n$ is $O(n^2)$ (prove this). Also the number of elements with $|\rho| \leq n$ is $O(n^2)$ (prove this). Hence elements other than the ρ_k contribute at most $O(n^2(1 + t - \delta)^n)$ to $\sum \Re\left[1 - (1 - 1/\rho)^{-n} \right]$. The remaining elements ρ_k contribute

$$K - (1 + t)^n \sum_{k=1}^{K} \cos(n\phi_k). \tag{3.102}$$

Consequently

$$\sum_{\rho \in \mathcal{R}} \Re \left[1 - (1 - 1/\rho)^{-n} \right] = K - (1 + t)^n \sum_{k=1}^{K} \cos(n\phi_k) + O(n^2(1 + t - \delta)^n). \tag{3.103}$$

By Exercise A.4.7 there are infinitely many n with $\sum_{k=1}^{K} \cos(n\phi_k)$ arbitrarily close to K. For such n the expression is negative and arbitrarily large in absolute value. This shows $3 \implies 1$. ☐

Exercise 3.2.44. *Complete the details.*

The paper of Bombieri-Lagarias contains the following interesting interpretation of Li's constants:

$$\lambda_n = -\sum_{j=1}^{n} \binom{n}{j} \frac{(-1)^{j-1}}{(j-1)!} \lim_{\epsilon \to 0+} \left\{ \sum_{m \leq \frac{1}{\epsilon}} \frac{\Lambda(m)(\log m)^{j-1}}{m} - \frac{1}{j} \left(\log \frac{1}{\epsilon} \right)^j \right\}$$

$$+ 1 - (\log 4\pi + \gamma) \frac{n}{2} - \sum_{j=2}^{n} (-1)^{j-1} \binom{n}{j} (1 - 2^{-j}) \zeta(j). \tag{3.104}$$

The observant reader will notice that as λ_n is a sum over the zeros of $\zeta(s)$, (3.104) very much resembles an explicit formula (see §3.2.5). In fact, the proof of (3.104) uses a variant of an explicit formula due to Weil. We advise the reader to consult the very interesting paper [BoLa] for a detailed proof.

Exercise 3.2.45. *Use the Bombieri-Lagarias' formula to evaluate* $\sum_\rho \frac{1}{\rho}$.

Remark 3.2.46. It would be very interesting to numerically test the values of Li's constants to check the validity of the Riemann Hypothesis. Recently, Voros [Vo] has found the following equivalent formulation of the Riemann Hypothesis:

$$\lambda_n \sim \frac{1}{2}n(\log n - \log 2\pi - 1 + \gamma) \tag{3.105}$$

as $n \to \infty$. He has verified that for $n < 3300$ the above formula agrees well with numerical data.

3.3 DIRICHLET CHARACTERS AND *L*-FUNCTIONS

We have seen that knowledge of $\zeta(s)$ translates to knowledge of primes, and that many questions concerning the distribution of primes are related to questions on the distribution of zeros of $\zeta(s)$. The zeta function is the first of many *L*-**functions**. To us, an *L*-function is a Dirichlet series that converges for $\Re s$ sufficiently large, has some type of Euler product expansion, and is "built" from arithmetic information. Similar to $\zeta(s)$, we can glean much from an analysis of these *L*-functions. In this section we concentrate on *L*-functions from Dirichlet characters with different moduli (to be defined below). Just as $\zeta(s)$ can be used to prove there are infinitely many primes and count how many there are less than x, the *L*-functions from Dirichlet characters give information about primes in arithmetic progressions. With the exception of *L*-functions of elliptic curves (see §4.2.2), we only investigate the Riemann zeta function and Dirichlet *L*-functions. See [Se] for an excellent introduction to additional *L*-functions and their connections to the arithmetic functions of Chapter 2.

Fix two positive integers m and a and look at all numbers congruent to a modulo m: $\{x : x = nm + a, n \in \mathbb{N}\}$. If m and a have a common divisor, there can be at most one prime congruent to a modulo m. If m and a are relatively prime, Dirichlet proved that not only are there infinitely many primes in this progression, but also that all such progressions have to first order the same number of primes in the limit. Explicitly, there are $\phi(m)$ numbers a that are relatively prime to m and $\pi(x)$ primes at most x. Let $\pi_{m,a}(x)$ denote the number of primes at most x congruent to a modulo m. Then

$$\pi_{m,a}(x) = \frac{1}{\phi(m)}\frac{x}{\log x} + o_a\left(\frac{1}{\phi(m)}\frac{x}{\log x}\right). \tag{3.106}$$

The main term is independent of a; how the error term varies with a, and what is its true dependence on m, is another story (see for example [Mon1, RubSa, Va]). Probabilistic arguments (see [Mon1] and the remark below) lead to natural conjectures for the m-dependence, and these have far reaching consequences for number theory; for one example, see §18.2.3.

Remark 3.3.1. We sketch some heuristics for the observations mentioned above. For ease of exposition, instead of keeping track of factors of $\log x$ we absorb these into an error x^ϵ. Assuming the Riemann Hypothesis (RH), the error in counting the number of primes at most x is of size $x^{\frac{1}{2}+\epsilon}$. For a fixed m, there are $\phi(m)$ residue classes. To first order, Dirichlet's Theorem asserts each residue class has the same number of primes, roughly $\frac{\pi(x)}{\phi(m)}$. By the Central Limit Theorem (see §8.4), if $\eta_1, \ldots, \eta_N \in \{-1, 1\}$, we expect a typical sum $\eta_1 + \cdots + \eta_N$ to be of size $\sqrt{N}$; the philosophy of square root cancellation asserts similar behavior happens in many situations (see also §4.4, §13.3.2 and Remark 13.3.7). If $\pi_{m,a}(x) = \frac{\pi(x)}{\phi(m)} + E_{m,a}(x)$, then the errors $E_{m,a}(x)$ can be positive or negative and assuming the RH sum to something of size $x^{\frac{1}{2}+\epsilon}$. If we assume they are roughly of the same size, say $\pm E_m$, then as there are $\phi(m)$ terms, the philosophy of square root cancellation predicts their sum is of size $\sqrt{\phi(m)}E_m$ Hence the errors should roughly be of size $x^{\frac{1}{2}+\epsilon}/\sqrt{\phi(m)}$. In many applications we are interested in results that hold for all residue classes as $m \to \infty$. Unfortunately, if even just one class had a much larger error, say of size $x^{\frac{1}{2}+\epsilon}/\phi(m)^{\frac{1}{4}}$ or even $o(x^{\frac{1}{2}+\epsilon})$, it would not noticeably affect the sum of the errors and hence cannot be ruled out. This would have profound consequences. So while we expect the typical $E_{m,a}(x)$ to be of size at most $x^{\frac{1}{2}+\epsilon}/\sqrt{\phi(m)}$, no bounds of this strength are known to hold uniformly in a for fixed m.

3.3.1 General Dirichlet Series

We will consider **Dirichlet series**

$$\sum_{n=1}^{\infty} \frac{a_n}{n^s} \tag{3.107}$$

where the a_n encode arithmetic information. We first prove some general results about convergence of such series, and then investigate Dirichlet series related to primes in arithmetic progressions.

Theorem 3.3.2. *If the above Dirichlet series converges for $s_0 = \sigma_0 + it_0$, then it converges for all s with $\Re s > \sigma_0$.*

Proof. The idea of the proof is to express the tail of the Dirichlet series at s in terms of the tail of the Dirichlet series at s_0 (which is known to converge). We put $C(n) = a(n)/n^{s_0}$, $h(x) = x^{-(s-s_0)}$ and $\sigma = \Re s$. Then

$$\sum_{n=1}^{\infty} \frac{a(n)}{n^s} = \sum_{n=1}^{\infty} C(n)h(n). \tag{3.108}$$

To show the Dirichlet series in (3.108) converges, it suffices (see for example [Rud]) to show that given any $\epsilon' > 0$ there exists an N such that for all $M > N$ then

$$\left| \sum_{n=N+1}^{M} \frac{a_n}{n^s} \right| = \epsilon'. \tag{3.109}$$

If (3.109) holds, we say the sequence satisfies the **Cauchy convergence condition**. Letting $M \to \infty$ shows the sum of the tail of the series can be made arbitrarily small. By Partial Summation (Theorem 2.2.6) we have

$$\sum_{n=1}^{K} \frac{a(n)}{n^s} = S(K)K^{-(s-s_0)} + (s-s_0)\int_{1}^{K} S(t)t^{-1-(s-s_0)}\,dt, \quad (3.110)$$

where $S(x) = \sum_{n \le x}\frac{a(n)}{n^{s_0}}$. Note $S(x)$ is a truncation of the Dirichlet series at s_0 (summing only those n up to x). We use the convergence at s_0 to obtain convergence at s. Subtracting (3.110) with $K = N$ from (3.110) with $K = M$ yields

$$\sum_{n=N+1}^{M} \frac{a(n)}{n^s} = (S(M) - S(N))M^{-(s-s_0)} + S(N)(M^{-(s-s_0)} - N^{-(s-s_0)})$$

$$+(s-s_0)\int_{N}^{M} S(t)t^{-1-(s-s_0)}\,dt. \quad (3.111)$$

By direct integration we see the second term in (3.111) equals

$$-(s-s_0)\int_{N}^{M} t^{-1-(s-s_0)}S(N)\,dt, \quad (3.112)$$

which implies

$$\sum_{n=N+1}^{M} \frac{a(n)}{n^s} = (S(M) - S(N))M^{-(s-s_0)}$$

$$+ (s-s_0)\int_{N}^{M} (S(t) - S(N))t^{-1-(s-s_0)}\,dt. \quad (3.113)$$

Since we have assumed that the series converges for s_0, for $\epsilon > 0$ there is an N_0 such that for $x \ge N \ge N_0$ we have $|S(x) - S(N)| < \epsilon$. Thus

$$\left|\sum_{n=N+1}^{M} \frac{a(n)}{n^s}\right| \le \left|(S(M) - S(N))M^{-(s-s_0)}\right| + \epsilon|s-s_0|\int_{N}^{M} \left|t^{-1-(s-s_0)}\right|\,dt.$$

$$(3.114)$$

For $\sigma > \sigma_0$,

$$\int_{N}^{M} \left|t^{-1-(s-s_0)}\right|\,dt = \int_{N}^{M} t^{-1-(\sigma-\sigma_0)}\,dt = \frac{N^{-(\sigma-\sigma_0)} - M^{-\sigma-\sigma_0}}{\sigma - \sigma_0}.$$

$$(3.115)$$

Since $|s - s_0| \ge \sigma - \sigma_0$, we get

$$\left|\sum_{n=N+1}^{M} \frac{a(n)}{n^s}\right| \le \epsilon\frac{|s-s_0|}{\sigma - \sigma_0}\left(M^{-(\sigma-\sigma_0)} + N^{-(\sigma-\sigma_0)} - M^{-(\sigma-\sigma_0)}\right)$$

$$= \epsilon\frac{|s-s_0|}{\sigma - \sigma_0}N^{-(\sigma-\sigma_0)}$$

$$\le \epsilon\frac{|s-s_0|}{\sigma - \sigma_0}. \quad (3.116)$$

Since ϵ is arbitrary, the convergence of the series in (3.108) for s now follows. $\square$

Exercise 3.3.3. *Assume a Dirichlet series converges absolutely for s_0. Prove this implies that for any $\epsilon > 0$ there is an N_0 such that if $x \geq N \geq N_0$ then $|S(x) - S(N)| < \epsilon$.*

Exercise 3.3.4 (Important). *Assume the partial sums of a_n are bounded. Prove that $\sum \frac{a_n}{n^s}$ converges for $\Re s > 0$. This exercise will be needed in our investigations of primes in arithmetic progressions.*

The following theorems assume some familiarity with uniform convergence and complex analysis (for a review, see Appendix A.3), and are not used in the remainder of the text.

Theorem 3.3.5. *Notation as in Theorem 3.3.2, the convergence is uniform inside any angle for which $|\arg(s - s_0)| \leq \frac{\pi}{2} - \delta$ for $0 < \delta < \frac{\pi}{2}$.*

Proof. Let s be a point in the angle. Then we have

$$\frac{|s - s_0|}{\sigma - \sigma_0} = \frac{1}{\sigma - \sigma_0} \frac{\sigma - \sigma_0}{\cos\arg|s - s_0|} \leq \frac{1}{\cos(\frac{\pi}{2} - \delta)} = \frac{1}{\sin\delta}. \qquad (3.117)$$

Arguing as in the proof of Theorem 3.3.2, we have

$$\left| \sum_{n=N+1}^{M} \frac{a(n)}{n^s} \right| \leq \frac{\epsilon}{\sin\delta}, \qquad (3.118)$$

and the convergence is thus uniform inside the angle. $\qquad\square$

Let R be the collection of all $\gamma \in \mathbb{R}$ with the property that the Dirichlet series converges for $\sigma > \gamma$. By Theorem 3.3.2, R must have an infimum (which may be $-\infty$ if the series converges for all s, or $+\infty$ if it never converges). Let α be this infimum, which is called the **abscissa of convergence**. The Dirichlet series converges if $\sigma > \alpha$ and diverges if $\sigma < \alpha$. As in the case of the convergence of power series on the boundary of the disk of convergence, the behavior of Dirichlet series for $\sigma = \alpha$ is in general undetermined. The following is an easy consequence of the above theorems:

Theorem 3.3.6. *The Dirichlet series*

$$\sum_{n=1}^{\infty} \frac{a(n)}{n^s} \qquad (3.119)$$

converges for $\Re s > \alpha$, and it defines an analytic function on that domain.

In general one must be very careful in dealing with conditionally convergent series, as the following exercise shows.

Exercise 3.3.7 (Rearrangement Theorem). *Let a_n be any series that conditionally converges but does not absolutely converge (thus $\sum a_n$ exists but $\sum |a_n|$ does not). Given $a < b$, show by rearranging the order of the a_n's (instead of $a_1, a_2, \ldots$, we now have the order $a_{n_1}, a_{n_2}, \ldots$), we can get the new partial sums arbitrarily close to a and b infinitely often.*

Exercise 3.3.8. *Let* $\{a_n\}$ *be a sequence of complex numbers. We say that the infinite product*

$$\prod_{n=1}^{\infty} a_n \tag{3.120}$$

converges if the sequence $\{p_m\}$ *defined by*

$$p_m = \prod_{n=1}^{m} a_n \tag{3.121}$$

converges; i.e., if $\lim_{m \to \infty} p_m$ *exists and is non-zero. We assume that* $a_n \neq 0$ *for all* n.

1. *State and prove a Cauchy convergence criterion for infinite products. If the infinite product (3.121) converges to a non-zero number, what is* $\lim_{n \to \infty} a_n$?

2. *Suppose for all* n, $a_n \neq -1$. *Prove that* $\prod_n (1 + a_n)$ *converges if and only if* $\sum_n a_n$ *converges.*

3. *Determine* $\prod_{n=1}^{\infty} \left(1 + \frac{1}{n}\right)$ *and* $\prod_{n=2}^{\infty} \left(1 - \frac{1}{n^2}\right)$.

The basic theory of Dirichlet series can be found in many references. We have used [Ay] in the above exposition.

3.3.2 Dirichlet Characters

Let m be a positive integer. A completely multiplicative (see Definition 2.1.2) arithmetic function with period m that is not identically zero is called a **Dirichlet character**. In other words, we have a function $f : \mathbb{Z} \to \mathbb{C}$ such that $f(xy) = f(x)f(y)$ and $f(x + m) = f(x)$ for all integers x, y. Often we call the period m the **conductor** or **modulus** of the character.

Exercise 3.3.9. *Let* χ *be a Dirichlet character with conductor* m. *Prove* $\chi(1) = 1$. *If* χ *is not identically 1, prove* $\chi(0) = 0$.

Because of the above exercise, we adopt the convention that a Dirichlet character has $\chi(0) = 0$. Otherwise, given any character, there is another character which differs only at 0.

A complex number z is a **root of unity** if there is some positive integer n such that $z^n = 1$. For example, numbers of the form $e^{2\pi i a/q}$ are roots of unity; if a is relatively prime to q, the smallest n that works is q, and we often say it is a qth **root of unity**. Let

$$\chi_0(n) = \begin{cases} 1 & \text{if } (n, m) = 1 \\ 0 & \text{otherwise.} \end{cases} \tag{3.122}$$

We call χ_0 the **trivial** or **principal character** (with conductor m); the remaining characters with conductor m are called **non-trivial** or **non-principal**.

Exercise 3.3.10. *Let χ be a non-trivial Dirichlet character with conductor m. Prove that if $(n, m) = 1$ then $\chi(n)$ is a root of unity, and if $(n, m) \neq 1$ then $\chi(n) = 0$.*

Theorem 3.3.11. *The number of Dirichlet characters with conductor m is $\phi(m)$.*

Proof. We prove the theorem for the special case when m equals a prime p. By Theorem 1.4.29 the group $(\mathbb{Z}/p\mathbb{Z})^*$ is cyclic, generated by some g of order $p - 1$. Thus any $x \in (\mathbb{Z}/p\mathbb{Z})^*$ is equivalent to g^k for some k depending on x. As $\chi(g^k) = \chi(g)^k$, once we have determined the Dirichlet character at a generator, its values are determined at all elements (of course, $\chi(0) = \chi(m) = 0$).

By Exercise 3.3.10, $\chi(g)$ is a root of unity. As $g^{p-1} \equiv 1 \bmod p$ and $\chi(1) = 1$, $\chi(g)^{p-1} = 1$. Therefore $\chi(g) = e^{2\pi i a/(p-1)}$ for some $a \in \{1, 2, \ldots, p - 1\}$. The proof is completed by noting each of these possible choices of a gives rise to a Dirichlet character, and all the characters are distinct (they have different values at g). $\square$

Not only have we proved (in the case of m prime) how many characters there are, but we have a recipe for them. If $a = p - 1$ in the above proof, we have the trivial character χ_0.

Exercise^(h) 3.3.12 (Important). *Let r and m be relatively prime. Prove that if n ranges over all elements of $\mathbb{Z}/m\mathbb{Z}$ then so does rn (except in a different order if $r \not\equiv 1 \bmod m$).*

Exercise 3.3.13. *If χ and χ' are Dirichlet characters with conductor m, so is $\chi'' = \chi\chi'$, given by $\chi''(n) = \chi(n)\chi'(n)$. Define $\overline{\chi}(n) = \overline{\chi(n)}$. Prove $\overline{\chi}$ is a Dirichlet character with conductor m, and $\overline{\chi}\chi = \chi_0$.*

Exercise^(h) 3.3.14 (Important). *Prove the Dirichlet characters with conductor m form a multiplicative group with $\phi(m)$ elements and identity element χ_0. In particular, if χ' is a fixed character with conductor m, if χ ranges over all Dirichlet characters with conductor m, so does $\chi'\chi$.*

The following lemma is often called the **orthogonality relations** for characters (orthogonal is another word for perpendicular). See Definition B.1.5 and §11.1 for other examples of orthogonality. By $\chi \bmod m$ we mean the set of all Dirichlet characters with conductor m.

Lemma 3.3.15 (Orthogonality Relations). *The Dirichlet characters with conductor m satisfy*

$$\sum_{n \bmod m} \chi(n) = \begin{cases} \phi(m) & \text{if } \chi = \chi_0 \\ 0 & \text{otherwise.} \end{cases} \tag{3.123}$$

$$\sum_{\chi \bmod m} \chi(n) = \begin{cases} \phi(m) & \text{if } n \equiv 1 \bmod m \\ 0 & \text{otherwise.} \end{cases} \tag{3.124}$$

Proof. We only prove (3.123) as the proof of (3.124) is similar. By $\chi \bmod m$ we mean χ ranges over all Dirichlet characters with conductor m. Let r be an integer with $(r, m) = 1$. Then

$$\chi(r) \sum_{n \bmod m} \chi(n) = \sum_{n \bmod m} \chi(rn) = \sum_{n \bmod m} \chi(n), \tag{3.125}$$

as when n ranges over a complete system of residues $\bmod m$, so does rn (Exercise 3.3.12). Consequently, denoting the sum in question by S, we have

$$\chi(r)S = S, \tag{3.126}$$

implying that $S = 0$ unless $\chi(r) = 1$ for all $(r, m) = 1$ (in this case, χ is the trivial character χ_0, and $S = \phi(m)$). This finishes the proof.   □

Exercise[h] **3.3.16.** *Prove* (3.124).

Exercise 3.3.17. *Give an alternate proof of* (3.123) *and* (3.124) *by using the explicit formulas for the characters χ with prime conductors. Specifically, for any character χ of prime conductor p with g a generator of $(\mathbb{Z}/p\mathbb{Z})^*$ there is an a such that $\chi(g) = e^{2\pi i a/(p-1)}$.*

Another useful form of the orthogonality relations is

Exercise 3.3.18 (Orthogonality Relations). *Show Lemma 3.3.15 implies*

$$\frac{1}{\phi(m)} \sum_{n \bmod m} \chi(n)\overline{\chi'}(n) = \begin{cases} 1 & \text{if } \chi' = \chi \\ 0 & \text{otherwise.} \end{cases} \tag{3.127}$$

To each character χ we can associate a vector of its values

$$\vec{\chi} \longleftrightarrow (\chi(1), \chi(2), \ldots, \chi(m-1), \chi(m)), \tag{3.128}$$

and we may interpret (3.127) *as saying $\vec{\chi}$ is perpendicular to $\vec{\chi'}$, where the dot product is*

$$\langle \vec{\chi}, \vec{\chi'} \rangle = \sum_{n \bmod m} \chi(n)\overline{\chi'}(n). \tag{3.129}$$

Exercise 3.3.19 (Important). *Given n and a integers, prove*

$$\frac{1}{\phi(m)} \sum_{\chi \bmod m} \overline{\chi}(a)\chi(n) = \begin{cases} 1 & \text{if } n \equiv a \bmod m \\ 0 & \text{otherwise.} \end{cases} \tag{3.130}$$

This exercise provides a way to determine if $a \equiv n \bmod m$, and is used below to find primes congruent to a modulo m.

3.3.3 *L*-functions and Primes in Arithmetic Progressions

The L-function of a general Dirichlet character with conductor m is defined by

$$L(s, \chi) = \sum_{n=1}^{\infty} \frac{\chi(n)}{n^s}. \tag{3.131}$$

Exercise[h] **3.3.20.** *Prove $L(s,\chi)$ converges for $\Re s > 1$. If $\chi \neq \chi_0$, prove that $L(s,\chi)$ can be extended to $\Re s > 0$.*

As in the case of the Riemann zeta function, because integers have unique factorization and because the Dirichlet characters are multiplicative, we have an **Euler product** defined for $\Re s > 1$:

$$L(s,\chi) = \prod_p \left(1 - \frac{\chi(p)}{p^s}\right)^{-1}. \qquad (3.132)$$

Arguing as in Exercise 3.1.9, for $\Re s > 1$, $L(s,\chi) \neq 0$ (again, this is not obvious from the series expansion).

Exercise[h] **3.3.21.** *Prove* (3.132).

We sketch how these L-functions can be used to investigate primes in arithmetic progressions (see [Da2, EE, Se] for complete details); another application is in counting solutions to congruence equations in §4.4. For example, say we wish to study primes congruent to a modulo m. Using Dirichlet characters modulo m, by Lemma 3.3.15 we have (at least for $\Re s > 1$)

$$\sum_{\chi \bmod m} \chi(a)L(s,\chi) = \sum_{n=1}^{\infty} \frac{1}{n^s} \sum_{\chi \bmod m} \chi(a)\chi(n)$$

$$= \sum_{\substack{n=1 \\ n \equiv a \bmod m}}^{\infty} \frac{\phi(m)}{n^s}. \qquad (3.133)$$

Thus, by using *all* the Dirichlet characters modulo m, we have obtained a sum over integers congruent to a modulo m. We want to study not integers but primes; thus, instead of studying $\chi(a)L(s,\chi)$ we study $\chi(a)\log L(s,\chi)$ (because of the Euler product, the logarithm of $L(s,\chi)$ will involve a sum over primes).

Similar to the Riemann zeta function, there is a Riemann Hypothesis, the **Generalized Riemann Hypothesis** (GRH), which asserts that all the non-trivial zeros of the L-function $L(s,\chi)$ lie on the line $\Re s = \frac{1}{2}$. This is, of course, beyond the reach of current technology, and if proven will have immense arithmetic implications. In fact, very interesting arithmetic information has already been obtained from progress towards GRH. The following exercise sketches one of these.

Exercise 3.3.22 (Dirichlet's Theorem on Primes in Arithmetic Progression). *The purpose of this exercise is sketch the ideas for Dirichlet's Theorem for primes in arithmetic progressions. Suppose for all Dirichlet characters $\chi \neq \chi_0$ modulo m, we have $L(1,\chi) \neq 0$. Then for any $(a,m) = 1$, there are infinitely many prime numbers $p \equiv a \bmod m$.*

1. *Using the Euler product for $L(s,\chi)$, and the Taylor series expansion for $\log(1-u)$ about $u = 0$, prove that for $\Re s > 1$,*

$$\log L(s,\chi) = \sum_p \sum_{k=1}^{\infty} \frac{\chi(p^k)}{kp^{ks}}. \qquad (3.134)$$

2. *Use Exercise 3.3.19 to show that*

$$\frac{1}{\phi(m)} \sum_\chi \overline{\chi}(a) \log L(s, \chi) = \sum_p \sum_{k=1}^{\infty} \sum_{p^k \equiv a \bmod m} \frac{1}{kp^{ks}}. \qquad (3.135)$$

3. *Show that the right hand side of (3.135) is*

$$\sum_{p \equiv a \bmod m} \frac{1}{p^s} + O(1) \qquad (3.136)$$

as $s \to 1$ from the right (i.e., $s > 1$ converges to 1, often denoted $s \to 1+$).

4. *Verify that*

$$L(s, \chi_0) = \zeta(s) \prod_{p|m} \left(1 - \frac{1}{p^s}\right), \qquad (3.137)$$

and conclude that $\lim_{s \to 1+} L(s, \chi_0) = +\infty$.

5. *Show that if for all $\chi \neq \chi_0$, $L(1, \chi) \neq 0$, then*

$$\lim_{s \to 1+} \sum_{p \equiv a \bmod m} \frac{1}{p^s} = \infty, \qquad (3.138)$$

which of course implies there are infinitely many primes congruent to a modulo m. Proving $L(1, \chi) \neq 0$ is the crux of Dirichlet's proof; see [Da2, EE, IR, Se] for details.

Note even if we are only interested in one residue class modulo m, in this proof we need to study $L(s, \chi)$ for all Dirichlet characters with conductor m.

Exercise 3.3.23. *The previous exercise allows us to reduce the question on whether or not there are infinitely many primes congruent to a modulo m to evaluating a finite number of L-functions at 1; thus any specific case can be checked. For $m = 4$ and $m = 6$, for each character χ use a good numerical approximation to $L(1, \chi)$ to show that it is non-zero. Note if one has a good bound on the tail of a series it is possible to numerically approximate an infinite sum and show it is non-zero; however, it is not possible to numerically prove an infinite sum is exactly zero.*

Exercise 3.3.24. *In the spirit of the previous problem, assume we know an infinite sum is rational and we know the denominator is at most Q. Prove that if we can show that $|\sum_{n=1}^{\infty} a_n - 0| < \frac{1}{Q}$ then this estimate improves itself to $\sum_{n=1}^{\infty} a_n = 0$. Unfortunately, it is difficult in practice to prove a sum is rational and to bound the denominator, though there are some instances involving L-functions attached to elliptic curves where this can be done. What is more common is to show a sum is a non-negative integer less than 1, which then implies the sum is 0. We shall see numerous applications of this in Chapter 5 (for example §5.4, where we prove e is irrational and transcendental).*

Exercise$^{(hr)}$ 3.3.25. *The difficult part of Dirichlet's proof is showing $L(1, \chi) \neq 0$ for real characters χ; we show how to handle the non-real characters (this means $\overline{\chi} \neq \chi$; for example, the Legendre symbol is a real character). Using $a = 1$ in Exercise 3.3.22, show*

$$\sum_\chi \log L(\sigma, \chi) \geq 0 \tag{3.139}$$

for real $\sigma \geq 1$; note this sum may be infinite. Therefore $\prod_\chi L(\sigma, \chi) \geq 1$ for $\sigma \geq 1$. Show that if $L(1, \chi) = 0$ so too does $L(1, \overline{\chi})$. Show for a non-real character χ that $L(1, \chi) \neq 0$.

Exercise 3.3.26. *We saw in Exercise 2.3.5 that for certain choices of m and a it is easy to prove there are infinitely many primes congruent to a modulo m. Modifying Euclid's argument (Theorem 2.3.1), prove there are infinitely many primes congruent to -1 modulo 4. Can you find an a modulo 5 (or 6 or 7) such that there are infinitely many primes? See [Mu1] for how far such elementary arguments can be pushed.*

Remark 3.3.27. One can show that, to first order, $\pi_{m,a}(x) \sim \frac{\pi(x)}{\phi(m)}$, where $\pi_{m,a}(x)$ is the number of primes at most x congruent to a modulo m. We can see evidence of this in (3.135). The left hand side of that equation depends very weakly on a. The contribution from the non-principal characters is finite as $s \to 1$; thus the main contribution comes from $\chi_0(a)L(s, \chi_0) = L(s, \chi_0)$. Therefore the main term in (3.136), $\sum_{p \equiv a \bmod q} p^{-s}$, has a similar $s \to 1$ limit for all a; specifically, the piece that diverges, diverges at the same rate for all a relatively prime to q. The behavior of the correction terms exhibit interesting behavior: certain congruence classes seem to have more primes. See [EE, RubSa] for details.

Exercise 3.3.28. *By Exercise 3.3.29 or 11.3.17,*

$$\frac{\pi}{4} = \sum_{n=1}^{\infty} \frac{(-1)^{n-1}}{2n-1}. \tag{3.140}$$

Note π is irrational (see [NZM], page 309). Define

$$\chi_4(n) = \begin{cases} (-1)^{(n-1)/2} & \text{if } n \text{ is odd} \\ 0 & \text{otherwise.} \end{cases} \tag{3.141}$$

Prove χ_4 is a Dirichlet character with conductor 4. By evaluating just $L(1, \chi_4)$ and noting π is irrational, show there are infinitely many primes; we sketch a proof of the irrationality of π^2 in Exercise 5.4.17. This is another special value proof and provides no information on the number of primes at most x. Using this and properties of $\zeta(s)$, can you deduce that there are infinitely many primes congruent to 1 modulo 4 or -1 modulo 4? Infinite products of rational numbers can be either rational or transcendental; see Exercise 5.6.9.

Exercise$^{(hr)}$ 3.3.29 (Gregory-Leibniz Formula). *Prove*

$$\frac{\pi}{4} = \sum_{n=1}^{\infty} \frac{(-1)^{n-1}}{2n-1}. \tag{3.142}$$

Exercise[h] **3.3.30** (Wallis' Formula). *Prove*

$$\frac{2}{\pi} = \frac{1}{2} \cdot \frac{3 \cdot 3}{2 \cdot 4} \cdot \frac{5 \cdot 5}{4 \cdot 6} \cdot \frac{7 \cdot 7}{6 \cdot 8} \cdots . \tag{3.143}$$

See Exercise 5.6.9 for more on infinite products and Chapter 11 of [BB] for more formulas for π. A good starting point is

$$\int_0^{\pi/2} (\sin x)^{2m} dx = \frac{1 \cdot 3 \cdot 5 \cdots (2m-1)}{2 \cdot 4 \cdot 6 \cdots 2m} \frac{\pi}{2}$$

$$\int_0^{\pi/2} (\sin x)^{2m+1} dx = \frac{2 \cdot 4 \cdot 6 \cdots 2m}{1 \cdot 3 \cdot 5 \cdots (2m+1)}. \tag{3.144}$$

3.3.4 Imprimitive Characters

We conclude with some remarks about imprimitive characters. Except for the definitions of primitive and imprimitive, the rest of this subsection may safely be skipped.

By definition a character χ modulo m is periodic. If the smallest period of χ is equal to m, χ is called a **primitive character**, otherwise it is **imprimitive**. Usually statements regarding characters and their L-functions are easier for primitive characters. All characters are built out of primitive ones:

Lemma 3.3.31. *Let χ be a non-principal imprimitive character modulo m. Then there is a divisor $m_1 \neq 1$ of m and a character ψ modulo m_1 such that*

$$\chi(n) = \begin{cases} \psi(n) & \text{if } (n,m) = 1 \\ 0 & \text{otherwise.} \end{cases} \tag{3.145}$$

Proof. Let m_1 be the least period of χ; $m_1 \neq 1$ because this would imply χ is periodic modulo 1, contradicting the assumption that χ is not principal. If $m_1 \nmid m$, then $(m_1, m) < m_1$ is also a period of χ, which contradicts the choice of m_1. We now construct the character ψ. If $(n, m) = 1$ we set $\psi(n) = \chi(n)$; if $(n, m_1) \neq 1$ we set $\psi(n) = 0$. The only remaining case is when $(n, m_1) = 1$ but $(n, m) \neq 1$. Here we choose an integer t with $(n+tm_1, m) = 1$, and we set $\psi(n) = \chi(n+tm_1)$. We leave it to the reader to verify that ψ is a well defined primitive character modulo m_1, and we say that ψ **induces** χ. □

Exercise 3.3.32. *Show that the Legendre symbol $\left(\frac{*}{p}\right)$ is a non-primitive character modulo p^α for $\alpha > 1$, which is induced by the same character modulo p. Here $\chi(n) = \left(\frac{n}{p}\right)$.*

Exercise 3.3.33. *Let χ be an imprimitive character modulo m, induced by a character ψ. Prove for $\Re s > 1$,*

$$L(s, \chi) = L(s, \psi) \prod_{p|m} \left(1 - \frac{\psi(p)}{p^s}\right). \tag{3.146}$$

3.3.5 Functional Equation

Let χ be a primitive character mod m. Similar to $\zeta(s)$, $L(s,\chi)$ satisfies a functional equation. Before proving this we need to define certain sums involving Dirichlet characters. Let the **Gauss Sum** $c(m,\chi)$ be defined by

$$c(m,\chi) \;=\; \sum_{k=0}^{m-1} \chi(k)e^{2\pi ik/m}. \tag{3.147}$$

Exercise[hr] **3.3.34.** *Prove that* $|c(m,\chi)| = m^{\frac{1}{2}}$. *See [BEW, IR] for additional results on Gauss and related sums.*

Exercise 3.3.35. *Prove that*

$$\chi(n) \;=\; \frac{1}{c(m,\overline{\chi})} \sum_{k=0}^{m-1} \overline{\chi}(k)e^{2\pi i\frac{kn}{m}}. \tag{3.148}$$

Note the right hand side is well defined for all $n \in \mathbb{R}$. This useful formula interpolates χ from a function on $\mathbb{Z}$ to one on $\mathbb{R}$.

Theorem 3.3.36 (Meromorphic Continuation of $L(s,\chi)$). *The function $L(s,\chi)$, originally defined only for $\Re s > 1$, has a meromorphic continuation to the entire complex plane. The meromorphic continuation of $L(s,\chi)$ has a unique pole at $s = 1$ when $\chi = \chi_0$; otherwise, it is entire. Furthermore, if we set*

$$\Lambda(s,\chi) \;=\; \left(\frac{m}{\pi}\right)^{\frac{1}{2}(s+\epsilon)} \Gamma\left(\frac{s+\epsilon}{2}\right) L(s,\chi), \tag{3.149}$$

where

$$\epsilon = \begin{cases} 0 & \text{if } \chi(-1) = 1 \\ 1 & \text{if } \chi(-1) = -1, \end{cases}$$

then we have

$$\Lambda(s,\chi) \;=\; (-i)^{\epsilon}\frac{c(m,\chi)}{m^{\frac{1}{2}}}\Lambda(1-s,\overline{\chi}). \tag{3.150}$$

Observe that

$$\left|(-i)^{\epsilon}\frac{c(m,\chi)}{m^{\frac{1}{2}}}\right| = 1. \tag{3.151}$$

The proof of this theorem is similar to the proof of the similar theorem for the Riemann zeta function as sketched above. We refer the reader to [Da2] for details.

Chapter Four

Solutions to Diophantine Equations

Often in investigations we encounter systems of equations with integer coefficients. Frequently what matters is not the solutions themselves, but rather the number of solutions. Thus we want to count the number of integer or rational solutions to these equations, or, failing that, at least estimate how many there are, or even prove a solution exists. Applications range from the combinatorial problems of §1.2.4, to the proof of Quadratic Reciprocity in §1.6.2, to using the rational solutions of elliptic curves in cryptology [Kob2, Kob3, M, Wa], to studying the eigenvalues of certain matrices (see §16.2.2 and [HM, MMS]), to writing integers as the sum of primes (Chapter 13). In some special cases we are able to not only count the number of rational solutions but also **parametrize** these solutions (i.e., derive a tractable expression which runs through these solutions). We assume the reader is familiar with the material from Chapter 1. See [BS, IR, NZM] for more on solving equations with integer coefficients.

4.1 DIOPHANTINE EQUATIONS

In loose terms, a **Diophantine Equation** is an algebraic equation with extra conditions on the set of acceptable solutions. For us, acceptable means of arithmetic interest (integral or rational). Diophantine equations are named in honor of the Greek mathematician Diophantus (approximately 200 to 280 A.D.), who studied equations of this form. In his famous book *Arithmetica*, Diophantus collected numerous puzzles and brain-teasers involving, among other things, the number of cows in a particular herd having various properties; as one cannot have fractional values for the number of cows in a herd, the solutions to these puzzles were integers. Diophantus' book was a favorite of Fermat; while studying this book Fermat formulated his so-called Last Theorem (see §4.1.3).

For example, consider the equation $x^2 + y^2 = 2$; see Figure 4.1. If we try to solve for $x, y \in \mathbb{C}$, we quickly find that not only are there infinitely many solutions, but given x we can easily determine y: $y = \pm\sqrt{2 - x^2}$. Next, instead of looking for complex solutions, we could restrict x and y to be in $\mathbb{R}$. In this case, if $|x| \leq \sqrt{2}$ the same argument works. If we restrict x and y to be integers, we find there are only four solutions: $(1, 1)$, $(1, -1)$, $(-1, 1)$ and $(-1, -1)$. If instead we allow $x, y \in \mathbb{Q}$, how many solutions are there? Can we describe them as easily as when x, y were in $\mathbb{R}$ or $\mathbb{C}$?

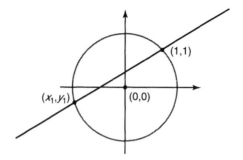

Figure 4.1 Solving $x^2 + y^2 = 2$

4.1.1 Solving $x^2 + y^2 = 2$ Over the Rationals

We know one rational solution, $(x_0, y_0) = (1, 1)$. The line through (x_0, y_0) with slope t is given by the equation

$$y = tx + (1 - t). \tag{4.1}$$

By construction, this line intersects $x^2 + y^2 = 2$ at the point $(1, 1)$; the other point of intersection is

$$(x_1(t),\ y_1(t)) = \left(\frac{t^2 - 2t - 1}{t^2 + 1},\ \frac{-t^2 - 2t + 1}{t^2 + 1} \right), \tag{4.2}$$

as demonstrated by an easy computation. It is clear that if t is any rational number, then $(x_1(t), y_1(t))$ is the point with rational coefficients solving $x^2 + y^2 = 2$. Furthermore, if (x_1, y_1) is an arbitrary rational point on the circle $x^2 + y^2 = 2$, the slope of the line connecting (x_0, y_0) and (x_1, y_1) is a rational number t_0, and $(x_1, y_1) = (x_1(t_0), y_1(t_0))$. We have therefore obtained a parametrization for all the rational points on the circle $x^2 + y^2 = 2$, given by

$$t \mapsto (x_1(t), y_1(t)) \tag{4.3}$$

which by the above remarks is **surjective** or **onto** (i.e., it hits all solutions).

It is clear that the same method works for any conic section, as long as we are able to find a rational point of reference on the curve. What about higher degree equations?

4.1.2 Pythagorean Triples

Consider Pythagorean triples $x^2 + y^2 = z^2$. If (x_0, y_0, z_0) is a solution, so is (ax_0, ay_0, az_0) for any non-zero a. Thus it suffices to understand the solutions for x, y and z coprime (no prime divides all three). The parametrization of Pythagorean triples will be useful in investigating a special case of Fermat's Last Theorem (see §4.1.3).

Exercise 4.1.1. *If $x^2 + y^2 = z^2$ and $x, y, z \in \mathbb{Z}$ are coprime, show exactly one of x and y is even and the other is odd. In particular, there are no solutions with both x and y odd.*

Lemma 4.1.2 (Pythagorean Triples). *Any positive integral solution to*

$$x^2 + y^2 = z^2 \tag{4.4}$$

with x, y relatively prime and x even must satisfy

$$x = 2ab, \quad y = a^2 - b^2, \quad z = a^2 + b^2 \tag{4.5}$$

for some coprime positive integers a, b, exactly one of which is even.

Proof. Since x is even and x and y are coprime, y must be odd. Hence z is odd. Therefore $\frac{1}{2}(z - y)$ and $\frac{1}{2}(z + y)$ are integral. By (4.4),

$$\left(\frac{x}{2}\right)^2 = \left(\frac{z+y}{2}\right)\left(\frac{z-y}{2}\right). \tag{4.6}$$

x and y coprime implies that y and z are coprime as well. Hence $\frac{1}{2}(z - y)$ and $\frac{1}{2}(z + y)$ are coprime (Exercise 4.1.3). Therefore, as their product is a square, they must both be squares (Exercise 4.1.4):

$$\frac{z+y}{2} = a^2, \quad \frac{z-y}{2} = b^2. \tag{4.7}$$

Thus a and b satisfy (4.5). Since y is odd, a and b are of opposite parity. □

Exercise 4.1.3. *Show that if y and z are coprime, so are $\frac{1}{2}(z - y)$ and $\frac{1}{2}(z + y)$.*

Exercise 4.1.4. *Prove that if uv is a square and u, v have no factors in common, then u and v are both squares.*

Remark 4.1.5. An alternative method of finding all solutions of the above equation is via the method of rational lines of the previous section. It is not hard to see that an integral solution of $x^2 + y^2 = z^2$ gives rise to a rational solution of $X^2 + Y^2 = 1$ and vice versa. Starting from the point $(1, 0)$, one gets the parametrization

$$(x(t), y(t)) = \left(\frac{t^2 - 1}{t^2 + 1}, \frac{2t}{t^2 + 1}\right) \tag{4.8}$$

for the rational points of $X^2 + Y^2 = 1$. Writing $t = \frac{a}{b}$, with $(a, b) = 1$, and clearing denominators gives the result.

4.1.3 Fermat's Last Theorem

We briefly discuss some special cases of Fermat's Last Theorem, which states there are no solutions to $x^n + y^n = z^n$ for $n \geq 3$ and x, y, z non-zero integers. See [Acz, Moz2, Moz3, vdP6] for details on the history of the problem, partial results, and summaries of Wiles' proof [Wi, TW]. For us, it is the method of proof in these special cases that is important, not the result.

Theorem 4.1.6 (Fermat). *There are no positive integers x, y and z such that*

$$x^4 + y^4 = z^2. \tag{4.9}$$

Proof. Suppose there are positive integral solutions to the given equation. We prove the theorem by showing that if there exists a solution then we can find a "smaller" one, and this process cannot continue indefinitely.

Order all solutions (x, y, z) according to the value of z. Let (x_0, y_0, z_0) be a solution with the smallest value of z; if there are several such solutions, choose any among them. As we are assuming the solution is positive, $z_0 > 0$. Clearly x_0 and y_0 are coprime: if they weren't, we could obtain a smaller solution by dividing x_0 and y_0 by $\gcd(x_0, y_0)$ and z_0 by $\gcd(x_0, y_0)^2$. We obtain a contradiction by finding another solution with strictly smaller z; this argument is known as the **Method of Descent**.

If x_0 and y_0 are both odd, then

$$x_0^4 \equiv 1 \bmod 4, \quad y_0^4 \equiv 1 \bmod 4 \tag{4.10}$$

and hence

$$z_0^2 = x_0^4 + y_0^4 \equiv 2 \bmod 4. \tag{4.11}$$

As no square can be congruent to 2 modulo 4 (Exercise 4.1.7), there are no integer solutions to (4.11). Thus our assumption that x_0 and y_0 are both odd is false, and either x_0 or y_0 is even. We can assume without loss of generality that x_0 is even. As we can write $x_0^4 + y_0^4 = z^2$ as

$$(x_0^2)^2 + (y_0^2)^2 = z_0^2, \tag{4.12}$$

by Lemma 4.1.2 we have

$$x_0^2 = 2ab, \quad y_0^2 = a^2 - b^2, \quad z_0 = a^2 + b^2, \tag{4.13}$$

with exactly one of a, b odd. If a were even and b odd, then y_0^2 would have to be congruent to 3 modulo 4, and this is impossible. Hence a is odd and b is even, and $b \neq 0$ (if $b = 0$ then $x_0 = 0$, contradicting $x_0 > 0$). Write $b = 2c$. Then

$$\left(\frac{x_0}{2}\right)^2 = \frac{x_0^2}{4} = \frac{2ab}{4} = \frac{4ac}{4} = ac. \tag{4.14}$$

Since a and b are coprime, a and c are coprime. Hence a and c must both be squares. Write $a = d^2, c = f^2$. Then (remembering $b = 2c = 2f^2$)

$$y^2 = a^2 - b^2 = d^4 - 4f^4, \tag{4.15}$$

and so

$$(2f^2)^2 + y^2 = (d^2)^2. \tag{4.16}$$

Applying Lemma 4.1.2 again, we obtain

$$2f^2 = 2lm, \quad d^2 = l^2 + m^2 \tag{4.17}$$

for some coprime positive integers l, m. Since $f^2 = lm$, both l and m are squares:

$$l = r^2, \quad m = s^2. \tag{4.18}$$

Therefore

$$d^2 = l^2 + m^2 \tag{4.19}$$

can be written as

$$d^2 = r^4 + s^4. \tag{4.20}$$

But as $b > 0$ we have

$$d \leq d^2 = a \leq a^2 < a^2 + b^2 = z_0, \tag{4.21}$$

and thus (4.20) is a solution to (4.9) with a value of z smaller than z_0, contradicting the claim that (x_0, y_0, z_0) had smallest z among integer solutions. $\square$

Exercise 4.1.7. *For any integer z, show that z^2 cannot be equivalent to 2 or 3 modulo 4.*

We say a solution of $x^4 + y^4 = z^2$ is **trivial** if x, y or z is zero, and **non-trivial** otherwise. The above argument proves that there are no non-trivial positive integer solutions, which immediately yields there are no non-trivial integer solutions. We can also conclude that the equation $x^4 + y^4 = z^4$ has no non-trivial integral solutions.

Exercise 4.1.8 (Fermat's Equation). *Assume $x^m + y^m = z^m$ has no non-trivial integer solutions for some $m \in \mathbb{N}$. Prove that for any $a, b, c \in \mathbb{N}$, $x^{am} + y^{bm} = z^{cm}$ has no non-trivial integer solutions. Thus it is enough to show there are no non-trivial solutions to Fermat's Equation for odd primes and for $m = 4$. Note we must do $m = 4$, as by Pythagoras there are solutions when $m = 2$.*

Exercise 4.1.9. *Classify all rational solutions to $x^{-2} + y^{-2} = z^{-2}$.*

Exercise 4.1.10. *Prove that if Fermat's Equation $x^m + y^m = z^m$ has a non-trivial rational solution then it has a non-trivial integer solution. Therefore $x^4 + y^4 = z^2$ has no non-trivial rational solutions.*

A prime p is a Germain prime if $\frac{p-1}{2}$ is also prime. Sophie Germain proved that if p is a Germain prime then the only integer solutions of $x^p + y^p = z^p$ have $p|xyz$; see Lecture III of [vdP6] for more details. We investigate the number of Germain primes in detail in Chapter 14.

4.2 ELLIPTIC CURVES

We give a brief introduction to some properties of elliptic curves. We refer the reader to the excellent books [IR, Kn, Milne, Sil1, ST] for full details. For us, an **elliptic curve** E is the set of solutions of an equation of the form

$$y^2 = x^3 + ax + b. \tag{4.22}$$

We usually assume that a, b are rational or integral. This section is meant only to be a brief introduction to an incredibly rich subject. The reader should consult [Sil1, ST] for complete results. A wonderful expository article on elliptic curves and their connections to modern number theory is [Maz3].

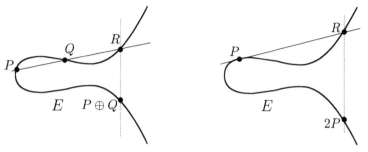

Addition of distinct points P and Q Adding a point P to itself

Figure 4.2 The addition law on an elliptic curve. In the second example the line is tangent to E at P.

4.2.1 Rational Solutions

We start with an exercise that shows the structure of the solutions to (4.22) is very different if $x, y \in \mathbb{C}$ or $x, y \in \mathbb{R}$ or $x, y \in \mathbb{Q}$.

Exercise 4.2.1. *Parametrize the solutions to (4.22) when $x, y \in \mathbb{C}$ or $x, y \in \mathbb{R}$ for $(a, b) = (-1, 0), (0, 1)$ and $(-7, 6)$.*

The set of rational points of the above curve, plus a point "at infinity," forms a group. We briefly describe the group operation, i.e., how to add points in the group; see Figure 4.2. Given two distinct points $P = (x_0, y_0)$ and $Q = (x_1, y_1)$ on the elliptic curve (i.e., satisfying (4.22)), we let $l(P, Q)$ be the line passing through the two points. We define $R(P, Q)$ to be the intersection of $l(P, Q)$ with E. We set $P + Q$ to be the reflection of $R(P, Q)$ along the x axis. If $P = Q$, we define $l(P, Q)$ to be the tangent line at P to the curve. Given a point P, we define $-P$ to be the reflection of P along the x axis. Also, formally, as our point at infinity we take the sum $P + (-P)$ as defined above; this will be the identity element of the elliptic curve.

Exercise 4.2.2. *Verify that addition of rational points on an elliptic curve is commutative.*

Suppose $P = (x_0, y_0)$ and $Q = (x_1, y_1)$ are two rational points on the curve $y^2 = x^3 + ax + b$, a and b rational. We write the following parametric equation for the line $l(P, Q)$:

$$P(t) = (x_0 + t(x_1 - x_0), y_0 + t(y_1 - y_0)), \qquad t \in \mathbb{R}. \qquad (4.23)$$

In order for $P(t)$ to be on E, we must have

$$(y_0 + t(y_1 - y_0))^2 = (x_0 + t(x_1 - x_0))^3 + a(x_0 + t(x_1 - x_0)) + b, \qquad (4.24)$$

which is a cubic equation in t. As $(x_0, y_0) = P(0)$ and $(x_1, y_1) = P(1)$ are on the elliptic curve, $t = 0, 1$ are two roots of the above cubic equation. Expanding the cubic equation in t in (4.24), we find all the coefficients of this cubic are rational. Therefore, the third root of (4.24) must also be rational. This implies that $P + Q$ has rational coordinates.

Note how similar the above arguments are to those in §4.1.1. There, starting with one rational solution we were able to generate all rational solutions. Here, given two rational solutions, we can generate a third. It is not clear from the above that all the points generated will be different, nor is it clear what is the additive identity (the point O such that $P + O = P$ for all P). Nevertheless, one can show

Lemma 4.2.3 (Advanced). *The collection of rational points of E, denoted $E(\mathbb{Q})$, forms a group. We have $E(\mathbb{Q})$ is isomorphic to $\mathbb{Z}^r \oplus \mathbb{T}$, where $\mathbb{T}$ is the torsion points (r and $\mathbb{T}$ depend on E).*

Remark 4.2.4. The hardest property to prove is associativity; the identity is the point at infinity alluded to above. The Mordell-Weil theorem asserts that this group is finitely generated. This should be contrasted to the case of degree two equations, where one rational point generates all rational solutions. Here one point may not suffice, but finitely many will. Mazur [Maz1, Maz2] has proven there are only 15 possibilities for the torsion part.

Definition 4.2.5. *We call $E(\mathbb{Q})$ the **Mordell-Weil group** of E, and r the **Mordell-Weil rank**.*

Exercise 4.2.6. *Prove that there at least four points with rational coordinates on the curve with equation*

$$y^2 = x^3 + x + 16. \tag{4.25}$$

Do you think there are infinitely many?

Exercise[h] **4.2.7.** *Find rational solutions for the elliptic curves in Exercise 4.2.1.*

Instead of looking for solutions with $x, y \in \mathbb{Q}$, for any prime p we can count the pairs of integers (x, y) such that

$$y^2 \equiv x^3 + ax + b \bmod p. \tag{4.26}$$

Here we restrict $x, y \in \{0, 1, \dots, p-1\}$, the finite field with p elements. In §4.4.1 we show how to use Legendre symbols and quadratic reciprocity (see §1.6) to quickly count the number of solutions. It is a remarkable fact that we can build a function $L(E, s) = \sum_n a_n/n^s$, where for p prime a_p is related to the number of solutions mod p. Properties of this function give information about the structure of the group of rational solutions. This continues a central theme in mathematics: L-functions are made up of local data that provide global information. This extends the results in Chapter 3, where we saw how $\zeta(s)$ and $L(s, \chi)$ yield information about primes and primes in arithmetic progression.

4.2.2 L-Functions

We assume the reader is familiar with Chapter 3. We briefly sketch how to attach an L-function $\sum_{n=1}^{\infty} \frac{a_n}{n^s}$ to an elliptic curve, and what information this function provides.

We have remarked that the rational solutions form a group $E(\mathbb{Q})$. In addition to theoretical interest, these groups are extremely important in cryptography

[Kob2, Kob3, M, Wa]. Standard algorithms are based on how difficult it is to factor numbers. Let $n = p_1 p_2$. The user must enter a factor of n to enter; thus, we
are working in $\mathbb{Z}/n\mathbb{Z}$. Similarly, one can design cryptosystems based on elliptic
curves, where now we need to do arithmetic on $E(\mathbb{Q})$.

Consider an elliptic curve E:

$$y^2 = f(x) = x^3 + ax + b, \ a, b \in \mathbb{Z}. \qquad (4.27)$$

Let $N_p(E)$ be the number of solutions to the above equation mod p:

$$N_p(E) = \#\{(x, y) : x, y \in \mathbb{Z}/p\mathbb{Z} \text{ and } y^2 \equiv f(x) \bmod p\}. \qquad (4.28)$$

We recall the Legendre symbol (see §1.6.1): for a prime p, $\left(\frac{a}{p}\right) = 0$ if $a \equiv 0 \bmod p$,
and otherwise

$$\left(\frac{a}{p}\right) = \begin{cases} +1 & x^2 \equiv a \bmod p \text{ has a solution} \\ -1 & x^2 \equiv a \bmod p \text{ does not have a solution.} \end{cases} \qquad (4.29)$$

Thus $\left(\frac{a}{p}\right)$ is 1 if a is a non-zero square mod p and -1 if a is not a square mod p.
One can show (see §4.4.1) that for a fixed x, $1 + \left(\frac{f(x)}{p}\right)$ is the number of y such that
$y^2 \equiv f(x) \bmod p$. As we expect $f(x)$ to be a square half of the time, we predict
$N_p(E)$ will be of size p. Let

$$a_p(E) = p - N_p(E). \qquad (4.30)$$

Building on the prime values, we can extend the above to define $a_n(E)$ for all n,
and attach an L-function to the elliptic curve E.

Exercise 4.2.8. *While we have not defined $a_n(E)$ except when n is prime, it can be
shown for any $\epsilon > 0$ that $|a_n(E)| \le C_\epsilon n^{\frac{1}{2}+\epsilon}$. Find an s_0 such that $\sum_{n=1}^{\infty} \frac{a_n(E)}{n^s}$
converges for $\Re s > s_0$.*

Similar to the Riemann zeta function and Dirichlet L-functions, by adding Γ-
factors we can build completed elliptic curve L-functions satisfying a functional
equation, now relating values at s to $2 - s$. By replacing $a_n(E)$ with $\frac{a_n(E)}{\sqrt{n}}$ we
obtain a new completed L-function $\Lambda(s, E)$ satisfying the **functional equation**

$$\Lambda(s, E) = \epsilon_E \Lambda(1 - s, E), \qquad (4.31)$$

where $\epsilon_E = \pm 1$ is the **sign** of the elliptic curve. If ϵ_E is even (resp., odd), then
$\Lambda(s, E)$ will have an even (resp., odd) number of zeros at $s = 1/2$.

The **Birch and Swinnerton-Dyer Conjecture** states that the Mordell-Weil rank
of the group $E(\mathbb{Q})$ is equal to the order of vanishing of (the analytically continued)
$\Lambda(s, E)$ at $s = 1/2$. Thus, knowledge of the zeros of $\Lambda(s, E)$ (here just at the
special point $s = 1/2$) yields information about the group $E(\mathbb{Q})$. If the Birch
and Swinnerton-Dyer Conjecture is true, then the group of rational solutions $E(\mathbb{Q})$
would have even or odd rank, depending on ϵ_E.

It is very hard to determine the order of vanishing of $\Lambda(s, E)$ at $s = 1/2$. The
best result to date is due to Gross and Zagier [GZ], who construct an elliptic curve
where $E(\mathbb{Q})$ is of rank 3 and $L(s, E)$ has three zeros at $s = 1/2$; see [Gol1] for
applications of such a curve. No curve E is known with $\Lambda(s, E)$ having more than

three zeros at $s = 1/2$; however, curves are known where the rank of $E(\mathbb{Q})$ is as large as 24 (due to Martin and McMillen [MaMc]).

Finally, we can consider a **one-parameter family** $\mathcal{E}$ of elliptic curves. Let $A(T)$ and $B(T) \in \mathbb{Z}[T]$ be two polynomials with integer coefficients. Consider

$$\mathcal{E} : y^2 = f_T(x) = x^3 + A(T)x + B(T). \tag{4.32}$$

For each $t \in \mathbb{Z}$, by evaluating $A(T)$ and $B(T)$ at $T = t$ we obtain an elliptic curve E_t. For example, for two such polynomials E_5 would be the curve $y^2 = x^3 + A(5)x + B(5)$. For given polynomials $A(T)$ and $B(T)$, there are many properties which hold for all E_t in the family. Similar to constructing $E(\mathbb{Q})$, we can construct a group $\mathcal{E}(\mathbb{Q}(T))$. Its rank will be related to the group of solutions $(x(T), y(T))$ of $y^2 = x^3 + A(T)x + B(T)$, where $x(T), y(T)$ are rational functions of T (i.e., in $\mathbb{Q}(T)$).

Remark 4.2.9. While it had been known for a long time that L-functions attached to certain elliptic curves satisfied functional equations, only recently was this shown to hold for all curves. In fact, proving the functional equations is related to proving Fermat's Last Theorem! See for example [vdP6].

Exercise 4.2.10. *Consider the elliptic curve* $\mathcal{E} : y^2 = x^3 - T^2 x + T^4$ *over* $\mathbb{Q}(T)$. *Find at least two points* $P_1(T) = (x(T), y(T))$ *and* $P_2(T) = (x(T), y(T))$ *of rational function solutions* $\mathcal{E}(\mathbb{Q}(T))$. *Does* $P_1(T)$ *ever equal* $P_2(T)$?

4.3 HEIGHT FUNCTIONS AND DIOPHANTINE EQUATIONS

One of the aspect of the modern theory of Diophantine equations is the qualitative study of the set of solutions of a given equation, rather than trying to write down all the solutions. Often, in fact, all we need to know is the number of solutions. If the equation has a finite number of solutions then this question is well defined; however, if the equation has an infinite number of solutions, one has to be more careful in formulating such questions. The set of all even numbers is an infinite set, and so is the set of all perfect squares. The latter is a much smaller set as far as arithmetic is concerned (the percentage of numbers at most N that are even is approximately $\frac{N/2}{N} = \frac{1}{2}$, whereas the percentage that are perfect squares is approximately $\frac{\sqrt{N}}{N} = \frac{1}{\sqrt{N}}$). We would like to generalize this intuition to Diophantine equations. In order to do this, we introduce a counting device, the **height function**. We shall also see height functions in our investigations in Chapter 6 on Roth's Theorem.

Explicitly, we compare even numbers at most N with perfect squares at most N. This is possible because the integers are ordered, and while there are infinitely many even numbers (and perfect squares), there are only finitely many less than a given N. There are infinitely many rationals in any interval, but there will only be finitely many of a given height.

4.3.1 Definition of the Height Function

For a non-zero rational number $x = \frac{p}{q}$ with $(p, q) = 1$, we define its **height** to be

$$H(x) = \max(|p|, |q|), \tag{4.33}$$

and we set $h(0) = 0$. The height of a rational number shows how arithmetically complicated that number is. Height functions can easily be generalized to n-tuples of rational numbers. Suppose $(x_1, x_2, \ldots, x_n)$ is an n-tuple of rational numbers. Set $x_0 = 1$ and choose an integer λ such that

1. $\lambda x_i \in \mathbb{Z}$ for $0 \leq i \leq n$;

2. $\mathrm{lcm}\,(\lambda x_0, \lambda x_1, \ldots, \lambda x_n) = 1$.

Set

$$H(x_1, x_2, \ldots, x_n) = \max_{0 \leq i \leq n} |\lambda x_i|. \tag{4.34}$$

For example, if $x_i = \frac{p_i}{q_i}$ (p_i, q_i coprime), let $d = \mathrm{lcm}\,(q_1, \ldots, q_n)$. Then $\frac{d}{q_i}$ is an integer for all i, and

$$H(x_1, \ldots, x_n) = \max\left(d, \left|p_1 \cdot \frac{d}{q_1}\right|, \left|p_2 \cdot \frac{d}{q_2}\right|, \ldots, \left|p_n \cdot \frac{d}{q_n}\right|\right). \tag{4.35}$$

This obviously gives the existence of λ.

Exercise 4.3.1. *Is λ unique? If not, is the height well defined (i.e., do different λ give the same height)?*

4.3.2 Rational Solutions of Bounded Height

Definition 4.3.2 ($N_n(B)$ and $N_F(B)$)**.** *Let $F(x_1, x_2, \ldots, x_n) = 0$ be a Diophantine equation with rational coefficients and denote the number of elements of a set S by $|S|$. Let $x = (x_1, \ldots, x_n)$ and define*

$$N_F(B) := |\{x \in \mathbb{Q}^n : F(x_1, \ldots, x_n) = 0, H(x_1, x_2, \ldots, x_n) < B\}|$$
$$N_n(B) := |\{x \in \mathbb{Q}^n : H(x_1, x_2, \ldots, x_n) < B\}|. \tag{4.36}$$

We are interested in rational solutions to $F(x_1, \ldots, x_n) = 0$ with the height of the solutions less than B. This is similar in spirit to counting how many primes are less than x. First we prove that this number is well defined: $N_F(B)$ is finite for any F and B.

Lemma 4.3.3. *Let F be a polynomial in n variables with rational coefficients. For all B, $N_n(B) < \infty$. As $N_F(B) \leq N_n(B)$, $N_F(B)$ is finite.*

Proof. Let $x_i = \frac{p_i}{q_i}$, with p_i, q_i coprime. Let $d = \mathrm{lcm}\,(q_1, \ldots, q_n)$. From (4.35)

$$H(r_1, \ldots, r_n) = \max\left(d, \left|p_1 \cdot \frac{d}{q_1}\right|, \ldots, \left|p_n \cdot \frac{d}{q_n}\right|\right). \tag{4.37}$$

If $H(x_1, \ldots, x_n) \leq B$, then $|p_i|, |q_i| \leq B$. We may assume $q_i > 0$. Therefore there are at most $(2B + 1)^n$ choices for the p_i's and B^n choices for the q_i's, for a total of at most $B^n(2B+1)^n$ tuples with height at most B. Thus $N_n(B)$ is finite. As the solutions to $F(x_1, \ldots, x_n) = 0$ with height less than B have height less than B, we have $N_F(B) \leq N_n(B)$.   □

We have shown $N_n(B) \leq B^n(2B+1)^n$; we prove a sharper bound, as the above crude estimates never used the fact that p_i and q_i are relatively prime. For example, let us consider the case of just one variable, and determine $N_1(B)$:

$$N_1(B) = |\{x \in \mathbb{Q} : H(x) < B\}|. \qquad (4.38)$$

Recall $\phi(n)$ is the number of integers in $\{1, \ldots, n-1\}$ relatively prime to n (see Exercise 1.4.5). Let $x = \frac{p}{q}$, p, q coprime and $q > 0$, be a non-zero rational number with $H(x) = k$. This means that $\max(|p|, |q|) = k$. If $k \neq 0, 1$, there are four possibilities:

1. $p, q > 0$, $p = k$: there are $\phi(k)$ such fractions (there is only one choice for p, and by definition $\phi(k)$ choices for q because $0 < q < k$ as $x \neq 1$).

2. $p < 0, q > 0$, $p = -k$: there are again $\phi(k)$ such fractions.

3. $p, q > 0$, $q = k$: there are again $\phi(k)$ such fractions.

4. $p < 0, q > 0$, $q = k$: there are again $\phi(k)$ such fractions.

Hence there are $4\phi(k)$ rational numbers x with $H(x) = k$, when $k > 1$. There are two rational numbers with $H(x) = 1$, $x = \pm 1$, and exactly one rational number with $H(x) = 0$. Consequently we have

$$N_1(B) = 3 + 4\sum_{k=2}^{B} \phi(k) = -1 + 4\sum_{k=1}^{B} \phi(k). \qquad (4.39)$$

We have reduced the problem to determining the average size of $\phi(k)$, which we solved in Example 2.2.18. Therefore, for B large,

$$N_1(B) = \frac{12}{\pi^2} B^2 + o(B^2). \qquad (4.40)$$

We have refined our original bound of $N_1(B) \leq 2B^2 + B$ to $N_1(B) \sim \frac{12}{\pi^2} B^2$. Typically we are interested in the asymptotic behavior of rational solutions, for example, what is the correct order of magnitude of $N_1(B)$ as $B \to \infty$. Note our crude arguments found the correct power of B (the number of solutions grows like B^2 and not B), but not the correct constant. In many investigations, such as the applications of Chebyshev's Theorem in §2.3.3, the most important part is determining the correct rate of growth. For more refined investigations the constant is needed as well (for such an example, see §4.3.3).

We apply the same type of analysis to counting Pythagorean triples. Finding rational solutions of $x^2 + y^2 = z^2$ is the same as finding rational solutions of $F_{\text{Pythag}}(x_1, x_2) = x_1^2 + x_2^2 - 1 = 0$, with $x_1 = x/z$ and $x_2 = y/z$. One can show a typical rational solution to this equation has the form

$$(x_1(t), x_2(t)) = \left(\frac{t^2 - 1}{t^2 + 1}, \frac{2t}{t^2 + 1} \right) \qquad (4.41)$$

for t rational. Let $t = \frac{a}{b}$, $(a,b) = 1$. Then we set

$$\delta(a,b) = \begin{cases} 2 & \text{if } a \text{ and } b \text{ are odd} \\ 1 & \text{otherwise.} \end{cases} \tag{4.42}$$

Then

$$H((x_1(t), x_2(t))) = \frac{1}{\delta(a,b)} \max(|a^2 - b^2|, 2|ab|, |a^2 + b^2|) = \frac{a^2 + b^2}{\delta(a,b)}. \tag{4.43}$$

Exercise[h] 4.3.4. *Verify the above statement.*

Here we will ignore the effect of $\delta(a,b)$ for simplicity. The corresponding counting function $N_{\text{Pythag}}(B)$ will be

$$N_{\text{Pythag}}(B) = \left| \{ (a,b) \in \mathbb{Z}^2 : (a,b) = 1, b > 0, a^2 + b^2 < B \} \right|. \tag{4.44}$$

Without the conditions $(a,b) = 1$ and $b > 0$, $N_{\text{Pythag}}(B)$ would be the number of points $(a,b) \in \mathbb{Z}^2$ in a circle of radius $\sqrt{B}$ about the origin. Gauss proved that

$$\left| \{ (a,b) \in \mathbb{Z}^2 : a^2 + b^2 < B \} \right| = \pi B + O(\sqrt{B}), \tag{4.45}$$

which implies that for B large, $N_{\text{Pythag}}(B) \sim cB$ for some constant c.

Exercise 4.3.5. *Determine the constant c.*

Exercise[h] 4.3.6. *Prove (4.45).*

Both $\mathbb{Q}$ and $F_{\text{Pythag}}(x_1, x_2) = x_1^2 + x_2^2 - 1 = 0$ have infinitely many rational points, but the two counting functions $N_1(B)$ and $N_{F_{\text{Pythag}}}(B)$ have very different asymptotic behaviors. In recent years, a great deal of work has been done on understanding the asymptotic behavior of $N_F(B)$ for a given Diophantine equation F in terms of the geometry of the variety defined by the equation. In the examples considered above, the determining factor is the degree of the equations.

Exercise 4.3.7. *Find an asymptotic formula as $B \to \infty$ for*

$$N_n(B) = \{ (x_1, \ldots, x_n) \in \mathbb{Q}^n : H(x_1, \ldots, x_n) < B \}. \tag{4.46}$$

4.3.3 Distribution of Rational Points

This subsection is more advanced and is independent of the rest of the book; however, similar problems are studied in Chapter 15. Consider the following question:

Question 4.3.8. *How are the rational solutions to $F(x_1, \ldots, x_n) = 0$ distributed?*

This question, too, has been the subject of much research. We concentrate on the example of §4.3.2, namely $F_{\text{zero}}(x_1) = 0$. We investigate whether there is a natural measure $d\mu$ with the property that for any continuous function f of compact support

$$\lim_{B \to \infty} \frac{\sum_{x \in \mathbb{Q}, H(x) \le B} f(x)}{|\{ x \in \mathbb{Q}, H(x) \le B \}|} = \int_{\mathbb{R}} f(x) \, d\mu. \tag{4.47}$$

In other words, we are asking whether rational numbers, ordered according to their height, are equidistributed (see Chapter 12) with respect to some natural measure $d\mu$. Let us investigate this question for rational points in the interval $[0, 1]$. By the Weirestrass Approximation Theorem (see Chapter 11), it suffices to consider polynomials $f(x) = x^n$. The number of rational points in $[0, 1]$ of height bounded by B is $\sum_{k=1}^{B} \phi(k)$ which is asymptotic to $\frac{3}{\pi^2} B^2$ for B large. Next we have

$$S := \sum_{\substack{x \in [0,1] \cap \mathbb{Q} \\ H(x) \le B}} x^n = \sum_{1 \le k \le B} \frac{1}{k^n} \sum_{\substack{(m,k) = 1 \\ 1 \le m \le k}} m^n. \tag{4.48}$$

We now use the identity

$$\sum_{d|n} \mu(d) = \begin{cases} 1 & \text{if } n = 1 \\ 0 & \text{otherwise.} \end{cases} \tag{4.49}$$

Then

$$S = \sum_{k=1}^{B} \sum_{m=1}^{n} \sum_{d|(m,k)} \frac{m^n \mu(d)}{k^n}$$

$$= \sum_{d=1}^{B} \sum_{y \le \frac{B}{d}} \sum_{x \le y} \frac{x^n \mu(d)}{y^n}$$

$$\sim \frac{1}{n+1} \sum_{d=1}^{B} \sum_{y \le \frac{B}{d}} y\mu(d)$$

$$\sim \frac{1}{2(n+1)} \sum_{d=1}^{B} \left[\frac{B}{d}\right] \left(\left[\frac{B}{d}\right] + 1\right)$$

$$\sim \frac{B^2}{2(n+1)} \sum_{d=1}^{B} \frac{\mu(d)}{d^2}, \tag{4.50}$$

as $B \to \infty$. Note

$$\lim_{B \to \infty} \sum_{d=1}^{B} \frac{\mu(d)}{d^2} = \prod_{p} \left(1 - \frac{1}{p^2}\right) = \frac{1}{\zeta(2)} = \frac{6}{\pi^2}; \tag{4.51}$$

for a proof of the last claim, see Exercise 11.3.15.

Exercise 4.3.9. *Fill in the details for the above proof.*

Using the methods of §4.3.2 we can show $\frac{B^2}{2} \sum_{d=1}^{B} \frac{\mu(d)}{d^2} \sim |\{x \in [0, 1] \cap \mathbb{Q}, H(x) \le B\}|$. Thus for $f(x) = x^n$ we have

$$\lim_{B \to \infty} \frac{\sum_{x \in \mathbb{Q}, H(x) \le B} f(x)}{|\{x \in \mathbb{Q}, H(x) \le B\}|} = \frac{1}{n+1}. \tag{4.52}$$

As $\int_0^1 x^n \, dx = \frac{1}{n+1}$, we conclude that the measure $d\mu$ is the ordinary Lebesgue measure.

Exercise 4.3.10. *Carry out similar computations for* $[0, 1]^k$, *the k-dimensional unit cube.*

Remark 4.3.11. In Chapter 15 we consider similar problems: how are the primes or zeros of L-functions distributed? See in particular Lemma 15.6.7.

4.3.4 Method of Descent

Consider our proof that there are no non-trivial rational solutions of $x^4 + y^4 = z^2$ (Theorem 4.1.6). Let (x_0, y_0, z_0) be a non-trivial integer solution. We reinterpret our proof in terms of height functions. The original argument we gave uses the height function $H_{\text{old}}(x_0, y_0, z_0) = z_0$. We could also use the height function

$$H_{\text{new}}(x, y, z) = \max(|x|, |y|). \tag{4.53}$$

Exercise 4.3.12. *Given any $B > 0$, prove there are only finitely many integer solutions of $x^4 + y^4 = z^2$ with $H(x, y, z) \leq B$ for $H = H_{\text{old}}$ or $H = H_{\text{new}}$. More generally, given any polynomial with integer coefficients $p(x, y, z) = 0$, prove there are only finitely many integer solutions with height less than B.*

Letting (x_0, y_0, z_0) be a non-trivial integer solution of $x^4 + y^4 = z^2$, our previous argument led to the existence of another integer solution

$$x_1^4 + y_1^4 = z_1^2. \tag{4.54}$$

Exercise 4.3.13. *Show $x_1, y_1 \neq 0$, which implies that (x_1, y_1, z_1) is a non-trivial integer solution. Prove $H(x_1, y_1, z_1) < H(x_0, y_0, z_0)$.*

Thus, given a non-trivial integer solution (x_0, y_0, z_0), we can always find another non-trivial integer solution with smaller height. We now apply the above construction to the non-trivial integer solution (x_1, y_1, z_1) and obtain another non-trivial integer solution (x_2, y_2, z_2) with *strictly smaller height*. We then apply the same construction to (x_2, y_2, z_2) and obtain another non-trivial integer solution (x_3, y_3, z_3) with *strictly smaller height*. And so on.

This is the Method of Descent: starting with one non-trivial integer solution, we construct an infinite sequence of non-trivial integer solutions, each solution with *strictly smaller height* than the previous. Here our concept of *smaller* comes from our height function. As each solution has strictly smaller height, and the initial height $H(x_0, y_0, z_0)$ was finite, we cannot continue constructing smaller solutions indefinitely; in fact, we can only proceed at most $H(x_0, y_0, z_0) + 1$ times.

Thus, as the Method of Descent gives infinitely many solutions, we reach a contradiction. Our only assumption was that there existed a non-trivial integer solution (x_0, y_0, z_0); therefore, there are no non-trivial integer solutions of $x^4 + y^4 = z^2$.

Height functions are a powerful tool in analyzing solutions to Diophantine equations. In fact, height functions are one of two key ingredients in proving the Mordell-Weil Theorem (the group of rational solutions of an elliptic curve is finitely generated; see [Kn, Sil1]).

4.4 COUNTING SOLUTIONS OF CONGRUENCES MODULO p

We consider the problem of solving

$$F(x_1, \ldots, x_n) \equiv 0 \bmod m, \tag{4.55}$$

where F is a polynomial with integer coefficients (for applications see [Kob1]). Any integral solution of $F(x_1, x_2, \ldots, x_n) = 0$ for integers $x_1, x_2, \ldots, x_n$ yields a solution of the congruence (4.55) for any integer m. It is often false, however, that one can "patch together" solutions of the above congruence for various m to obtain integral solution of the equation (though see Exercise 1.3.10 for patching solutions to a different problem). For example, the congruence equation

$$(x^2 - 3)(x^2 - 5)(x^2 - 15) \equiv 0 \bmod m \tag{4.56}$$

has solutions for all m and for $x \in \mathbb{R}$, but the equation

$$(x^2 - 3)(x^2 - 5)(x^2 - 15) = 0 \tag{4.57}$$

has no integral solutions.

Exercise[hr] **4.4.1.** *Prove there are no integral solutions to (4.56).*

For some equations a reverse engineering process exists. Hasse and Minkowski proved that if F is a homogeneous quadratic polynomial with integral coefficients (i.e., $F(x_1, \ldots, x_n) = \sum_{i \leq j} a_{ij} x_i x_j$ with $a_{ij} \in \mathbb{Z}$) for which (4.55) is solvable for every modulus m and there is a real solution to $F(x_1, \ldots, x_n) = 0$, then there is an integral solution.

Remark 4.4.2. It is not enough to assume a solution exists to (4.55) for each m. Consider

$$F(x_1, x_2, x_3, x_4) = x_1^2 + x_2^2 + x_3^2 + x_4^2 + 1. \tag{4.58}$$

It is a theorem of Lagrange, and perhaps Fermat, that every integer is a sum of at most four squares. Thus there are solutions to $F(x) \equiv 0 \bmod m$ for every m (choose x_i so that $x_1^2 + \cdots + x_4^2 = m - 1$); however, there are no real solutions (and if there are no real solutions, there cannot be any integer solutions).

Exercise 4.4.3. *Find a polynomial f in one variable with integer coefficients such that $f(x) \equiv 0 \bmod m$ has a solution for all m but $f(x) = 0$ has no solutions for $x \in \mathbb{R}$ (and therefore $f(x) = 0$ has no integral solutions).*

4.4.1 Elliptic Curves Modulo p

Let $y^2 = x^3 + ax + b$, $a, b \in \mathbb{Z}$, be an elliptic curve. Consider for p prime

$$y^2 \equiv x^3 + ax + b \bmod p. \tag{4.59}$$

We show how the Legendre symbol (§1.6.1) yields a closed form expression for the number of solutions (x, y), $x, y \in \{0, 1, \ldots, p-1\}$. For any x, there are three possibilities:

1. If $x^3 + ax + b \equiv 0 \bmod p$ then the only solution is $y = 0$. In this case $\left(\frac{x^3+ax+b}{p}\right) = 0$.

2. If x^3+ax+b is a non-zero square mod p, say d^2, then there are two solutions: (x, d) and $(x, p - d)$. In this case $\left(\frac{x^3+ax+b}{p}\right) = 1$.

3. If $x^3 + ax + b$ is not a square mod p, then there are no solutions. In this case, $\left(\frac{x^3+ax+b}{p}\right) = -1$.

Thus, for a fixed x, the number of y such that (x, y) solves (4.59) is $1 + \left(\frac{x^3+ax+b}{p}\right)$. Therefore, if N_p is the number of solutions mod p, then

$$N_p = \sum_{x=0}^{p-1}\left[1 + \left(\frac{x^3 + ax + b}{p}\right)\right] = p + \sum_{x=0}^{p-1}\left(\frac{x^3 + ax + b}{p}\right). \quad (4.60)$$

We have reduced the problem to evaluating a sum of Legendre symbols. Hasse proved the deep bound that

$$\left|\sum_{x=0}^{p-1}\left(\frac{x^3 + ax + b}{p}\right)\right| < 2\sqrt{p}; \quad (4.61)$$

by the **philosophy of square root cancellation** (see §8.4.1), such a result is not surprising as we are summing p terms that are each ± 1. Thus

$$|N_p - p| < 2\sqrt{p}. \quad (4.62)$$

As $p \to \infty$ the number solutions is approximately p with corrections at most $2\sqrt{p}$. Equation (4.62) implies the existence of an angle $\theta_p \in [0, 2\pi]$ such that

$$\frac{N_p - p}{2\sqrt{p}} = \cos\theta_p. \quad (4.63)$$

A very interesting arithmetic problem is the study of the family of angles $\{\theta_p\}$ as p varies. The Sato-Tate conjecture predicts the behavior of the $\{\theta_p\}$ for a large class of elliptic curves. We refer the reader to [MM], §4.7 for the statement, a generalization due to Serre, and some recent progress.

In many cases we are able to evaluate the sum appearing in (4.61) directly.

Exercise[h] 4.4.4. *Consider $y^2 \equiv x^3 + ax \bmod p$, $a \not\equiv 0$. If $p \equiv 3 \bmod 4$, show $N_p = p$.*

Exercise[h] 4.4.5. *Consider $y^2 \equiv x^3 + b \bmod p$, $b \not\equiv 0$. If $p \equiv 2 \bmod 3$, show $N_p = p$.*

Exercise 4.4.6. *For $a \not\equiv 0 \bmod p$, show*

$$\sum_{t \bmod p}\left(\frac{at + b}{p}\right) = 0. \quad (4.64)$$

For $p > 2$, show

$$\sum_{t \bmod p}\left(\frac{at^2 + bt + c}{p}\right) = \begin{cases} (p-1)\left(\frac{a}{p}\right) & \text{if } p \text{ divides } b^2 - 4ac \\ -\left(\frac{a}{p}\right) & \text{otherwise.} \end{cases} \quad (4.65)$$

Using these linear and quadratic sums, one can build elliptic curves with many rational solutions (see [ALM]).

The bound $|N_p - p| < 2\sqrt{p}$ has been generalized to more general equations. In order to state a generalization we need a definition:

Definition 4.4.7 (Irreducible). *A polynomial $F(x_1, \ldots, x_n)$ with rational coefficients is called absolutely irreducible if it cannot be written as a product of polynomials with complex coefficients in any nontrivial way (each polynomial must involve at least one of the variables). If it cannot be written as a product over polynomials with rational (resp., integer) coefficients, it is irreducible over $\mathbb{Q}$ (resp., over $\mathbb{Z}$).*

For $n = 1$ the Fundamental Theorem of Algebra implies the only absolutely irreducible polynomials are linear polynomials, i.e., polynomials of degree 1; see Appendix A of [NZM] for a proof of the Fundamental Theorem of Algebra. The generalization of the bound for elliptic curves is the following:

Theorem 4.4.8. *For $n \geq 2$, if $F(x_1, \ldots, x_n)$ is an absolutely irreducible polynomial with integral coefficients then the number $N_F(p)$ of solutions of the congruence*

$$F(x_1, \ldots, x_n) \equiv 0 \bmod p \tag{4.66}$$

satisfies

$$|N_F(p) - p^{n-1}| \leq C(F)p^{n-1-\frac{1}{2}}, \tag{4.67}$$

with the constant $C(F)$ depending only on F and not on p.

The most important consequence of this theorem is that given an absolutely irreducible polynomial F with integer coefficients, the above congruence equation has non-trivial solutions for p large. For elliptic curves $n = 2$ and $C(F) = 2$. We refer the reader to the magnificent book of [BS] for an account of the history of the above fundamental theorem. Our exposition here is heavily based on that book. The proof of the above theorem is beyond the scope of this text. Below we prove the theorem in some special cases using the method of exponential sums.

Remark 4.4.9. Note we prove that for all large p there must be a solution to (4.66) by showing there are *many* solutions! This is an extremely common technique for proving the existence of solutions; we meet it again in Chapters 13 and 14 in our Circle Method investigations. There we provide evidence towards Goldbach's conjecture (every even number is the sum of two primes) by showing that every even number should be expressible as a sum of two primes in *many* ways, and therefore for each n there should be a solution to $2n = p_1 + p_2$.

4.4.2 Exponential Sums and Polynomial Congruences

The starting point for counting integer solutions is deceptively simple, namely the geometric series formula with $r = e^{2\pi i x/p}$.

Lemma 4.4.10. *Let x be an integer. Then*

$$\sum_{m=0}^{p-1} e^{2\pi i m x/p} = \begin{cases} p & \text{if } x \equiv 0 \bmod p \\ 0 & \text{otherwise.} \end{cases} \tag{4.68}$$

Proof. If $x \equiv 0 \bmod p$, each summand is 1 and the claim is clear. If $x \not\equiv 0 \bmod p$, $e^{2\pi i x/p} \neq 1$ and we have a geometric series:

$$\sum_{m=0}^{p-1} e^{2\pi i m x/p} = \sum_{m=0}^{p-1} \left(e^{\frac{2\pi i x}{p}} \right)^m = \frac{1 - e^{2\pi i x p/p}}{1 - e^{2\pi i x/p}} = 0, \qquad (4.69)$$

where we used Exercise 3.1.4 to sum the geometric series. $\square$

Consider the polynomial congruence

$$F(x_1, \ldots, x_n) \equiv 0 \bmod p. \qquad (4.70)$$

Assume at least one coefficient of F is non-zero. We want to count the number of solutions to (4.70) with $x_1, \ldots, x_n \in \{0, \ldots, p-1\}$.

Lemma 4.4.11. *The number of solutions to* (4.70) *is given by*

$$N_p = \frac{1}{p} \sum_{x_1, \ldots, x_n = 0}^{p-1} \left[\sum_{m=0}^{p-1} e^{2\pi i m F(x_1, \ldots, x_n)/p} \right]. \qquad (4.71)$$

Proof. By Lemma 4.4.10, for each n-tuple $(x_1, \ldots, x_n)$ the m-sum is p if $F(x_1, \ldots, x_n) \equiv 0 \bmod p$ and 0 otherwise. The claim follows by summing over all n-tuples and dividing by p. $\square$

The terms with $m = 0$ contribute p^{n-1} (each x_i-sum gives p and then we divide by p). Consequently Theorem 4.4.8 is in fact an estimate for

$$N_p' = \frac{1}{p} \sum_{x_1, \ldots, x_n = 0}^{p-1} \sum_{m=1}^{p-1} e^{2\pi i m F(x_1, \ldots, x_n)/p}, \qquad (4.72)$$

under the conditions of the theorem. While we have written down a closed form expansion for our problem, we have not made any progress unless we can estimate (4.72). This is similar in spirit to the reformulations we see in writing a formula for the number of primes at most x (Exercise 2.3.18) or in studying the Circle Method (Chapter 13).

Remark 4.4.12. While (4.71) shows $N_p \leq p^n$, we expect it to be of size p^{n-1} (choose $n-1$ variables freely and the last is determined). It is worth commenting that we have found a new expression which counts the number of solutions. We have split it into two parts. First, the terms with $m = 0 \bmod p$; the contribution from these terms is p^{n-1} and is expected to be the main term. The proof is completed by showing the contribution from the remaining terms N_p' is smaller than the main term (at least for p large).

Exercise 4.4.13. *For sums of squares (each $r_i = 2$), prove Theorem 4.4.8 for $n = 2$.*

4.4.3 Proof of Theorem 4.4.8 for Sums of Powers

In this section we assume the reader is familiar with group theory at the level of §1.3 and §1.4. In particular, by Theorem 1.4.29 for p prime, $(\mathbb{Z}/p\mathbb{Z})^*$ is cyclic of order $p-1$. For example, for $p = 7$ we have

$$\{1, 2, 3, 4, 5, 6\} \equiv \{3^6, 3^2, 3^1, 3^4, 3^5, 3^3\}, \qquad (4.73)$$

so 3 is a generator of $(\mathbb{Z}/7\mathbb{Z})^*$. Fix a generator g of $(\mathbb{Z}/p\mathbb{Z})^*$. As this group is cyclic, every $x \not\equiv 0 \bmod p$ can be written as g^k for a unique $k \in \{1, \dots, p-1\}$. If $x \equiv 0 \bmod p$, let $k = 0$. We call k the **index** of x (relative to g).

We prove Theorem 4.4.8 for polynomials of the form

$$F(x_1, \dots, x_n) = a_1 x_1^{r_1} + \cdots + a_n x_n^{r_n}, \tag{4.74}$$

with a_i integers not divisible by p, and $n \geq 2$. Because of Fermat's Little Theorem (Theorem 1.4.19), the number of integer tuples with $a_1 x_1^{r_1} + \cdots + a_n x_n^{r_n} \equiv 0 \bmod p$ should depend on divisibility relations between the r_i's and $p-1$. The reason for this is that for any integer $x \not\equiv 0 \bmod p$, $x^{p-1} \equiv 1 \bmod p$.

For example, assume each $r_i \equiv 1 \bmod p - 1$; this would happen if each $r_i = 1$. Then our equation reduces to solving

$$a_1 x_1 + \cdots + a_n x_n \equiv 0 \bmod p. \tag{4.75}$$

As each $a_i \not\equiv 0 \bmod p$, we can choose any integer values for x_1 through x_{n-1}, and then x_n is determined:

$$x_n \equiv -(a_1 x_1 + \cdots + a_{n-1} x_{n-1}) \cdot a_n^{-1} \bmod p. \tag{4.76}$$

Thus the number of solutions is *exactly* p^{n-1}; there is no error and we may take $C(F) = 0$ in Theorem 4.4.8.

This is of course a special case, although a very instructive one. Another good case to consider is a prime $p > 3$, each $r_i = 3$ and 3 does not divide $p - 1$ (i.e., $p \equiv 2 \bmod 3$). The reason this case is interesting is that the map sending x to $x^3 \bmod p$ is an isomorphism of $\mathbb{Z}/p\mathbb{Z}$. What this means is that this map simply permutes the order of the elements $0, 1, \dots, p-1$. For example, if $p = 5$ we have

$$\{0^3, 1^3, 2^3, 3^3, 4^3\} \equiv \{0, 1, 3, 2, 4\} \bmod 5. \tag{4.77}$$

The important point to note is that as x_i ranges over $\{0, 1, \dots, p-1\}$, so too does x_i^3, each taking on each value once and only once. Thus for $p \equiv 2 \bmod 3$ we again see that it suffices to count solutions to

$$a_1 x_1 + \cdots + a_n x_n \equiv 0 \bmod p, \tag{4.78}$$

and again we see we may take $C(F) = 0$.

What goes wrong if $p \equiv 1 \bmod 3$? In this case the map $x \to x^3 \bmod p$ is not an isomorphism (i.e., it does not just permute the elements). For example, if $p = 7$ we have

$$\{0^3, 1^3, 2^3, 3^3, 4^3, 5^3, 6^3\} \equiv \{0, 1, 1, 6, 1, 6, 6\} \bmod 7. \tag{4.79}$$

The point of these examples is that whenever an r_i is relatively prime to $p - 1$, then we may replace $a_i x_i^{r_i}$ with $a_i x_i$. In fact, this leads to the following result:

Lemma 4.4.14 (Special Case of Theorem 4.4.8 for Sums of Squares). *If some r_i is relatively prime to $p - 1$ then the number of solutions to*

$$a_1 x^{r_1} + \cdots + a_n x^{r_n} \equiv 0 \bmod p \tag{4.80}$$

is exactly p^{n-1}.

Proof. Without loss of generality, assume r_n is relatively prime to $p - 1$. Then as x_n ranges over $\{0, 1, \ldots, p - 1\}$, so too does $x_n^{r_n}$. Therefore the number of solutions of (4.80) is the same as the number of solutions of

$$a_1 x_1^{r_1} + \cdots + a_{n-1} x_{n-1}^{r_{n-1}} + a_n x_n \equiv 0 \bmod p. \qquad (4.81)$$

We may choose $x_1, \ldots, x_{n-1}$ arbitrarily in $\{0, 1, \ldots, p-1\}$, and then x_n is uniquely determined:

$$x_n \equiv (a_1 x_1^{r_1} + \cdots + a_{n-1} x_{n-1}^{r_{n-1}}) \cdot a_n^{-1} \bmod p. \qquad (4.82)$$

The proof is completed by noting there are p^{n-1} choices for $x_1, \ldots, x_{n-1} \in \{0, 1, \ldots, p - 1\}$. $\qquad\square$

Exercise 4.4.15. *Show the above arguments imply that if we write each r_i as $a_i b_i$ with b_i relatively prime to $p - 1$, then the number of integer solutions is independent of b_i. Specifically, show that $x_i \to x_i^{b_i} \bmod p$ is an isomorphism, permuting $\{0, 1, \ldots, p - 1\}$.*

We now return to the proof of Theorem 4.4.8 for sums of squares. By (4.72), we need a good bound for

$$N_p' = \frac{1}{p} \sum_{x_1, \ldots, x_n = 0}^{p-1} \sum_{m=1}^{p-1} e^{2\pi i m (a_1 x_1^{r_1} + \cdots + a_n x_n^{r_n})/p}$$

$$= \frac{1}{p} \sum_{m=1}^{p-1} \prod_{i=1}^{n} \sum_{x_i=0}^{p-1} e^{2\pi i m a_i x_i^{r_i}/p}. \qquad (4.83)$$

This leads us to consider, for $a \not\equiv 0 \bmod p$, sums of the form

$$\sum_{y=0}^{p-1} e^{2\pi i a y^r/p}. \qquad (4.84)$$

It is clear that if $m(x)$ is the number of solutions of the congruence $y^r \equiv x \bmod p$, then

$$\sum_{y=0}^{p-1} e^{2\pi i a y^r/p} = \sum_{x=0}^{p-1} m(x) e^{2\pi i a x/p}. \qquad (4.85)$$

Exercise 4.4.16. *Let $d = (r, p - 1)$ and k the index of x relative to g; remember that we defined the index of 0 to be 0. Then*

$$m(x) = \begin{cases} 1 & \text{if } k = 0 \\ d & \text{if } k \neq 0, \ k \equiv 0 \bmod d \\ 0 & \text{if } k \neq 0, \ k \not\equiv 0 \bmod d. \end{cases} \qquad (4.86)$$

Show the answer is independent of the generator g.

If we want, we may invoke Lemma 4.4.14 to show that it suffices to consider the case when each $r_i | (p - 1)$. This follows from Fermat's Little Theorem (as $x^p \equiv x \bmod p$, we may assume each r_i is at most $p - 1$) and because if there is an

r_i relatively prime to $p - 1$, then $x_i \to x_i^{r_i} \bmod p$ is an isomorphism, permuting $\{0, 1, \ldots, p - 1\}$; this leads to there are exactly p^{n-1} solutions. We do not assume this for now. We argue in greater generality to highlight key features of arguments like this.

We present a different formula for $m(x)$. For $x \not\equiv 0 \bmod p$ with x of index k and d as above, define a character $\chi : (\mathbb{Z}/p\mathbb{Z})^* \to \mathbb{C}$ by

$$\chi(x) = e^{2\pi i k / d}. \tag{4.87}$$

For $s \in \{0, 1, \ldots, d - 1\}$, set

$$\chi_s(x) = \chi(x)^s. \tag{4.88}$$

So far we have only defined the character for x that are invertible modulo p. We extend the definition of χ_s, s as above, by

$$\chi_s(x) = \begin{cases} 0 & \text{if } x \equiv 0 \bmod p \text{ and } s \neq 0 \\ 1 & \text{if } x \equiv 0 \bmod p \text{ and } s = 0. \end{cases} \tag{4.89}$$

The χ_s are **multiplicative characters** (see Definition 2.1.2), which means that for all x, y:

$$\chi_s(xy) = \chi_s(x)\chi_s(y). \tag{4.90}$$

Note χ_0 is the constant function and is identically 1; while this conflicts a bit with earlier notation of Dirichlet characters, it is notationally convenient to denote the constant function by χ_0 here. We can reconcile the χ_s with $s > 0$ with the explicit expressions for Dirichlet characters with prime conductors from Theorem 3.3.11. As $d = (r, p - 1)$ and k is the index of x relative to a generator g, we have

$$\chi_s(x) = e^{2\pi i k s / d} = e^{2\pi i k s \frac{p-1}{d}/(p-1)}. \tag{4.91}$$

Note we use sums over characters to count solutions to congruences (in this case, to $y^r \equiv x \bmod p$). This is just one of many uses of characters.

Exercise 4.4.17. *Prove* (4.90).

For more properties of these characters, see §3.3.2.

Lemma 4.4.18. *We have*

$$m(x) = \sum_{s=0}^{d-1} \chi_s(x). \tag{4.92}$$

Proof. Remember $m(x)$ is the number of solutions to $y^r \equiv x \bmod p$. Assume first that $x \not\equiv 0 \bmod p$. If $k \equiv 0 \bmod d$, then $e^{2\pi i k / d} = 1$ and the sum is equal to d. If $k \not\equiv 0 \bmod d$, then $e^{2\pi i k / d} \neq 1$, and therefore

$$\sum_{s=0}^{d-1} \chi(x)^s = \frac{\chi(x)^d - 1}{\chi(x) - 1} = 0. \tag{4.93}$$

If $x \equiv 0 \bmod p$, as $m(x) = 1$ and $\chi_0(0) = 1$ while all other $\chi_s(0) = 0$, the formula still holds. $\qquad\square$

Using (4.85) and Lemma 4.4.18 we may write

$$\sum_{y=0}^{p-1} e^{2\pi i a y^r / p} = \sum_{x=0}^{p-1} m(x) e^{2\pi i a x / p}$$

$$= \sum_{x=0}^{p-1} \sum_{s=0}^{d-1} \chi_s(x) e^{2\pi i a x / p}$$

$$= \sum_{x=0}^{p-1} \sum_{s=1}^{d-1} \chi_s(x) e^{2\pi i a x / p} + \sum_{x=0}^{p-1} \chi_0(x) e^{2\pi i a x / p}; \quad (4.94)$$

of course if $d = 1$ then the sum $\sum_{s=1}^{d-1}$ is vacuous and does not exist; we shall still write the sum as above, but in this case we interpret the sum to be zero. By (4.87), (4.88) and (4.89), $\chi_0(x) = 1$ for all x. Therefore the $s = 0$ terms contribute

$$\sum_{x=0}^{p-1} \chi_0(x) e^{2\pi i a x / p} = \sum_{x=0}^{p-1} e^{2\pi i a x / p} = \frac{e^{2\pi i a \cdot 0 / p} - e^{2\pi i a \cdot p / p}}{1 - e^{2\pi i a / p}} = 0, \quad (4.95)$$

because this is a geometric series and $a \not\equiv 0 \bmod p$. Thus

$$\sum_{y=0}^{p-1} e^{2\pi i a y^r / p} = \sum_{x=0}^{p-1} \sum_{s=1}^{d-1} \chi_s(x) e^{2\pi i a x / p} = \sum_{s=1}^{d-1} \sum_{x=0}^{p-1} \chi_s(x) e^{2\pi i a x / p}, \quad (4.96)$$

and again if $d = 1$ then the sum is vacuous and hence zero. Set

$$c_a(p, \chi_s) = \sum_{x=0}^{p-1} \chi_s(x) e^{2\pi i a x / p}. \quad (4.97)$$

We encountered these sums in the special case of $a = 1$ in §3.3.5. We call the $c_a(p, \chi_s)$ a **Gauss sum**. Therefore

$$\sum_{y=0}^{p-1} e^{2\pi i a y^r / p} = \sum_{s=1}^{d-1} c_a(p, \chi_s). \quad (4.98)$$

Combining all the pieces, we have

$$N_p' = \frac{1}{p} \sum_{m=1}^{p-1} \prod_{i=1}^{n} \sum_{s=1}^{d_i - 1} c_{m a_i}(p, \chi_{i,s}) \quad (4.99)$$

with $d_i = (r_i, p - 1)$ and the character $\chi_{i,s}$ is defined as above but with d_i instead of d. It is then clear that in order to establish the bound of Theorem 4.4.8, we just need to find appropriate upper bounds for $|c_a(p, \chi_s)|$. By Lemma 4.4.14, if any $r_i \equiv 0 \bmod p - 1$ then the corresponding $c_{m a_i}(p) = 0$ (because $d_i = 1$ and the vacuous sum is just zero), proving Theorem 4.4.8 with $C(F) = 0$. In general we have

Theorem 4.4.19. Let $d = (r, p - 1) \geq 2$; thus r and $p - 1$ are not relatively prime. For $s \in \{0, 1, \ldots, d - 1\}$, if $s \neq 0$ and $a \not\equiv 0 \bmod p$ then

$$|c_a(p, \chi_s)| = \sqrt{p}. \quad (4.100)$$

We saw a similar sum in Exercise 3.3.34. Note there are p terms in $c_a(p, \chi)$, each term of size 1; however, the sum exhibits square root cancellation and is of size $\sqrt{p}$. This is a general phenomenon, arising in many places (see, for example, the Central Limit Theorem in Chapter 8 and the generating functions in the Circle Method problems of Chapters 13 and 14).

We shall prove this result in §4.4.4. Using the bound from Theorem 4.4.19 yields

$$|N_p'| \leq \frac{1}{p} \sum_{m=1}^{p-1} \prod_{i=1}^{n} \sum_{s=1}^{d_i-1} |c_{ma_i}(p, \chi_{i,s})|$$

$$= \frac{1}{p}(p-1) \prod_{i=1}^{n} (d_i - 1)\sqrt{p}$$

$$= (p-1)p^{\frac{n}{2}-1} \prod_{i=1}^{n} (d_i - 1). \tag{4.101}$$

As always, if *any* of the $d_i = 1$ then the corresponding sum is vacuous and hence zero (which we see above as then the factor $d_i - 1$ is zero); in this case the error term can be improved to zero!

We have thus proved the following theorem:

Theorem 4.4.20. *If* $F(x_1, \ldots, x_n) = a_1 x_1^{r_1} + \cdots + a_n x_n^{r_n}$ *is a polynomial with integral coefficients, and* p *a prime number with* $a_i \not\equiv 0 \bmod p$, $i = 1, \ldots, n$, *then the number* $N_F(p)$ *of solutions of the congruence*

$$F(x_1, \ldots, x_n) \equiv 0 \bmod p \tag{4.102}$$

satisfies

$$|N_F(p) - p^{n-1}| \leq C(f)p^{\frac{n}{2}}, \tag{4.103}$$

with $C(f) = (d_1 - 1) \cdots (d_n - 1)$, $d_i = (r_i, p - 1)$.

Remark 4.4.21. For a shorter proof of Theorem 4.4.20, see [BS], exercises 9-12, page 18 (also see the important paper [Weil], where a similar method has been used). Following [BS] we decided to include the proof using Gauss sums. Gauss sums are important objects; we have already encountered them when studying the functional equations of Dirichlet L-series (Theorem 3.3.36), and they surface in many other places as well. Note the square root cancellation in the sums above: there are p terms of absolute value 1 in each Gauss sum, but the sum is of size $\sqrt{p}$.

Exercise 4.4.22. Let $F(x_1, x_2) = ax_1^2 + bx_1 x_2 + cx_2^2$, $a, b, c \in \mathbb{Z}$. We say F is *homogenous* of degree 2. If $p \neq 2$, show that we can change variables and obtain a diagonal polynomial $F(x_1', x_2') = a'x_1^2 + c'x_2^2$, and hence our previous bounds to count solutions of $F(x) \equiv 0 \bmod p$ are applicable. Does this method generalize to degree 2 homogenous polynomials with integer coefficients in n variables?

The following exercise recasts our results in the terminology of Linear Algebra.

Exercise 4.4.23. Let $F(x_1, \ldots, x_n)$ be a homogenous polynomial of degree 2 with integer coefficients:

$$F(x_1, \ldots, x_n) = \sum_{1 \leq i \leq j \leq n} c_{ij} x_i x_j. \tag{4.104}$$

To each F we can associate a real symmetric matrix A_F with $a_{ii} = c_{ii}$ and $a_{ij} = c_{ij}/2$ if $i \neq j$. Solving $F(x) \equiv 0 \bmod p$ is the same as solving $x^T A_F x \equiv 0 \bmod p$. We have bounded the number of solutions when A is diagonal. Assume we have an F whose A_F is not diagonal. For which A_F can we change basis and apply our previous results?

4.4.4 Size of Gauss Sums

Proof of Theorem 4.4.19. We are studying

$$c_a(p, \chi_s) = \sum_{x=0}^{p-1} \chi_s(x) e^{2\pi i a x/p}, \tag{4.105}$$

where $s > 0$ and the χ_s are multiplicative Dirichlet characters with conductor p. A common approach to obtaining estimates of complex-valued functions is to consider the absolute value square, namely multiply by the complex conjugate. This yields

$$|c_a(p, \chi_s)|^2 = c_a(p, \chi_s)\overline{c_a(p, \chi_s)}$$

$$= \sum_{x=0}^{p-1} \chi_s(x) e^{2\pi i a x/p} \sum_{y=0}^{p-1} \overline{\chi_s}(y) e^{-2\pi i a y/p}. \tag{4.106}$$

We expect the main contribution is from the $x = y$ terms. Note if $x = 0$ then $\chi_s(x) = 0$ (as we are assuming $s > 0$); thus we may assume $x \not\equiv 0 \bmod p$. For a fixed x, change variables in the y-sum by sending y to $xy \bmod p$; this is permissible as $x \not\equiv 0 \bmod p$; note $x = y$ terms now become the terms with $y = 1$. This yields

$$|c_a(p, \chi_s)|^2 = \sum_{x=1}^{p-1}\sum_{y=0}^{p-1} \chi_s(x)\overline{\chi_s(xy)} e^{2\pi i a(x-xy)/p}$$

$$= \sum_{x=1}^{p-1}\sum_{y=0}^{p-1} \chi_s(x)\overline{\chi_s(x)}\,\overline{\chi_s(y)} e^{2\pi i a(x-xy)/p}$$

$$= \sum_{x=1}^{p-1}\sum_{y=0}^{p-1} \overline{\chi_s(y)} e^{2\pi i a x(1-y)/p}. \tag{4.107}$$

If $y \equiv 1 \bmod p$ then the x-sum is just $p - 1$ (each term is just 1). We switch orders of summation, summing now on x first and then y. Interchanging orders is a common method, as often it is easier to handle one sum first rather than the other. Therefore

$$|c_a(p, \chi_s)|^2 = p - 1 + \sum_{y \not\equiv 1 \bmod p} \sum_{x=1}^{p-1} \overline{\chi_s(y)} e^{2\pi i a x(1-y)/p}$$

$$= p - 1 + \sum_{y \not\equiv 1 \bmod p} \sum_{x=0}^{p-1} \overline{\chi_s(y)} e^{2\pi i a x(1-y)/p} - \sum_{y \not\equiv 1 \bmod p} \overline{\chi_s(y)}$$

$$= p - 1 + 0 - \sum_{y \not\equiv 1 \bmod p} \overline{\chi_s(y)}, \tag{4.108}$$

as the x-sum vanishes by the geometric series formula. Notice how we have added and subtracted terms to make our sums full (i.e., to remove the annoying restrictions of summing over just some of the x's or y's modulo p). Arguing as above gives

$$|c_a(p, \chi_s)|^2 = p - 1 - \sum_{y \bmod p} \overline{\chi_s(y)} + \chi_s(1) = p, \qquad (4.109)$$

where the y-sum vanishes by the Orthogonality Relations of Dirichlet characters (see Lemma 3.3.15). As $|c_a(p, \chi_s)|^2 = p$ we finally obtain $|c_a(p, \chi_s)| = \sqrt{p}$. $\quad\square$

4.5 RESEARCH PROJECTS

Research Project 4.5.1. Let $f(x, y, z, t)$ be an irreducible homogeneous polynomial with integer coefficients. We are basically interested in finding asymptotic expressions for

$$N_f(B) := \#\left\{ (x, y, z, t) \in \mathbb{Z}^4 \;\middle|\; \begin{array}{c} \gcd(x, y, z, t) = 1 \\ |x|, |y|, |z|, |t| \le B, \; f(x, y, z, t) = 0 \end{array} \right\}$$
$$(4.110)$$

as $B \to \infty$. It is, however, often necessary to throw out those tuples (x, y, z, t) for which $f(x, y, z, t) = 0$ for trivial reasons. The reason is often in the limit "most" solutions arise from trivial reasons, and if we did not remove these we would not see the interesting features specific to the particular problem. For example, if

$$f(x, y, z, t) = xyz - t^3, \qquad (4.111)$$

then we will throw out those (x, y, z, t) which are of the form $(x, y, 0, 0)$, $(0, y, z, 0)$, $(x, 0, z, 0)$ or (x, x, x, x). The above problem as stated is, in the current state of knowledge, hopeless, but one can explore specific examples. A nice list of possibilities is the following collection of *singular cubic surfaces*:

1. $xyz - t^3$

2. $(x + y + z)^2 t - xyz$

3. $t(xy + yz + zx) - xyz$

4. $x^2 t + yt^2 - z^3$.

In each case one expects an asymptotic formula of the form $C_f B(\log B)^6$. Except for the first cubic this is not known, though there are partial results available in the literature; see [BBD] for recent progress on the fourth surface. Further, for the first cubic there are at least B^2 trivial solutions; we now see why we removed these, as their number dwarfs the expected number of non-trivial solutions. For various results and conjectures see [BT, deBr, Bro, Fou, He] and the references therein, as well as the undergraduate research project [Ci]. *Warning:* it is very difficult to see logarithmic factors in numerical investigations. Thus, for small B, it is hard to distinguish $C_f B(\log B)^6$ from $C_f' B^{1+\epsilon}$ for some small ϵ, and without theoretical support for a conjecture it is easy to be mislead. See Exercise 2.3.16 for an example of a slowly diverging sum.

The Circle Method (Chapters 13 and 14) can be used to predict the answer for questions similar to the following project; see those chapters as well as [Ci].

Research Project 4.5.2. Another interesting counting problem is to consider the distribution of rational point of bounded height on the cubic

$$x_0^3 + x_1^3 + x_2^3 = x_3^3 + x_4^3 + x_5^3 \tag{4.112}$$

after throwing out obvious "accumulating subsets." The reader should consult the website http://euler.free.fr/ for many numerical examples and questions. See [Ci] and the references therein for integer solutions.

PART 2
Continued Fractions and Approximations

Chapter Five

Algebraic and Transcendental Numbers

We have the following inclusions: the natural numbers $\mathbb{N} = \{0, 1, 2, 3, \dots\}$ are a subset of the integers $\mathbb{Z} = \{\dots, -1, 0, 1, \dots\}$ are a subset of the rationals $\mathbb{Q} = \{\frac{p}{q} : p, q \in \mathbb{Z}, q \neq 0\}$ are a subset of the real numbers $\mathbb{R}$ are a subset of the complex numbers $\mathbb{C}$. The notation $\mathbb{Z}$ comes from the German zahl (number) and $\mathbb{Q}$ comes from quotient. Are most real numbers rational? We show that, not only are rational numbers "scarce," but irrational numbers like $\sqrt{n}$ or $\sqrt[m]{n}$ are also scarce.

Definition 5.0.3 (Algebraic Number). *An $\alpha \in \mathbb{C}$ is an algebraic number if it is a root of a polynomial with finite degree and integer coefficients.*

Definition 5.0.4 (Transcendental Number). *An $\alpha \in \mathbb{C}$ is a transcendental number if it is not algebraic.*

Later (Chapters 7, 9 and 12) we see many properties of numbers depend on whether or not a number is algebraic or transcendental. We prove in this chapter that most real numbers are transcendental *without ever constructing a transcendental number!* We then show that e is transcendental but only later in §5.6.2 will we explicitly construct infinitely many transcendental numbers.

The main theme of this chapter is to describe a way to compare sets with infinitely many elements. In Chapter 2 we compared subsets of the natural numbers. For any set A, let $A_N = A \cap \{1, 2, \dots, N\}$, and consider $\lim_{N \to \infty} \frac{A_N}{N}$. Such comparisons allowed us to show that in the limit zero percent of all integers are prime (see Chebyshev's Theorem, Theorem 2.3.9), but there are far more primes than perfect squares. While such limiting arguments work well for subsets of the integers, they completely fail for other infinite sets and we need a new notion of size.

For example, consider the closed intervals $[0, 1]$ and $[0, 2]$. In one sense the second set is larger as the first is a proper subset. In another sense they are the same size as each element $x \in [0, 2]$ can be paired with a unique element $y = \frac{x}{2} \in [0, 1]$. The idea of defining size through such correspondences has interesting consequences. While there are as many perfect squares as primes as integers as algebraic numbers, such numbers are rare and in fact essentially all numbers are transcendental.

5.1 RUSSELL'S PARADOX AND THE BANACH-TARSKI PARADOX

The previous example, where in some sense the sets $[0, 1]$ and $[0, 2]$ have the same number of elements, shows that we must be careful with our definition of counting. To motivate our definitions we give some examples of paradoxes in set theory, which emphasize why we must be so careful to put our arguments on solid mathematical ground.

Russell's Paradox: Assume for any property P the collection of all elements having property P is a set. Consider $\mathcal{R} = \{x : x \notin x\}$; thus $x \in \mathcal{R}$ if and only if $x \notin x$. Most objects are not elements of themselves; for example, $\mathbb{N} \notin \mathbb{N}$ because the set of natural numbers is not a natural number. If $\mathcal{R}$ exists, it is natural to ask whether or not $\mathcal{R} \in \mathcal{R}$. Unwinding the definition, we see $\mathcal{R} \in \mathcal{R}$ if and only if $\mathcal{R} \notin \mathcal{R}$! Thus the collection of all objects satisfying a given property is not always a set. This strange situation led mathematicians to reformulate set theory. See, for example, [HJ, Je].

Banach-Tarski Paradox: Consider a solid unit sphere in $\mathbb{R}^3$. It is possible to divide the sphere into 5 disjoint pieces such that, by simply translating and rotating the 5 pieces, we can assemble 3 into a solid unit sphere and the other 2 into a disjoint solid unit sphere. But translating and rotating should not change volumes, yet we have doubled the volume of our sphere! This construction depends on the (Uncountable) Axiom of Choice (see §5.3.4). See, for example, [Be, Str].

Again, the point of these paradoxes is to remind ourselves that plausible statements need not be true, and one must be careful to build on firm foundations.

5.2 DEFINITIONS

We now define the terms we will use in our counting investigations. We assume some familiarity with set theory; we will not prove all the technical details (see [HJ] for complete details).

A function $f : A \to B$ is **one-to-one** (or **injective**) if $f(x) = f(y)$ implies $x = y$; f is **onto** (or **surjective**) if given any $b \in B$ there exists $a \in A$ with $f(a) = b$. A **bijection** is a one-to-one and onto function.

Exercise 5.2.1. *Show* $f : \mathbb{R} \to \mathbb{R}$ *given by* $f(x) = x^2$ *is not a bijection, but* $g : [0, \infty) \to \mathbb{R}$ *given by* $g(x) = x^2$ *is. If* $f : A \to B$ *is a bijection, prove there exists a bijection* $h : B \to A$. *We usually write* f^{-1} *for* h.

We say two sets A and B **have the same cardinality** (i.e., are the same size) if there is a bijection $f : A \to B$. We denote the common cardinality by $|A| = |B|$. If A has finitely many elements (say n elements), then there is a bijection from A to $\{1, \ldots, n\}$. We say A is **finite** and $|A| = n < \infty$.

Exercise 5.2.2. *Show two finite sets have the same cardinality if and only if they have the same number of elements.*

Exercise 5.2.3. *Suppose* A *and* B *are two sets such that there are onto maps* $f : A \to B$ *and* $g : B \to A$. *Prove* $|A| = |B|$.

Exercise 5.2.4. *A set A is said to be infinite if there is a one-to-one map $f : A \to A$ which is not onto. Using this definition, show that the sets $\mathbb{N}$ and $\mathbb{Z}$ are infinite sets. In other words, prove that an infinite set has infinitely many elements.*

Exercise 5.2.5. *Show that the cardinality of the positive even integers is the same as the cardinality of the positive integers is the same as the cardinality of the perfect squares is the same as the cardinality of the primes.*

Remark 5.2.6. Exercise 5.2.5 is surprising. Let E_N be all positive even integers at most N. The percentage of positive integers less than $2M$ and even is $\frac{M}{2M} = \frac{1}{2}$, yet the even numbers have the same cardinality as $\mathbb{N}$. If S_N is all perfect squares up to N, one can similarly show the percentage of perfect squares up to N is approximately $\frac{1}{\sqrt{N}}$, which goes to zero as $N \to \infty$. Hence in one sense there are a lot more even numbers or integers than perfect squares, but in another sense these sets are the same size.

A is **countable** if there is a bijection between A and the integers $\mathbb{Z}$. A is **at most countable** if A is either finite or countable. A is **uncountable** if A is not at most countable

Definition 5.2.7 (Equivalence Relation). *Let R be a binary relation (taking values true and false) on a set S. We say R is an equivalence relation if the following properties hold:*

1. Reflexive: $\forall x \in S$, $R(x,x)$ is true;

2. Symmetric: $\forall x, y \in S$, $R(x,y)$ is true if and only if $R(y,x)$ is true;

3. Transitive: $\forall x, y, z \in S$, $R(x,y)$ and $R(y,z)$ are true imply $R(x,z)$ is true.

Exercise 5.2.8.

1. *Let S be any set, and let $R(x,y)$ be $x = y$. Prove that R is an equivalence relation.*

2. *Let $S = \mathbb{Z}$ and let $R(x,y)$ be $x \equiv y \bmod n$. Prove R is an equivalence relation.*

3. *Let $S = (\mathbb{Z}/m\mathbb{Z})^*$ and let $R(x,y)$ be xy is a quadratic residue modulo m. Is R an equivalence relation?*

If A and B are sets, the **Cartesian product** $A \times B$ is $\{(a,b) : a \in A, b \in B\}$.

Exercise 5.2.9. *Let $S = \mathbb{N} \times (\mathbb{N} - \{0\})$. For $(a,b), (c,d) \in S$, we define $R((a,b),(c,d))$ to be true if $ad = bc$ and false otherwise. Prove that R is an equivalence relation. What type of number does a pair (a,b) represent?*

Exercise 5.2.10. *Let x, y, z be subsets of X (for example, $X = \mathbb{Q}, \mathbb{R}, \mathbb{C}, \mathbb{R}^n$, et cetera). Define $R(x,y)$ to be true if $|x| = |y|$ (the two sets have the same cardinality), and false otherwise. Prove R is an equivalence relation.*

5.3 COUNTABLE AND UNCOUNTABLE SETS

We show that several common sets are countable. Consider the set of whole numbers $\mathbb{W} = \{1, 2, 3, \dots\}$. Define $f : \mathbb{W} \to \mathbb{Z}$ by $f(2n) = n-1$, $f(2n+1) = -n-1$. By inspection, we see f gives the desired bijection between $\mathbb{W}$ and $\mathbb{Z}$. Similarly, we can construct a bijection from $\mathbb{N}$ to $\mathbb{Z}$, where $\mathbb{N} = \{0, 1, 2, \dots\}$. Thus, we have proved

Lemma 5.3.1. *To show a set S is countable, it is sufficient to find a bijection from S to either $\mathbb{W}$ or $\mathbb{N}$ or $\mathbb{Z}$.*

We need the intuitively plausible

Lemma 5.3.2. *If $A \subset B$, then $|A| \leq |B|$.*

Lemma 5.3.3. *If $f : A \to C$ is a one-to-one function (not necessarily onto), then $|A| \leq |C|$. Further, if $C \subset A$ then $|A| = |C|$.*

Theorem 5.3.4 (Cantor-Bernstein). *If $|A| \leq |B|$ and $|B| \leq |A|$, then $|A| = |B|$.*

Exercise 5.3.5. *Prove Lemmas 5.3.2 and 5.3.3 and Theorem 5.3.4.*

Theorem 5.3.6. *If A and B are countable then so is $A \cup B$ and $A \times B$.*

Proof. We have bijections $f : \mathbb{N} \to A$ and $g : \mathbb{N} \to B$. Thus we can label the elements of A and B by

$$A = \{a_0, a_1, a_2, a_3, \dots\}$$
$$B = \{b_0, b_1, b_2, b_3, \dots\}. \tag{5.1}$$

Assume $A \cap B$ is empty. Define $h : \mathbb{N} \to A \cup B$ by $h(2n) = a_n$ and $h(2n+1) = b_n$. As h is a bijection from $\mathbb{N}$ to $A \cup B$, this proves $A \cup B$ is countable. We leave to the reader the case when $A \cap B$ is not empty. To prove $A \times B$ is countable, consider the following function $h : \mathbb{N} \to A \times B$ (see Figure 5.1):

$h(1) = (a_0, b_0)$
$h(2) = (a_1, b_0), h(3) = (a_1, b_1), h(4) = (a_0, b_1)$
$h(5) = (a_2, b_0), h(6) = (a_2, b_1), h(7) = (a_2, b_2), h(8) = (a_1, b_2), h(9) = (a_0, b_2)$

and so on. For example, at the nth stage we have

$$h(n^2 + 1) = (a_n, b_0), h(n^2 + 2) = (a_n, b_{n-1}), \dots$$
$$h(n^2 + n + 1) = (a_n, b_n), h(n^2 + n + 2) = (a_{n-1}, b_n), \dots$$
$$\dots, h((n+1)^2) = (a_0, b_n).$$

We are looking at all pairs of integers (a_x, b_y) in the first quadrant (including those on the axes). The above function h starts at $(0, 0)$, and then moves through the first quadrant, hitting each pair once and only once, by going up and over and then restarting on the x-axis. $\square$

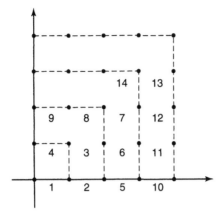

Figure 5.1 $A \times B$ is countable

Corollary 5.3.7. *Let $(A_i)_{i \in \mathbb{N}}$ be a collection of sets such that A_i is countable for all $i \in \mathbb{N}$. Then for any n, $A_1 \cup \cdots \cup A_n$ and $A_1 \times \cdots \times A_n$ are countable, where the last set is all n-tuples $(a_1, \ldots, a_n)$, $a_i \in A_i$. Further $\cup_{i=0}^{\infty} A_i$ is countable. If each A_i is at most countable, then $\cup_{i=0}^{\infty} A_i$ is at most countable.*

Exercise[(h)] **5.3.8.** *Prove Corollary 5.3.7.*

As the natural numbers, integers and rationals are countable, by taking each $A_i = \mathbb{N}$, $\mathbb{Z}$ or $\mathbb{Q}$ we immediately obtain

Corollary 5.3.9. $\mathbb{N}^n$, $\mathbb{Z}^n$ *and* $\mathbb{Q}^n$ *are countable.*

Proof. Proceed by induction; for example write $\mathbb{Q}^{n+1}$ as $\mathbb{Q}^n \times \mathbb{Q}$. $\square$

Exercise 5.3.10. *Prove that there are countably many rationals in the interval $[0, 1]$.*

5.3.1 Irrational Numbers

If $\alpha \notin \mathbb{Q}$, we say α is **irrational**. Clearly, not all numbers are rational (for example, $\sqrt{-1}$). Are there any real irrational numbers? The following example disturbed the ancient Greeks:

Theorem 5.3.11. *The square root of two is irrational.*

Proof. Assume not. Then we have $\sqrt{2} = \frac{p}{q}$, and we may assume p and q are relatively prime. Then $2q^2 = p^2$. We claim that $2|p^2$. While this appears obvious, this must be proved. If p is even, this is clear. If p is odd, we may write $p = 2m+1$. Then $p^2 = 4m^2 + 4m + 1 = 2(2m^2 + 2m) + 1$, which is clearly not divisible by 2. Thus p is even, say $p = 2p_1$. Then $2q^2 = p^2$ becomes $2q^2 = 4p_1^2$, and a similar argument yields q is even. Hence p and q have a common factor, contradicting our assumption. $\square$

This construction was disturbing for the following reason: consider an isosceles right triangle with bases of length 1. By the Pythagorean theorem, the hypotenuse has length $\sqrt{2}$. Thus, using a straight edge and compass, one easily constructs a non-rational length from rational sides and a right angle.

The above proof would be faster if we appealed to unique factorization: any positive integer can be written uniquely as a product of powers of primes. If one does not use unique factorization, then for $\sqrt{3}$ one must check p of the form $3m, 3m+1$ and $3m+2$.

Exercise 5.3.12. *If n is a non-square positive integer, prove $\sqrt{n}$ is irrational.*

Exercise 5.3.13. *Using a straight edge and compass, given two segments (one of unit length, one of length r with $r \in \mathbb{Q}$), construct a segment of length $\sqrt{r}$.*

Exercise[(h)] **5.3.14.** *Prove the Pythagorean theorem: if a right triangle has bases of length a and b and hypotenuse c then $a^2 + b^2 = c^2$.*

5.3.2 Algebraic Numbers

Let $f(x)$ be a polynomial with rational coefficients. By multiplying by the least common multiple of the denominators, we can clear the fractions. Thus without loss of generality it suffices to consider polynomials with integer coefficients.

The set of **algebraic numbers** $\mathcal{A}$ is the set of all $x \in \mathbb{C}$ such that there is a polynomial of finite degree and integer coefficients (depending on x, of course) such that $f(x) = 0$. The remaining complex numbers are the **transcendentals**. The set of **algebraic numbers of degree** n, $\mathcal{A}_n$, is the set of all $x \in \mathcal{A}$ such that

1. there exists a polynomial with integer coefficients of degree n such that $f(x) = 0$;

2. there is no polynomial g with integer coefficients and degree less than n with $g(x) = 0$.

Thus $\mathcal{A}_n$ is the subset of algebraic numbers x where for each $x \in \mathcal{A}_n$ the degree of the smallest polynomial f with integer coefficients and $f(x) = 0$ is n.

Exercise 5.3.15. *Show the following are algebraic: any rational number, the square root of any rational number, the cube root of any rational number, $r^{\frac{p}{q}}$ where $r, p, q \in \mathbb{Q}$, $i = \sqrt{-1}$, $\sqrt{3\sqrt{2} - 5}$.*

Theorem 5.3.16. *The algebraic numbers are countable.*

Proof. If we show each $\mathcal{A}_n$ is at most countable, then as $\mathcal{A} = \cup_{n=1}^{\infty} \mathcal{A}_n$ by Corollary 5.3.7 $\mathcal{A}$ is at most countable. The proof proceeds by finding a bijection from the set of all roots of polynomials of degree n with a subset of the countable set $\mathbb{Z}^n$.

Recall the **Fundamental Theorem of Algebra:** Let $f(x)$ be a polynomial of degree n with complex coefficients. Then $f(x)$ has n (not necessarily distinct) roots. Actually, we only need a weaker version, namely that a polynomials with integer coefficients has at most countably many roots.

Fix an $n \in \mathbb{N}$. We show $\mathcal{A}_n$ is at most countable. We can represent every integral polynomial $f(x) = a_n x^n + \cdots + a_0$ by an $(n+1)$-tuple $(a_0, \ldots, a_n)$. By Corollary 5.3.9, the set of all $(n+1)$-tuples with integer coefficients ($\mathbb{Z}^{n+1}$) is countable. Thus there is a bijection from $\mathbb{N}$ to $\mathbb{Z}^{n+1}$ and we can index each $(n+1)$-tuple $a \in \mathbb{Z}^{n+1}$

$$\{a : a \in \mathbb{Z}^{n+1}\} = \bigcup_{i=1}^{\infty} \{\alpha_i\}, \tag{5.2}$$

where each $\alpha_i \in \mathbb{Z}^{n+1}$. For each tuple α_i (or $a \in \mathbb{Z}^{n+1}$), there are n roots to the corresponding polynomial. Let R_{α_i} be the set of roots of the integer polynomial associated to α_i. The roots in R_{α_i} need not be distinct, and the roots may solve an integer polynomial of smaller degree. For example, $f(x) = (x^2 - 1)^4$ is a degree 8 polynomial. It has two roots, $x = 1$ with multiplicity 4 and $x = -1$ with multiplicity 4, and each root is a root of a degree 1 polynomial.

Let $P_n = \{x \in \mathbb{C} : x \text{ is a root of a degree } n \text{ polynomial}\}$. One can show that

$$P_n = \bigcup_{i=1}^{\infty} R_{\alpha_i} \supset \mathcal{A}_n. \tag{5.3}$$

By Lemma 5.3.7, P_n is at most countable. Thus by Lemma 5.3.2, as P_n is at most countable, $\mathcal{A}_n$ is at most countable. By Corollary 5.3.7, $\mathcal{A}$ is at most countable. As $\mathcal{A}_1 \supset \mathbb{Q}$ (given $\frac{p}{q} \in \mathbb{Q}$ consider $qx - p = 0$), $\mathcal{A}_1$ is countable. As $\mathcal{A}$ is at most countable, this implies $\mathcal{A}$ is countable. $\square$

Exercise 5.3.17. *Show the full force of the Fundamental Theorem of Algebra is not needed in the above proof; namely, it is enough that every polynomial have finitely many (or even countably many!) roots.*

Exercise 5.3.18. *Prove $R_n \supset \mathcal{A}_n$.*

Exercise 5.3.19. *Prove any real polynomial of odd degree has a real root.*

Remark 5.3.20. The following argument allows us to avoid using the Fundamental Theorem of Algebra. Let $f(x)$ be a polynomial of degree n with real coefficients. If $\alpha \in \mathbb{C}$ is such that $f(\alpha) = 0$, prove $f(\overline{\alpha}) = 0$, where $\overline{\alpha}$ is the complex conjugate of α ($\alpha = x + iy$, $\overline{\alpha} = x - iy$). Using polynomial long division, divide $f(x)$ by $h(x) = (x - \alpha)$ if $\alpha \in \mathbb{R}$ and $h(x) = (x - \alpha)(x - \overline{\alpha})$ otherwise. As both of these polynomials are real, $\frac{f(x)}{h(x)} = g(x) + \frac{r(x)}{h(x)}$ has all real coefficients, and the degree of $r(x)$ is less than the degree of $h(x)$. As $f(x)$ and $h(x)$ are zero for $x = \alpha$ and $\overline{\alpha}$, $r(x)$ is identically zero. We now have a polynomial of degree $n - 1$ (or $n - 2$). Proceeding by induction, we see f has at most n roots. Note we have not proved f *has* n roots. Note also the use of the Euclidean algorithm (see §1.2.3) in the proof.

Exercise 5.3.21 (Divide and Conquer). *For $f(x)$ continuous, if $f(x_l) < 0 < f(x_r)$ then there must be a root between x_l and x_r (Intermediate Value Theorem, Theorem A.2.1); look at the midpoint $x_m = \frac{x_l + x_r}{2}$. If $f(x_m) = 0$ we have found the root; if $f(x_m) < 0 \ (> 0)$ the root is between x_m and x_r (x_m and x_l). Continue subdividing the interval. Prove the division points converge to a root.*

Remark 5.3.22. By completing the square, one can show that the roots of $ax^2 + bx + c = 0$ are given by $x = \frac{-b \pm \sqrt{b^2 - 4ac}}{2a}$. More complicated formulas exist for the general cubic and quartic; however, there is no such formula which gives the roots of a general degree 5 (or higher) polynomial in terms of its coefficients (see [Art]). While we can use Newton's Method (see §1.2.4) or Divide and Conquer to approximate a root, we do not have a procedure in general to give an exact answer involving radicals and the coefficients of the polynomial.

Exercise 5.3.23 (Rational Root Test). *Let $f(x) = a_n x^n + \cdots + a_0$ be a polynomial with integer coefficients, $a_n, a_0 \neq 0$ and coprime. Let $p, q \in \mathbb{Z}$, $q \neq 0$. If $f(p/q) = 0$, show $q|a_n$ and $p|a_0$. Thus given a polynomial one can determine all the rational roots in a finite amount of time. Generalize this by finding a criterion for numbers of the form $\sqrt{p/q}$ to be a root. Does this work for higher powers, such as $\sqrt[m]{p/q}$? Does this contradict the claim in Remark 5.3.22 about degree 5 and higher polynomials?*

5.3.3 Transcendental Numbers

A set is **uncountable** if it is infinite and there is no bijection between it and the rationals (or the integers, or any countable set). We prove

Theorem 5.3.24 (Cantor). *The set of all real numbers is uncountable.*

Cantor's Theorem is an immediate consequence of

Lemma 5.3.25. *Let $\mathcal{S}$ be the set of all sequences $(y_i)_{i \in \mathbb{N}}$ with $y_i \in \{0, 1\}$. Then $\mathcal{S}$ is uncountable.*

Proof. We proceed by contradiction. Suppose there is a bijection $f : \mathcal{S} \to \mathbb{N}$. It is clear that this is equivalent to listing of the elements of $\mathcal{S}$:

$$
\begin{aligned}
x_1 &= .x_{11}x_{12}x_{13}x_{14}\cdots \\
x_2 &= .x_{21}x_{22}x_{23}x_{24}\cdots \\
x_3 &= .x_{31}x_{32}x_{33}x_{34}\cdots \\
&\vdots \\
x_n &= .x_{n1}x_{n2}x_{n3}x_{n4}\cdots x_{nn}\cdots \\
&\vdots
\end{aligned}
$$

(5.4)

Define an element $\theta = (\theta_i)_{i \in \mathbb{N}} \in \mathcal{S}$ by $\theta_i = 1 - x_{ii}$. Note θ cannot be in the list; it is not x_N because $1 - x_{NN} \neq x_{NN}$. But our list was supposed to be a complete enumeration of $\mathcal{S}$, contradiction. $\qquad \square$

Proof of Cantor's Theorem. Consider all numbers in the interval $[0, 1]$ whose decimal expansion (see §7.1 or §7.1) consists entirely of 0's and 1's. There is a bijection between this subset of $\mathbb{R}$ and the set $\mathcal{S}$. We have established that $\mathcal{S}$ is uncountable. Consequently $\mathbb{R}$ has an uncountable subset, and is uncountable. $\qquad \square$

Exercise 5.3.26. *Instead of using decimal expansions one could use binary expansions. Unfortunately there is the problem that some rationals have two expansions, a finite terminating and an infinite non-terminating expansion. For example,* $.001 = .0001111111\ldots$ *in base two, or* $.1 = .0999\cdots$ *in base ten. Using binary expansions, prove there are uncountably many reals. Prove* $.001 = .0001111111\ldots$ *in base two.*

Exercise 5.3.27. *Prove* $\|[0,1]\| = |\mathbb{R}| = |\mathbb{R}^n| = |\mathbb{C}^n|$. *Find a set with strictly larger cardinality than* $\mathbb{R}$.

The above proof is due to Cantor (1873–1874), and is known as **Cantor's Diagonalization Argument**. Note Cantor's proof shows that *most* numbers are transcendental, though it does not tell us *which* numbers are transcendental. We can easily show many numbers (such as $\sqrt{3} + \sqrt[5]{2^3} \sqrt[11]{5} + \sqrt{7}$) are algebraic. What of other numbers, such as π and e?

Lambert (1761), Legendre (1794), Hermite (1873) and others proved π irrational and Lindemann (1882) proved π transcendental (see [HW, NZM]); in Exercise 3.1.7, we showed that $\pi^2 \notin \mathbb{Q}$ implies there are infinitely many primes! What about e? Euler (1737) proved that e and e^2 are irrational, Liouville (1844) proved e is not an algebraic number of degree 2, and Hermite (1873) proved e is transcendental. Liouville (1851) gave a construction for an infinite (in fact, uncountable) family of transcendental numbers; see Theorem 5.6.1 as well as Exercise 5.6.9.

5.3.4 Axiom of Choice and the Continuum Hypothesis

Let $\aleph_0 = |\mathbb{Q}|$. Cantor's diagonalization argument can be interpreted as saying that $2^{\aleph_0} = |\mathbb{R}|$. As there are more reals than rationals, $\aleph_0 < 2^{\aleph_0}$. Does there exist a subset of $\mathbb{R}$ with strictly larger cardinality than the rationals, yet strictly smaller cardinality than the reals? Cantor's **Continuum Hypothesis** says that there are no subsets of intermediate size, or, equivalently, that $\aleph_1 = 2^{\aleph_0}$ (the reals are often called the continuum, and the $\aleph_i$ are called cardinal numbers).

The standard axioms of set theory are known as the Zermelo-Fraenkel axioms. A more controversial axiom is the **Axiom of Choice**, which states given any collection of sets $(A_x)_{x \in J}$ indexed by some set J, then there is a function f from J to the disjoint union of the A_x with $f(x) \in A_x$ for all x. Equivalently, this means we can form a new set by choosing an element a_x from each A_x; f is our choice function. If we have a countable collection of sets this is quite reasonable: a countable set is in a one-to-one correspondence with $\mathbb{N}$, and "walking through" the sets we know exactly when we will reach a given set to choose a representative. If we have an uncountable collection of sets, however, it is not clear "when" we would reach a given set to choose an element.

Exercise 5.3.28. *The construction of the sets in the Banach-Tarski Paradox uses the Axiom of Choice; we sketch the set $\mathcal{R}$ that arises. For $x, y \in [0, 1]$ we say x and y are equivalent if $x - y \in \mathbb{Q}$. Let $[x]$ denote all elements equivalent to x. We form a set of representatives $\mathcal{R}$ by choosing one element from each equivalence class. Prove there are uncountably many distinct equivalence classes.*

Kurt Gödel [Gö] showed that if the standard axioms of set theory are consistent, so too are the resulting axioms where the Continuum Hypothesis is assumed true; Paul Cohen [Coh] showed that the same is true if the negation of the Continuum Hypothesis is assumed. These two results imply that the Continuum Hypothesis is independent of the other standard axioms of set theory! See [HJ] for more details.

Exercise 5.3.29. *The cardinal numbers have strange multiplication properties. Prove* $\aleph_0^{\aleph_0} = 2^{\aleph_0}$ *by interpreting the two sides in terms of operations on sets.*

5.4 PROPERTIES OF e

In this section we study some of the basic properties of the number e (see [Rud] for more properties and proofs). One of the many ways to define the number e, the base of the natural logarithm, is to write it as the sum of the following infinite series:

$$e = \sum_{n=0}^{\infty} \frac{1}{n!}. \tag{5.5}$$

Denote the partial sums of the above series by

$$s_m = \sum_{n=0}^{m} \frac{1}{n!}. \tag{5.6}$$

Hence e is the limit of the convergent sequence s_m. This representation is one of the main tool in analyzing the nature of e.

Exercise 5.4.1. *Define*

$$e^x = \sum_{n=0}^{\infty} \frac{x^n}{n!}. \tag{5.7}$$

Prove $e^{x+y} = e^x e^y$. *Show this series converges for all* $x \in \mathbb{R}$; *in fact, it makes sense for* $x \in \mathbb{C}$ *as well. One can define* a^b *by* $e^{b \ln a}$.

Exercise 5.4.2. *An alternate definition of* e^x *is*

$$e^x = \lim_{n \to \infty} \left(1 + \frac{x}{n}\right)^n. \tag{5.8}$$

Show this definition agrees with the series expansion. This formulation is useful for growth problems such as compound interest or radioactive decay; see for example [BoDi].

Exercise 5.4.3. *Prove* $\frac{d}{dx} e^x = e^x$. *As* $e^{\ln x} = x$, *the chain rule implies* $\frac{d}{dx} \ln x = \frac{1}{x}$ (ln x *is the inverse function to* e^x).

From the functions e^x and $\ln x$, we can interpret a^b for any $a > 0$ and $b \in \mathbb{R}$: $a^b = e^{b \ln a}$. Note the series expansion for e^x makes sense for all x, thus we have a well defined process to determine numbers such as $3^{\sqrt{2}}$. We cannot compute $3^{\sqrt{2}}$ directly because we do not know what it means to raise 3 to the $\sqrt{2}$-power; we can only raise numbers to *rational* powers.

Exercise 5.4.4. *Split* 100 *into smaller integers such that each integer is two or more and the product of all these integers is as large as possible.*

Suppose now N is a large number and we wish to split N into smaller pieces, but all we require is that each piece be positive. How should we break up a large N?

5.4.1 Irrationality of e

Theorem 5.4.5 (Euler, 1737). *The number e is irrational.*

Proof. Assume $e \in \mathbb{Q}$. Then we can write $e = \frac{p}{q}$, where p, q are relatively prime positive integers. Now

$$
\begin{aligned}
e - s_m &= \sum_{n=m+1}^{\infty} \frac{1}{n!} \\
&= \frac{1}{(m+1)!}\left(1 + \frac{1}{m+1} + \frac{1}{(m+1)(m+2)} + \cdots\right) \\
&< \frac{1}{(m+1)!}\left(1 + \frac{1}{m+1} + \frac{1}{(m+1)^2} + \frac{1}{(m+1)^3} + \cdots\right) \\
&= \frac{1}{(m+1)!}\frac{1}{1 - \frac{1}{m+1}} = \frac{1}{m!m}.
\end{aligned}
\tag{5.9}
$$

Hence we obtain

$$
0 < e - s_m < \frac{1}{m!m}.
\tag{5.10}
$$

In particular, taking $m = q$ we and multiplying (5.10) by $q!$ yields

$$
0 < q!e - q!s_q < 1,
\tag{5.11}
$$

which is clearly impossible since $q!e - q!s_q$ would have to be an integer between 0 and 1. This contradicts our assumption that e was rational. $\qquad\square$

The key idea in the above proof is the simple fact that there are no integers between 0 and 1. We use a variant of this argument to prove e is transcendental.

5.4.2 Transcendence of e

We know there are more transcendental numbers than algebraic; we finally show a specific number is transcendental.

Theorem 5.4.6 (Hermite, 1873). *The number e is transcendental.*

Proof. The proof is again by contradiction. Assume e is algebraic. Then it must satisfy a polynomial equation

$$
a_n X^n + \cdots + a_1 X + a_0 = 0,
\tag{5.12}
$$

where $a_0, a_1, \ldots, a_n$ are integers. The existence of such a polynomial leads to an integer greater than zero but less than one; and this contradiction proves the theorem. This is a common technique for proving such results; see also Remark 1.6.14.

Exercise 5.4.7. *Prove one may assume without loss of generality that $a_0, a_n \neq 0$.*

Consider a polynomial $f(X)$ of degree r, and associate to it the following linear combination of its derivatives:

$$F(X) = f(X) + f'(X) + \cdots + f^{(r)}(X). \tag{5.13}$$

Exercise 5.4.8. *Prove the polynomial $F(X)$ has the property that*

$$\frac{d}{dx}\left[e^{-x}F(x)\right] = -e^{-x}f(x). \tag{5.14}$$

As $F(X)$ is differentiable, applying the Mean Value Theorem (Theorem A.2.2) to $e^{-x}F(X)$ on the interval $[0, k]$ for k any integer gives

$$e^{-k}F(k) - F(0) = -ke^{-c_k}f(c_k) \text{ for some } c_k \in (0, k), \tag{5.15}$$

or equivalently

$$F(k) - e^k F(0) = -ke^{k-c_k}f(c_k) = \epsilon_k. \tag{5.16}$$

Substituting $k = 0, 1, \ldots, n$ into (5.16), we obtain the following system of equations:

$$
\begin{array}{rcccccl}
F(0) & - & F(0) & = & 0 & = & \epsilon_0 \\
F(1) & - & eF(0) & = & -e^{1-c_1}f(c_1) & = & \epsilon_1 \\
F(2) & - & e^2 F(0) & = & -2e^{2-c_2}f(c_2) & = & \epsilon_2 \\
& & & \vdots & & & \\
F(n) & - & e^n F(0) & = & -ne^{n-c_n}f(c_n) & = & \epsilon_n.
\end{array} \tag{5.17}
$$

We multiply the first equation by a_0, the second by $a_1, \ldots,$ the $(n+1)^{st}$ by a_n. Adding the resulting equations gives

$$\sum_{k=0}^{n} a_k F(k) - \left(\sum_{k=0}^{n} a_k e^k\right) F(0) = \sum_{k=0}^{n} a_k \epsilon_k. \tag{5.18}$$

Notice that on the left hand side we have exactly the polynomial that we assume e satisfies:

$$\sum_{k=0}^{n} a_k e^k = 0; \tag{5.19}$$

this is the key step: we have now incorporated the (fictitious) polynomial. Hence (5.18) reduces to

$$\sum_{k=0}^{n} a_k F(k) = \sum_{k=0}^{n} a_k \epsilon_k. \tag{5.20}$$

We have used the hypothetical algebraicity of e to prove a certain integral combination of its powers vanish.

So far we had complete freedom in our choice of f, and (5.20) always holds for its associate F. In what follows we choose a special polynomial f in order to reach a contradiction. Choose a prime p large enough so that $p > |a_0|$ and $p > n$. Let f equal

$$
\begin{aligned}
f(X) &= \frac{1}{(p-1)!} X^{p-1}(1-X)^p(2-X)^p \cdots (n-X)^p \\
&= \frac{1}{(p-1)!} \left((n!)^p X^{p-1} + \text{higher order terms} \right) \\
&= \frac{a_{p-1}X^{p-1} + a_p X^p + \cdots + a_r X^r}{(p-1)!}.
\end{aligned} \tag{5.21}
$$

Though it plays no role in the proof, we note that the degree of f is $r = (n+1)p-1$. We prove a number of results which help us finish the proof. Recall that $p\mathbb{Z}$ denotes the set of integer multiples of p.

Claim 5.4.9. *Let p be a prime number and m any positive integer. Then $(p-1)(p-2) \cdots 2 \cdot 1$ divides $(p-1+m)(p-2+m) \cdots (2+m)(1+m)$.*

Warning: It is clearly not true that any consecutive set of $p-1$ numbers divides any larger consecutive set of $p-1$ numbers. For example, $7 \cdot 6 \cdot 5 \cdot 4$ does not divide $9 \cdot 8 \cdot 7 \cdot 6$, and $8 \cdot 7 \cdot 6 \cdot 5$ does not divide $14 \cdot 13 \cdot 12 \cdot 11$. In the first example we have 5 divides the smaller term but not the larger; in the second we have 2^4 divides the smaller term but only 2^3 divides the larger.

Proof of Claim 5.4.9. Let $x = (p-1)!$ and $y = (p-1+m) \cdots (1+m)$. The claim follows by showing for each prime $q < p$ that if $q^a | x$ then $q^a | y$. Let k be the largest integer such that $q^k \leq p-1$ and $\lfloor z \rfloor$ be the greatest integer at most z. Then there are $\lfloor \frac{p-1}{q} \rfloor$ factors of x divisible by q once, $\lfloor \frac{p-1}{q^2} \rfloor$ factors of x divisible by q twice, and so on up to $\lfloor \frac{p-1}{q^k} \rfloor$ factors of x divisible by q a total of k times. Thus the exponent of q dividing x is $\sum_{\ell=1}^{k} \lfloor \frac{p-1}{q^\ell} \rfloor$. The proof is completed by showing that for each $\ell \in \{1, \dots, k\}$ we have as many terms in y divisible by q^ℓ as we do in x; it is possible to have more of course (let $q = 5$, $x = 6 \cdots 1$ and $y = 10 \cdots 5$). Clearly it is enough to prove this for $m < (p-1)!$; we leave the remaining details to the reader. $\square$

Claim 5.4.10. *For $i \geq p$, $\forall j \in \mathbb{N}$, $f^{(i)}(j) \in p\mathbb{Z}$.*

Proof. Differentiate (5.21) $i \geq p$ times. Consider any term which survives, say $\frac{a_k X^k}{(p-1)!}$ with $k \geq i$. After differentiating this term becomes $\frac{k(k-1) \cdots (k-(i-1)) a_k X^{k-i}}{(p-1)!}$. By Claim 5.4.9 we have $(p-1)! | k(k-1) \cdots (k-(i-1))$. Further, $p | k(k-1) \cdots (k-(i-1))$ as we differentiated at least p times and any product of p consecutive numbers is divisible by p. As p does not divide $(p-1)!$, we see that all surviving terms are multiplied by p. $\square$

Claim 5.4.11. *For $0 \leq i < p$ and $j \in \{1, \dots, n\}$, $f^{(i)}(j) = 0$.*

Proof. The multiplicity of a root of a polynomial gives the order of vanishing of the polynomial at that particular root. As $j = 1, 2, \dots, n$ are roots of $f(X)$ of multiplicity p, differentiating $f(x)$ less than p times yields a polynomial which still vanishes at these j. $\square$

Claim 5.4.12. *Let F be the polynomial associated to f. Then $F(1)$, $F(2)$, ..., $F(n) \in p\mathbb{Z}$.*

Proof. Recall that $F(j) = f(j) + f'(j) + \cdots + f^{(r)}(j)$. By the first claim, $f^{(i)}(j)$ is a multiple of p for $i \geq p$ and any integer j. By the second claim, $f^{(i)}(j) = 0$ for $0 \leq i < p$ and $j = 1, 2, \ldots, n$. Thus $F(j)$ is a multiple of p for these j. $\qquad\square$

Claim 5.4.13. *For $0 \leq i \leq p - 2$, we have $f^{(i)}(0) = 0$.*

Proof. Similar to the second claim, we note that $f^{(i)}(0) = 0$ for $0 \leq i < p - 2$, because 0 is a root of $f(x)$ of multiplicity $p - 1$. $\qquad\square$

Claim 5.4.14. *$F(0)$ is not a multiple of p.*

Proof. By the first claim, $f^{(i)}(0)$ is a multiple of p for $i \geq p$; by the fourth claim, $f^{(i)}(0) = 0$ for $0 \leq i \leq p - 2$. Since

$$F(0) = f(0) + f'(0) + \cdots + f^{(p-2)}(0) + f^{(p-1)}(0) + f^{(p)}(0) + \cdots + f^{(r)}(0), \tag{5.22}$$

to prove $F(0)$ is a not multiple of p it is sufficient to prove $f^{(p-1)}(0)$ is not multiple of p because all the other terms *are* multiples of p. However, from the Taylor series expansion (see §A.2.3) of f in (5.21), we see that

$$f^{(p-1)}(0) = (n!)^p + \text{ terms that are multiples of } p. \tag{5.23}$$

Since we chose $p > n$, $n!$ is not divisible by p, proving the claim. $\qquad\square$

We resume the proof of the transcendence of e. Remember we also chose p such that a_0 is not divisible by p. This fact plus the above claims imply first that $\sum_k a_k F(k)$ is an integer, and second that

$$\sum_{k=0}^{n} a_k F(k) \equiv a_0 F(0) \not\equiv 0 \bmod p. \tag{5.24}$$

Thus $\sum_k a_k F(k)$ is a non-zero integer. Recall (5.20):

$$\sum_{k=0}^{n} a_k F(k) = a_1 \epsilon_1 + \cdots + a_n \epsilon_n. \tag{5.25}$$

We have already proved that the left hand side is a non-zero integer. We analyze the sum on the right hand side. We have

$$\epsilon_k = -k e^{k-c_k} f(c_k) = \frac{-k e^{k-c_k} c_k^{p-1} (1 - c_k)^p \cdots (n - c_k)^p}{(p-1)!}. \tag{5.26}$$

As $0 \leq c_k \leq k \leq n$ we obtain

$$|\epsilon_k| \leq \frac{e^k k^p (1 \cdot 2 \cdots n)^p}{(p-1)!} \leq \frac{e^n (n! n)^p}{(p-1)!} \longrightarrow 0 \text{ as } p \to \infty. \tag{5.27}$$

Exercise 5.4.15. *For fixed n, prove that as $p \to \infty$, $\frac{(n! n)^p}{(p-1)!} \to 0$. See Lemma 8.4.5.*

Recall that n is fixed, as are the constants $a_0, \ldots, a_n$ (they define the polynomial equation supposedly satisfied by e); in our argument only the prime number p varies. Hence, by choosing p sufficiently large, we can make sure that all ϵ_k's are uniformly small. In particular, we can make them small enough such that the following holds:

$$\left| \sum_{k=1}^{n} a_k \epsilon_k \right| < 1. \tag{5.28}$$

To be more precise, we only have to choose a prime p such that $p > n$, $|a_0|$ and

$$\frac{e^n (n!n)^p}{(p-1)!} < \frac{1}{\sum_{k=0}^{n} |a_k|}. \tag{5.29}$$

In this way we reach a contradiction in the identity (5.20) where the left hand side is a non-zero integer, while the right hand side is a real number of absolute value less than 1. □

This proof illustrates two of the key features of these types of arguments: considering properties of the "fictitious" polynomial, and finding an integer between 0 and 1. It is very hard to prove a given number is transcendental. Note this proof heavily uses special properties of e, in particular the derivative of e^x is e^x. The reader is invited to see Theorem 205 of [HW] where the transcendence of π is proved. It is known that $\zeta(k) = \sum_{n=1}^{\infty} \frac{1}{n^k}$ is transcendental for k even (in fact, it is a rational multiple of π^k); very little is known if k is odd. If $k = 3$, Apery [Ap] proved $\zeta(3)$ is irrational (see also [Mill]), though it is not known if it is transcendental. For infinitely many odd k, $\zeta(k)$ is irrational ([BR]), and at least one of $\zeta(5), \zeta(7), \zeta(9)$ or $\zeta(11)$ is irrational [Zu]. See also §3.1.3.

In field theory, one shows that if α, β are algebraic then so are $\alpha + \beta$ and $\alpha\beta$; if both are transcendental, at least one of $\alpha + \beta$ and $\alpha\beta$ is transcendental. Hence, while we expect both $e + \pi$ and $e\pi$ to be transcendental, all we know is at least one is! In §5.6.2 we construct uncountably many transcendentals. In §A.5.2 we show the Cantor set is uncountable, hence "most" of its elements are transcendental.

Exercise 5.4.16. *Complete the proof of Claim 5.4.9.*

Arguing similarly as in the proof of the transcendence of e, we can show π is transcendental. We content ourselves with proving π^2 is irrational, which we have seen (Exercises 3.1.7 and 3.3.28) implies there are infinitely many primes. For more on such proofs, see Chapter 11 of [BB] (specifically pages 352 to 356, where the following exercise is drawn from).

Exercise 5.4.17 (Irrationality of π^2). *Fix a large n (how large n must be will be determined later). Let $f(x) = \frac{x^n (1-x)^n}{n!}$. Show f attains its maximum at $x = \frac{1}{2}$, for $x \in (0,1)$, $0 < f(x) < \frac{1}{n!}$, and all the derivatives of f evaluated at 0 or 1 are integers. Assume π^2 is rational; thus we may write $\pi^2 = \frac{a}{b}$ for integers a, b. Consider*

$$G(x) = b^n \sum_{k=0}^{n} (-1)^k f^{(2k)}(x) \pi^{2n-2k}. \tag{5.30}$$

Show $G(0)$ and $G(1)$ are integers and

$$\frac{d}{dx}[G'(x)\sin(\pi x) - \pi G(x)\cos(\pi x)] = \pi^2 a^n f(x)\sin(\pi x). \qquad (5.31)$$

Deduce a contradiction (to the rationality of π^2) by showing that

$$\pi \int_0^1 a^n f(x)\sin(\pi x)dx = G(0) + G(1), \qquad (5.32)$$

which cannot hold for n sufficiently large. The contradiction is the usual one, namely the integral on the left is in $(0,1)$ and the right hand side is an integer.

5.5 EXPONENT (OR ORDER) OF APPROXIMATION

Let α be a real number. We desire a rational number $\frac{p}{q}$ such that $\left|\alpha - \frac{p}{q}\right|$ is small. Some explanation is needed. In some sense, the size of the denominator q measures the "cost" of approximating α, and we want an error that is small relative to q. For example, we could approximate π by $314159/100000$, which is accurate to 5 decimal places (about the size of q), or we could use $103993/33102$, which uses a smaller denominator and is accurate to 9 decimal places (about twice the size of q)! This ratio comes from the continued fraction expansion of π (see Chapter 7). We will see later (present chapter and Chapters 9 and 12) that many properties of numbers are related to how well they can be approximated by rationals. We start with a definition.

Definition 5.5.1 (Approximation Exponent). *The real number ξ has approximation order (or exponent) $\tau(\xi)$ if $\tau(\xi)$ is the smallest number such that for all $e > \tau(\xi)$ the inequality*

$$\left|\xi - \frac{p}{q}\right| < \frac{1}{q^e} \qquad (5.33)$$

has only finitely many solutions.

In Theorem 12.2.8 we shall see how the approximation exponent yields information about the distribution of the fractional parts of $n^k\alpha$ for fixed k and α. In particular, if α has approximation exponent greater than 4 then the sequence $n^k\alpha \bmod 1$ comes arbitrarily close to all numbers in $[0,1]$.

The following exercise gives an alternate definition for the approximation exponent. The definition below is more convenient for constructing transcendental numbers (Theorem 5.6.1).

Exercise[h] **5.5.2** (Approximation Exponent). *Show ξ has approximation exponent $\tau(\xi)$ if and only if for any fixed $C > 0$ and $e > \tau(\xi)$ the inequality*

$$\left|\xi - \frac{p}{q}\right| < \frac{C}{q^e} \qquad (5.34)$$

has only finitely many solutions with p, q relatively prime.

5.5.1 Bounds on the Order of Real Numbers

Lemma 5.5.3. *A rational number has approximation exponent 1.*

Proof. If $\xi = \frac{a}{b}$ and $r = \frac{s}{t} \neq \frac{a}{b}$, then $sb - at \neq 0$. Thus $|sb - at| \geq 1$ (as it is integral). This implies

$$\left| \xi - \frac{s}{t} \right| = \left| \frac{a}{b} - \frac{s}{t} \right| = \frac{|sb - at|}{bt} \geq \frac{1}{bt}. \qquad (5.35)$$

If the rational ξ had approximation exponent $e > 1$ we would find

$$\left| \xi - \frac{s}{t} \right| < \frac{1}{t^e}, \quad \text{which implies} \quad \frac{1}{t^e} > \frac{1}{bt}. \qquad (5.36)$$

Therefore $t^{e-1} < b$. Since b is fixed, there are only finitely many such t. $\qquad\square$

Theorem 5.5.4 (Dirichlet). *An irrational number has approximation exponent at least 2.*

Proof. It is enough to prove this for $\xi \in (0, 1)$. Let $Q > 1$ be an integer. Divide the interval $(0, 1)$ into Q equal intervals, say $[\frac{k}{Q}, \frac{k+1}{Q})$. Consider the $Q + 1$ numbers inside the interval $(0, 1)$:

$$\{\xi\}, \{2\xi\}, \ldots, \{(Q + 1)\xi\}, \qquad (5.37)$$

where $\{x\}$ denotes the fractional part of x. Letting $[x]$ denote the greatest integer less than or equal to x, we have $x = [x] + \{x\}$. As $\xi \notin \mathbb{Q}$, the $Q + 1$ fractional parts are all different.

By Dirichlet's Pigeon-Hole Principle (§A.4), at least two of these numbers, say $\{q_1\xi\}$ and $\{q_2\xi\}$, belong to a common interval of length $\frac{1}{Q}$. Without loss of generality we may take $1 \leq q_1 < q_2 \leq Q + 1$. Hence

$$|\{q_2\xi\} - \{q_1\xi\}| \leq \frac{1}{Q} \qquad (5.38)$$

and

$$|(q_2\xi - n_2) - (q_1\xi - n_1)| \leq \frac{1}{Q}, \quad n_i = [q_i\xi]. \qquad (5.39)$$

Now let $q = q_1 - q_2 \in \mathbb{Z}$ and $p = n_1 - n_2 \in \mathbb{Z}$. Note $1 \leq q \leq Q$ and

$$|q\xi - p| \leq \frac{1}{Q} \qquad (5.40)$$

and hence

$$\left| \xi - \frac{p}{q} \right| \leq \frac{1}{qQ} \leq \frac{1}{q^2}. \qquad (5.41)$$

We leave the rest of the proof to the reader (Exercise 5.5.5). $\qquad\square$

Exercise[(h)] 5.5.5. *Show the above argument leads to an infinite sequence of q with $q \to \infty$; thus there are infinitely many solutions to $\left| \xi - \frac{p}{q} \right| \leq \frac{1}{q^2}$. Further, as $\frac{p}{q} \in \mathbb{Q}$ and $\xi \notin \mathbb{Q}$, we may replace the $\leq$ with $<$, and ξ has approximation exponent at least 2.*

Exercise$^{(h)}$ 5.5.6. *Use Exercises 5.5.5 and 5.4.17 (where we prove π is irrational) to show that $\sum_{n=1}^{\infty}(\cos n)^n$ diverges; the argument of the cosine function is in radians. Harder: what about $\sum_{n=1}^{\infty}(\sin n)^n$?*

Exercise 5.5.7. *In Theorem 5.5.4, what goes wrong if $\xi \in \mathbb{Q}$? Is the theorem true for $\xi \in \mathbb{Q}$?*

Later we give various improvements to Dirichlet's theorem. For example, we use continued fractions to give constructions for the rational numbers $\frac{p}{q}$ (see the proof of Theorem 7.9.4). Further, we show that any number $\frac{p}{q}$ that satisfies Dirichlet's theorem for an irrational ξ has to be a continued fraction convergent of ξ (§7.9). We also ask whether the exponent two can be improved. Our first answer to this question is Liouville's theorem (Theorem 5.6.1), which states that a real algebraic number of degree n *cannot* be approximated to order larger than n. In other words, if ξ satisfies a polynomial equation with integer coefficients of degree n, then $\tau(\xi) \leq n$. Liouville's theorem provides us with a simple method to construct transcendental numbers: if a number can be approximated by rational numbers too well, it will have to be transcendental. We work out a classical example in 5.6.2.

Liouville's theorem combined with Dirichlet's theorem implies the interesting fact that a quadratic irrational number ξ has approximation exponent exactly 2. Roth's spectacular theorem (Theorem 5.7.1) asserts that this is in fact the case for all algebraic numbers: the approximation exponent of any real algebraic number is equal to two, regardless of the degree! We will see that the order of approximation of numbers has many applications, for example in digit bias of sequences (Chapter 9) and Poissonian behavior of the fractional parts of $n^k \alpha$ (Chapter 12).

Exercise 5.5.8. *Let α (respectively β) be approximated to order n (respectively m). What is the order of approximation of $\alpha + \frac{a}{b}$ ($\frac{a}{b} \in \mathbb{Q}$), $\alpha + \beta$, $\alpha \cdot \beta$, and $\frac{\alpha}{\beta}$.*

5.5.2 Measure of Well Approximated Numbers

We assume the reader is familiar with the notions of lengths or measures of sets; see §A.5. In loose terms, the following theorem states that almost all numbers have approximation exponent equal to two.

Theorem 5.5.9. *Let C, ϵ be positive constants. Let S be the set of all points $x \in [0,1]$ such that there are infinitely many relatively prime integers p, q with*

$$\left| x - \frac{p}{q} \right| \leq \frac{C}{q^{2+\epsilon}}. \tag{5.42}$$

Then the length (or measure) of S, denoted $|S|$, equals 0.

Proof. Let $N > 0$. Let S_N be the set of all points $x \in [0,1]$ such that there are $p, q \in \mathbb{Z}$, $q > N$ for which

$$\left| x - \frac{p}{q} \right| \leq \frac{C}{q^{2+\epsilon}}. \tag{5.43}$$

If $x \in S$ then $x \in S_N$ for every N. Thus if we can show that the measure of the sets S_N becomes arbitrarily small as $N \to \infty$, then the measure of S must be zero.

How large can S_N be? For a given q there are at most q choices for p. Given a pair (p, q), we investigate how many x's are within $\frac{C}{q^{2+\epsilon}}$ of $\frac{p}{q}$. Clearly the set of such points is the interval

$$I_{p,q} = \left(\frac{p}{q} - \frac{C}{q^{2+\epsilon}}, \frac{p}{q} + \frac{C}{q^{2+\epsilon}} \right). \tag{5.44}$$

Note that the length of $I_{p,q}$ is $\frac{2C}{q^{2+\epsilon}}$. Let I_q be the set of all x in $[0, 1]$ that are within $\frac{C}{q^{2+\epsilon}}$ of a rational number with denominator q. Then

$$I_q \subset \bigcup_{p=0}^{q} I_{p,q} \tag{5.45}$$

and therefore

$$|I_q| \leq \sum_{p=0}^{q} |I_{p,q}| = (q+1) \cdot \frac{2C}{q^{2+\epsilon}} = \frac{q+1}{q} \frac{2C}{q^{1+\epsilon}} < \frac{4C}{q^{1+\epsilon}}. \tag{5.46}$$

Hence

$$|S_N| \leq \sum_{q>N} |I_q| = \sum_{q>N} \frac{4C}{q^{1+\epsilon}} < \frac{4C}{1+\epsilon} N^{-\epsilon}. \tag{5.47}$$

Thus, as N goes to infinity, $|S_N|$ goes to zero. As $S \subset S_N$, $|S| = 0$. $\quad\square$

Remark 5.5.10. It follows from Roth's Theorem (Theorem 5.7.1) that the set S consists entirely of transcendental numbers; however, in terms of length, it is a small set of transcendentals.

Exercise 5.5.11. *Instead of working with* $\left| x - \frac{p}{q} \right| \leq \frac{C}{q^{2+\epsilon}}$, *show the same argument works for* $\left| x - \frac{p}{q} \right| \leq \frac{C}{f(q)}$, *where* $\sum \frac{q}{f(q)} < \infty$.

Exercise 5.5.12. *Another natural question to ask is what is the measure of all $x \in [0, 1]$ such that each digit of its continued fraction is at most K? In Theorem 10.2.1 we show this set also has length zero. This should be contrasted with Theorem 10.2.4, where we show if $\sum_{n=1}^{\infty} \frac{1}{k_n}$ converges, then the set $\{x \in [0, 1] : a_i(x) \leq k_i\}$ has positive measure. What is the length of $x \in [0, 1]$ such that there are no 9's in x's decimal expansion?*

Exercise 5.5.13 (Hard). *For a given C, what is the measure of the set of $\xi \in (0, 1)$ such that*

$$\left| \xi - \frac{p}{q} \right| < \frac{C}{q^2} \tag{5.48}$$

holds only finitely often? What if $C < 1$? More generally, instead of $\frac{C}{q^2}$ we could have $\frac{1}{q^2 \log q}$ or any such expression. Warning: The authors are not aware of a solution to this problem!

5.6 LIOUVILLE'S THEOREM

5.6.1 Proof of Lioville's Theorem

Theorem 5.6.1 (Liouville's Theorem). *Let α be a real algebraic number of degree d. Then α is approximated by rationals to order at most d.*

Proof. Let

$$f(x) = a_d x^d + \cdots + a_1 x + a_0 \tag{5.49}$$

be the polynomial with relatively prime integer coefficients of smallest degree (called the **minimal polynomial**) such that $f(\alpha) = 0$. The condition of minimality implies that $f(x)$ is irreducible over $\mathbb{Z}$.

Exercise[h] **5.6.2.** *Show that a polynomial irreducible over $\mathbb{Z}$ is irreducible over $\mathbb{Q}$.*

In particular, as $f(x)$ is irreducible over $\mathbb{Q}$, $f(x)$ does not have any rational roots. If it did then $f(x)$ would be divisible by a linear polynomial $(x - \frac{a}{b})$. Therefore f is non-zero at every rational. Our plan is to show the existence of a rational number $\frac{p}{q}$ such that $f(\frac{p}{q}) = 0$. Let $\frac{p}{q}$ be such a candidate. Substituting gives

$$f\left(\frac{p}{q}\right) = \frac{N}{q^d}, \quad N \in \mathbb{Z}. \tag{5.50}$$

Note the integer N depends on p, q and the a_i's. To emphasize this dependence we write $N(p, q; \alpha)$. As usual, the proof proceeds by showing $|N(p, q; \alpha)| < 1$, which then forces $N(p, q; \alpha)$ to be zero; this contradicts f is irreducible over $\mathbb{Q}$.

We find an upper bound for $N(p, q; \alpha)$ by considering the Taylor expansion of f about $x = \alpha$. As $f(\alpha) = 0$, there is no constant term in the Taylor expansion. We may assume $\frac{p}{q}$ satisfies $|\alpha - \frac{p}{q}| < 1$. Then

$$f(x) = \sum_{i=1}^{d} \frac{1}{i!} \frac{d^i f}{dx^i}(\alpha) \cdot (x - \alpha)^i. \tag{5.51}$$

Consequently

$$\left| f\left(\frac{p}{q}\right) \right| = \left| \frac{N(p, q; \alpha)}{q^d} \right| \leq \left| \frac{p}{q} - \alpha \right| \cdot \sum_{i=1}^{d} \left| \frac{1}{i!} \frac{d^i f}{dx^i}(\alpha) \right| \cdot \left| \frac{p}{q} - \alpha \right|^{i-1}$$

$$\leq \left| \frac{p}{q} - \alpha \right| \cdot d \cdot \max_i \left| \frac{1}{i!} \frac{d^i f}{dx^i}(\alpha) \cdot 1^{i-1} \right|$$

$$\leq \left| \frac{p}{q} - \alpha \right| \cdot A(\alpha), \tag{5.52}$$

where $A(\alpha) = d \cdot \max_i \left| \frac{1}{i!} \frac{d^i f}{dx^i}(\alpha) \right|$. If α were approximated by rationals to order greater than d, then (Exercise 5.5.2) for some $\epsilon > 0$ there would exist a constant $B(\alpha)$ and infinitely many $\frac{p}{q}$ such that

$$\left| \frac{p}{q} - \alpha \right| \leq \frac{B(\alpha)}{q^{d+\epsilon}}. \tag{5.53}$$

Combining yields

$$\left| f\left(\frac{p}{q}\right) \right| \leq \frac{A(\alpha)B(\alpha)}{q^{d+\epsilon}}. \tag{5.54}$$

Therefore

$$|N(p, q; \alpha)| \leq \frac{A(\alpha)B(\alpha)}{q^{\epsilon}}. \tag{5.55}$$

For q sufficiently large, $A(\alpha)B(\alpha) < q^{\epsilon}$. As we may take q arbitrarily large, for sufficiently large q we have $|N(p, q; \alpha)| < 1$. As the only non-negative integer less than 1 is 0, we find for q large that $f\left(\frac{p}{q}\right) = 0$, contradicting f is irreducible over $\mathbb{Q}$. □

Exercise 5.6.3. *Justify the fact that if $\{\frac{p_i}{q_i}\}_{i\geq 1}$ is a sequence of rational approximations to order $n \geq 1$ of x then $q_i \to \infty$.*

5.6.2 Constructing Transcendental Numbers

We have seen that the order to which an algebraic number can be approximated by rationals is bounded by its degree. Hence if a real, irrational number α can be approximated by rationals to an arbitrarily large order, then α must be transcendental! This provides us with a recipe for constructing transcendental numbers. Note the reverse need not be true: if a number x can be approximated to order at most n, it does not follow that x is algebraic of degree at most n (see Theorem 5.7.1); for example, Hata [Hata] showed the approximation exponent of π is at most 8.02; see Chapter 11 of [BB] for bounds on the approximation exponent for e, π, $\zeta(3)$ and $\log 2$. We use the definition of approximation exponent from Exercise 5.5.2.

Theorem 5.6.4 (Liouville). *The number*

$$\alpha = \sum_{m=1}^{\infty} \frac{1}{10^{m!}} \tag{5.56}$$

is transcendental.

Proof. The series defining α is convergent, since it is dominated by the geometric series $\sum \frac{1}{10^m}$. In fact the series converges very rapidly, and it is this high rate of convergence that makes x transcendental. Fix N large and choose $n > N$. Write

$$\frac{p_n}{q_n} = \sum_{m=1}^{n} \frac{1}{10^{m!}} \tag{5.57}$$

with $p_n, q_n > 0$ and $(p_n, q_n) = 1$. Then $\{\frac{p_n}{q_n}\}_{n\geq 1}$ is a monotone increasing sequence converging to α. In particular, all these rational numbers are distinct. Note also that q_n must divide $10^{n!}$, which implies that $q_n \leq 10^{n!}$. Using this, and the fact that $10^{-(n+1+k)!} < 10^{-(n+1)!}10^{-k}$, we obtain

$$0 < \alpha - \frac{p_n}{q_n} = \sum_{m>n} \frac{1}{10^{m!}}$$

$$< \frac{1}{10^{(n+1)!}} \left(1 + \frac{1}{10} + \frac{1}{10^2} + \cdots \right)$$

$$= \frac{1}{10^{(n+1)!}} \cdot \frac{10}{9}$$

$$< \frac{2}{(10^{n!})^{n+1}}$$

$$< \frac{2}{q_n^{n+1}} \leq \frac{2}{q_n^N}. \tag{5.58}$$

This gives an approximation by rationals of order N of α, in fact infinitely many such approximations (one for each $n > N$). Since N can be chosen arbitrarily large, this implies that α can be approximated by rationals to arbitrary order. By Theorem 5.6.1, if α were algebraic of degree m it could be approximated by rationals to order at most m; thus, α is transcendental. □

Exercise 5.6.5. *Consider the binary expansion for $x \in [0, 1)$, namely*

$$x = \sum_{n=1}^{\infty} \frac{b_n(x)}{2^n}, \quad b_n(x) \in \{0, 1\}. \tag{5.59}$$

For irrational x this expansion is unique. Consider the function

$$M(x) = \sum_{n=1}^{\infty} 10^{-(b_n(x)+1)n!}. \tag{5.60}$$

Prove for irrational x that $M(x)$ is transcendental. Thus the above is an explicit construction for uncountably many transcendentals! Investigate the properties of this function. Is it continuous or differentiable (everywhere or at some points)? What is the measure of these numbers? These are "special" transcendental numbers (compare these numbers to Theorem 5.5.9). See also Remark 8.3.3.

The following example uses some results concerning continued fraction studied in the next chapter. The reader should return to this theorem after studying Chapter 7.

Theorem 5.6.6. *The number*

$$\beta = [10^{1!}, 10^{2!}, \dots] \tag{5.61}$$

is transcendental.

Proof. Let $\frac{p_n}{q_n}$ be the continued fraction of $[10^{1!}, \dots, 10^{n!}]$. Then

$$\left| \beta - \frac{p_n}{q_n} \right| = \frac{1}{q_n q'_{n+1}} = \frac{1}{q_n(a'_{n+1}q_n + q_{n-1})} < \frac{1}{a_{n+1}} = \frac{1}{10^{(n+1)!}}. \tag{5.62}$$

Since $q_k = a_k q_{k-1} + q_{k-2}$, we have $q_k > q_{k-1}$ Also $q_{k+1} = a_{k+1}q_k + q_{k-1}$, so we obtain

$$\frac{q_{k+1}}{q_k} = a_{k+1} + \frac{q_{k-1}}{q_k} < a_{k+1} + 1. \tag{5.63}$$

Writing this inequality for $k = 1, \ldots, n - 1$ and multiplying yields

$$q_n = q_1 \frac{q_2}{q_1} \frac{q_3}{q_2} \cdots \frac{q_n}{q_{n-1}} < (a_1 + 1)(a_2 + 1) \cdots (a_n + 1)$$

$$= \left(1 + \frac{1}{a_1}\right) \cdots \left(1 + \frac{1}{a_n}\right) a_1 \cdots a_n$$

$$< 2^n a_1 \cdots a_n = 2^n 10^{1! + \cdots + n!}$$

$$< 10^{2 \cdot n!} = a_n^2. \tag{5.64}$$

Combining (5.62) and (5.64) gives

$$\left| \beta - \frac{p_n}{q_n} \right| < \frac{1}{a_{n+1}} = \frac{1}{a_n^{n+1}} < \left(\frac{1}{a_n^2}\right)^{\frac{n}{2}} < \left(\frac{1}{q_n^2}\right)^{\frac{n}{2}} = \frac{1}{q_n^n}. \tag{5.65}$$

In this way we get, just as in Liouville's Theorem, an approximation of β by rationals to arbitrary order. This proves that β is transcendental.   □

Exercise 5.6.7. *Without using the factorial function, construct transcendental numbers (either by series expansion or by continued fractions). Can you do this using a function $f(n)$ which grows slower than $n!$?*

The following exercises construct transcendental numbers by investigating infinite products of rational numbers; see Exercise 3.3.8 for a review of infinite products. algebraic and which are transcendental.

Exercise 5.6.8. *Let a_n be a sequence of positive numbers such that $\sum_{n=1}^{\infty} a_n$ converges. Assume also for all $N > 1$ that $a_N > \sum_{n=N+1}^{\infty} a_n$. Let $(n_1, n_2, \ldots)$ and $(m_1, m_2, \ldots)$ be any two distinct infinite sequences of increasing positive integers; this means that there is at least one k such that $n_k \neq m_k$. Prove*

$$\sum_{k=1}^{\infty} a_{n_k} \neq \sum_{k=1}^{\infty} a_{m_k}, \tag{5.66}$$

and find three different sequences $\{a_n\}_{n=1}^{\infty}$ satisfying the conditions of this problem.

Exercise 5.6.9. *Prove*

$$\prod_{n=2}^{\infty} \frac{n^2 - 1}{n^2} = \prod_{n=2}^{\infty} \left(1 - \frac{1}{n^2}\right) = \frac{1}{2}. \tag{5.67}$$

For each $\alpha \in [0, 1]$, let $\alpha(n)$ be the nth of α's binary expansion; if α has two expansions take the finite one. Consider the function

$$f(\alpha) = \prod_{n=2}^{\infty} \left(1 - \frac{\alpha(n)}{n^2}\right). \tag{5.68}$$

Prove $f(\alpha)$ takes on countably many distinct rational values and uncountably many distinct transcendental values. Hint: one approach is to use the previous exercise. For a generic $\alpha \in [0, 1]$, do you expect $f(\alpha)$ to be algebraic or transcendental? Note if $\alpha(n) = 1$ for n prime and 0 otherwise we get $\frac{6}{\pi^2}$; see Exercise 3.1.7 and 11.3.15.

5.7 ROTH'S THEOREM

As we saw earlier, Liouville's Theorem asserts that there is a limit to the accuracy with which algebraic numbers can be approximated by rational numbers. There is a long list of improvements associated with Liouville's Theorem. More precise and more profound results were proved by Thue in 1908, Siegel in 1921, Dyson in 1947 and Roth in 1955, to mention but a few of the improvements. Thue proved that the exponent n can be replaced by $\frac{n}{2} + 1$; Siegel proved

$$\min_{1 \le s \le n-1} \left(s + \frac{n}{s+1} \right) \tag{5.69}$$

works, and Dyson showed that $\sqrt{2n}$ is sufficient. It was, however, conjectured by Siegel that for any $\epsilon > 0$, $2 + \epsilon$ is enough! Proving Siegel's conjecture was Roth's remarkable achievement that earned him a Fields medal in 1958. For an enlightening historical analysis of the work that led to Roth's Theorem see [Gel], Chapter I.

Theorem 5.7.1 (Roth's Theorem). *Let α be a real algebraic number (a root of a polynomial equation with integer coefficients). Given any $\epsilon > 0$ there are only finitely many relatively prime pairs of integers (p, q) such that*

$$\left| \alpha - \frac{p}{q} \right| < \frac{1}{q^{2+\epsilon}}. \tag{5.70}$$

Remark 5.7.2. We have seen for $\alpha \notin \mathbb{Q}$ that there are infinitely many pairs of relatively prime integers (p, q) such that

$$\left| \alpha - \frac{p}{q} \right| < \frac{1}{q^2}. \tag{5.71}$$

Therefore any non-rational algebraic number has approximation exponent exactly 2.

Roth's Theorem has been generalized to more general settings. For a generalization due to Lang, and other historical remarks, see [HS]. For another generalization due to Schmidt see [B].

The remainder of this chapter is devoted to various applications of this fundamental theorem. For a proof, see Chapter 6.

5.7.1 Applications of Roth's Theorem to Transcendental Numbers

In this section we indicate, without proof, some miscellaneous applications of Roth's Theorem to constructing transcendental numbers. From this theorem follows a sufficient, but not necessary, condition for transcendency: let ξ and $\tau > 2$ be real numbers. If there exists an infinite sequence of distinct rational numbers $\frac{p_1}{q_1}, \frac{p_2}{q_2}, \frac{p_3}{q_3}, \dots$ satisfying

$$0 < \left| \xi - \frac{p_r}{q_r} \right| \le \frac{1}{q_r^\tau} \tag{5.72}$$

for $r = 1, 2, 3, \dots$, then ξ is transcendental.

Exercise 5.7.3. *Verify that the collection of all such ξ is an uncountable set of measure zero.*

The first application is a theorem due to Mahler which was originally proved by an improvement of Thue's result mentioned above. One can of course prove the same result using Roth's Theorem; the proof is easier, but still non-trivial. Let $P(x)$ be a polynomial with integral coefficients with the property that $P(n) > 0$ if $n > 0$. Let $q > 1$ be a positive integer. For any number n we let $l_q(n)$ be the string of numbers obtained from writing n in base q. Then Mahler's theorem [Mah] asserts that the number

$$\alpha(P;q) = 0.l_q(P(1))l_q(P(2))l_q(P(3))\cdots$$

$$= \sum_{n=1}^{\infty} \frac{P(n)}{\prod_{k=1}^{n} q^{\lceil \log_q P(k) \rceil}} \tag{5.73}$$

is transcendental (see [Gel], page 6). For example, when $P(x) = x$ and $q = 10$, we obtain Champernowne's constant

$$0.12345678910111213141516171819\ldots. \tag{5.74}$$

Exercise 5.7.4. *Prove, using elementary methods, that the above number is irrational. Can you prove this particular number is transcendental?*

Another application is the transcendence of various continued fractions expansions (see Chapter 7 for properties of continued fractions). As an illustration we state the following theorem due to Okano [Ok]: let $\gamma > 16$ and suppose $A = [a_1, a_2, a_3, \ldots]$ and $B = [b_1, b_2, b_3, \ldots]$ are two simple continued fractions with $a_n > b_n > a_{n-1}^{\gamma(n-1)}$ for n sufficiently large. Then $A, B, A \pm B$ and $AB^{\pm}$ are transcendental. The transcendence of A, B easily follows from Liouville's theorem, but the remaining assertions rely on Roth's Theorem.

5.7.2 Applications of Roth's Theorem to Diophantine Equations

Here we collect a few applications of Roth's Theorem to Diophantine equations (mostly following [Hua], Chapter 17). Before stating any hard theorems, however, we illustrate the general idea with an example (see pages 244–245 of [Sil1]).

Example 5.7.5. *There are only finitely many integer solutions $(x, y) \in \mathbb{Z}^2$ to*

$$x^3 - 2y^3 = a. \tag{5.75}$$

In order to see this, we proceed as follows. Let $\rho = e^{2\pi i/3} = (-1)^{1/3} = -\frac{1}{2} + i\frac{\sqrt{3}}{2}$. Then

$$x^3 - 2y^3 = (x - 2^{1/3}y)(x - \rho 2^{1/3}y)(x - \rho^2 2^{1/3}y), \tag{5.76}$$

and therefore

$$\left| \frac{a}{y^3} \right| = \left| \frac{x}{y} - 2^{1/3} \right| \left| \frac{x}{y} - \rho 2^{1/3} \right| \left| \frac{x}{y} - \rho^2 2^{1/3} \right|$$

$$\geq \left| \frac{x}{y} - 2^{1/3} \right| \left| \Im(\rho 2^{1/3}) \right| \left| \Im(\rho^2 2^{1/3}) \right|$$

$$= \frac{3}{2^{4/3}} \left| \frac{x}{y} - 2^{1/3} \right|. \tag{5.77}$$

Hence every integer solution (x, y) to $x^3 - 2y^3 = a$ is a solution to

$$\left| 2^{1/3} - \frac{x}{y} \right| \le \frac{3 \cdot 2^{-4/3}}{|y|^3}. \tag{5.78}$$

By Roth's Theorem there are only finitely many such solutions.

Note Liouville's Theorem is *not strong enough to allow us to conclude there are only finitely many integer solutions. As $2^{1/3}$ is an algebraic number of degree 3, Liouville's Theorem says $2^{1/3}$ can be approximated by rationals to order at most 3. Thus the possibility that $2^{1/3}$ can be approximated by rationals to order 3 is not ruled out by Liouville's Theorem.*

Remark 5.7.6. The reader should keep in mind that "finite" does not mean "a small number"; 10^{456} is still a finite number! In general, Roth's Theorem and other finiteness results of the same nature do not provide effective bounds. In some sense this is similar to the special value proofs of the infinitude of primes: $\pi^2 \notin \mathbb{Q}$ implies there are infinitely many primes, but gives no information on how many primes there are at most x (see Exercise 3.1.7).

Building on the above example, we state the following important theorem.

Theorem 5.7.7. *Let $n \ge 3$ and let $f(x, y)$ be an irreducible homogeneous polynomial of degree n with integer coefficients. Suppose that $g(x, y)$ is a polynomial with rational coefficients of degree at most $n - 3$. Then the equation*

$$f(x, y) = g(x, y) \tag{5.79}$$

has only finitely many solutions in integers (x, y).

Proof. Let us assume $a_0 \ne 0$. Without loss of generality we may also assume $|x| \le |y|$. Suppose $y > 0$, the other cases being similar or trivial. Let $\alpha_1, \ldots, \alpha_n$ be the roots of the equation $f(x, 1) = 0$, and let G be the maximum of the absolute values of the coefficients of $g(x, y)$. Then (5.79) implies

$$\begin{aligned} |a_0(x - \alpha_1 y) \ldots (x - \alpha_n y)| &\le G(1 + 2|y| + \cdots + (n - 2)|y|^{n-3}) \\ &< n^2 G|y|^{n-3}. \end{aligned} \tag{5.80}$$

Exercise 5.7.8. *Prove the above inequalities.*

Consequently

$$|(x - \alpha_1 y) \ldots (x - \alpha_n y)| < \frac{n^2 G}{a_0} |y|^{n-3}. \tag{5.81}$$

As on the left hand side there are n factors, at least one the factors must be strictly less than the right hand side raised to the power $\frac{1}{n}$; there exist an index ν such that

$$|x - \alpha_\nu y| < \left(\frac{n^2 G}{a_0} \right)^{\frac{1}{n}} |y|^{1 - \frac{3}{n}}. \tag{5.82}$$

Since there are infinitely many solutions (x, y), it is a consequence of the Pigeonhole Principle that infinitely many of the pairs of solutions correspond to the same index ν. We fix one such index and denote it again by ν. Next let $\mu \neq \nu$ and $|y| > N$, N a large positive number whose size we will determine in a moment. Then

$$|x - \alpha_\mu y| = |(\alpha_\nu - \alpha_\mu)y + (x - \alpha_\nu y)|$$

$$> |(\alpha_\nu - \alpha_\mu)| \cdot |y| - \left(\frac{n^2 G}{a_0}\right)^{\frac{1}{n}} \cdot |y|^{1-\frac{3}{n}}$$

$$> \frac{1}{2}|(\alpha_\nu - \alpha_\mu)| \cdot |y| \qquad (5.83)$$

for N sufficiently large. Next, 5.80 and 5.81 imply that for $|y| > N$ we have

$$\frac{n^2 G}{a_0}|y|^{n-3} > \left[\prod_{\mu \neq \nu} \frac{1}{2}|\alpha_\nu - \alpha_\mu|\right] \cdot |y|^{n-1} |x - \alpha_\nu y|. \qquad (5.84)$$

Hence

$$\left|\frac{x}{y} - \alpha_\nu\right| < \frac{K}{|y|^3} \qquad (5.85)$$

for infinitely many pairs of integers (x, y) for a fixed explicitly computable constant K. By Roth's Theorem, this contradicts the algebraicity of α_ν. □

Exercise 5.7.9. *In the proof of Theorem 5.7.7, handle the cases where $|x| > |y|$.*

Remark 5.7.10. In the proof of the above theorem, and also the example preceding it, we used the following simple, but extremely useful, observation: If $a_1, \ldots, a_n, B$ are positive quantities subject to $a_1 \ldots a_n < B$, then for some i, we have $a_i < B^{\frac{1}{n}}$.

An immediate corollary is the following:

Corollary 5.7.11 (Thue). *Let $n \geq 3$ and let f be as above. Then for any integer a the equation*

$$f(x, y) = a \qquad (5.86)$$

has only finitely many solutions.

Exercise 5.7.12 (Thue). *Show that if $a \neq 0$ and $f(x, y)$ is not the nth power of a linear form or the $\frac{n}{2}^{th}$ power of a quadratic form, then the conclusion of the corollary still holds.*

Example 5.7.13. *Consider Pell's Equation $x^2 - dy^2 = 1$ where d is not a perfect square. We know that if $d > 0$ this equation has infinitely many solutions in integers (x, y). Given integers d and n, we can consider the generalized Pell's Equation $x^n - dy^n = 1$. Exercise 5.7.12 shows that if $n \geq 3$ the generalized Pell's Equation can have at most finitely many solutions. See §7.6.3 for more on Pell's Equation.*

Example 5.7.14. *We can apply the same idea to Fermat's equation $x^n + y^n = z^n$. Again, Exercise 5.7.12 shows that if $n \geq 3$ there are at most a finite number* *of solutions* (x, y, z), *provided that we require one of the variables to be a fixed integer. For example, the equation* $x^n + y^n = 1$ *cannot have an infinite number of integer solutions* (x, y). *This is of course not hard to prove directly (exercise!). Fermat's Last Theorem states that there are no rational solutions to the equation* $x^n + y^n = 1$ *for* n *larger than two except when* $xy = 0$ *(if* x *or* y *is zero, we say the solution is trivial). A deep result of Faltings, originally conjectured by Mordell, implies that for any given* $n \geq 3$ *there are at most a finite number of rational solutions to the equation. Incidentally, the proof of Faltings' theorem uses a generalization of Roth's Theorem. Unfortunately, Faltings' theorem does not rule out the possibility of the existence of non-trivial solutions as conjectured by Fermat. This was finally proved by Wiles in 1995; see [Acz, Maz3, Wi].*

Exercise 5.7.15 (Hua). *Let* $n \geq 3$, $b^2 - 4ac \neq 0$, $a \neq 0$, $d \neq 0$. *Then a theorem of Landau, Ostrowski, and Thue states that the equation*

$$ay^2 + by + c = dx^n \tag{5.87}$$

has only finitely many solutions. Assuming this statement, prove the following two assertions:

1. *Let* n *be an odd integer greater than 1. Arrange the integers which are either a square or an* n^{th} *power into an increasing sequence* (z_r). *Prove that* $z_{r+1} - z_r \to \infty$ *as* $r \to \infty$.

2. *Let* $\langle \xi \rangle = \min(\xi - [\xi], [\xi] + 1 - \xi)$. *Prove that*

$$\lim_{x \to \infty, x \neq k^2} x^{\frac{n}{2}} \langle x^{\frac{n}{2}} \rangle = \infty, \tag{5.88}$$

where the conditions on the limit mean $x \to \infty$ *and* x *is never a perfect square.*

Chapter Six

The Proof of Roth's Theorem

In this chapter we prove Roth's Theorem in its original form, following the argument presented in [Ro]. Our presentation is motivated by [HS, Ra]. A nice treatment can also be found in [La1].

There are no easy proofs of Roth's Theorem. The proof presented here, while technical, is elementary in terms of the needed mathematical prerequisites. The difficulty is keeping the main idea clear, as the details and complications are long and involved. In all arguments below we assume the algebraic numbers are irrational, as the rational case is straightforward.

Theorem 6.0.16 (Roth's Theorem). *Let α be a real algebraic number (a root of a polynomial equation with integer coefficients). Given any $\epsilon > 0$ there are only finitely many relatively prime pairs of integers (p, q) such that*

$$\left| \alpha - \frac{p}{q} \right| < \frac{1}{q^{2+\epsilon}}. \tag{6.1}$$

6.1 LIOUVILLE'S THEOREM AND ROTH'S THEOREM

As we shall see below, the proof of Roth's Theorem follows from an argument that generalizes that of Liouville's Theorem. Given a real algebraic number α of degree $d > 1$, Liouville (see §5.6.1) showed that for any $\epsilon > 0$, except for finitely many $\frac{p}{q}$,

$$\left| \alpha - \frac{p}{q} \right| \geq \frac{1}{q^{d+\epsilon}}. \tag{6.2}$$

In other words, a real algebraic number of degree d can be approximated by rationals to at most order d. Recall that the proof proceeds in two steps:

1. We construct the minimal irreducible polynomial of α, $f(x)$; by construction $f(\alpha) = 0$.

2. Assuming α has approximation exponent greater than d we prove for some $\frac{p}{q} \in \mathbb{Q}$ that $f\left(\frac{p}{q}\right) = 0$; this contradicts the irreducibility of f and completes the proof.

The basic idea in proving Roth's Theorem is to generalize the construction of the polynomial f. Thue's improvement of the exponent in Liouville's Theorem to $\frac{d}{2} + 1$ uses a similar method to Liouville's method but with a polynomial of two variables of the form

$$f(x, y) = (y - \alpha)f_1(x, y) + (x - \alpha)^m f_2(x, y). \tag{6.3}$$

The proof of Roth's Theorem proceeds along the same lines. Suppose Roth's Theorem is false for some algebraic number α. Then for some $\epsilon > 0$ there is an infinite sequence of rational numbers $\beta_i = \frac{h_i}{q_i}$ with $\left| \alpha - \frac{h_i}{q_i} \right| < \frac{1}{q_i^{2+\epsilon}}$. We can choose the sequence so that q_1 is very large and for each i, q_i is much larger than q_{i-1}. Let m be a large natural number; how large will depend on α. In the course of the proof we construct a non-zero polynomial $Q(x_1, \ldots, x_m)$ with $Q(\alpha, \ldots, \alpha) = 0$ and $Q(\beta_1, \ldots, \beta_m) = 0$. As in the proof of Liouville's Theorem, we will know enough about the polynomial Q so that these vanishing statements imply Q is the zero polynomial, giving a contradiction. The vanishing of Q at $(\beta_1, \ldots, \beta_m)$ is really the heart of the proof of Roth's Theorem, and is the most technically demanding part. Things are very different in higher dimensions as the following simple example shows.

Example 6.1.1. *Consider the irreducible polynomial of four variables $xy - zw$. This polynomial is zero infinitely often; in fact, it is zero for infinitely many rational (x, y, z, w). Consider, for example, $(0, m, 0, n)$, $m, n \in \mathbb{Q}$.*

We will explain in 6.3 the procedure for proving Roth's Theorem. Before doing that, however, we need some preparation.

6.2 EQUIVALENT FORMULATION OF ROTH'S THEOREM

The goal of this section is to show that in order to prove Roth's Theorem for all algebraic numbers, we just need to prove it for a subclass of algebraic numbers, the **algebraic integers**. Recall

Definition 6.2.1. *An algebraic integer α is a number $\alpha \in \mathbb{C}$ which satisfies an equation*

$$a_d \alpha^d + \cdots + a_0 = 0, \tag{6.4}$$

with $a_i \in \mathbb{Z}$, d a positive integer and $a_d = 1$.

Exercise 6.2.2. *Show that $\alpha = \frac{1+\sqrt{5}}{2}$ is an algebraic integer. Is $\frac{1}{\alpha}$ an algebraic integer?*

The basic idea is quite simple. If α is an algebraic number, then there is a integer D such that $D\alpha$ is an algebraic integer. Then if β is a good rational approximation to α, $D\beta$ must be a good rational approximation to $D\alpha$ (see the proof of Lemma 6.2.10 for details). We start with a definition.

Definition 6.2.3 (Height). *Assume p and q are relatively prime integers. We define the height of $\frac{p}{q}$ to be*

$$H\left(\frac{p}{q}\right) = \max\left(|p|, |q|\right). \tag{6.5}$$

See §4.3 for more on height functions and connections with Diophantine equations.

Exercise 6.2.4. *Prove for any $\beta \in \mathbb{Q}$ and $D \in \mathbb{Z}$ that $H(D\beta) \le H(D)H(\beta)$.*

Definition 6.2.5 (Ineq$_*(\alpha, \beta; \epsilon)$). *Let*

- α *be a real algebraic number;*

- β *be a rational number with $|\beta - \alpha| \le 1$;*

- $\epsilon > 0$;

- $H(\beta)$ *be the height of β.*

We write Ineq$_*(\alpha, \beta; \epsilon)$ if α and β satisfy

$$|\alpha - \beta| \le \frac{1}{H(\beta)^{2+\epsilon}}. \tag{6.6}$$

We say there are only finitely many solutions to Ineq$_*(\alpha, \beta; \epsilon)$ if for a fixed algebraic α and a fixed $\epsilon > 0$ there are only finitely many $\beta \in \mathbb{Q}$ such that (6.6) holds.

Definition 6.2.6. *Let α be an irrational real algebraic number. We say Roth's Theorem holds, or is valid for α, if for all $\epsilon > 0$ there are only finitely many rational numbers β such that Ineq$_*(\alpha, \beta; \epsilon)$; otherwise, we say Roth's Theorem fails for α.*

The following two lemmas give an equivalent formulation to Roth's Theorem. As for each $\epsilon > 0$ the inequality (6.6) is to hold for only finitely many $\beta \in \mathbb{Q}$, we can replace the 1 in the numerator of (6.6) with any fixed $c > 0$. We then relate these inequalities to Roth's Theorem:

Lemma 6.2.7. *For an $\epsilon > 0$ there are only finitely many solutions to Ineq$_*(\alpha, \beta; \epsilon)$ if and only if for any fixed c, for any $\epsilon' > 0$ there are only finitely many solutions to*

$$|\alpha - \beta| \le \frac{c}{H(\beta)^{2+\epsilon'}}. \tag{6.7}$$

Proof. Assume for any ϵ there are only finitely many solutions to (6.6). Given a c and an ϵ', chose $\epsilon < \epsilon'$. Assume α, β satisfy the inequality in (6.7). Then

$$|\alpha - \beta| \le \frac{c}{H(\beta)^{2+\epsilon'}} = \frac{c}{H(\beta)^{\epsilon'-\epsilon}} \frac{1}{H(\beta)^{2+\epsilon}}. \tag{6.8}$$

For all but finitely many β, $H(\beta)^{\epsilon'-\epsilon} > c$, therefore except for finitely many β,

$$|\alpha - \beta| < \frac{1}{H(\beta)^{2+\epsilon}}. \tag{6.9}$$

By assumption, there are only finitely many solutions to (6.9); therefore there are only finitely many solutions to

$$|\alpha - \beta| \le \frac{c}{H(\beta)^{2+\epsilon'}}. \tag{6.10}$$

Conversely, for any c and any $\epsilon' > 0$ assume there are only finitely many solutions to (6.7). Choose $\epsilon > \epsilon'$. Assume α, β satisfy the inequality in (6.6). Then

$$|\alpha - \beta| \le \frac{1}{H(\beta)^{2+\epsilon}} = \frac{1}{H(\beta)^{\epsilon-\epsilon'}} \frac{1}{H(\beta)^{2+\epsilon'}}. \tag{6.11}$$

For all but finitely many β, $\frac{1}{H(\beta)^{\epsilon - \epsilon'}} < c$, therefore except for finitely many β,

$$|\alpha - \beta| < \frac{c}{H(\beta)^{2+\epsilon'}}. \tag{6.12}$$

By assumption, there are only finitely many solutions to (6.12); therefore there are only finitely many solutions to

$$|\alpha - \beta| \leq \frac{1}{H(\beta)^{2+\epsilon'}}. \tag{6.13}$$

$\square$

Lemma 6.2.8. *Given $\alpha \notin \mathbb{Q}$, $Ineq_*(\alpha, \beta; \epsilon)$ has finitely many solutions for every $\epsilon > 0$ if and only if Roth's Theorem is true for α.*

We show that if Roth's Theorem is false for $\alpha \notin \mathbb{Q}$ and some $\epsilon > 0$ then $Ineq_*(\alpha, \beta; \epsilon)$ has infinitely many solutions for that ϵ, and if $Ineq_*(\alpha, \beta; \epsilon)$ has infinitely many solutions for an $\epsilon > 0$ then Roth's Theorem fails for $\alpha \notin \mathbb{Q}$ and that ϵ. Why will proving this suffice? If Roth's Theorem is true for α and all ϵ we must show for all ϵ that $Ineq_*(\alpha, \beta; \epsilon)$ has only finitely many solutions. Assume $Ineq_*(\alpha, \beta; \epsilon)$ has infinitely many solutions for some ϵ, say ϵ'. Then we can conclude that Roth's Theorem fails for α and that ϵ', contradicting our assumption that Roth's Theorem is true for α for all ϵ. A similar argument shows that if for all ϵ, $Ineq_*(\alpha, \beta; \epsilon)$ has finitely many solutions then Roth's Theorem is true for α. We repeatedly use Lemma 6.2.7 below.

Proof. Assume for some $\epsilon > 0$ that Roth's Theorem is false for $\alpha \notin \mathbb{Q}$. Then there are infinitely many relatively prime p, q with

$$\left| \alpha - \frac{p}{q} \right| \leq \frac{1}{q^{2+\epsilon}}. \tag{6.14}$$

Exercise 6.2.9. *Use the triangle inequality to show that $\left| \alpha - \frac{p}{q} \right| \leq 1$ implies $|p| \leq |q| \cdot (|\alpha| + 1)$.*

As $|p| \leq |q| \cdot (|\alpha| + 1)$, we have

$$\begin{aligned} \left| \alpha - \frac{p}{q} \right| &\leq \frac{1}{q^{2+\epsilon}} \\ &\leq \frac{(|\alpha| + 1)^{2+\epsilon}}{(|\alpha| + 1)^{2+\epsilon} q^{2+\epsilon}} \\ &\leq \frac{(|\alpha| + 1)^{2+\epsilon}}{|p|^{2+\epsilon}}. \end{aligned} \tag{6.15}$$

As $\left| \alpha - \frac{p}{q} \right| < \frac{1}{q^{2+\epsilon}}$, combining the this inequality with (6.15) gives

$$\begin{aligned} \left| \alpha - \frac{p}{q} \right| &\leq \frac{(|\alpha| + 1)^{2+\epsilon}}{\max(|p|, |q|)^{2+\epsilon}} \\ &= \frac{(|\alpha| + 1)^{2+\epsilon}}{H\left(\frac{p}{q}\right)}. \end{aligned} \tag{6.16}$$

Thus if for a given $\epsilon > 0$, Roth's Theorem fails for α, then $\text{Ineq}_*(\alpha, \beta; \epsilon)$ holds infinitely often.

Conversely, assume $\text{Ineq}_*(\alpha, \beta; \epsilon)$ holds infinitely often. Let $\beta = \frac{p}{q}$. If $q > |p|$ then $H(\beta) = q$ and we immediately find

$$\left| \alpha - \frac{p}{q} \right| \leq \frac{1}{q^{2+\epsilon}}. \tag{6.17}$$

Therefore it is enough to consider the case when $\text{Ineq}_*(\alpha, \beta; \epsilon)$ holds and $|p| > q$. Thus

$$\left| \alpha - \frac{p}{q} \right| \leq \frac{1}{|p|^{2+\epsilon}}. \tag{6.18}$$

As $|p| > q$ and $H(\beta) = |p|$, we have

$$\frac{1}{H(\beta)} = \frac{1}{|p|} \leq \frac{1}{|q|}. \tag{6.19}$$

Therefore

$$\left| \alpha - \frac{p}{q} \right| \leq \frac{1}{H(\beta)^{2+\epsilon}} \leq \frac{1}{q^{2+\epsilon}}. \tag{6.20}$$

Thus if $\text{Ineq}_*(\alpha, \beta; \epsilon)$ has infinitely many solutions for a given $\epsilon > 0$, then Roth's Theorem fails for α with this ϵ. $\square$

Lemma 6.2.10 (Reduction Lemma). *If Roth's Theorem is true for all algebraic integers α, then it is true for all algebraic numbers.*

Proof. Suppose α is an algebraic number and suppose Roth's Theorem fails for α. We find an algebraic integer for which Roth's Theorem fails. Our assumption means there are infinitely many $\beta \in \mathbb{Q}$ such that

$$|\alpha - \beta| \leq \frac{1}{H(\beta)^{2+\epsilon}}. \tag{6.21}$$

We find an integer D such that $D\alpha$ is an algebraic integer, and then show that $D\beta$ provide good approximations to $D\alpha$. As α is an algebraic number, there are integers a_i such that

$$a_d \alpha^d + a_{d-1} \alpha^{d-1} + a_{d-2} \alpha^{d-2} + \cdots + a_0 = 0. \tag{6.22}$$

Multiply the above equation by a_d^{d-1}. Regrouping we find

$$(a_d \alpha)^d + a_{d-1}(a_d \alpha)^{d-1} + a_{d-2} a_d (a_d \alpha)^{d-2} + \cdots + a_0 a_d^{d-1} = 0. \tag{6.23}$$

Hence $a_d \alpha$ satisfies the equation

$$X^d + a_{d-1} X^{d-1} + a_{d-2} a_d X^{d-2} + \cdots + a_0 a_d^{d-1} = 0. \tag{6.24}$$

As all the coefficients are integers, $a_d \alpha$ is an algebraic integer. Note $D = a_d \in \mathbb{Z}$ and $H(D) = |D|$. For all but finitely many β we have $H(\beta) > H(D)^{1+\frac{6}{\epsilon}}$ (why?). By Exercise 6.2.4, $H(D\beta) \leq H(D)H(\beta)$.

We are assuming Roth's Theorem fails for the algebraic number α; we now show Roth's Theorem fails for the algebraic integer $D\alpha = a_d\alpha$. We have

$$
\begin{aligned}
|D\alpha - D\beta| &= |D| \cdot |\alpha - \beta| \\
&\leq \frac{|D|}{H(\beta)^{2+\epsilon}} \\
&\leq \frac{H(D)}{H(\beta)^{2+\epsilon}} \\
&= \frac{H(D)}{H(\beta)^{2+\frac{\epsilon}{2}}} \cdot \frac{1}{H(\beta)^{\frac{\epsilon}{2}}} \\
&\leq \frac{H(D)}{\left(\frac{H(D\beta)}{H(D)}\right)^{2+\frac{\epsilon}{2}}} \cdot \frac{1}{\left(H(D)^{1+\frac{6}{\epsilon}}\right)^{\frac{\epsilon}{2}}} \\
&= \frac{1}{H(D\beta)^{2+\frac{\epsilon}{2}}}.
\end{aligned}
\tag{6.25}
$$

Thus if Roth's Theorem fails for an algebraic number then it also fails for some algebraic integer. $\qquad\square$

The advantage of the above theorem is that we may restrict our investigations to algebraic integers. This gives a little more information about the characteristic polynomial (namely, the leading coefficient must be 1).

6.3 ROTH'S MAIN LEMMA

In this section we state Roth's Lemma, the most technical part of the proof, and then derive Roth's Theorem. Before doing that, however, we would like to take a moment to motivate Roth's Lemma. Suppose that an algebraic number α is given, and suppose there is a number $u > 2$ and infinitely many rational fractions $\frac{h}{q}$, $(h, q) = 1$ and $q > 0$, such that

$$
\left|\alpha - \frac{h}{q}\right| < \frac{1}{q^u}.
\tag{6.26}
$$

We choose m rational numbers $\frac{h_1}{q_1}, \ldots, \frac{h_m}{q_m}$ satisfying (6.26). Then for any polynomial $Q(x_1, x_2, \ldots, x_m) \in \mathbb{Z}[x_1, x_2, \ldots, x_m]$ of degree at most r_j in x_j ($1 \leq j \leq m$) which does not vanish at the point $\left(\frac{h_1}{q_1}, \ldots, \frac{h_m}{q_m}\right)$, from a multidimensional Taylor series expansion we have

$$
\frac{1}{q_1^{r_1} \ldots q_m^{r_m}} \leq \left| Q\left(\frac{h_1}{q_1}, \ldots, \frac{h_m}{q_m}\right) \right|
\tag{6.27}
$$

$$
\leq \left| \sum_{i_m=0}^{r_m} \cdots \sum_{i_1=1}^{r_1} Q_{i_1,\ldots,i_m}(\alpha, \ldots, \alpha) \prod_{k=1}^{m} \left(\frac{h_k}{q_k} - \alpha\right)^{i_k} \right|
$$

$$
< B 2^{r_1 + \cdots + r_m} \max_{i_1,\ldots,i_m} \prod_{k=1}^{m} q_k^{-u i_k};
\tag{6.28}
$$

the maximum is taken over all m-tuples $(i_1, \ldots, i_m)$ such that $Q_{i_1,\ldots,i_m}(\alpha, \ldots, \alpha) \neq 0$. Here,

$$Q_{i_1,\ldots,i_m}(x_1, \ldots, x_m) = \frac{1}{i_1! i_2! \cdots i_m!} \frac{\partial^{i_1 + \cdots + i_m}}{\partial x_1^{i_1} \cdots \partial x_m^{i_m}} Q(x_1, \ldots, x_m) \quad (6.29)$$

and

$$B = \max_{i_1,\ldots,i_m} |Q_{i_1,\ldots,i_m}(\alpha, \ldots, \alpha)|. \quad (6.30)$$

The inequality (6.27) is valid, because the right hand side is a rational number, whose denominator divides $q_1^{r_1} \cdots q_m^{r_m}$; any such number is at least as large as the left hand side. The content of Roth's Lemma is the existence of a polynomial Q such that the right hand side of (6.28) is too small for the string of inequalities ending with (6.28) to be valid. Naively, one can do the following. Rewrite the last inequality as

$$B^{-1} 2^{-(r_1 + \cdots + r_m)} \leq \max_{i_1,\ldots,i_m} \prod_{k=1}^{m} q_k^{-u i_k + r_k}. \quad (6.31)$$

Now suppose for each $(i_1, \ldots, i_m)$ for which $Q_{i_1,\ldots,i_m}(\alpha, \ldots, \alpha) \neq 0$ we have $u i_k - r_k > 0$, and also that $q_j^{r_j} \geq q_1^{r_1}$. Then the right hand side of (6.31) is bounded by

$$\max \prod_{k=1}^{m} \frac{1}{q_k^{r_k(u \frac{i_k}{r_k} - 1)}} \leq \max \prod_{k=1}^{m} \frac{1}{q_1^{r_1(u \frac{i_k}{r_k} - 1)}} \leq \frac{q_1^{r_1 m}}{q_1^{u \min \frac{i_1}{r_1} + \cdots + \frac{i_m}{r_m}}}.$$

Here again the maximum is taken over all tuples $(i_1, \ldots, i_m)$ such that $Q_{i_1,\ldots,i_m}(\alpha, \ldots, \alpha) \neq 0$. Motivated by the above computation, we make the following definition:

Definition 6.3.1 (Index). *Let $P(x_1, \ldots, x_p)$ be a non-identically zero polynomial in p variables. Let $\alpha_1, \ldots, \alpha_p$ be real numbers and $r_1, \ldots, r_p$ be positive integers. The index θ of P at $(\alpha_1, \ldots, \alpha_p)$ relative to $(r_1, \ldots, r_p)$ is the smallest value of*

$$\frac{i_1}{r_1} + \cdots + \frac{i_p}{r_p} \quad (6.32)$$

such that

$$\frac{1}{i_1! \cdots i_p!} \frac{\partial^{i_1 + \cdots + i_p}}{\partial x_1^{i_1} \cdots \partial x_p^{i_p}} P(\alpha_1, \ldots, \alpha_p) \neq 0. \quad (6.33)$$

We denote the index described above by $\mathrm{ind}(P)(\alpha_1, \ldots, \alpha_p; r_1, \ldots, r_p)$. *For brevity we often write* $\mathrm{ind}(P)$ *when the argument is clear.*

Note $\theta \geq 0$. Further, $\theta = 0$ implies $P(\alpha_1, \ldots, \alpha_p) = 0$.

Lemma 6.3.2. *Let P and Q be polynomials. Then*

1. $\mathrm{ind}(P + Q) \geq \min(\mathrm{ind}(P), \mathrm{ind}(Q))$,

2. $\mathrm{ind}(PQ) = \mathrm{ind}(P) + \mathrm{ind}(Q)$,

3. ind $\left(\dfrac{\partial^{i_1 + \cdots + i_p}}{\partial x_1^{i_1} \cdots \partial x_p^{i_p}} P \right) \geq$ ind $P - \left(\dfrac{i_1}{r_1} + \cdots + \dfrac{i_p}{r_p} \right)$.

Lemma 6.3.3. *Assume* $F = U(x_1, \ldots, x_{p-1}))V(x_p)$. *Then*

$$\text{ind}(F)(\alpha_1, \ldots, \alpha_p; r_1, \ldots, r_p)$$
$$= \text{ind}(U)(\alpha_1, \ldots, \alpha_{p-1}; r_1, \ldots, r_{p-1}) + \text{ind}(V)(\alpha_p, r_p). \tag{6.34}$$

Exercise 6.3.4. *Prove Lemmas 6.3.2 and 6.3.3.*

Exercise 6.3.5. *State and prove a general multi-dimensional Taylor expansion formula for polynomials in several variables.*

We now return to Roth's Lemma. Suppose we have an algebraic integer α satisfying

$$x^n + a_1 x^{n-1} + \cdots + a_{n-1}x + a_n = 0, \quad a_i \in \mathbb{Z}. \tag{6.35}$$

Let

$$A = \max\left(1, |a_1|, \ldots, |a_n|\right). \tag{6.36}$$

There are a lot of parameters and relations in the proof. We collect the key equations in (6.37) through (6.45). Let $m, \delta, q_1, \ldots, q_m, r_1, \ldots, r_m$ satisfy

$$0 < \delta < \frac{1}{m}, \tag{6.37}$$

$$10^m \delta^{\frac{m}{2}} + 2(1 + 3\delta)nm^{\frac{1}{2}} < \frac{m}{2}, \tag{6.38}$$

$$r_m > \frac{10}{\delta}, \quad \frac{r_{j-1}}{r_j} > \frac{1}{\delta}, \quad j = 2, \ldots, m, \tag{6.39}$$

$$\log q_1 > \frac{1}{\delta^2} \left[2m + 1 + 2m \log(1 + A) + 2m \log(1 + |\alpha|) \right], \tag{6.40}$$

$$q_j^{r_j} \geq q_1^{r_1}. \tag{6.41}$$

Given m, it is easy to choose δ to satisfy (6.37); in order for equation (6.38) to be satisfied δ has to be extremely small and m very large. For (6.39) to be valid, r_m has to be a large number, and for each j, r_{j-1} much bigger than r_j; in particular, r_1 is the largest number among $r_1, r_2, \ldots, r_m$. Equations (6.40) and (6.41) tell us how to choose $q_1, \ldots, q_m$. Here we urge the reader to study again the inequalities ending with (6.28) before moving on, and keep coming back to them to keep the ideas clear. It will be clear from the proof of the theorem that the above inequalities are to some extent quite arbitrary; we have chosen these particular ones for historical reasons: They were the inequalities that Roth used in his original paper [Ro].

For ease of reference below, define λ, γ, m, B_1 by

$$\lambda = 4(1 + 3\delta)nm^{\frac{1}{2}}, \tag{6.42}$$

$$\gamma = \frac{m - \lambda}{2}, \tag{6.43}$$

$$\eta = 10^m \delta^{\frac{m}{2}}, \tag{6.44}$$

$$B_1 = \left[q_1^{\delta r_1} \right], \tag{6.45}$$

where $[z]$ denotes the greatest integer at most z. Note (6.38) is equivalent to $\eta < \gamma$. The reader will observe that the specific shapes and values of the above parameters are irrelevant.

We can now state the lemma which will imply Roth's Theorem. The proof of Roth's Lemma is given in §6.3.6.

Lemma 6.3.6 (Roth's Main Lemma). *Suppose (6.37) through (6.42) are satisfied. Let $h_1, \ldots, h_m$ be given with $(h_i, q_i) = 1$ for $i = 1, \ldots, m$. Then there is a polynomial $Q(x_1, \ldots, x_m)$ with integer coefficients of degree at most r_j in x_j with the following properties:*

1. $\mathrm{ind}(Q)(\alpha, \ldots, \alpha; r_1, \ldots, r_m)$ is at least $\gamma - \eta$,

2. $Q\left(\frac{h_1}{q_1}, \ldots, \frac{h_m}{q_m} \right) \neq 0$,

3. $\left| \frac{1}{i_1! \cdots i_m!} \left(\frac{\partial}{\partial x_1} \right)^{i_1} \cdots \left(\frac{\partial}{\partial x_m} \right)^{i_m} Q(\alpha, \ldots, \alpha) \right| < B_1^{1+3\delta}.$

We now show how Roth's Theorem follows from Lemma 6.3.6.

Proof of Roth's Theorem. Let $k > 2$ and assume

$$\left| \alpha - \frac{h}{q} \right| < \frac{1}{q^k} \tag{6.46}$$

has infinitely many solutions with h, q relatively prime. Choose an m such that $m > 16n^2$ and

$$\frac{2m}{m - 4nm^{1/2}} < k. \tag{6.47}$$

Recall (6.38) and (6.44):

$$m - 4(1 + 3\delta)nm^{1/2} - 2\eta > 0. \tag{6.48}$$

Note that by (6.44), $\eta \to 0$ as $\delta \to 0$. After we choose δ small enough to satisfy the above, we further choose δ small enough so that

$$\frac{2m(1 + 4\delta)}{m - 4(1 + 3\delta)nm^{1/2} - 2\eta} < k. \tag{6.49}$$

This is possible as $k > 2$. This gives

$$\frac{m(1 + 4\delta)}{\gamma - \eta} < k. \tag{6.50}$$

Let $\frac{h_1}{q_1}$, $(h_1, q_1) = 1$, be a rational number satisfying (6.46) with q_1 *very* large so that (6.40) is satisfied; such a rational number exists as we are assuming that there are infinitely many rational numbers $\frac{h}{q}$ satisfying (6.46). Similarly, choose $\frac{h_2}{q_2}, \ldots, \frac{h_m}{q_m}$ satisfying (6.46), with $(h_j, q_j) = 1$ for $j = 2, \ldots, m$ such that $\frac{\log q_j}{\log q_{j-1}} > \frac{2}{\delta}$. Choose

$$r_1 > \frac{10}{\delta} \frac{\log q_m}{\log q_1}. \tag{6.51}$$

Now choose $r_2, \ldots, r_m$ recursively such that

$$\frac{r_1 \log q_1}{\log q_j} \le r_j < 1 + \frac{r_1 \log q_1}{q_j}. \tag{6.52}$$

Exercise 6.3.7. *Show that (6.41) is satisfied.*

Now we take

$$\frac{r_j \log q_j}{r_1 \log q_1} < 1 + \frac{\log q_j}{r_1 \log q_1} \le 1 + \frac{\log q_m}{r_1 \log q_1} < 1 + \frac{\delta}{10}, \tag{6.53}$$

where the last bit follows from (6.51). We now have

$$r_m \ge \frac{r_1 \log q_1}{\log q_m} \ge \frac{10}{\delta}$$

$$\frac{r_{j-1}}{r_j} > \frac{\log q_j}{\log q_{j-1}} \left(1 + \frac{\delta}{10}\right)^{-1} > \frac{1}{\delta}. \tag{6.54}$$

Roth's Main Lemma (Lemma 6.3.6) gives

$$Q\left(\frac{h_1}{q_1}, \ldots, \frac{h_m}{q_m}\right) \ne 0. \tag{6.55}$$

The polynomial Q has integer coefficients and is of degree at most r_j in x_j. We therefore have

$$\left|Q\left(\frac{h_1}{q_1}, \ldots, \frac{h_m}{q_m}\right)\right| \ge \frac{1}{q_1^{r_1}} \cdots \frac{1}{q_m^{r_m}}$$

$$> \frac{1}{q_1^{mr_1(1+\delta)}}, \tag{6.56}$$

where the last follows from (6.53). What we are saying is

$$r_1 + \cdots + r_m < mr_1(1+\delta). \tag{6.57}$$

On the other hand,

$$Q\left(\frac{h_1}{q_1}, \ldots, \frac{h_m}{q_m}\right) = \sum_{i_1=0}^{r_1} \cdots \sum_{i_m=0}^{r_m} \frac{1}{i_1! \cdots i_m!} \frac{\partial^{i_1 + \cdots + i_m} Q}{\partial x_1^{i_1} \cdots \partial x_m^{i_m}} (\alpha, \ldots, \alpha)$$

$$\times \left(\frac{h_1}{q_1} - \alpha\right)^{i_1} \cdots \left(\frac{h_m}{q_m} - \alpha\right)^{i_m}. \tag{6.58}$$

By the first part of Roth's Lemma, we have that if

$$\frac{i_1}{r_1} + \cdots + \frac{i_m}{r_m} < \gamma - \eta, \tag{6.59}$$

then the coefficient $Q_{i_1,\ldots,i_m}(\alpha,\alpha,\ldots,\alpha)$ vanishes. In every other term

$$\left| Q_{i_1,\ldots,i_m}(\alpha,\ldots,\alpha) \cdot \left(\frac{h_1}{q_1} - \alpha\right)^{i_1} \cdots \left(\frac{h_m}{q_m} - \alpha\right)^{i_m} \right| < \frac{1}{q_1^{ki_1} \cdots q_m^{ki_m}}$$

$$= \frac{1}{(q_1^{i_1} \cdots q_m^{i_m})^k}$$

$$\leq \frac{1}{q_1^{r_1(\gamma-\eta)k}}, \quad (6.60)$$

where $q_j \geq q_1^{\frac{r_1}{r_j}}$ as $\frac{r_1 \log q_1}{\log q_j} \leq r_j$. Thus

$$\frac{1}{q^{mr_1(1+\delta)}} \leq \left| Q\left(\frac{h_1}{q_1},\ldots,\frac{h_m}{q_m}\right) \right|$$

$$\leq q^{-r_1(\gamma-\eta)k}(1 + r_1)\cdots(1 + r_m)B_1^{1+3\delta}$$

$$\leq B_1^{1+4\delta}q_1^{-r_1(\gamma-\eta)k}, \quad (6.61)$$

because as r_1 is the largest of the r_i's,

$$(1 + r_1)\cdots(1 + r_m) \leq 2^{r_1+\cdots+r_m} < 2^{mr_1} < B_1^\delta. \quad (6.62)$$

Therefore we find

$$\frac{1}{q^{mr_1(1+\delta)}} < q^{(1+4\delta)\delta r_1 - r_1(\gamma-\eta)k}, \quad (6.63)$$

with

$$k < \frac{m(1 + 4\delta)}{\gamma - \eta}, \quad (6.64)$$

which contradicts (6.50). This contradiction proves the theorem. $\square$

The reader is advised to compare the above argument with the inequalities ending with (6.28). The remainder of this chapter is devoted to giving the proof from [Ro] of Roth's Lemma (Lemma 6.3.6).

6.4 PRELIMINARIES TO PROVING ROTH'S LEMMA

Before we can prove Roth's Lemma, we need some preliminaries. We introduce Wronskians and their basic properties as they will be useful in the proof of Roth's Lemma. Lemma 6.4.1 below is a needed combinatorial statement whose proof is independent of everything else in this chapter.

For a first reading of this section, you should just study the statement of Lemma 6.4.1, familiarize yourself with Definition 6.4.13, and then skip to the statement of Lemma 6.4.20; after this, you can safely move to §6.5 where the proof of Roth's Main Lemma is presented.

6.4.1 A Combinatorial Lemma

The following lemma will be used to bound the number of elements of a certain set $\mathcal{D}$ defined in (6.113).

Lemma 6.4.1. *Let $r_1, \ldots, r_m$ be any positive integers, $\lambda > 0$. The number of integer m-tuples $(j_1, \ldots, j_m)$ with $0 \le j_i \le r_i$ that satisfy*

$$\frac{j_1}{r_1} + \cdots + \frac{j_m}{r_m} \le \frac{m - \lambda}{2} \tag{6.65}$$

is at most

$$2 \frac{\sqrt{m}}{\lambda} (r_1 + 1) \cdots (r_m + 1). \tag{6.66}$$

Proof. We proceed by induction. The result holds for $m = 1$. As the claim is trivial if $\lambda \le 2m^{\frac{1}{2}}$, we may assume $\lambda > 2m^{\frac{1}{2}}$. For any fixed value of j_m the conditions become $0 \le j_i \le r_i$ for $i \le m - 1$ and

$$\frac{j_1}{r_1} + \cdots + \frac{j_{m-1}}{r_{m-1}} \le \frac{m - 1 - \lambda'}{2} \tag{6.67}$$

with $\lambda' = \lambda - 1 + 2\frac{j_m}{r_m}$. Because we assumed $\lambda > 2m^{\frac{1}{2}}$, for $0 \le j_m \le r_m$ we have $\lambda' > 0$. Consequently, by the inductive assumption the number of solutions to (6.65) does not exceed

$$\sum_{j_m=1}^{r_m} 2 \frac{\sqrt{m-1}}{\lambda - 1 + 2\frac{j_m}{r_m}} (r_1 + 1) \cdots (r_{m-1} + 1). \tag{6.68}$$

Hence it suffices to prove that

$$\sum_{j=0}^{r} \left(\lambda - 1 + 2\frac{j}{r}\right)^{-1} < \lambda^{-1}(m-1)^{-\frac{1}{2}} m^{\frac{1}{2}} (r + 1), \tag{6.69}$$

for any positive integer r and m and $\lambda > 2m^{\frac{1}{2}}$. We leave this for the reader as a pleasant exercise. $\square$

Exercise 6.4.2. *Prove (6.69) with r, m and λ as above.*

6.4.2 Wronskians

Let $\phi_0(x), \ldots, \phi_{l-1}(x)$ be a collection of polynomials. Consider the determinant of the matrix

$$\left(\frac{1}{\mu!} \frac{d^\mu}{dx^\mu} \phi_\nu(x)\right)_{\mu,\nu=0,\ldots,l-1}. \tag{6.70}$$

For example, for $l - 1 = 2$ we have

$$\begin{vmatrix} \phi_0(x) & \phi_1(x) & \phi_2(x) \\ \phi_0'(x) & \phi_1'(x) & \phi_2'(x) \\ \frac{1}{2!}\phi_0''(x) & \frac{1}{2!}\phi_1''(x) & \frac{1}{2!}\phi_2''(x) \end{vmatrix}. \tag{6.71}$$

The following lemma is standard and explains the utility of the Wronskian:

Lemma 6.4.3. *Let* $\phi_0(x), \ldots, \phi_{l-1}(x)$ *have rational coefficients. Then they are linearly independent (see Definition B.1.7) if and only if the Wronskian is non-zero.*

See [BoDi] for more on ordinary Wronskians and their basic properties and applications to differential equations. We generalize the Wronskian to polynomials in several variables. Consider

$$\Delta = \frac{1}{i_1! \cdots i_p!} \left(\frac{\partial}{\partial x_1} \right)^{i_1} \cdots \left(\frac{\partial}{\partial x_p} \right)^{i_p}, \tag{6.72}$$

with $i_1 + \cdots + i_p$ the order of the symbol Δ. Such a symbol acts on the space of polynomials by differentiation.

By Δ_ν we mean some operator with tuple $(i_1, \ldots, i_p)$ such that $i_1 + \cdots i_p \leq \nu$. Thus the order of Δ_ν is at most ν. Often we do not care which tuple $(i_1, \ldots, i_p)$ we have; we often only care about the order, which is $i_1 + \cdots + i_p$. For notational convenience we write Δ_ν for such an operator, although $\Delta_{i_1, \ldots, i_p}$ would be more accurate.

Definition 6.4.4 (Generalized Wronskian). *Let* $\Delta_0, \ldots, \Delta_{l-1}$ *be as above. If* ϕ_0, *...* ϕ_{l-1} *are* l *polynomials in* $x_1, \ldots, x_p$, *we set*

$$G(x_1, \ldots, x_p) = \det \left(\Delta_\mu \phi_\nu (x_1, \ldots, x_p) \right)_{\mu, \nu = 0, \ldots, l-1}. \tag{6.73}$$

Exercise 6.4.5. *Verify that if* $\phi_0, \ldots, \phi_{l-1}$ *are linearly dependent (see Definition B.1.7) over* $\mathbb{Q}$, *then all of their generalized Wronskians are identically zero.*

Lemma 6.4.6. *If* $\phi_0, \ldots, \phi_{l-1}$ *are polynomials with rational coefficients which are linearly independent over* $\mathbb{Q}$, *then at least one of their generalized Wronskians is not identically zero.*

Proof. Let $k \in \mathbb{N}$ be larger than all the exponents of the individual x_i's, and consider the following l one-variable polynomials

$$P_\nu(t) = \phi_\nu \left(t, t^k, t^{k^2}, \ldots, t^{k^{p-1}} \right) \qquad (0 \leq \nu \leq l-1). \tag{6.74}$$

The polynomials $P_\nu(t)$ are linearly independent.

Exercise 6.4.7. *Prove the above statement.*

Hence the *standard Wronskian* will be non-zero. The Wronskian is

$$W(t) = \det \left(\frac{1}{\mu!} \frac{d^\mu}{dt^\mu} \phi_\nu(t, t^k, \ldots, t^{k^{p-1}}) \right)_{0 \leq \mu, \nu \leq l-1}. \tag{6.75}$$

Now

$$\frac{dP_\nu}{dt} = \frac{\partial \phi_\nu}{\partial x_1} \bigg|_{x_1 = t} + k t^{k-1} \frac{\partial \phi_\nu}{\partial x_2} \bigg|_{x_2 = t^k} + \cdots + k^{p-1} t^{k^{p-1}-1} \frac{\partial \phi_\nu}{\partial x_p} \phi_\nu \bigg|_{x_p = t^{k^{p-1}}}. \tag{6.76}$$

Thus

$$\frac{d^\mu}{dt^\mu} \tag{6.77}$$

is a linear combination of differential operators of several variables on $x_1, \ldots, x_p$, of orders not exceeding μ, say

$$\frac{d^\mu}{dt^\mu} = f_1(t)\Delta_1 + \cdots + f_r(t)\Delta_r, \tag{6.78}$$

where r depends only on μ and p, $\Delta_1, \ldots, \Delta_r$ are operators of orders not exceeding μ of the form (6.72) and $f_1, \ldots, f_r$ are polynomials with rational coefficients. Inserting this in the definition of the standard Wronskian and expressing the determinant in terms of other determinants, we obtain an expression for W of the form

$$W(t) = g_1(t)G_1(t, \ldots, t^{k^{p-1}}) + \cdots + g_s(t)G_s(t, \ldots, t^{k^{p-1}}), \tag{6.79}$$

where $G_1, \ldots, G_s$ are generalized Wronskians of $\phi_0, \ldots, \phi_{l-1}$ and $g_1(t), \ldots, g_s(t)$ are polynomials in t.

Exercise 6.4.8. *Verify the details of the last statement.*

Since $W(t)$ does not vanish identically, there is some i for which $G_i(t, \ldots, t^{k^{p-1}})$ does not vanish identically, and obviously $G_i(x_1, \ldots, x_p)$ does not vanish identically. □

6.4.3 More Properties

This lemma will be used in the induction step of Lemma 6.4.18.

Lemma 6.4.9. *Let $R(x_1, \ldots, x_p)$ be a polynomial in $p \geq 2$ variables with integral coefficients that is not identically zero. Let R be of degree at most r_j in the variable x_j, $1 \leq j \leq p$. Then there exists an integer l satisfying $1 \leq l \leq r_p + 1$ and differential operators $\Delta_0, \ldots, \Delta_{l-1}$ on $x_1, \ldots, x_{p-1}$ (with the order of Δ_ν at most ν) such that if*

$$F(x_1, \ldots, x_p) = \det\left(\Delta_\mu \frac{1}{\nu!}\frac{\partial^\nu}{\partial x_p^\nu}R\right)_{0 \leq \mu, \nu \leq l-1}, \tag{6.80}$$

then

1. *F has integral coefficients and F is not identically zero.*

2. *$F(x_1, \ldots, x_p) = u(x_1, \ldots, x_{p-1})v(x_p)$, u and v have integral coefficients with the degree of u at most lr_j ($1 \leq j \leq p-1$) and the degree of v is at most lr_p.*

Proof. Consider all representations of R of the form

$$R(x_1, \ldots, x_p) = \phi_0(x_p)\psi_0(x_1, \ldots, x_{p-1}) + \cdots + \phi_{l-1}(x_p)\psi_{l-1}(x_1, \ldots, x_{p-1}) \tag{6.81}$$

in such a way that for each $0 \leq i \leq l-1$ ϕ_i, ψ_i has rational coefficients and each ϕ_ν is of degree at most r_p and ψ_ν is of degree at most r_j in each x_j for $1 \leq j \leq p-1$. Such a representation is possible. We can collect common powers of x_p and factor out. In this case, $\phi_\nu(x_p) = x_p^\nu$, and clearly $l \leq r_p + 1$.

Choose the smallest l where we have a representation as in (6.81). We have seen this method of proof before, namely in Exercise 1.2.17 where we gave a non-constructive proof of the Euclidean algorithm. The reader is encouraged to review that exercise.

Claim 6.4.10. $\phi_0(x_p), \ldots, \phi_{l-1}(x_p)$ *are linearly independent, and* $\psi_0, \ldots, \psi_{l-1}$ *are linearly independent.*

Proof of the claim. Assume the ϕ_i are linearly dependent. By relabeling if necessary, we may assume ϕ_{l-1} can be written in terms of the others:

$$\phi_{l-1}(x_p) = d_0 \phi_0(x_p) + \cdots + d_{l-2} \phi_{l-2}(x_p), \quad d_i \in \mathbb{Q}. \quad (6.82)$$

Then

$$R(x_1, \ldots, x_p) = \phi_0(x_p) \left[\psi_0(x_1, \ldots, x_{p-1}) + d_0 \psi_{l-1}(x_1, \ldots, x_{p-1}) \right] + \cdots$$
$$+ \phi_{l-2}(x_p) \left[\psi_{l-2}(x_1, \ldots, x_{p-1}) + d_{l-2} \psi_{l-1}(x_1, \ldots, x_{p-1}) \right]. \quad (6.83)$$

We have written R in terms of $l-1$ summands, which contradicts the definition of l. A similar argument gives the linear independence of the ψ_i, proving the claim. $\square$

We now finish the proof of Lemma 6.4.9. Let $W(x_p)$ be the Wronskian of $\phi_0(x_p), \ldots, \phi_{l-1}(x_p)$. This is non-zero, has rational coefficients and can be expressed as a determinant. Let $G(x_1, \ldots, x_{p-1})$ be some non-vanishing generalized Wronskian of $\psi_0, \ldots, \psi_{l-1}$, say

$$G(x_1, \ldots, x_{p-1}) = \det \left(\Delta_\mu \psi_n u(x_1, \ldots, x_{p-1}) \right). \quad (6.84)$$

If A and B are any two $l \times l$ matrices then $\det(AB) = \det A \cdot \det B$. Applying this identity to G and W (which are determinants) and using matrix multiplication formulas to handle AB, we find

$$G(x_1, \ldots, x_{p-1})W(x_p) = \det \left(\Delta_\mu \psi_n u(x_1, \ldots, x_{p-1}) \right) \det \left(\frac{1}{\mu!} \frac{d^\mu}{dx^\mu} \phi_\nu(x) \right)$$
$$= \det \left(\sum_{\rho=0}^{l-1} \Delta_\mu \frac{1}{\nu!} \frac{\partial^\nu}{\partial x_p^\nu} \phi_\rho(x_p) \psi_\rho(x_1, \ldots, x_{p-1}) \right)$$
$$= \det \left(\Delta_\mu \frac{1}{\nu!} \frac{\partial^\nu}{\partial x_p^\nu} R \right) \not\equiv 0. \quad (6.85)$$

Since $G(x_1, \ldots, x_{p-1})W(x_p)$ is integral, there exists a rational number q such that $u = qG$ and $v = q^{-1}W$ both have integral coefficients. We leave it to the reader to verify the assertion about the degrees. $\square$

Lemma 6.4.11. *Let R be as above and suppose all the coefficients of R have absolute value bounded by B. Then all the coefficients of F are bounded by*

$$[(r_1 + 1) \cdots (r_p + 1)]^l \, l! B^l 2^{(r_1 + \cdots + r_p)l}, \quad l \le r + p + 1. \quad (6.86)$$

Exercise 6.4.12. *Prove the above lemma.*

6.4.4 The Set R_m and the Number Θ_m

The reader is advised to review the definition of index (see §6.3). Fix $r_1, \ldots, r_m$ and $B \geq 1$.

Definition 6.4.13 (R_m). *Let*

$$R_m = R_m(B; r_1, \ldots, r_m) \tag{6.87}$$

be the set of all polynomials $R(x_1, \ldots, x_m)$ such that

1. *R has integral coefficients and is* not *identically zero;*

2. *R is of degree at most r_j in x_j, $1 \leq j \leq m$;*

3. *the coefficients of R have absolute value at most B.*

Definition 6.4.14. *Let $q_1, \ldots, q_m$ be positive integers, $h_1, \ldots, h_m$ numbers relatively prime to $q_1, \ldots, q_m$, respectively. Let $\theta(R)$ denote the index of $R(x_1, \ldots, x_m)$ at the point $\left(\frac{h_1}{q_1}, \ldots, \frac{h_m}{q_m} \right)$ relative to $r_1, \ldots, r_m$. Set $\Theta_m(B, q_1, \ldots, q_m; r_1, \ldots, r_m)$ to be the supremum over $R \in R_m$ of $\theta(R)$ over all choices of $h_1, \ldots, h_m$.*

The goal of this section is to find nice bounds for the number Θ_m. We need the following result from algebra (see [Art, J, La3]):

Lemma 6.4.15 (Gauss' Lemma). *Let $f(x)$ be a polynomial with integer coefficients. Then we can factor $f(x)$ as a product of two polynomials of lower degree with rational coefficients if and only if we can factor f as a product of integer polynomials of lower degree.*

We state three technical lemmas to be used in the proof of Roth's Lemma (Lemma 6.3.6) in §6.5. We advise the reader to move directly to the statement of Lemma 6.4.20 in the first reading and skip over the details of proofs.

Lemma 6.4.16.

$$\Theta_1(B; q_1, r_1) \leq \frac{\log B}{r_1 \log q_1}. \tag{6.88}$$

Proof. From the definition of θ, $R(x_1)$ vanishes at $x_1 = \frac{h_1}{q_1}$ to order $\theta(R) r_1$. This implies $\left(x_1 - \frac{h_1}{q_1} \right)^{\theta(R) r_1}$ is a rational factor of $R(x_1)$. Therefore, as $(h_1, q_1) = 1$, by Gauss' lemma there is a polynomial $Q(x_1)$ with integer coefficients such that

$$R(x_1) = (q_1 x_1 - h_1)^{\theta r_1} Q(x_1). \tag{6.89}$$

This implies that the coefficient of the highest term in $R(x_1)$, which is bounded in absolute value by B, is a multiple of $q_1^{\theta r_1}$; thus

$$q_1^{\theta r_1} \leq B, \tag{6.90}$$

which yields the claim.   □

Exercise 6.4.17. *Prove that $(h_1, q_1) = 1$ allows us to conclude Q has integer coefficients.*

We will then prove a result which will be used in the proof of Lemma 6.4.20. The significance of this lemma, and its corollary Lemma 6.4.20, is that it will enable us to find bounds for the indices of a large class of polynomials of several variables. Observe that the statement of the lemma relates Θ_p to Θ_1 and Θ_{p-1} in an inductive fashion. The bounds obtained in Lemma 6.4.20 will then be used in the proof of Roth's Main Lemma (Lemma 6.3.6).

Lemma 6.4.18. *Let $p \geq 2$ be a positive integer. Let $r_1, \ldots, r_p$ be positive integers satisfying*

$$r_p > \frac{10}{\delta}, \quad \frac{r_{j-1}}{r_j} > \frac{1}{\delta}, \quad j = 2, \ldots, p, \tag{6.91}$$

where $0 < \delta < 1$ and $q_1, \ldots, q_p$ are positive integers. Then

$$\Theta_p(B; q_1, \ldots, q_p; r_1, \ldots, r_p) \leq 2 \max_l \left(\Phi + \Phi^{\frac{1}{2}} + \delta^{\frac{1}{2}} \right) \tag{6.92}$$

where $1 \leq l \leq r_p + 1$ and

$$\Phi = \Theta_1(M; q_p; lr_p) + \Theta_{p-1}(M; q_1, \ldots, q_{p-1}; lr_1, \ldots, lr_{p-1}),$$
$$M = (r_1 + 1)^{p^l} l! B^l 2^{p l r_1}. \tag{6.93}$$

Proof. Let R be a polynomial in $R_p(B; r_1, \cdots, r_p)$, and if $h_1, \ldots, h_p$ are integers relatively prime to $q_1, \ldots, q_p$, respectively, then we need to show that the index θ of R at $\left(\frac{h_1}{q_1}, \ldots, \frac{h_p}{q_p} \right)$ does not exceed the right hand side of (6.92). Construct F, u, v as in Lemma 6.4.9.

Exercise 6.4.19. *Use Lemma 6.4.11 to verify that the coefficients of F, u, v do not exceed M.*

The polynomial $u(x_1, \ldots, x_{p-1})$ has degree at most lr_j in x_j for $1 \leq j \leq p - 1$. In fact, $u \in R_{p-1}(M; lr_1, lr_2, \ldots, lr_{p-1})$. Hence its index at the point $\left(\frac{h_1}{q_1}, \ldots, \frac{h_p}{q_p} \right)$ relative to $(lr_1, \ldots, lr_{p-1})$ will not exceed

$$\Theta_{p-1}(M; q_1, \ldots, q_{p-1}; lr_1, \ldots, lr_{p-1}). \tag{6.94}$$

Consequently the index of u at the same point relative to $(r_1, \ldots, r_{p-1})$ will not exceed

$$l\Theta_{p-1}(M; q_1, \ldots, q_{p-1}; lr_1, \ldots, lr_{p-1}). \tag{6.95}$$

Similarly the index of $v(x_p)$ at $\frac{h_p}{q_p}$ relative to r_p does not exceed $l\Theta_1(M; q_p; lr_p)$. Therefore by Lemma 6.3.3

$$\text{ind}(F) = \text{ind}(u) + \text{ind}(v) \leq l\Phi, \tag{6.96}$$

with Φ as in the statement of the lemma. Now we need to deduce a bound for the index θ of R; for this, we use the determinant expression of F. Consider a differential operator of the form

$$\Delta = \frac{1}{i_1! \cdots i_p!} \left(\frac{\partial}{\partial x_1} \right)^{i_1} \cdots \left(\frac{\partial}{\partial x_p} \right)^{i_p} \tag{6.97}$$

of order $w = i_1 + \cdots + i_{p-1} \leq l - 1$. If the polynomial

$$\Delta \frac{1}{\nu!} \left(\frac{\partial}{\partial x_p} \right)^\nu R \tag{6.98}$$

is not identically zero, its index at $\left(\frac{h_1}{q_1}, \ldots, \frac{h_p}{q_p} \right)$ relative to $r_1, \ldots, r_p$ is at least

$$\theta - \sum_{j=1}^{p-1} \frac{i_j}{r_j} - \frac{\nu}{r_p} \geq \theta - \frac{w}{r_{p-1}} - \frac{\nu}{r_p}. \tag{6.99}$$

Next,

$$\frac{w}{r_{p-1}} \leq \frac{l-1}{r_{p-1}} \leq \frac{r_p}{r_{p-1}} < \delta \tag{6.100}$$

by assumption. Consequently, as the index is never negative, by Lemma 6.3.2 the index of the differentiated polynomial (6.98) must be at least

$$\max \left(0, \theta - \frac{\nu}{r_p} - \delta \right) \geq \max \left(0, \theta - \frac{\nu}{r_p} \right) - \delta. \tag{6.101}$$

If we expand the determinant expression of F, we find that F is a sum of $l!$ terms, each of which is of the form

$$\pm (\Delta_{\mu_0} R) \left(\Delta_{\mu_1} \frac{\partial}{\partial x_p} R \right) \cdots \left(\Delta_{\mu_{l-1}} \frac{1}{(l-1)!} \left(\frac{\partial}{\partial x_p} \right)^{l-1} R \right) \tag{6.102}$$

where Δ_{μ_i} are differential operators of orders at most $l - 1$. By Lemma 6.3.2, for any two polynomials P, Q we have $\operatorname{ind}(P + Q) \geq \min(\operatorname{ind}(P), \operatorname{ind}(Q))$. We therefore conclude that the index of F is bounded below by the index of any of the above terms, if non-zero. By the above arguments and basic properties of the index, we have

$$\operatorname{ind}(F) \geq \sum_{v=0}^{l-1} \max(0, \theta - \frac{\nu}{r_p}) - l\delta. \tag{6.103}$$

If $\theta \leq 10r_p^{-1}$ then as $10r_p^{-1} < \delta < 2\delta^{\frac{1}{2}}$ we get $\theta < 2\delta^{\frac{1}{2}}$, which is what we need to prove. So we may assume $\theta > 10r_p^{-1}$. If $\theta r_p < l$, we have

$$\sum_{v=0}^{l-1} \max \left(0, \theta - \frac{\nu}{r_p} \right) = r_p^{-1} \sum_{0 \leq \nu \leq r_p \theta} (r_p \theta - \nu)$$

$$\geq \frac{1}{2} r_p^{-1} [\theta r_p]^2$$

$$> \frac{1}{3} r_p \theta^2. \tag{6.104}$$

If $\theta r_p \geq l$ we have

$$\sum_{v=0}^{l-1} \max \left(0, \theta - \frac{\nu}{r_p} \right) = \sum_{v=0}^{l-1} \left(\theta - \frac{\nu}{r_p} \right) \geq \frac{1}{2} l\theta. \tag{6.105}$$

Combining everything we have proved so far, we obtain

$$\mathrm{ind}(F) \geq \min\left(\frac{1}{2}l\theta, \frac{1}{3}r_p\theta^2\right) - l\delta, \tag{6.106}$$

and consequently

$$\min\left(\frac{1}{2}l\theta, \frac{1}{3}r_p\theta^2\right) \leq l(\Phi + \delta). \tag{6.107}$$

We leave it to the reader to finish the proof of the lemma. $\qquad\square$

Lemma 6.4.18 is needed as the induction step in the proof of Lemma 6.4.20 below. Roth thought of Lemma 6.4.20 as the novel part of his proof (see [Ro], end of paragraph 1).

Lemma 6.4.20. *Let m be a positive integer, $0 < \delta < m^{-1}$, $r_1, \ldots, r_m$ integers with $r_m > \frac{10}{\delta}$ and for $j = 2, \ldots, m$ assume $\frac{r_{j-1}}{r_j} > \frac{1}{\delta}$. Also let $q_1, \ldots, q_m$ be positive integers with $\log q_1 > \frac{m(2m+1)}{\delta}$ and for $j = 2, \ldots, m$, $r_j \log q_j > r_1 \log q_1$. Then*

$$\Theta_m(q_1^{\delta r_1}; q_1, \ldots, q_m; r_1, \ldots, r_m) < 10^m \delta^{(\frac{1}{2})^m}. \tag{6.108}$$

Proof. We proceed by induction. The first step $m = 1$ follows from Lemma 6.4.16. Now suppose $p \geq 2$ is an integer, and the inequality of the lemma is valid for $m = p - 1$. The claim follows from Lemma 6.4.18, which is applicable. We leave the details to the reader. $\qquad\square$

Exercise 6.4.21. *Complete the proof of Lemmas 6.4.18 and 6.4.20.*

6.5 PROOF OF ROTH'S LEMMA

We are now (finally) able to prove Roth's Lemma (Lemma 6.3.6); we have already shown how Roth's Theorem immediately follows. Remember α is a non-rational algebraic integer and $f(x), A, m, \delta, q_1, h_1, \ldots, q_m, h_m$ satisfy (6.37) to (6.45).

Proof of Roth's Lemma. Let $\mathcal{W}$ be the set of all polynomials W of the form

$$W(x_1, \ldots, x_m) = \sum_{s_1=0}^{r_1} \cdots \sum_{s_m=0}^{r_m} c(s_1, \ldots, s_m)x_1^{s_1} \cdots x_m^{s_m} \tag{6.109}$$

with integer coefficients $c(s_1, \ldots, s_m)$ satisfying

$$0 \leq c(s_1, \ldots, s_m) \leq B_1. \tag{6.110}$$

The number of such polynomials equals $(1 + B_1)^r$, with

$$r = (r_1 + 1) \cdots (r_m + 1). \tag{6.111}$$

Consider the partial derivatives of W:

$$W_{j_1, \ldots, j_m}(x_1, \ldots, x_m) = \frac{1}{j_1! \ldots j_m!}\left(\frac{\partial}{\partial x_1}\right)^{j_1} \cdots \left(\frac{\partial}{\partial x_m}\right)^{j_m} W(x_1, \ldots, x_m). \tag{6.112}$$

Let $\mathcal{D}$ be the collection of all m-tuples of integers $j_1, \ldots, j_m$ such that

$$0 \le j_i \le r_i, \quad \frac{j_1}{r_1} + \cdots + \frac{j_m}{r_m} \le \gamma. \tag{6.113}$$

Let $D = |\mathcal{D}|$. By Lemma 6.4.1, $D \le \frac{2\sqrt{m}}{\lambda} r$.

Let $T_{j_1,\ldots,j_m}(W; x)$ be the remainder from dividing $W_{j_1,\ldots,j_m}(x, \ldots, x)$ by $f(x)$. The coefficients of W are bounded by B_1 and the coefficients of $W_{j_1,\ldots,j_m}(x_1, \ldots, x_m)$ are bounded by $2^{r_1 + \cdots + r_m} B_1 \le 2^{mr_1} B_1$, which is at most $B_1^{1+\delta}$. This follows from (6.40). When we put $x_1 = x_2 = \cdots = x_m = x$, we have a bound of $rB_1^{1+\delta} < B_1^{1+2\delta}$ for the coefficients of $W_{j_1,\ldots,j_m}(x, \ldots, x)$.

Exercise 6.5.1. *Verify this bound.*

We now find upper bounds for the size of the coefficients of $T_{j_1,\ldots,j_m}$. Write

$$W_{j_1,\ldots,j_m}(x, \ldots, x) = w_s x^s + w_{s-1} x^{s-1} + \cdots + w_0. \tag{6.114}$$

We divide $W_{j_1,\ldots,j_m}(x, \ldots, x)$ by $f(x)$. Recall f is the minimal polynomial of α. As we assume α is an algebraic integer, the leading coefficient of $f(x)$ is 1. Thus $f(x) = x^n + \cdots$. We may assume $s \ge n$ (as otherwise there is nothing to do). The first step in dividing $W_{j_1,\ldots,j_m}$ by $f(x)$ is to subtract $w_s x^{s-n} f(x)$ from $W_{j_1,\ldots,j_m}$; this gives a polynomial whose coefficients are of form $w_k - a_{s-k} w_s$ or w_k. Hence the absolute values of the coefficients of

$$W_{j_1,\ldots,j_m}(x, \ldots, x) - w_s x^{s-n} f(x) \tag{6.115}$$

are at most $B_1^{1+2\delta}(1 + A)$, where A is given by (6.36). Repeating the same procedure, combined with mathematical induction, implies that the coefficients of $T_{j_1,\ldots,j_m}(W; x)$ are bounded by

$$(1 + A)^{s-n+1} B_1^{1+2\delta}. \tag{6.116}$$

By equations (6.40) and (6.45), any such number is at most $B_1^{1+3\delta}$.

We now bound the number of possibilities for polynomials $T_{j_1,\ldots,j_m}(W; x)$ for all choices of $(j_1, \ldots, j_m) \in \mathcal{D}$. We know that each polynomial of the form $T_{j_1,\ldots,j_m}$ is a polynomial of degree at most $n - 1$, all of whose coefficients are integers bounded in absolute value by $B_1^{1+3\delta}$. It is not hard to see that the number of all polynomials of degree at most $n - 1$ all of whose coefficients are integers bounded in absolute value by a positive number K is at most $(1 + 2K)^n$. This simple fact implies that there is a set $\mathcal{S}$ of at most $\left(1 + 2B_1^{1+3\delta}\right)^n$ polynomials such that $T_{j_1,\ldots,j_m}(W; x) \in \mathcal{S}$, for all $W \in \mathcal{W}$ and all $(j_1, \ldots, j_m) \in \mathcal{D}$.

Next, we recall the definition of D immediately following (6.113). We have a map

$$\partial : \mathcal{W} \to \mathcal{S}^D \tag{6.117}$$

given by

$$W \mapsto (T_{j_1,\ldots,j_m}(W; x))_{(j_1,\ldots,j_m) \in \mathcal{D}}, \tag{6.118}$$

for $W \in \mathcal{W}$.

We have $|\mathcal{W}| = (1 + B_1)^r$ and $|\mathcal{S}^D| \leq (1 + 2B_1^{1+3\delta})^{nD}$. We then have

$$\left(1 + 2B_1^{1+3\delta}\right)^{nD} < (1 + B_1)^r. \tag{6.119}$$

By the Pigeon-Hole Principle (see §A.4), there are $W_1, W_2 \in \mathcal{W}$ such that $\partial(W_1) = \partial(W_2)$; i.e., all admissible derivatives of W_1 and W_2 have equal remainders when divided by $f(x)$. Let $W^* = W_1 - W_2$. Then $f | W^*_{j_1,\ldots,j_m}$ for all $(j_1, \ldots, j_m) \in \mathcal{D}$.

Exercise 6.5.2. *Prove* $W^* \in R_m(q_1^{\delta r_1}; r_1, \ldots, r_m)$.

By Lemma 6.4.20, we have a bound for the indices of the elements of R_m. We get that the index of W^* at $\left(\frac{h_1}{q_1}, \ldots, \frac{h_m}{q_m}\right)$ is at most η, defined by (6.44). Hence W^* possesses some derivative

$$Q(x_1, \ldots, x_m) = \frac{1}{k_1! \cdots k_m!} \left(\frac{\partial}{\partial x_1}\right)^{k_1} \cdots \left(\frac{\partial}{\partial x_m}\right)^{k_m} W^*(x_1, \ldots, x_m) \tag{6.120}$$

with

$$\frac{k_1}{r_1} + \cdots + \frac{k_m}{r_m} < \eta \tag{6.121}$$

such that

$$Q\left(\frac{h_1}{q_1}, \ldots, \frac{h_m}{q_m}\right) \neq 0. \tag{6.122}$$

Exercise 6.5.3. *Show that the polynomial Q satisfies the conditions of Lemma 6.3.6.*

This completes the proof of Roth's Lemma. $\square$

Exercise 6.5.4 (Hard). *Pick an algebraic integer with a simple minimal polynomial, e.g. $\sqrt[3]{2}$. Work out the details of the proof of Roth's theorem and lemma for this specific number. Can you simplify any part of the proof for this number? Fix an small positive number ϵ, say $\frac{1}{100}$. Give an upper bound for the number of rational numbers $\frac{p}{q}$, $(p, q) = 1$, that satisfy*

$$\left|\sqrt[3]{2} - \frac{p}{q}\right| < \frac{1}{q^{2 + \frac{1}{100}}}. \tag{6.123}$$

Chapter Seven

Introduction to Continued Fractions

For many problems (such as approximations by rationals and algebraicity), the continued fraction expansion of a number provides information that is hidden in the binary or decimal expansion. We will see many applications of this knowledge, for example in our investigations of digit bias in some sequences (Chapter 9) and in the behavior of the fractional parts of $n^k \alpha$ (Chapter 12). Unlike binary or decimal expansions where the digits are chosen from a finite set, *any* positive integer can be a digit in a continued fraction. While we expect a generic real decimal to have as many 1's as 2's as 3's (and so on), clearly such a result cannot be true for continued fractions; see Theorem 10.3.1 for how the digits are distributed. Many numbers (for example $\sqrt{2}$ and e) which have complicated decimal expansions have very simple and elegant continued fraction expansions. We develop the necessary machinery and then tackle these and other problems, culminating in several open problems in §7.10 and §10.6. Our treatment of the subject is heavily influenced by the excellent books [HW, La2]; see also [Kh, vdP1].

7.1 DECIMAL EXPANSIONS

There are many ways to represent numbers. A common way is to use decimal or base 10 expansions. For a positive real number x,

$$x = x_n 10^n + x_{n-1} 10^{n-1} + \cdots + x_1 10^1 + x_0 + x_{-1} 10^{-1} + x_{-2} 10^{-2} + \cdots$$
$$x_i \in \{0, 1, \ldots, 9\}. \tag{7.1}$$

Exercise[hr] **7.1.1.** *Let x have a periodic decimal expansion from some point onward, that is, there exists $N_0 \in \mathbb{N}$ and $a_1, \ldots, a_n \in \{0, \ldots, 9\}$ such that*

$$x = x_m x_{m-1} \cdots x_1 x_0 . x_{-1} \cdots x_{N_0+1} x_{N_0} a_1 \cdots a_n a_1 \cdots a_n a_1 \cdots a_n \cdots$$
$$= x_m x_{m-1} \cdots x_1 x_0 . x_{-1} \cdots x_{N_0+1} x_{N_0} \overline{a_1 \cdots a_n}. \tag{7.2}$$

Prove that x is rational, and bound the size of the denominator. Conversely, prove that if x is rational then its decimal expansion is either finite or eventually periodic. Thus x is rational if and only if its decimal expansion is finite or eventually periodic.

Recall $[x]$ is the largest integer less than or equal to x.

Exercise 7.1.2. *Find $[x]$ for $x = -2, 2.9, \pi$ and $\frac{29}{5}$. Does $[x + y] = [x] + [y]$? Does $[xy] = [x] \cdot [y]$?*

We calculate the decimal (or base 10) expansion of $x = 9.75$. As $[x] = [9.75] = 9$, the first digit is $x_0 = 9$. To retrieve the next digit, we study the difference $x - x_0$.

This is .75; if we multiply by 10, we get 7.5, and we note that the greatest integer less than or equal to 7.5 is 7. The next digit x_1 is $[10(x - x_0)] = 7$, and then $[10(10(x - x_0) - x_1)]$ gives the final digit, $x_2 = 5$. This construction is similar to Horner's algorithm (see §1.2.2), and a generalization of this method will lead to a new expansion for numbers, continued fractions.

Exercise 7.1.3. *Formally write down the procedure to find the base ten expansion of a positive number x. Discuss the modifications needed if x is negative.*

7.2 DEFINITION OF CONTINUED FRACTIONS

7.2.1 Introductory Remarks

A **Finite Continued Fraction** is a number of the form

$$a_0 + \cfrac{1}{a_1 + \cfrac{1}{a_2 + \cfrac{1}{\ddots + \cfrac{1}{a_n}}}}, \qquad a_i \in \mathbb{R}. \qquad (7.3)$$

As n is finite, the above expression makes sense provided we never divide by 0. Since this notation is cumbersome to write, we introduce the following shorthand notations. The first is

$$a_0 + \frac{1}{a_1 +} \frac{1}{a_2 +} \cdots \frac{1}{a_n}. \qquad (7.4)$$

A more common notation, which we often use, is

$$[a_0, a_1, \ldots, a_n]. \qquad (7.5)$$

Remark 7.2.1. Later we discuss infinite continued fractions. While we can formally consider expansions such as

$$[a_0, a_1, a_2, \ldots], \qquad (7.6)$$

we need to show such expressions converge to a unique number, and any number has a unique expansion. Note the "numerators" are always one (but see §7.8.2 for generalizations).

Exercise 7.2.2. *Show $[a_0] = a_0$, $[a_0, a_1] = a_0 + \frac{1}{a_1} = \frac{a_0 a_1 + 1}{a_1}$, and $[a_0, a_1, a_2] = \frac{a_0(a_1 a_2 + 1) + a_2}{a_1 a_2 + 1}$.*

Definition 7.2.3 (Positive Continued Fraction). *A continued fraction $[a_0, \ldots, a_n]$ is positive if each $a_i > 0$ for $i \geq 1$.*

Definition 7.2.4 (Digits). *If $\alpha = [a_0, \ldots, a_n]$ we call the a_i the digits of the continued fraction. Note some books call a_i the ith partial quotient of α.*

Definition 7.2.5 (Simple Continued Fraction). *A continued fraction is simple if for each $i \geq 1$, a_i is a positive integer.*

Below we mostly concern ourselves with simple continued fractions; however, in truncating infinite simple continued fractions we encounter expansions which are simple except for the last digit.

Definition 7.2.6 (Convergents). *Let $x = [a_0, a_1, \ldots, a_n]$. For $m \le n$, set $x_m = [a_0, \ldots, a_m]$. Then x_m can be written as $\frac{p_m}{q_m}$, where p_m and q_m are polynomials in $a_0, a_1, \ldots, a_m$. The fraction $x_m = \frac{p_m}{q_m}$ is the mth convergent of x.*

We will see that continued fractions have many advantages over decimal expansion, and some disadvantages. Clearly any finite continued fraction with integer digits is a rational number. The surprising fact is that the converse of this statement is also valid, that is, rational numbers have finite continued fractions. In contrast with periodic decimal expansions (which are just rational numbers), periodic continued fractions encapsulate more information: they are solutions of quadratic equations with rational coefficients (and the converse of this statement is also true).

Many complicated numbers, such as e, have very simple continued fraction expansions. Using continued fractions we obtain very interesting results on how to approximate numbers by rational numbers. If we truncate the decimal expansion of a number, the resulting rational number is an approximation to the original number. We can do this with a simple continued fraction as well: we truncate it at some point and obtain a rational number, and use that rational to approximate the original number. We will see that this gives the best rational approximation; we will, of course, quantify what we mean by best approximation. This is very similar to Fourier series (see Chapter 11) and Taylor series (see §A.2.3): for a given expansion, the first n terms of a Fourier Series (or Taylor series) give the best approximation of a certain order to the given function. But how does one find the continued fraction expansion of $x \in \mathbb{R}$?

7.2.2 Calculating Continued Fractions

We first describe the formal process to calculate simple continued fractions. We want to find integers a_i (all positive except possibly for a_0) such that

$$x = a_0 + \cfrac{1}{a_1 + \cfrac{1}{a_2 + \cdots}}. \tag{7.7}$$

Obviously $a_0 = [x]$, the greatest integer at most x. Then

$$x - [x] = \cfrac{1}{a_1 + \cfrac{1}{a_2 + \cdots}}, \tag{7.8}$$

and the inverse is

$$x_1 = \cfrac{1}{x - [x]} = a_1 + \cfrac{1}{a_2 + \cfrac{1}{a_3 + \cdots}}. \tag{7.9}$$

Therefore the next digit of the continued fraction expansion is $[x_1] = a_1$. Then $x_2 = \frac{1}{x_1 - [x_1]}$, and $[x_2] = a_2$, and so on.

Exercise 7.2.7. *Formally write down the procedure to find the continued fraction expansion of $x \in \mathbb{R}$. Note: By introducing two operators $I(x) = \frac{1}{x}$ and $F(x) = [x]$, one can write the procedure in a manner similar to Horner's algorithm (see §1.2.2 and §7.1).*

Exercise 7.2.8. *Find the first few terms in the continued fraction expansions of $\sqrt{2}$, $\sqrt{3}$, π and e. How accurately does one need to know these numbers? In Theorem 7.7.1 we give an efficient construction for many numbers.*

Exercise(hr) 7.2.9. *Use the Euclidean algorithm (see §1.2.3) to quickly determine the continued fraction of $\frac{p}{q} \in \mathbb{Q}$. Is it finite? If so, how does the number of digits depend on p and q? Prove x is rational if and only if the continued fraction expansion is finite. Modify the method to handle additional numbers such as $\sqrt{n}$.*

Exercise 7.2.10. *Show that any positive integer k may occur as the value of a_1 in a continued fraction expansion. Similarly it may occur as a_2 (or a_3, a_4, ...). What is the probability that the first digit of a continued fraction is k? If the first digit is 1, what is the probability the second digit is k? Harder: what is the probability that the second digit is k without any restrictions on the first digit? Compare with similar statements for decimal digits.*

We will discuss the last exercise and its generalizations in great detail in Chapter 10. In §12.8.1 we show a connection between large square factors in the convergents arising from the continued fraction expansion of α and the distribution of $n^2\alpha \bmod 1$.

7.3 REPRESENTATION OF NUMBERS BY CONTINUED FRACTIONS

Decimal expansions are not unique. For example, 3.49999... = 3.5. We adopt the convention that we throw away any decimal expansion ending with all 9's and replace it with the appropriate expansion. A similar phenomenon occurs with continued fractions. In fact for continued fractions the situation is slightly more interesting. While ambiguities for decimal expansion can occur only when dealing with infinite expansions, for continued fractions such ambiguities occur only for finite continued fractions.

Lemma 7.3.1. *Let $x = [a_0, \dots, a_N]$ be a simple continued fraction. If N is odd there is another simple continued fraction which also equals x but with an even number of terms (and vice versa).*

Proof. If $a_N > 1$, $[a_0, \dots, a_N] = [a_0, \dots, a_N - 1, 1]$. For example, if $a_1 > 1$,

$$a_0 + \frac{1}{a_1} = a_0 + \cfrac{1}{(a_1 - 1) + \cfrac{1}{1}}. \tag{7.10}$$

If $a_N = 1$, then $[a_0, \dots, a_N] = [a_0, \dots, a_{N-1} + 1]$.   □

Consider $[a_0, a_1, a_2, \ldots, a_N]$. Define $a_n' = [a_n, \ldots, a_N]$, the tail of the continued fraction. Then $[a_0, \ldots, a_N] = [a_0, \ldots, a_{n-1}, a_n']$; however, the second continued fraction is positive but not necessarily simple (as a_n' need not be an integer).

Theorem 7.3.2. *Suppose $[a_0, \ldots, a_N]$ is positive and simple. Then $[a_n'] = a_n$ except when both $n = N - 1$ and $a_N = 1$, in which case $[a_{N-1}'] = a_{N-1} - 1$.*

Proof. a_n' is a continued fraction given by

$$a_n' = a_n + \cfrac{1}{a_{n+1} + \cfrac{1}{\ddots}}. \tag{7.11}$$

We just need to make sure that

$$\cfrac{1}{a_{n+1} + \cfrac{1}{\ddots}} < 1. \tag{7.12}$$

How could this equal 1 or more? The only possibility is if $a_{n+1} = 1$ and the sum of the remaining terms is 0. This happens only if both $n = N - 1$ and $a_N = 1$. $\square$

Uniqueness Assumption (Notation): whenever we write a finite continued fraction, we assume $a_N \neq 1$, where N corresponds to the last digit. This is similar to notation from decimal expansions.

Lemma 7.3.3. *Let $[a_0, \ldots, a_n]$ be a continued fraction. Then*

1. $[a_0, \ldots, a_n] = \left[a_0, \ldots, a_{n-2}, a_{n-1} + \frac{1}{a_n}\right].$

2. $[a_0, \ldots, a_n] = [a_0, \ldots, a_{m-1}, [a_m, \ldots, a_n]].$

These are the most basic properties of continued fractions, and will be used frequently below.

Exercise 7.3.4. *Prove Lemma 7.3.3.*

7.3.1 Convergents to a Continued Fraction

Direct calculation shows that $[a_0] = a_0$, $[a_0, a_1] = \frac{a_0 a_1 + 1}{a_1}$, and $[a_0, a_1, a_2] = \frac{a_0(a_1 a_2 + 1) + a_2}{a_1 a_2 + 1}$. In general we have $\frac{p_n}{q_n} = \frac{p_n(a_0, \ldots, a_n)}{q_n(a_0, \ldots, a_n)}$ where p_n and q_n are polynomials with integer coefficients of $a_0, a_1, \ldots, a_n$.

Theorem 7.3.5 (Recursion Relations). *For any $m \in \{2, \ldots, n\}$ we have*

1. $p_0 = a_0$, $p_1 = a_0 a_1 + 1$, and $p_m = a_m p_{m-1} + p_{m-2}$ for $m \geq 2$.

2. $q_0 = 1$, $q_1 = a_1$, and $q_m = a_m q_{m-1} + q_{m-2}$ for $m \geq 2$.

Proof. We proceed by induction (see §A.1 for a review of proofs by induction). First we check the basis case; we actually need to check $n = 0$ and $n = 1$ for the induction because the recurrence relations for m involve *two* previous terms; we check $n = 2$ as well to illuminate the pattern.

By definition $[a_0] = \frac{a_0}{1}$ which is $\frac{p_0}{q_0}$; $[a_0, a_1] = \frac{a_0 a_1 + 1}{a_1}$ and this agrees with $\frac{p_1}{q_1}$. We must show $[a_0, a_1, a_2] = \frac{a_2 p_1 + p_0}{a_2 q_1 + q_0}$. As $p_1 = a_0 a_1 + 1$ and $q_1 = a_1$, direct substitution gives $\frac{a_0(a_0 a_1 + 1) + a_2}{a_2 a_1 + 1}$.

We now proceed by induction on the number of digits. We must show that if

$$[a_0, \ldots, a_m] = \frac{p_m}{q_m} = \frac{a_m p_{m-1} + p_{m-2}}{a_m q_{m-1} + q_{m-2}} \tag{7.13}$$

then

$$[a_0, \ldots, a_{m+1}] = \frac{p_{m+1}}{q_{m+1}} = \frac{a_{m+1} p_m + p_{m-1}}{a_{m+1} q_m + q_{m-1}}. \tag{7.14}$$

We calculate the continued fraction of $x = [a_0, \ldots, a_m, a_{m+1}]$. By Lemma 7.3.3, this is the same as the continued fraction of $y = [a_0, \ldots, a_{m-1}, a_m + \frac{1}{a_{m+1}}]$. Note, of course, that $x = y$; we use different letters to emphasize that x has a continued fraction expansion with $m + 2$ digits, and y has a continued fraction expansion with $m + 1$ digits (remember we start counting with a_0).

We consider the continued fraction of y; it will have its own expansion with numerator P_m and denominator Q_m. By induction (we are assuming we know the theorem for all continued fractions with $m + 1$ digits), $y = \frac{P_m}{Q_m}$. Therefore

$$\frac{P_m}{Q_m} = \frac{\left(a_m + \frac{1}{a_{m+1}}\right) P_{m-1} + P_{m-2}}{\left(a_m + \frac{1}{a_{m+1}}\right) Q_{m-1} + Q_{m-2}}. \tag{7.15}$$

The first m digits $(a_0, \ldots, a_{m-1})$ of y are the same as those of x. Thus $P_{m-1} = p_{m-1}$, and similarly for Q_{m-1}, P_{m-2}, and Q_{m-2}. Substituting gives

$$\frac{P_m}{Q_m} = \frac{\left(a_m + \frac{1}{a_{m+1}}\right) p_{m-1} + p_{m-2}}{\left(a_m + \frac{1}{a_{m+1}}\right) q_{m-1} + q_{m-2}}. \tag{7.16}$$

Simple algebra yields

$$\frac{P_m}{Q_m} = \frac{(a_m a_{m+1} + 1) p_{m-1} + p_{m-2} a_{m+1}}{(a_m a_{m+1} + 1) q_{m-1} + q_{m-2} a_{m+1}}. \tag{7.17}$$

This is the same as

$$\frac{a_{m+1}(a_m p_{m-1} + p_{m-2}) + p_{m-1}}{a_{m+1}(a_m q_{m-1} + q_{m-2}) + q_{m-1}} = \frac{a_{m+1} p_m + p_{m-1}}{a_{m+1} q_m + q_{m-1}}, \tag{7.18}$$

where the last step (substituting in with p_m and q_m) follows from the inductive assumption. This completes the proof. $\square$

Example 7.3.6. *An important example is* $[1, 1, \ldots, 1] = \frac{p_n}{q_n}$, *where we have* $n + 1$ *ones. What are the* p_i's *and the* q_i's? *We have* $p_0 = 1$, $p_1 = 2$, $p_m = p_{m-1} + p_{m-2}$. *Similarly, we have* $q_0 = 1$, $q_1 = 1$, *and* $q_m = q_{m-1} + q_{m-2}$. *Let* F_m *be the* m^{th} *Fibonacci number:* $F_0 = 1$, $F_1 = 1$, $F_2 = 2$, $F_3 = 3$, *and* $F_m = F_{m-1} + F_{m-2}$. *Thus* $[1, 1, \ldots, 1] = \frac{F_{n+1}}{F_n}$. *As* $n \to \infty$, $\frac{p_n}{q_n} \to \frac{1+\sqrt{5}}{2}$, *the golden mean. Notice how* *beautiful continued fractions are. A simple expression like this captures the golden mean, which has many deep, interesting properties (see Exercise 7.9.6). In base ten, .111111 ... is the not so very interesting $\frac{1}{9}$. See [Kos] for much more on the Fibonacci numbers.*

Exercise 7.3.7. *Let $[a_0, a_1, \ldots, a_n]$ be a continued fraction. Prove*

$$\begin{pmatrix} a_0 & 1 \\ 1 & 0 \end{pmatrix} \begin{pmatrix} a_1 & 1 \\ 1 & 0 \end{pmatrix} \cdots \begin{pmatrix} a_n & 1 \\ 1 & 0 \end{pmatrix} = \begin{pmatrix} p_n & p_{n-1} \\ q_n & q_{n-1} \end{pmatrix}. \tag{7.19}$$

What is the determinant?

Exercise 7.3.8. *Notation as in the previous exercise, prove that*

$$\frac{p_n}{q_n} = a_0 + \sum_{k=1}^{n} \frac{(-1)^{k-1}}{q_{k-1} q_k}. \tag{7.20}$$

Exercise 7.3.9 (Recurrence Relations). *Let $\alpha_0, \ldots, \alpha_{k-1}$ be fixed integers and consider the recurrence relation of order k*

$$x_{n+k} = \alpha_{k-1} x_{n+k-1} + \alpha_{k-2} x_{n+k-2} + \cdots + \alpha_1 x_{n+1} + \alpha_0 x_n. \tag{7.21}$$

Show once k values of x_m are specified, all values of x_n are determined. Let

$$f(r) = r^k - \alpha_{k-1} r^{k-1} - \cdots - \alpha_0; \tag{7.22}$$

we call this the characteristic polynomial of the recurrence relation. Show if $f(\rho) = 0$ then $x_n = c\rho^n$ satisfies the recurrence relation for any $c \in \mathbb{C}$. If $f(r)$ has k distinct roots $r_1, \ldots, r_k$, show that any solution of the recurrence equation can be represented as

$$x_n = c_1 r_1^n + \cdots + c_k r_k^n \tag{7.23}$$

for some $c_i \in \mathbb{C}$. The Initial Value Problem is when k values of x_n are specified; using linear algebra, this determines the values of $c_1, \ldots, c_k$. Investigate the cases where the characteristic polynomial has repeated roots. For more on recursive relations, see §9.3, Exercise 8.1.18 and [GKP], §7.3.

Exercise 7.3.10. *Solve the Fibonacci recurrence relation $F_{n+2} = F_{n+1} + F_n$, given $F_0 = F_1 = 1$. Show F_n grows exponentially, i.e., F_n is of size r^n for some $r > 1$. What is r? Let $r_n = \frac{F_{n+1}}{F_n}$. Show that the even terms r_{2m} are increasing and the odd terms r_{2m+1} are decreasing. Investigate $\lim_{n \to \infty} r_n$ for the Fibonacci numbers. Show r_n converges to the golden mean, $\frac{1+\sqrt{5}}{2}$. See [PS2] for a continued fraction involving Fibonacci numbers.*

Exercise 7.3.11 (Binet's Formula). *For F_n as in the previous exercise, prove*

$$F_{n-1} = \frac{1}{\sqrt{5}} \left[\left(\frac{1+\sqrt{5}}{2} \right)^n - \left(\frac{1-\sqrt{5}}{2} \right)^n \right]. \tag{7.24}$$

This formula should be surprising at first: F_n is an integer, but the expression on the right involves irrational numbers and division by 2. More generally, for which positive integers m is

$$\frac{1}{\sqrt{m}} \left[\left(\frac{1+\sqrt{m}}{2} \right)^n - \left(\frac{1-\sqrt{m}}{2} \right)^n \right] \tag{7.25}$$

an integer for any positive integer n?

Exercise 7.3.12. *Let $x = [a_0, \ldots, a_n]$ be a simple continued fraction. For $m \geq 1$, show $q_m \geq F_m$; therefore, the q_m's grow exponentially. Find a number $c > 1$ such that for any simple continued fraction, $q_n \geq c^n$. Note the same observations apply to the p_m's.*

7.3.2 Fundamental Recurrence Relation

In Chapter 5 we investigated how well numbers can be approximated by rational $\frac{p}{q}$ with p, q relatively prime. One of the most important properties of continued fractions is that the convergents of x are good rational approximations to x. We know $\frac{p_n}{q_n} = \frac{a_n p_{n-1} + p_{n-2}}{a_n q_{n-1} + q_{n-2}}$. In Exercise 7.3.17 we show p_n and q_n are relatively prime.

To do so we need some preliminaries which are useful and interesting in their own right. Consider the difference $p_n q_{n-1} - p_{n-1} q_n$. Using the recursion relations in Theorem 7.3.5, this difference also equals

$$(a_n p_{n-1} + p_{n-2}) q_{n-1} - p_{n-1}(a_n q_{n-1} + q_{n-2}). \tag{7.26}$$

This is the same (expand and cancel) as $p_{n-2} q_{n-1} - p_{n-1} q_{n-2}$.

The key observation is that $p_n q_{n-1} - p_{n-1} q_n = -(p_{n-1} q_{n-2} - p_{n-2} q_{n-1})$. The index has been reduced by one, and there has been a sign change. Repeat, and we obtain $(-1)^2 (p_{n-2} q_{n-3} - p_{n-3} q_{n-2})$. After $n - 1$ iterations we find $(-1)^{n-1}(p_1 q_0 - p_0 q_1)$. Substituting $p_1 = a_0 a_1 + 1$, $q_1 = a_1$, $p_0 = a_0$ and $q_0 = 1$ yields

Lemma 7.3.13 (Fundamental Recurrence Relation).

$$p_n q_{n-1} - p_{n-1} q_n = (-1)^{n-1}. \tag{7.27}$$

So, even though a priori this difference should depend on a_0 through a_n, it is in fact just -1 to a power. See Exercise 7.3.7 for an alternate proof. Similarly, one can show

Lemma 7.3.14.

$$p_n q_{n-2} - p_{n-2} q_n = (-1)^n a_n. \tag{7.28}$$

Exercise 7.3.15. *Prove Lemma 7.3.14. Is there a nice formula for $p_n q_{n-k} - p_{n-k} q_n$?*

Notice that the consecutive convergents to the continued fraction $\frac{p_{n-2}}{q_{n-2}}, \frac{p_{n-1}}{q_{n-1}}$ and $\frac{p_n}{q_n}$ satisfy

Lemma 7.3.16. *We have*

$$\frac{p_n}{q_n} - \frac{p_{n-1}}{q_{n-1}} = \frac{(-1)^{n-1}}{q_n q_{n-1}} \tag{7.29}$$

and

$$\frac{p_n}{q_n} - \frac{p_{n-2}}{q_{n-2}} = \frac{(-1)^n a_n}{q_n q_{n-2}}. \tag{7.30}$$

Proof. Divide (7.27) by $q_n q_{n-1}$ and (7.28) by $q_n q_{n-2}$. □

Exercise 7.3.17 (Important). *Show p_n and q_n are relatively prime. If each a_i is a positive integer, show q_n is at least as large as the nth Fibonacci number. Thus the denominators corresponding to simple continued fractions grow at least exponentially.*

7.3.3 Continued Fractions with Positive Terms

Theorem 7.3.18. *For $x = [a_0, \dots, a_n]$, let $x_m = \frac{p_m}{q_m}$ be the mth convergent. If the digits are positive, then $\{x_{2m}\}$ is an increasing sequence, $\{x_{2m+1}\}$ is a decreasing sequence, and for every m, $x_{2m} < x < x_{2m+1}$ (if $n \neq 2m$ or $2m+1$).*

Proof. Note $\{x_{2m}\}$ increasing means $x_0 < x_2 < x_4 < \dots$. By Lemma 7.3.16,

$$x_{2(m+1)} - x_{2m} = \frac{(-1)^{2m} a_{2m}}{q_{2m} q_{2(m+1)}}. \tag{7.31}$$

Everything on the right hand side is positive, so $x_{2(m+1)} > x_{2m}$. The result for the odd terms is proved similarly; there we will have $(-1)^{2m+1}$ instead of $(-1)^{2m}$.

We assume $n = 2k$; the proof when $n = 2k+1$ follows similarly. Clearly $x_2 < x_4 < \dots < x_{2k} = x_n$. By Lemma 7.3.16 (with $n = 2k$ in the lemma), $x_{2k} < x_{2k-1}$. As $x_{2k-1} < x_{2k-3} < \dots < x_1$, we have $x_{2m} \leq x_n < x_{2m+1}$ (with strict inequality if $2m \neq n$). □

7.3.4 Uniqueness of Continued Fraction Expansions

Theorem 7.3.19 (Uniqueness of Continued Fraction Expansion). *Let $x = [a_0, \dots, a_N] = [b_0, \dots, b_M]$ be continued fractions with $a_N, b_M > 1$. Then $N = M$ and $a_i = b_i$.*

Proof. We proceed by induction on $N + M$. If N or M is 1 the claim is clear. Thus we may assume $N, M \geq 2$. By definition $a_0 = [x]$, $b_0 = [x]$. As $[a_0, \dots, a_N] = [b_0, \dots, b_M]$,

$$[[x], [a_1, \dots, a_N]] = [[x], [b_1, \dots, b_M]]. \tag{7.32}$$

Then

$$[x] + \frac{1}{[a_1, \dots, a_N]} = [x] + \frac{1}{[b_1, \dots, b_M]}. \tag{7.33}$$

Thus $[a_1, \dots, a_N] = [b_1, \dots, b_M]$. As $(N-1) + (M-1) < N+M$, the proof follows by induction. □

Theorem 7.3.20. *A number is rational if and only if its continued fraction expansion is finite.*

Proof. Clearly if the continued fraction expansion is finite then the number is rational. We use the Euclidean algorithm (see §1.2.3) for the other direction.

Without loss of generality we may assume $x = \frac{p}{q} > 0$. By the Euclidean algorithm, $p = a_0 q + r_1$ with $a_0, r_1 \in \mathbb{N}$ and $0 \leq r_1 < q$. Thus

$$x = \frac{a_0 q + r_1}{q} = a_0 + \frac{r_1}{q} = a_0 + \frac{1}{q/r_1}. \tag{7.34}$$

Applying the Euclidean algorithm again yields $q = a_1 r_1 + r_2$ with $a_1, r_1 \in \mathbb{N}$ and $0 \le r_2 < r_1$. Similar arguments give

$$\frac{q}{r_1} = a_1 + \frac{1}{r_1/r_2}, \tag{7.35}$$

which implies

$$x = a_0 + \cfrac{1}{a_1 + \cfrac{1}{r_1/r_2}}. \tag{7.36}$$

Note the numerator (resp., denominator) of $\frac{r_1}{r_2}$ is less than the numerator (resp., denominator) of $\frac{p}{q}$. By Lemma 1.2.14 the Euclidean algorithm terminates after finitely many steps. There are two possibilities: either eventually an $r_k = 0$, in which case $x = [a_0, a_1, \ldots, a_{k-1}]$, or eventually an $r_k = 1$, in which case we have $x = [a_0, a_1, \ldots, a_{k-1}, r_{k-1}]$. $\qquad\square$

Remark 7.3.21 (Important). Note the above gives a constructive proof that a rational number has a finite continued fraction expansion. In practice, for numbers approximated by truncating infinite series expansions it is often difficult to determine if the number is rational or not because with any truncation there will be a small but finite error. For example, $\frac{1}{\pi^2}\sum_{n=1}^{\infty}\frac{1}{n^2} = \frac{1}{6}$; see Exercises 3.1.7 and 11.3.15. One indicator for rationality is to look at the continued fraction expansion of the observed approximation. If the true number is rational and the approximate value has a large digit in its continued fraction expansion, replacing the large digit with infinity suggests a rational value for the number. See [BP, QS1] and the references there for more details.

Exercise[h] 7.3.22. *Let $x = \frac{p}{q}$ be rational. What can one say about its decimal expansion? If it is periodic, what is the relation between the length of the period and q? See also Exercise 7.1.1.*

7.4 INFINITE CONTINUED FRACTIONS

Definition 7.4.1. *By $[a_0, a_1, \ldots]$ we mean*

$$\lim_{n\to\infty} [a_0, a_1, \ldots, a_n] = \lim_{n\to\infty} \frac{p_n}{q_n}. \tag{7.37}$$

If the a_i are positive integers and the limit exists, we say the above is an **(infinite) simple continued fraction**. We show the limit exists for *any* sequence of positive integers a_i.

Exercise 7.4.2. *Prove*

1. *The even convergents $\frac{p_{2n}}{q_{2n}}$ are increasing and bounded.*

2. *The odd convergents $\frac{p_{2n+1}}{q_{2n+1}}$ are decreasing and bounded.*

Theorem 7.4.3. *Let* $[a_0, \ldots, a_n]$ *be a simple continued fraction. Then*

1. $q_n \geq q_{n-1}$ *for all* $n \geq 1$, *and* $q_n > q_{n-1}$ *if* $n > 1$.

2. $q_n \geq n$, *with strict inequality if* $n > 3$.

Proof. Recall $q_0 = 1$, $q_1 = a_1 \geq 1$, $q_n = a_n q_{n-1} + q_{n-2}$. Each $a_n > 0$ and is an integer. Thus $a_n \geq 1$ and $a_n q_{n-1} + q_{n-2} \geq q_{n-1}$, yielding $q_n \geq q_{n-1}$. If $n > 1$ then $q_{n-2} > 0$, giving a strict inequality. We prove the other claim by induction. Suppose $q_{n-1} \geq n - 1$. Then $q_n = a_n q_{n-1} + q_{n-2} \geq q_{n-1} + q_{n-2} \geq (n-1) + 1 = n$. If at one point the inequality is strict, it is strict from that point onward. $\square$

Theorem 7.4.4. *Let* $[a_0, \ldots, a_n]$ *be a simple continued fraction with quotient* $\frac{p_n}{q_n}$. *Then* $\frac{p_n}{q_n}$ *is reduced; i.e.,* p_n *and* q_n *are relatively prime. Further,* $q_n^2 > 2^{n-1}$ *for* $n > 1$.

Proof. Assume not, and let $d|p_n$ and $d|q_n$. Then $d|(p_n q_{n-1} - q_n p_{n-1})$. By Lemma 7.3.13, $p_n q_{n-1} - q_n p_{n-1} = (-1)^{n-1}$. Thus $d|(-1)^{n-1}$, which implies $d = \pm 1$ and $\frac{p_n}{q_n}$ is reduced.

For the second claim we proceed by induction, recalling that $q_0 = 1$, $q_1 = a_1$, and $q_m = a_m q_{m-1} + q_{m-2}$. Thus $q_2 \geq 1q_1 + q_0 \geq 2$, proving the basis case.

Assume $q_{n-1}^2 > 2^{n-1}$. As $q_{n+1} = a_{n+1} q_n + q_{n-1} \geq q_n + q_{n-1} > 2q_{n-1}$, we have

$$q_{n+1}^2 > (2q_{n-1})^2 > 2^2 2^{(n-1)-1} = 2^n, \tag{7.38}$$

completing the proof; see also Exercise 7.3.12. $\square$

Theorem 7.4.5. *The limit*

$$\lim_{n \to \infty} [a_0, a_1, \ldots, a_n] \tag{7.39}$$

exists for any sequence $\{a_i\}_{i=1}^{\infty}$ *of positive integers.*

Proof. By Exercise 7.4.2 the even convergents $\frac{p_{2n}}{q_{2n}}$ converge (say to A) and the odd convergents $\frac{p_{2n-1}}{q_{2n-1}}$ converge (say to B). We need only show $A = B$, or, equivalently,

$$\left| \frac{p_{2n}}{q_{2n}} - \frac{p_{2n-1}}{q_{2n-1}} \right| \longrightarrow 0. \tag{7.40}$$

By Lemma 7.3.13, $p_{2n} q_{2n-1} - p_{2n-1} q_{2n} = (-1)^{2n-1}$. Therefore

$$\left| \frac{p_{2n}}{q_{2n}} - \frac{p_{2n-1}}{q_{2n-1}} \right| = \left| \frac{p_{2n} q_{2n-1} - p_{2n-1} q_{2n}}{q_{2n} q_{2n-1}} \right| = \frac{1}{q_{2n} q_{2n-1}} \leq \frac{1}{2n(2n-1)} \longrightarrow 0, \tag{7.41}$$

as by Lemma 7.4.3, $q_n \geq n$. Thus $A = B$. $\square$

Theorem 7.4.6 (Existence and Uniqueness Theorem). *Let* $[a_0, a_1, \ldots]$ *be the continued fraction expansion of* x. *Then* $x = [a_0, a_1, \ldots]$. *If* x *also equals* $[b_0, b_1, \ldots]$, *then* $a_i = b_i$.

As there is no final digit, we do not need to worry about the ambiguity in the last digit. This is markedly different from the slight non-uniqueness in finite continued fraction expansions.

Exercise 7.4.7. *Prove Theorem 7.4.6.*

7.5 POSITIVE SIMPLE CONVERGENTS AND CONVERGENCE

Similar to an infinite decimal expansion, the infinite continued fraction of x converges to x. We investigate the rate of convergence. The speed of the convergence is one of the great assets of continued fractions. For simplicity we assume we have an infinite continued fraction so we do not need to worry about trivial modifications at the last digit. Thus $x \notin \mathbb{Q}$.

Let $x = [a_0, \ldots, a_n, a_{n+1}, \ldots] = [a_0, a_1, \ldots, a_n, a'_{n+1}]$; a'_{n+1} arises from truncating the continued fraction and is at least 1. Then

$$x = \frac{a'_{n+1} p_n + p_{n-1}}{a'_{n+1} q_n + q_{n-1}}. \tag{7.42}$$

How large is $\left| x - \frac{p_n}{q_n} \right|$, the difference between x and the nth convergent?

$$\left| x - \frac{p_n}{q_n} \right| = \frac{a'_{n+1} p_n + p_{n-1}}{a'_{n+1} q_n + q_{n-1}} - \frac{p_n}{q_n}$$

$$= \frac{p_{n-1} q_n - p_n q_{n-1}}{q_n (a'_{n+1} q_n + q_{n-1})}$$

$$= \frac{(-1)^n}{q_n q'_{n+1}} \tag{7.43}$$

as $q'_1 = a'_1$, $q'_n = a'_n q_{n-1} + q_{n-2}$, and by Lemma 7.3.13, $p_{n-1} q_n - p_n q_{n-1} = (-1)^n$. We now determine the size of q'_{n+1}. Note $a_{n+1} < a'_{n+1} < a_{n+1} + 1$ and the inequalities are strict as x is irrational. Thus

$$q'_{n+1} = a'_{n+1} q_n + q_{n-1} > a_{n+1} q_n + q_{n-1} = q_{n+1} \tag{7.44}$$

and

$$q'_{n+1} < (a_{n+1} + 1) q_n + q_{n-1}$$

$$= a_{n+1} q_n + q_{n-1} + q_n$$

$$= q_{n+1} + q_n$$

$$\leq a_{n+2} q_{n+1} + q_n = q_{n+2}, \tag{7.45}$$

as a_{n+2} is a positive integer. We have shown

Theorem 7.5.1. *For an irrational x the convergents satisfy*

$$\frac{1}{q_n q_{n+2}} < \left| x - \frac{p_n}{q_n} \right| < \frac{1}{q_n q_{n+1}}, \tag{7.46}$$

or equivalently

$$\frac{1}{q_{n+2}} < |p_n - q_n x| < \frac{1}{q_{n+1}}. \tag{7.47}$$

By Theorem 7.4.4, $\frac{1}{q_{n+1}} < 2^{-n/2}$. Thus, even when multiplied by a huge number like the exponentially growing q_n, these differences tend to zero exponentially fast. The terrific approximation power of the convergents of a continued fraction is one of their most important properties, which we shall explore in more detail below.

Exercise 7.5.2. *Show* $\left| x - \frac{p_n}{q_n} \right| \leq \frac{1}{a_{n+1}q_n^2}$.

7.6 PERIODIC CONTINUED FRACTIONS AND QUADRATIC IRRATIONALS

In this section we study periodic continued fractions and an interesting application to Pell's Equation. Our main references are [HW, Da1]; see also [BuP, vdP4].

7.6.1 A Theorem on Periodic Continued Fractions

By a periodic continued fraction we mean a continued fraction of the form

$$[a_0, a_1, \ldots, a_k, \ldots, a_{k+m}, a_k, \ldots, a_{k+m}, a_k, \ldots, a_{k+m}, \ldots]. \tag{7.48}$$

For example,

$$[1, 2, 3, 4, 5, 6, 7, 8, 9, 7, 8, 9, 7, 8, 9, 7, 8, 9, \ldots]. \tag{7.49}$$

Theorem 7.6.1 (Lagrange). *A number $x \in \mathbb{R}$ has a periodic continued fraction if and only if it satisfies an irreducible quadratic equation; i.e., there exist $A, B, C \in \mathbb{Z}$ such that $Ax^2 + Bx + C = 0$, $A \neq 0$, and x does not satisfy a linear equation with integer coefficients.*

Recall that a quadratic polynomial $f(x)$ with integer (resp., rational) coefficients is **irreducible** over $\mathbb{Z}$ (resp., over $\mathbb{Q}$) if it is not divisible by a linear factor with integer (resp., rational) coefficients. We prove Lagrange's Theorem in Lemmas 7.6.2 and 7.6.5; we start with the easier direction.

Lemma 7.6.2. *If x has a periodic continued fraction, then x satisfies a quadratic equation with integer coefficients.*

Proof. We first prove the result when $x = [a_0, \ldots, a_k, a_0, \ldots, a_k, \ldots] = [a_0, \ldots, a_k, x]$. Such numbers are said to be **purely periodic**, as the repeating block begins immediately. We have

$$x = [a_0, \ldots, a_k, x]$$

$$\frac{1}{x - a_0} = [a_1, \ldots, a_k, x]$$

$$\frac{x - a_0}{1 - (x - a_0)a_1} = [a_2, \ldots, a_k, x]. \tag{7.50}$$

Note the left hand side is the ratio of two linear polynomials in x, say $\frac{p_1(x)}{q_1(x)}$. The next stage gives

$$\frac{p_2(x)}{q_2(x)} = \frac{q_1(x)}{p_1(x) - a_2 q_1(x)} = [a_3, \ldots, a_k, x]. \tag{7.51}$$

We will eventually have

$$\frac{p_k(x)}{q_k(x)} = x, \tag{7.52}$$

implying x satisfies the quadratic equation $xq_k(x) - p_k(x) = 0$. Note the linear functions $p_i(x)$ and $q_i(x)$ are never constant; for $j \geq 1$ the coefficient of x is positive for $p_{2j}(x)$ and negative for $p_{2j+1}(x)$ (and the reverse for $q_j(x)$). If the expansion of x does not begin with its repeating block, one argues similarly until we reach the repeating block.   □

Exercise 7.6.3. *Prove the claim about the coefficients of x in $p_i(x)$ and $q_i(x)$.*

Exercise 7.6.4. *Prove Lemma 7.6.2 when x is not purely periodic.*

Lemma 7.6.5. *If x satisfies an irreducible quadratic equation with integer coefficients and leading term non-zero, then x has a periodic continued fraction.*

Proof. Note if the quadratic factored over $\mathbb{Z}$ or $\mathbb{Q}$, then x would satisfy a linear equation and therefore be rational. This direction is significantly harder, and we proceed in stages. Assume x solves

$$ax^2 + bx + c = 0. \tag{7.53}$$

As the quadratic equation is irreducible over $\mathbb{Z}$ or $\mathbb{Q}$ (i.e., it does not split into the product of two linear equations), (7.53) has no rational roots. We must show x has a periodic continued fraction expansion. We write $x = [a_0, a_1, \dots]$. We may write $x = [a_0, \dots, a_{n-1}, a'_n]$ and obtain

$$x = \frac{p_{n-1}a'_n + p_{n-2}}{q_{n-1}a'_n + q_{n-2}}. \tag{7.54}$$

Substitute $\frac{p_{n-1}a'_n + p_{n-2}}{q_{n-1}a'_n + q_{n-2}}$ for x in (7.53). Clear denominators by multiplying through by $(q_{n-1}a'_n + q_{n-2})^2$. We find a'_n satisfies the following quadratic equation

$$A_n(a'_n)^2 + B_n a'_n + C_n = 0, \tag{7.55}$$

where a messy but straightforward calculation gives

$$\begin{aligned}
A_n &= ap_{n-1}^2 + bp_{n-1}q_{n-1} + cq_{n-1}^2 \\
B_n &= 2ap_{n-1}q_{n-2} + b(p_{n-1}q_{n-2} + p_{n-2}q_{n-1}) + 2cq_{n-1}q_{n-2} \\
C_n &= ap_{n-2}^2 + bp_{n-2}q_{n-2} + cq_{n-2}^2.
\end{aligned} \tag{7.56}$$

Claim 7.6.6. $A_n \neq 0$.

Proof of claim. If $A_n = 0$, then dividing the expression for A_n by q_{n-1}^2 gives $\frac{p_{n-1}}{q_{n-1}}$ satisfies (7.53); however, (7.53) has no rational solutions (this is where we use the quadratic is irreducible).   □

Returning to the proof of Lemma 7.6.5, we have

$$A_n y^2 + B_n y + C_n = 0, \quad y = a'_n \text{ is a solution}, \quad A_n \neq 0. \tag{7.57}$$

The discriminant of the above quadratic is (another messy but straightforward calculation)

$$\Delta \;=\; B_n^2 - 4A_nC_n \;=\; b^2 - 4ac. \tag{7.58}$$

By Theorem 7.5.1 (use $\frac{q_n}{q_{n+1}} < 1$),

$$x q_{n-1} - p_{n-1} \;=\; \frac{\delta_{n-1}}{q_{n-1}}, \quad |\delta_{n-1}| < 1. \tag{7.59}$$

Thus

$$A_n \;=\; a\left(x q_{n-1} + \frac{\delta_{n-1}}{q_{n-1}}\right)^2 + b q_{n-1}\left(x q_{n-1} + \frac{\delta_{n-1}}{q_{n-1}}\right) + c q_{n-1}^2. \tag{7.60}$$

Taking absolute values and remembering that $ax^2 + bx + c = 0$ gives

$$\left|(ax^2 + bx + c)q_{n-1}^2 + 2axq_{n-1}\delta_{n-1} + a\frac{\delta_{n-1}^2}{q_{n-1}^2} + b\delta_{n-1}\right| \leq 2\cdot|a|\cdot|x| + |b| + |a|. \tag{7.61}$$

As $C_n = A_{n-1}$ we find that

$$|C_n| \;\leq\; 2\cdot|a|\cdot|x| + |b| + |a|$$
$$B_n^2 - 4A_nC_n \;=\; b^2 - 4ac$$
$$B_n \;\leq\; \sqrt{|4A_nC_n| + |b^2 - 4ac|}$$
$$<\; \sqrt{4\left(2\cdot|a|\cdot|x| + |b| + |a|\right)^2 + |b^2 - 4ac|}. \tag{7.62}$$

We have shown

Lemma 7.6.7. *There is an M such that, for all n,*

$$|A_n|, \;|B_n|, \;|C_n| \;<\; M. \tag{7.63}$$

We now complete the proof of Lemma 7.6.5. By Dirichlet's Pigeon-Hole Principle (§A.4), we can find three triples such that

$$(A_{n_1}, B_{n_1}, C_{n_1}) \;=\; (A_{n_2}, B_{n_2}, C_{n_2}) \;=\; (A_{n_3}, B_{n_3}, C_{n_3}). \tag{7.64}$$

We obtain three numbers a'_{n_1}, a'_{n_2} and a'_{n_3} which all solve the same quadratic equation (7.55), and the polynomial *is not* the zero polynomial as $A_{n_i} \neq 0$. As any non-zero polynomial has at most two distinct roots, two of the three a'_{n_i} are equal. Without loss of generality assume $a'_{n_1} = a'_{n_2}$. This implies periodicity, because

$$[a_{n_1}, a_{n_1+1}, \ldots, a_{n_2}, \ldots] \;=\; [a_{n_1}, a_{n_1+1}, \ldots, a'_{n_2}]$$
$$=\; [a_{n_1}, a_{n_1+1}, \ldots, a'_{n_1}]. \tag{7.65}$$

This concludes the proof of Lemma 7.6.5. $\qquad\qquad\qquad\qquad\qquad\quad\square$

Notice we have not determined where the periodicity begins. The above arguments use a'_n to convert an infinite continued fraction $[a_0, a_1, \ldots]$ to a finite continued fraction $[a_0, a_1, \ldots, a'_n]$, which we then notice is periodic.

Exercise 7.6.8. *Show* $\sqrt{2} = [1, 2, 2, 2, \ldots]$.

Exercise 7.6.9. *Show* $\sqrt{3} = [1, 1, 2, 1, 2, 1, 2, \ldots]$.

Exercise 7.6.10. *Give an explicit upper bound for the constant* M *in the proof of Lemma 7.6.5; the bound should be a function of the coefficients of the quadratic. Use this bound to determine an* N *such that we can find three numbers* $a_{n_1}, a_{n_2}, a_{n_3}$ *as above with* $n_i \leq N$. *Deduce a bound for where the periodicity must begin. Similarly, deduce a bound for the length of the period.*

Exercise 7.6.11. *For* a, N *relatively prime, consider the continued fraction of* $\frac{a}{N}$. *Obtain an upper bound for* $l(a, N)$, *the length of the continued fraction (which is finite as* $\frac{a}{N} \in \mathbb{Q}$). *Heillbronn [Hei] proved that*

$$\sum_{\substack{1 \leq a < N \\ (a,N)=1}} l(a, N) = \frac{12 \log 2}{\pi^2} \phi(N) \log N + O(N\sigma_{-1}^3(N)), \qquad (7.66)$$

where $\sigma_{-1}(N) = \sum_{d|N} \frac{1}{d}$. *See [Sz] for a good estimate for individual* $l(a, N)$.

Remark 7.6.12. Some of the results of this section have very interesting interpretations in the theory of dynamical systems; for some of these and much more see [Corl]. For a recent more advanced survey on the connections between dynamical systems and number theory, see [KSS]. A wonderful reference for the *Gauss Map* is [CFS].

7.6.2 Purely Periodic Continued Fractions

Let $\alpha = \overline{[a_0, a_1, \ldots, a_n]}$ be a purely periodic continued fraction with $a_0 > 0$. We know that a number with such a continued fraction expansion has to be a quadratic irrational. What else can be said about such an α? It is clear that $\alpha > 1$. Define a new continued fraction β by

$$\beta = \overline{[a_n, a_{n-1}, \ldots, a_1, a_0]}. \qquad (7.67)$$

Hence β is obtained from α by reversing the period.

Exercise 7.6.13. *Let* α' *be the second root of the quadratic equation satisfied by* α. *If* $\alpha = \overline{[a_0, a_1, \ldots, a_n]}$ *and* β *is as above, prove that*

$$\alpha' = -\frac{1}{\beta}. \qquad (7.68)$$

Observe that as $\beta > 1$, we have $-1 < \alpha' < 0$.

Definition 7.6.14. *We call a quadratic irrational number* α *a reduced quadratic if* $\alpha > 1$ *and* $-1 < \alpha' < 0$.

Exercise 7.6.15. *Let* D *be a positive integer that is not a perfect square. Prove that if*

$$\alpha = \frac{P + \sqrt{D}}{Q} \qquad (7.69)$$

is reduced, then P, Q *are natural numbers satisfying*

1. $P < \sqrt{D}$;

2. $Q < 2\sqrt{D}$;

3. $Q | P^2 - D$.

Show that these conditions are not *sufficient to force* α *to be reduced.*

Exercise 7.6.16. *Let* α *be reduced and let* $a_0 = [\alpha]$. *Define* α_1 *by*

$$\alpha = a_0 + \frac{1}{\alpha_1}. \tag{7.70}$$

Show that α_1 *is reduced.*

For example, $2 + \sqrt{5}$ is reduced. The above discussion then implies that if the continued fraction of a quadratic number α is purely periodic, then α is a reduced quadratic. The converse is also true.

Theorem 7.6.17 (Galois). *A quadratic number* α *has a purely periodic continued fraction expansion if and only if it is a reduced quadratic number.*

Sketch of the proof. We have already done one direction. Let

$$\alpha = \frac{P + \sqrt{D}}{Q} \tag{7.71}$$

be reduced. We calculate α_1 as in Exercise 7.6.16. We have

$$\alpha_1 = \frac{1}{\alpha - a_0} = \frac{P_1 + \sqrt{D}}{Q_1} \tag{7.72}$$

with $P_1 = a_0 Q - P$ and $Q_1 = \frac{D - P_1^2}{Q}$. By Exercise 7.6.15 P_1 and Q_1 are natural numbers with $P_1 < \sqrt{D}$ and $Q_1 < 2\sqrt{D}$. If we repeat the process we obtain two sequences $\{P_n\}$ and $\{Q_n\}$ of natural numbers with $P_n < \sqrt{D}$ and $Q_n < 2\sqrt{D}$. By Dirichlet's Pigeon-Hole Principle there are n, m with $P_n = P_m$ and $Q_n = Q_m$. This implies that $\alpha_n = \alpha_m$, with the obvious notation. We leave it to the reader to verify that this implies $\alpha_{n-1} = \alpha_{m-1}$ which yields the continued fraction of α is purely periodic. $\qquad\qquad\square$

Exercise 7.6.18. *Fill in the details of the proof.*

Example 7.6.19. *The continued fraction of any number of the form* $\sqrt{d}$, *d not a perfect square, is of the form*

$$[a_0, \overline{a_1, \dots, a_{n-1}, 2a_0}] \tag{7.73}$$

for positive integers $a_0, a_1, \dots, a_{n-1}$. *In order to see this, we let* $\alpha = [\sqrt{d}] + \sqrt{d}$. *We determine the continued fraction of* α, *and from this we will deduce the expansion for* $\sqrt{d}$. *As* $\alpha > 1$ *and* $-1 < \alpha' < 0$, *we may apply Theorem 7.6.17 and we conclude that the continued fraction expansion of* α *is purely periodic. Then*

$$[\alpha] = \left[[\sqrt{d}] + \sqrt{d} \right] = 2[\sqrt{d}]. \tag{7.74}$$

Setting $a_0 = \sqrt{d}$, *the continued fraction of* α *must be of the form* $\overline{[2a_0, a_1, \dots, a_{n-1}]}$. *As a result the continued fraction of* $\sqrt{d}$ *will be*

$$\overline{[2a_0, a_1, \dots, a_{n-1}]} - a_0 = [a_0, \overline{a_1, \dots, a_{n-1}, 2a_0}]. \tag{7.75}$$

Exercise 7.6.20. *Verify the last identity.*

Exercise 7.6.21. *Compute the continued fraction expansion of $\sqrt{N^2 + 1}$, N a natural number.*

For other interesting properties of the continued fraction expansion of $\sqrt{d}$ we refer the reader to [Cohn] and Module 8 of [Bur].

7.6.3 Applications to Pell's Equation

Let d be a positive integer. The Diophantine equation

$$X^2 - dY^2 = \pm 1 \tag{7.76}$$

is referred to as *Pell's Equation.* We refer the reader to [HW], Notes on Chapter XIV, for references and historical comments; see also [Bur, vdP1]. For more on Diophantine equations, see Chapter 4. For now, assume the positive sign holds in (7.76).

We may assume that d is not a perfect square. If $d = n^2$ we have $X^2 - (nY)^2 = 1$, or $X^2 = (nY)^2 + 1^2$. This is a specific Pythagorean equation; in §4.1.2 we found all solutions.

It is not hard to solve Pell's Equation $X^2 - dY^2 = 1$ over the rational numbers. One can show that every rational solution of the equation is of the form

$$(X, Y) = \left(\frac{t^2 + d}{t^2 - d}, \frac{2t}{t^2 - d} \right) \tag{7.77}$$

for t rational.

Exercise[h] **7.6.22.** *Prove the parametrization of solutions in (7.77).*

Solving the equation over the integers, however, poses a much bigger challenge. There is a natural relationship between solutions to Pell's Equation and continued fraction as the following lemma shows.

Lemma 7.6.23. *Let (X_0, Y_0) be a solution of (7.76) of positive integers. Then $\frac{X_0}{Y_0}$ is a convergent of the continued fraction expansion of $\sqrt{d}$.*

Proof. We first consider the case $X^2 - dY^2 = 1$. We have

$$\left| \frac{X_0}{Y_0} - \sqrt{d} \right| = \frac{1}{Y_0(X_0 + Y_0\sqrt{d})} < \frac{1}{2Y_0^2 \sqrt{d}}, \tag{7.78}$$

as $X_0 > Y_0\sqrt{d}$. In Theorem 7.9.8 we show that (7.78) implies that $\frac{X_0}{Y_0}$ must be a convergent of $\sqrt{d}$. The reason is that (7.78) gives a very good approximation to $\sqrt{d}$ by rationals, and the only approximations that are this good come from convergents in the continued fraction of $\sqrt{d}$.

We now investigate $X^2 - dY^2 = -1$. We now have

$$0 < \sqrt{d} - \frac{X_0}{Y_0} = \frac{1}{Y_0(Y_0\sqrt{d} + \sqrt{dY_0^2 - 1})}. \tag{7.79}$$

If we could show that the last expression is less than $\frac{1}{2Y_0^2}$ then another application of Theorem 7.9.8 would finish the proof. Equivalently, we need to prove

$$Y_0\sqrt{d} + \sqrt{dY_0^2 - 1} > 2Y_0. \tag{7.80}$$

If $d \geq 5$, then $\sqrt{d} > 2$ and the inequality is trivial. If $d < 5$, as d is not a square then $d = 2$ or 3; these cases can be checked by hand. $\square$

Remark 7.6.24. This proof should remind the reader of the arguments used in §5.7.2.

We can use our results on periodic continued fractions to solve Pell's Equation for any non-square integer d. We need the following lemma:

Lemma 7.6.25. *Let $\frac{p_m}{q_m}$ be the convergents to the continued fraction expansion of $\sqrt{d}$. Let n, as in Example 7.6.19, be the period of the continued fraction of $\sqrt{d}$. Then for any positive integer k, we have*

$$p_{kn-1}^2 - dq_{kn-1}^2 = (-1)^{kn}. \tag{7.81}$$

Proof. We adopt the notation of Exercise 7.6.19, except that we set $a_n = 2a_0$. For any positive integer k we have

$$\sqrt{d} = [a_0, a_1, \ldots, a_{kn-1}, r_{kn}] \tag{7.82}$$

with

$$r_{kn} = [2a_0, \overline{a_1, \ldots, a_n}] = a_0 + \sqrt{d}. \tag{7.83}$$

By Theorem 7.3.5, we have

$$\sqrt{d} = \frac{r_{kn}p_{kn-1} + p_{kn-2}}{r_{kn}q_{kn-1} + q_{kn-2}}. \tag{7.84}$$

Substituting $r_{kn} = \sqrt{d} + a_0$ in (7.84) and simplifying gives

$$\sqrt{d}(a_0q_{kn-1} + q_{kn-2} - p_{kn-1}) = a_0p_{kn-1} + p_{kn-2} - dq_{kn-1}. \tag{7.85}$$

Since the right hand side is a rational number and the left hand side a rational multiple of $\sqrt{d}$, which we are assuming to be irrational, both sides must be 0; consequently,

$$a_0q_{kn-1} + q_{kn-2} = p_{kn-1},$$
$$a_0p_{kn-1} + p_{kn-2} = dq_{kn-1}. \tag{7.86}$$

Next, we have

$$\begin{aligned}
p_{kn-1}^2 - dq_{kn-1}^2 &= p_{kn-1}(a_0q_{kn-1} + q_{kn-2}) - q_{kn-1}(a_0p_{kn-1} + p_{kn-2}) \\
&= p_{kn-1}q_{kn-2} - q_{kn-1}p_{kn-2} \\
&= (-1)^{kn-2} \\
&= (-1)^{kn}
\end{aligned}$$

$$\tag{7.87}$$

by Lemma 7.3.13, which completes the proof. $\square$

Exercise 7.6.26. *Prove that if $p_m^2 - dq_m^2 = \pm 1$, then $m = kn - 1$ for some integer k.*

The following result is immediate from the lemmas and exercises above:

Theorem 7.6.27. *Every integral solution of $X^2 - dY^2 = \pm 1$ is of the form*

$$(X, Y) = (p_{kn-1}, q_{kn-1}) \tag{7.88}$$

for some integer k.

Corollary 7.6.28. *The equation $X^2 - dY^2 = 1$ always has infinitely many solutions. The equation $X^2 - dY^2 = -1$ has no solutions if and only if the length of the period of the continued fraction of $\sqrt{d}$ is an even number.*

Exercise 7.6.29. *Find all integral solutions to the following equations:*

1. $X^2 - 2Y^2 = 1$;

2. $X^2 - 3Y^2 = -1$.

7.7 COMPUTING ALGEBRAIC NUMBERS' CONTINUED FRACTIONS

We give a construction to find the continued fraction expansion of many algebraic numbers. We follow the exposition in [LT]; see also [BP, BPR, Bry, La2, NT, RDM]. See Theorem 5.6.6 for the continued fraction expansion of a transcendental number.

Theorem 7.7.1. *Consider a positive real algebraic number α of degree d. Let $P_0(x)$ be its minimal polynomial. Assume*

1. *the leading coefficient of $P_0(x)$ is positive;*

2. *$P_0(x)$ has exactly one positive simple root α (of degree d), and $\alpha > 1$. Thus $P_0(\alpha) = 0$ and $P_0'(\alpha) \neq 0$.*

Then we can construct the continued fraction expansion of α just by looking at a sequence of polynomials of degree d which are explicitly determined by α.

Proof. We construct the sequence of polynomials by induction. In the course of proving the inductive step the basis step will be implicitly handled. Note for $d = 1$ this reduces to generating continued fractions by Euclid's algorithm.

Assume we have constructed polynomials $P_0(x), P_1(x), \ldots, P_n(x)$ satisfying the above conditions. By induction $P_n(x)$ has positive leading coefficient and only one positive root, which is greater than 1 and of degree d. We want to construct $P_{n+1}(x)$ with the same properties.

Let y_n be the sole positive root of $P_n(x)$. Let $a_n = [y_n]$, the greatest integer less than or equal to y_n. If $d > 1$ then $P_n(a_n) \neq 0$, as y_n irrational implies $[y_n] < y_n$; if $d = 1$ and $[y_n] = y_n$ then the algorithm terminates. We analyze the case $d > 1$ below, which implies $a_n < y_n$.

Since the leading coefficient of P_n is positive, $P_n(x) \to \infty$ as $x \to \infty$. Since y_n is the only positive root, it is a simple root (i.e., it is not a multiple root). If $P_n(a_n) > 0$, y_n cannot be the only positive root. Hence $P_n(a_n) \le 0$. Define

$$Q_n(x) = P_n(x + a_n)$$
$$P_{n+1}(x) = -x^d Q_n(x^{-1}). \tag{7.89}$$

Then $Q_n(x)$ has a root of degree d strictly between 0 and 1, and no other positive roots. It follows that $P_{n+1}(x)$ has only one positive root, y_{n+1}, satisfying $y_{n+1} > 1$; this root is also of degree d. The constant term of $Q_n(x)$ is $P_n(a_n) < 0$. Hence $-P_n(a_n)$, the leading coefficient of $P_{n+1}(x)$, is positive. One can show y_{n+1} is a simple root of $P_{n+1}(x)$; therefore, $P_{n+1}(x)$ satisfies the desired conditions.

We now show that the a_n's give the continued fraction expansion of α. Recall $a_n = [y_n]$. The root y_{n+1} satisfies $Q_n(y_{n+1}^{-1}) = 0$. Therefore

$$P_n(y_{n+1}^{-1} + a_n) = 0, \tag{7.90}$$

which implies

$$y_{n+1}^{-1} + a_n = y_n$$
$$y_{n+1} = \frac{1}{y_n - a_n}$$
$$a_{n+1} = \left[\frac{1}{y_n - a_n} \right] = \left[\frac{1}{y_n - [y_n]} \right]. \tag{7.91}$$

Recalling how we construct the continued fraction expansion of a number, we see that the a_n's are just the coefficients of the continued fraction expansion of α. □

We discuss applications and advantages of this construction. First we must make sure that the given polynomial is irreducible, with exactly one positive root which is greater than 1. Let $P_n(y_n) = 0$. In each iteration we need to find $a_n = [y_n]$. Thus we need a ballpark approximation for the roots y_n. One method is divide and conquer, looking for sign changes of $P_n(x)$ for $x > 1$ (see Exercise 5.3.21). Another approach is to apply Newton's Method (§1.2.4). Finally, note this method is not based on calculating a lengthy decimal expansion and then determining the continued fraction of the decimal approximation. Such a construction would involve many technical issues, such as how many decimal digits does one need to ensure N continued fraction digits are correct, and how does N depend on our number (see Exercise 7.7.9). At each stage we need only store $d + 1$ numbers (as we have a polynomial of degree d). One great advantage of this is that if we have calculated a million digits and later decide we needed more, we can resume the calculation where we left off easily by saving the final polynomial. See [BPR] for a comparison of various algorithms for determining continued fraction expansions.

Exercise 7.7.2. *Prove the claim that if $P(a_n) > 0$, then there are at least two positive roots.*

Exercise 7.7.3. *Find continued fraction expansions for $\sqrt{n}$ for some rational $n > 1$ (or $\sqrt[m]{n}$).*

Definition 7.7.4 (Modular Group). *The modular group $SL_2(\mathbb{Z})$ is the set of 2×2 matrices with integer entries and determinant $+1$. This group is also called the Special Linear group (hence the notation).*

Lemma 7.7.5. *The group $SL_2(\mathbb{Z})$ is generated by the matrices $T = \left(\begin{smallmatrix} 1 & 1 \\ 0 & 1 \end{smallmatrix}\right)$ and $S = \left(\begin{smallmatrix} 0 & -1 \\ 1 & 0 \end{smallmatrix}\right)$. Thus any $\gamma \in SL_2(\mathbb{Z})$ can be written as a finite product of T and S and their inverses.*

Proof. See for example [Se]. □

Definition 7.7.6 (Equivalent Numbers). *Consider invertible matrices $\gamma = \left(\begin{smallmatrix} a & b \\ c & d \end{smallmatrix}\right)$, with $a, b, c, d \in \mathbb{Z}$. We define an action of these matrices on complex numbers by*

$$\gamma z = \begin{pmatrix} a & b \\ c & d \end{pmatrix} z = \frac{az + b}{cz + d}. \tag{7.92}$$

If $x, y \in \mathbb{R}$ satisfy $x = \gamma y$ with $\det(\gamma) = ad - bc = n$, we say x and y are n-equivalent.

Exercise[h] 7.7.7. *Let x be 1-equivalent to y, neither rational. Prove that there exist $k, N \in \mathbb{N}$ such that for all $n \geq N$, $a_n(x) = a_{n+k}(y)$ where $a_n(x)$ is the nth digit of x's continued fraction expansion. In other words, the two continued fractions are eventually the same, although one of them might need to be shifted.*

Exercise 7.7.8. *Prove that for any real algebraic number x there is an integer n and an algebraic number y such that x is n-equivalent to y and y satisfies the conditions of Theorem 7.7.1. Simple continued fractions have all 1's as the numerators. If one allows any positive integer, how are the continued fractions of x and an n-equivalent y related?*

Exercise 7.7.9. *Let $\alpha \in \mathbb{R}$. Let $D_d(\alpha)$ give the first d digits of α's decimal expansion. For any fixed $N > 1$ and any d, show there is an α such that $D_d(\alpha)$ does not provides enough information to calculate the first N digits of α's continued fraction expansion.*

7.8 FAMOUS CONTINUED FRACTION EXPANSIONS

We give some closed form expansions for the golden mean and similar numbers, as well as a few famous continued fraction expansions. For more expansions, see for example [BP, PS1, PS2, vdP5].

7.8.1 Closed Form Expressions for Quadratic Irrationals

We describe an algorithm to generate closed form expressions for powers of the golden mean (which we denote by ϕ below), and then discuss its generalizations. Let F_n denote the nth Fibonacci number. They satisfy the recurrence $F_{n+1} = F_n + F_{n-1}$, with $F_0 = F_1 = 1$. By Exercise 7.3.9, $\phi = \lim_{n\to\infty} \frac{F_n}{F_{n-1}} = \frac{1+\sqrt{5}}{2}$. Similarly

$$\lim_{n\to\infty} \frac{F_n}{F_{n-2}} = \lim_{n\to\infty} \frac{F_n}{F_{n-1}} \frac{F_{n-1}}{F_{n-2}} = \phi^2, \tag{7.93}$$

and similar arguments show that $\lim_{n\to\infty} \frac{F_n}{F_{n-k}} = \phi^k$. In Example 7.3.6 we proved $\phi = [\overline{1}] = [1,1,1,\dots]$. For ϕ^2 we have

$$\phi^2 = \lim_{n\to\infty} \frac{F_n}{F_{n-2}} = \lim_{n\to\infty} \frac{F_{n-1}+F_{n-2}}{F_{n-2}} = 1 + \lim_{n\to\infty} \frac{F_{n-1}}{F_{n-2}} = 1+\phi. \quad (7.94)$$

This implies

$$\phi^2 = [2,\overline{1}] = [2,1,1,1,\dots]. \quad (7.95)$$

The first interesting power is ϕ^3. The proof of this case generalizes nicely to the remaining powers.

$$
\begin{aligned}
\frac{F_n}{F_{n-3}} &= \frac{F_{n-1}+F_{n-2}}{F_{n-3}} \\
&= \frac{2F_{n-2}+F_{n-3}}{F_{n-3}} \\
&= 1 + \frac{2F_{n-2}}{F_{n-3}} \\
&= 1 + \frac{2F_{n-3}+2F_{n-4}}{F_{n-3}} \\
&= 3 + \frac{2F_{n-4}}{F_{n-3}} \\
&= 3 + \frac{F_{n-4}+F_{n-3}-F_{n-5}}{F_{n-3}} \\
&= 4 + \frac{F_{n-4}-F_{n-5}}{F_{n-3}} \\
&= 4 + \frac{F_{n-6}}{F_{n-3}} \\
&= 4 + \frac{1}{\dfrac{F_{n-3}}{F_{n-6}}}. \quad (7.96)
\end{aligned}
$$

We have obtained a repeating block, as $\lim_n \frac{F_n}{F_{n-3}} = \lim_n \frac{F_{n-3}}{F_{n-6}} = \phi^3$. Thus,

$$\phi^3 = [\overline{4}] = [4,4,4,\dots]. \quad (7.97)$$

Any integer power of ϕ can be determined by a similar argument. The two key facts in the proof are

1. $F_n = f_k F_{n-k} + f_{k-1} F_{n-k-1}$, where the f_k's are also Fibonacci numbers, with $f_0 = f_1 = 1$.

2. $f_{k-2}f_k - (f_{k-1})^2 = (-1)^{k-1}$.

For positive k, these two facts lead to

$$\phi^k = \begin{cases} [f_k + f_{k-2} - 1, \overline{1, f_k + f_{k-2} - 2}] & k \text{ even} \\ [\overline{f_k + f_{k-2}}] & k \text{ odd}. \end{cases} \quad (7.98)$$

Exercise 7.8.1. *Prove the above results. Find continued fractions for* ϕ^{-k}.

The method generalizes to handle slightly more general x. Consider a recurrence relation $G_{n+1} = mG_n + G_{n-1}$, where m is a fixed positive integer. By Exercise 7.3.9, $x_{m,\pm} = \frac{m \pm \sqrt{m^2+4}}{2}$. Note $x_{m,+} \cdot x_{m,-} = -1$; this property is very important in the proof. Similar to the Fibonacci numbers, one can calculate $x_{m,+}^k$. For details see [Fi].

7.8.2 Famous Continued Fractions

Often a special number whose decimal expansion seems random has a continued fraction expansion with a very rich structure. For example, compare the first 25 digits for e:

$$e = 2.7182818284590452353602 87 \ldots$$
$$= [2, 1, 2, 1, 1, 4, 1, 1, 6, 1, 1, 8, 1, 1, 10, 1, 1, 12, 1, 1, 14, 1, 1, 16, 1, \ldots].$$

For a proof see [La2, vdP5], where expansions are given for many expressions involving e which exhibit similar striking patterns. For π, the positive simple continued fraction does not look particularly illuminating:

$$\pi = 3.14159265358979323846 2643 \ldots$$
$$= [3, 7, 15, 1, 292, 1, 1, 1, 2, 1, 3, 1, 14, 2, 1, 1, 2, 2, 2, 2, 1, 84, 2, 1, 1, \ldots].$$

If, however, we drop the requirement that the expansions are simple, the story is quite different. One nice expression for π is

$$\frac{4}{\pi} = 1 + \cfrac{1^2}{2 + \cfrac{3^2}{2 + \cfrac{5^2}{2 + \cdots}}}. \tag{7.99}$$

There are many different types of non-simple expansions, leading to some of the most beautiful formulas in mathematics. For example,

$$e = 2 + \cfrac{1}{1 + \cfrac{1}{2 + \cfrac{2}{3 + \cfrac{3}{4 + \cdots}}}}. \tag{7.100}$$

For some nice articles and simple and non-simple continued fraction expansions, see [We] (in particular, the entries on π and e).

Exercise 7.8.2. *Show the large digit of 292 in the continued fraction expansion of π leads to a good rational approximation to π with a small denominator.*

Exercise 7.8.3. *Try to generalize as many properties as possible from simple continued fractions to non-simple. Clearly, numbers do not have unique expansions* unless *we specify exactly what the "numerators" of the non-simple expansions must be. One often writes such expansions in the more economical notation*

$$\cfrac{a}{\alpha+}\ \cfrac{b}{\beta+}\ \cfrac{c}{\gamma+}\ \cfrac{d}{\delta+}\cdots. \tag{7.101}$$

For what choices of $a, b, c \ldots$ and $\alpha, \beta, \gamma, \ldots$ will the above converge? How rapidly will it converge? Are there generalizations of the recurrence relations? How rapidly do the numerators and denominators of the rationals formed by truncating these expansions grow? See also Exercise 7.7.8.

7.9 CONTINUED FRACTIONS AND APPROXIMATIONS

We continue our survey of continued fractions, concentrating on how well a continued fraction approximates a number. These results were used in §7.6.3 to solve Pell's Equation.

7.9.1 Convergents Give the Best Approximations

Theorem 7.9.1. *Let $x = [a_0, a_1, \ldots]$ with convergents $\frac{p_n}{q_n}$. Then for $0 < q \le q_n$, if $\frac{p}{q} \ne \frac{p_n}{q_n}$ then*

$$\left| x - \frac{p_n}{q_n} \right| < \left| x - \frac{p}{q} \right|. \tag{7.102}$$

The above theorem states that among numbers with bounded denominators, the continued fraction is the best approximation. The theorem clearly follows from

Theorem 7.9.2. *Under the same assumptions as in Theorem 7.9.1,*

$$|p_n - q_n x| < |p - qx|. \tag{7.103}$$

Proof. Suppose p and q are relatively prime. By Theorem 7.5.1

$$|p_n - q_n x| < |p_{n-1} - q_{n-1} x|. \tag{7.104}$$

Thus it suffices to investigate $q_{n-1} < q \le q_n$.

Case One: $q = q_n$: If $q = q_n$ we must have $p \ne p_n$ as otherwise $\frac{p}{q} = \frac{p_n}{q_n}$. Therefore

$$\left| \frac{p_n}{q_n} - \frac{p}{q} \right| \ge \frac{1}{q_n} \tag{7.105}$$

if $p \ne p_n$ (as $q = q_n$, and $|p - p_n| \ge 1$). Thus $\frac{p_n}{q_n}$ is not too close to $\frac{p}{q}$. Again by Theorem 7.5.1,

$$\left| \frac{p_n}{q_n} - x \right| \le \frac{1}{q_n q_{n+1}} \le \frac{1}{(n+1)q_n} \tag{7.106}$$

(we are assuming $n > 1$, and by Theorem 7.4.3, $q_{n+1} \ge n + 1$). Thus $\frac{p_n}{q_n}$ is close to x. Consider the interval of radius $\frac{1}{(n+.5)q_n}$ about $\frac{p_n}{q_n}$. Then x is in this interval;

however, $\frac{p}{q}$ is *not* in the interval because $\frac{p}{q}$ is at least $\frac{1}{q_n}$ units from $\frac{p_n}{q_n}$. Therefore the closest x can be to $\frac{p}{q}$ is $\frac{n-.5}{n+.5}\frac{1}{q_n}$, as

$$\left| x - \frac{p}{q} \right| > \left| \frac{p_n}{q_n} - \frac{p}{q} \right| - \left| x - \frac{p_n}{q_n} \right| > \frac{n-.5}{n+.5}\frac{1}{q_n}. \tag{7.107}$$

As $\left| x - \frac{p_n}{q_n} \right| < \frac{1}{(n+.5)q_n}$, we have

$$\left| x - \frac{p_n}{q_n} \right| < \left| x - \frac{p}{q} \right|, \tag{7.108}$$

which completes the proof.

Case Two: $q_{n-1} < q < q_n$: By our assumptions on q, $\frac{p}{q} \neq \frac{p_n}{q_n}$ or $\frac{p_{n-1}}{q_{n-1}}$. We will find integers μ and ν such that

$$\mu p_n + \nu p_{n-1} = p, \quad \mu q_n + \nu q_{n-1} = q. \tag{7.109}$$

Consider the above relation; we show that if μ and ν solve the above then they are integers. Multiplying the first by q_{n-1} and the second by p_{n-1} yields

$$\mu(p_n q_{n-1} - p_{n-1}q_n) = pq_{n-1} - qp_{n-1}$$
$$\mu = (-1)^{n-1}(pq_{n-1} - qp_{n-1}), \tag{7.110}$$

where we used Lemma 7.3.13 to get $p_n q_{n-1} - p_{n-1}q_n = (-1)^{n-1}$. Similarly we find that

$$\nu = -(-1)^{n-1}(pq_n - qp_{n-1}). \tag{7.111}$$

Thus we can find integers μ and ν such that (7.109) is true. As $q = \mu q_n + \nu q_{n-1} < q_n$, μ and ν must have opposite signs. Further, we know $p_n - q_n x$ and $p_{n-1} - q_{n-1}x$ have opposite signs (the even convergents are increasing, the odd convergents are decreasing; see Exercise 7.4.2). Therefore $\mu(p_n - q_n x)$ and $\nu(p_{n-1} - q_{n-1}x)$ have the same sign. But

$$p - qx = \mu(p_n - q_n x) + \nu(p_{n-1} - q_{n-1}x). \tag{7.112}$$

Thus

$$|p - qx| > |p_{n-1} - q_{n-1}x| > |p_n - q_n x|. \tag{7.113}$$

The above is the desired inequality.   □

Remark 7.9.3. One could use linear algebra to solve for μ and ν. To find μ, ν such that

$$\mu p_n + \nu p_{n-1} = p$$
$$\mu q_n + \nu q_{n-1} = q, \tag{7.114}$$

note

$$\begin{vmatrix} p_n & p_{n-1} \\ q_n & q_{n-1} \end{vmatrix} = (-1)^{n-1}, \tag{7.115}$$

and by Cramer's rule we can find such a μ and ν (or take the inverse of this matrix). As we want integer solutions, it is essential that the determinant is ± 1.

7.9.2 Approximation Properties of Convergents

Now that we know the convergents give the best rational approximations, we investigate *how well* they approximate. The statement of the following theorem does not involve continued fractions; the proof, however, uses continued fractions in an essential fashion.

Theorem 7.9.4 (Hurwitz). *Let $x \in [0, 1]$ be an irrational number. Then there are infinitely many relatively prime integers p, q such that*

$$\left| x - \frac{p}{q} \right| \le \frac{1}{q^2\sqrt{5}}. \tag{7.116}$$

Proof. As x is rational if and only if its continued fraction is finite, we may assume x has an infinite continued fraction expansion. We show that, of any three consecutive quotients $\frac{p_{n-1}}{q_{n-1}}, \frac{p_n}{q_n}, \frac{p_{n+1}}{q_{n+1}}$ to x coming from the continued fraction expansion, at least one satisfies (7.116). Showing one of three consecutive terms has the desired property is a common technique; see also Remark 1.2.19.

Let

$$b_{i+1} = \frac{q_{i-1}}{q_i}. \tag{7.117}$$

Then

$$\left| \frac{p_i}{q_i} - x \right| = \frac{1}{q_i q'_{i+1}} = \frac{1}{q_i(a'_{i+1}q_i + q_{i-1})}$$
$$= \frac{1}{q_i(a'_{i+1}q_i + b_{i+1}q_i)}$$
$$= \frac{1}{q_i^2} \frac{1}{a'_{i+1} + b_{i+1}}. \tag{7.118}$$

It is thus enough to prove that

$$a'_i + b_i \ge \sqrt{5} \tag{7.119}$$

for at least one of any three consecutive values $n-1, n, n+1$ of i. Assume

$$a'_{n-1} + b_{n-1} \le \sqrt{5}$$
$$a'_n + b_n \le \sqrt{5}. \tag{7.120}$$

By definition

$$a'_{n-1} = a_{n-1} + \frac{1}{a'_n} \tag{7.121}$$

and

$$\frac{1}{b_n} = \frac{q_{n-1}}{q_{n-2}} = \frac{a_{n-1}q_{n-2} + q_{n-3}}{q_{n-2}} = a_{n-1} + \frac{q_{n-3}}{q_{n-2}} = a_{n-1} + b_{n-1}. \tag{7.122}$$

Hence

$$\frac{1}{a'_n} + \frac{1}{b_n} = a'_{n-1} + b_{n-1} \le \sqrt{5}. \tag{7.123}$$

Therefore

$$1 = a'_n \frac{1}{a'_n} \le a'_n \left(\sqrt{5} - \frac{1}{b_n} \right) \le (\sqrt{5} - b_n) \left(\sqrt{5} - \frac{1}{b_n} \right)$$

$$= 6 - \sqrt{5} \left(b_n + \frac{1}{b_n} \right). \tag{7.124}$$

In other words

$$b_n + \frac{1}{b_n} \le \sqrt{5}. \tag{7.125}$$

Since b_n is rational, the inequality must be strict. Completing the square we obtain

$$b_n > \frac{1}{2}(\sqrt{5} - 1). \tag{7.126}$$

Now suppose

$$a'_{m-1} + b_{m-1} \le \sqrt{5}$$
$$a'_m + b_m \le \sqrt{5}$$
$$a'_{m+1} + b_{m+1} \le \sqrt{5}. \tag{7.127}$$

Applying the above reasoning that led to (7.126) to $n = m$ and $n = m + 1$, we obtain

$$b_m > \frac{1}{2}(\sqrt{5} - 1)$$

$$b_{m+1} > \frac{1}{2}(\sqrt{5} - 1). \tag{7.128}$$

By (7.121) with $n = m + 1$ and (7.122) and (7.123) with $n = m$,

$$a_m = \frac{1}{b_{m+1}} - b_m$$

$$< \frac{1}{b_{m+1}} - \frac{1}{2}(\sqrt{5} - 1)$$

$$< \frac{1}{\frac{1}{2}(\sqrt{5} - 1)} - \frac{1}{2}(\sqrt{5} - 1)$$

$$= \frac{1}{2}(\sqrt{5} + 1) - \frac{1}{2}(\sqrt{5} - 1) = 1. \tag{7.129}$$

However, a_m is a positive integer, and there are no positive integers less than 1. This contradiction completes the proof. $\qquad\square$

Remark 7.9.5. From the above we see that the approximation is often better than $\frac{1}{\sqrt{5}q^2}$. For example, if our continued fraction expansion has infinitely many 3's in its expansion we can do at least as well as $\frac{1}{3q^2}$ infinitely often.

Exercise 7.9.6. *Show that $\frac{1}{\sqrt{5}q^2}$ is the best one can have for all irrationals by studying the golden mean, $\frac{1+\sqrt{5}}{2} = [1, 1, 1, \ldots]$.*

Exercise 7.9.7. *Let x be any irrational other than the golden mean. How well can x be approximated? See for example [HW].*

Theorem 7.9.8. *Let $x \notin \mathbb{Q}$ and p and q be relatively prime integers. If $\left| x - \frac{p}{q} \right| < \frac{1}{2q^2}$ then $\frac{p}{q}$ is a convergent of x. Thus for some n, $p = p_n$ and $q = q_n$.*

Exercise 7.9.9. *Prove the above theorem. Is it true for $\left| x - \frac{p}{q} \right| < \frac{1}{q^2}$?*

Theorem 7.9.10. *Of any two consecutive convergents, one will satisfy*

$$\left| x - \frac{p}{q} \right| < \frac{1}{2q^2}. \tag{7.130}$$

Proof. See [HW], Theorem 183.   □

Lemma 7.9.11. *If $x = \frac{P\zeta + R}{Q\zeta + S}$ with $\zeta > 1$ and P, Q, R, S are integers such that $Q > S > 0$ and $PS - QR = \pm 1$, then $\frac{R}{S}$ and $\frac{P}{Q}$ are two consecutive convergents to x.*

Exercise 7.9.12. *Prove Lemma 7.9.11.*

See Remark 10.2.2 for better approximations that work for almost all numbers.

7.10 RESEARCH PROJECTS

We end this chapter with several partial results and interesting open conjectures and problems for the reader to pursue. See Chapter 10 for projects concerning the distribution of digits of continued fractions.

We have shown that x is a quadratic irrational if and only if its continued fraction is periodic from some point onward. Thus, given any repeating block we can find a quadratic irrational. In some sense this means we completely understand these numbers; however, depending on how we transverse countable sets we can see greatly different behavior.

For example, consider the following ordered subsets of $\mathbb{N}$:

$$S_1 = \{1, \mathbf{2}, 3, \mathbf{4}, 5, \mathbf{6}, 7, \mathbf{8}, 9, \mathbf{10}, 11, \mathbf{12}, \dots \}$$
$$S_2 = \{1, 3, \mathbf{2}, 5, 7, \mathbf{4}, 9, 11, \mathbf{6}, 13, 15, \mathbf{8}, \dots \}. \tag{7.131}$$

For N large, in the first set the even numbers make up about half of the first N numbers, while in the second set, only one-third. Simply by reordering the terms, we can adjust certain types of behavior. What this means is that, depending on how we transverse a set, we can see different limiting behaviors.

Exercise 7.10.1 (Rearrangement Theorem). *Consider a sequence of real numbers a_n that is conditionally convergent but not absolutely convergent: $\sum_{n=1}^{\infty} a_n$ exists and is finite, but $\sum_{n=1}^{\infty} |a_n| = \infty$; for example, $a_n = \frac{(-1)^n}{n}$. Prove by re-arranging the order of the a_n's one can obtain a new series which converges to any desired real number! Moreover, one can design a new sequence that oscillated between any two real numbers.*

Therefore, when we decide to investigate quadratic irrationals, we need to specify how the set is ordered. This is similar to our use of height functions to investigate rational numbers (see §4.3). One interesting set is $\mathcal{F}_N = \{\sqrt{n} : n \leq N\}$;
another is $\mathcal{G}_N = \{x : ax^2 + bx + c = 0, |a|, |b|, |c| \leq N\}$. We could fix a quadratic
irrational x and study its powers $\mathcal{H}_N = \{x^n : 0 < |n| \leq N\}$ or its multiples
$\mathcal{I}_N = \{nx : 0 < |n| \leq N\}$ or ratios $\mathcal{J}_N = \{\frac{x}{n} : 0 < |n| \leq N\}$.

Remark 7.10.2 (Dyadic intervals). In many applications, instead of considering
$0 < n \leq N$ one investigates $N \leq n \leq 2N$. There are many advantage to such
studies. For N large, all elements are of a comparable magnitude. Additionally,
often there are low number phenomena which do not persist at larger values: by
starting the count at 1, these low values could pollute the conclusions. For example,
looking at

$$\{1, 2, 3, 4, 5, 6, 7, 8, 9, 10\}, \tag{7.132}$$

we conclude 40% of numbers are prime, and 50% of primes p also have $p + 2$
prime (i.e., start a twin prime pair); further, these percentages hold if we extend
to $\{1, \ldots, 20\}$! Both these conclusions are false. We have seen in Theorem 2.3.9
that the proportion of numbers less than x that are prime is like $\frac{1}{\log x}$, and heuristics
(Chapters 13 and 14) indicate the proportion that are twin primes is $\frac{2C_2}{\log^2 x}$, where
$C_2 \approx .66016$ is the Hardy-Littlewood twin prime constant.

One must be very careful about extrapolations from data. A terrific example
is Skewes' number. Let $\pi(x)$ equal the number of primes at most x. A good
approximation to $\pi(x)$ is $\text{Li}(x) = \int_2^x \frac{dt}{\log t}$; note to first order, this integral is
$\frac{x}{\log x}$. By studying tables of primes, mathematicians were led to the conjecture
that $\pi(x) < \text{Li}(x)$. While simulations supported this claim, Littlewood proved that
which of the two functions is larger changes infinitely often; his student Skewes
[Sk] proved the first change occurs by $x = 10^{10^{10^{10^3}}}$. This bound has been significantly improved; however, one expects the first change to occur around 10^{250}.
Numbers this large are beyond the realm of experimentation. The moral is: for
phenomena whose natural scale is logarithmic (or log-logarithmic, and so on), numerics can be very misleading.

Research Project 7.10.3. Determine if possible simple closed formulas for the
sets $\mathcal{H}_N, \mathcal{I}_N$ and $\mathcal{J}_N$ arising from ϕ (the golden mean) and $x_{m,+}$ (see §7.8.1). In
particular, what can one say about $nx_{m,+}^k$ or $x_{m,+}^k/n$? How are the lengths of the
periods related to (m, k, n), and what digits occur in these sets (say for fixed m and
k, $0 < n \leq N$)? If $m = 1$, $x_{1,+} = \phi$, the golden mean. For $\frac{p}{q} \in \mathbb{Q}$, note $\frac{p}{q}x^k$ can
be written as $\frac{p_1}{q_1}\sqrt{5} + \frac{p_2}{q_2}$. Thus, in some sense, it is sufficient to study $\frac{p_1}{q_1}\sqrt{5}$. See
also §7.8.1.

Remark 7.10.4 (Important). Many of the formulas for the continued fraction expansions were first seen in numerical experiments. The insights that can be gained
by investigating cases on the computer cannot be underestimated. Often *seeing* the
continued fraction for some cases leads to an idea of how to do the calculations,

and at least as importantly what calculations may be interesting and worthwhile. For example,

$$\frac{\sqrt{5}}{4} = [0, 1, \overline{1, 3, 1, 2}]$$

$$\frac{\sqrt{5}}{8} = [0, 3, \overline{1, 1, 2, 1, 2, 1, 1, 6}]$$

$$\frac{\sqrt{5}}{16} = [0, 7, \overline{6, 2, 3, 3, 3, 2, 6, 14, 6, 2, 3, 3, 3, 2, 6, 14}]$$

$$\frac{\sqrt{5}}{10} = [0, 4, \overline{2, 8}]$$

$$\frac{\sqrt{5}}{6} = [0, 2, \overline{1, 2, 6, 2, 1, 4}]$$

$$\frac{\sqrt{5}}{12} = [0, 5, \overline{2, 1, 2, 1, 2, 10}]$$

$$\frac{\sqrt{5}}{14} = [0, 6, \overline{3, 1, 4, 1, 14, 1, 4, 1, 3, 12}]$$

$$\frac{\sqrt{5}}{28} = [0, 12, \overline{1, 1, 10, 1, 6, 1, 10, 1, 1, 24}]$$

$$\frac{\sqrt{5}}{42} = [0, 18, \overline{1, 3, 1, 1, 1, 1, 4, 1, 1, 1, 1, 3, 1, 36}]. \tag{7.133}$$

Research Project 7.10.5. The data in Remark 7.10.4 seem to indicate a pattern between the length of the repeating block and the factorization of the denominator, as well as what the largest digit is. Discover and prove interesting relations. How are the digits distributed (i.e., how many are 1's, 2's, 3's and so on; see Chapter 10 for more on the digits of a continued fraction). Also, the periodic expansions are almost symmetric (if one removes the final digit, the remaining piece is of the form $abc \ldots xyzyx \ldots cba$). Is this always true? What happens if we divide by other n, say odd n?

Research Project 7.10.6. How are the continued fractions of n-equivalent numbers related? We have seen quadratic irrationals have periodic continued fractions. Consider the following generalization. Fix functions $f_1, \ldots, f_k$, and study numbers of the form

$$[f_1(1), \ldots, f_k(1), f_1(2), \ldots, f_k(2), f_1(3), \ldots, f_k(3), f_1(4), \ldots]. \tag{7.134}$$

Which numbers have such expansions (say if the f_i's are linear)? See [Di] for some results. For results on multiplying continued fractions by rationals see [vdP1], and see [PS1, PS2, vdP3] for connections between power series and continued fractions.

Research Project 7.10.7. For more on the lengths of the period of $\sqrt{n}$ or $\sqrt{p}$, as well as additional topics to investigate, see [Be, Gl]. For a generalization to what has been called "linearly periodic" expansions, see [Di].

Research Project 7.10.8 (Davenport). Determine whether the digits of the continued fraction expansion of $\sqrt[3]{2} = [1, 3, 1, 5, 1, 1, 4, 1, \ldots]$ are bounded or not. This problem appears on page 107 of [Da1].

PART 3
Probabilistic Methods and Equidistribution

Chapter Eight

Introduction to Probability

In this chapter we give a quick introduction to the basic elements of Probability Theory, which we use to describe the limiting behavior of many different systems; for more details see [Du, Fe, Kel]. Consider all numbers in $[0,1]$. Let $p_{10,n}(k)$ be the probability that the nth decimal (base 10) digit is k for $k \in \{0,\dots,9\}$. It is natural to expect that each digit is equally likely. This leads us to conjecture that $p_{10,n}(k) = \frac{1}{10}$ for all n. There is nothing special about base 10 — the universe does not care that we have ten fingers on our hands. Thus if we were to write our numbers in base b, then $k \in \{0,1,\dots,b-1\}$ and it is natural to conjecture that $p_{b,n}(k) = \frac{1}{b}$. These statements can be easily proved. If we look at the nth digit of 10 million randomly chosen numbers, we expect to see about 1 million ones, 1 million twos, and so on; we will, of course, have to specify what we mean by randomly. What about the fluctuations about the expected values? Would we be surprised if we see $1,000,053$ ones? If we see $1,093,127$? The answer is given by the Central Limit Theorem, stated in §8.4 and proved in §11.5.

Instead of choosing numbers randomly in $[0,1]$, what if we consider special sequences? For example, how is the *first* digit of 2^n base 10 distributed? The possible digit values are $1,\dots,9$. Are all numbers equally likely to be the first digit of 2^n? We see in Chapter 9 that the answer is a resounding no. Another possible experiment is to investigate the nth decimal digit of $\sqrt{p}$ as p varies through the primes. Do we expect as $n \to \infty$ that each number 0 through 9 occurs equally often? Do numerical experiments support our conjecture? Building on this chapter, in Chapter 9 we discuss how to analyze such data.

The probability of observing a digit depends on the base we use. What if we instead write the continued fraction expansion (see Chapter 7) of numbers in $[0,1]$? The advantage of this expansion is that it does not depend on a base *as there is no base!* What is the probability that the nth digit of the continued fraction expansion equals k, $k \in \{1,2,\dots\}$? How likely is it that the nth digit is large, say more than a million? Small? We can already answer this question for certain numbers α. If α is rational then it has a finite continued fraction expansion; if α is a quadratic irrational, it has a periodic expansion. What is true about the expansions of the other $\alpha \in (0,1)$? We answer such questions in Chapter 10.

Let $\{x\}$ denote the fractional part of x. Thus $\{x\} = x \bmod 1$. Consider an irrational number $\alpha \in (0,1)$. For each N look at the N numbers $\{1\alpha\}, \{2\alpha\}, \dots, \{N\alpha\}$. Rearrange the above $\{n\alpha\}$ in increasing order, and for definiteness label them $\beta_1, \dots, \beta_N$:

$$0 \le \beta_1 \le \beta_2 \le \cdots \le \beta_N. \tag{8.1}$$

As we have N numbers in $[0,1]$, the average distance between numbers is about

$\frac{1}{N}$. What does the spacing between adjacent β_i's look like? How often are two adjacent β_i's twice the average spacing apart? Half the average spacing apart? We prove some results and describe open problems in Chapter 12, and then in Part 5 we investigate the spacings between eigenvalues of matrices, energy levels of heavy nuclei like Uranium and zeros of L-functions, showing connections between these very different systems!

8.1 PROBABILITIES OF DISCRETE EVENTS

We begin by studying the probabilities of discrete sets; for example, subsets of the integers or rationals or any finite set. Many interesting systems are discrete. One common example is flipping a coin a finite number of times; in this case we are often interested in the number of heads or tails. Another is to have time discrete; for example, people waiting in line at a bank, and every minute there is a chance a teller will serve the next person in line.

In the last example, if instead of measuring time in minutes we measured time in seconds or tenths of a second, for all practical purposes we would have a continuous process. While discrete sets are often good approximations to continuous processes, sometimes we actually need the continuous case; we describe continuous probability distributions in §8.2.3. We assume the reader is familiar with elementary set operations and countable sets (see §5.2).

8.1.1 Introduction

Definition 8.1.1 (Outcome Space, Outcomes). *Let* $\Omega = \{\omega_1, \omega_2, \omega_3, \dots\}$ *be an at most countable set. We call* Ω *the sample (or outcome) space, and the elements* $\omega \in \Omega$ *the outcomes.*

Thus, the outcome space is the collection of possible outcomes.

Example 8.1.2. *Flip a coin 3 times. The possible outcomes are*

$$\Omega = \{HHH, HHT, HTH, THH, HTT, THT, TTH, TTT\}. \qquad (8.2)$$

If we flip a coin three times, how many heads do we expect to see? What is the probability we observe exactly three heads? Exactly two heads? The answer depends on the coin. If the coin is fair, for each flip we have a 50% chance of getting a head and a 50% chance of getting a tail. The coin, however, need not be fair. It could have some probability p of landing on heads, and then probability $1-p$ of landing on tails. For many investigations, we need more than just a collection of possible outcomes: we need to know how likely each possible outcome is.

Definition 8.1.3 (Probability Function). *We say* $p(\omega)$ *is a (**discrete**) probability function or distribution on* Ω *if*

1. $0 \le p(\omega_i) \le 1$ *for all* $\omega_i \in \Omega$.

2. $\sum_i p(\omega_i) = 1$.

The first statement says that each outcome has a non-negative probability of occurring, and nothing can have a probability greater than 1 (a probability of 1 of
happening means the event happens); the second statement quantifies the observation that something definitely happens.

We call $p(\omega)$ the probability of the outcome ω. Given an outcome space with a
probability function, we can investigate functions of the outcomes.

Definition 8.1.4 (Random Variable). *Let X be a function from Ω to $\mathbb{R}$. That is, for
each outcome $\omega \in \Omega$ we attach a real number $X(\omega)$. We call X a random variable.*

A random variable is essentially a function of the outcomes, assigning a number
to each outcome. As there are many functions that could convert outcomes to
numbers, for any outcome space there are many random variables. With the same
outcome space from Example 8.1.2, one possible random variable is $X(\omega)$ equals
the number of heads in ω. Thus, $X(HHT) = 2$ and $X(TTT) = 0$. Additionally,
for $i \in \{1, 2, 3\}$ let

$$X_i(\omega) = \begin{cases} 1 & \text{if the } i^{\text{th}} \text{ toss is a head} \\ 0 & \text{if the } i^{\text{th}} \text{ toss is a tail.} \end{cases} \tag{8.3}$$

Note that

$$X(\omega) = X_1(\omega) + X_2(\omega) + X_3(\omega). \tag{8.4}$$

Remark 8.1.5 (Important). The following situation occurs frequently. Consider the
case when $\Omega \subset \mathbb{R}$ and X is a random variable. We often adjust our notation and
write x for $\omega \in \Omega$; thus a capital letter denotes a random variable and a lowercase
letter denotes a value it attains. For example, consider a roll of a fair die. The
outcome space is $\Omega = \{1, 2, 3, 4, 5, 6\}$, and the probability of each $\omega \in \Omega$ is $\frac{1}{6}$.
Let X be the number rolled on the die. Then $X(1) = 1$, $X(2) = 2$, and so on.
In this example, it is very convenient to call the outcome space the number rolled.
The outcomes are the numbers 1, 2 and so on, rather then "the dice is a 1," "the
dice is a 2"; X is the random variable that is the number rolled, taking on values
$x \in \{1, \ldots, 6\}$. We shall mostly use $X : \Omega \to \mathbb{R}$ to represent a random variable
and emphasize that the outcome space need not be a subset of $\mathbb{R}$, though the reader
should be aware of both notations.

Example 8.1.6 (Important). *Given an outcome space Ω with events ω with probability function p, p is a random variable.*

The terminology can be confusing, as a given random variable X is clearly not
random — it is what it is! The point is we can attach many different random
variable to a given Ω.

8.1.2 Events

Definition 8.1.7 (Events). *We call a subset $A \subset \Omega$ an event, and we write*

$$\text{Prob}(A) = \sum_{\omega \in A} p(\omega). \tag{8.5}$$

Note each outcome is also an event.

Definition 8.1.8 (Range of X). *The range of a random variable X is the set of values it attains, denoted $X(\Omega)$:*

$$X(\Omega) \ = \ \{r \in \mathbb{R} : \exists \omega \in \Omega \text{ with } X(\omega) = r\}. \tag{8.6}$$

Note $X(\Omega)$ is the set of values attained by $X(\omega)$ as we vary $\omega \in \Omega$. Given a set $S \subset X(\Omega)$, we let $X^{-1}(S) = \{\omega \in \Omega : X(\omega) \in S\}$. This is the set of all outcomes where the random variable assigns a number in S.

Exercise 8.1.9. *Let Ω be the space of all tosses of a fair coin where all but the last toss are tails, and the last is a head. Thus $\Omega = \{H, TH, TTH, TTTH, \dots\}$. One possible random variable is X equals the number of tails; another is Y equals the number of the flip which is a head. Calculate the probabilities of the following outcomes in Ω. What is the probability that $X(\omega) \leq 3$? What is the probability that $Y(\omega) > 3$? What events do these correspond to?*

In general, we can associate events to any random variable. Let Ω be an outcome space with outcomes ω, and let X be a random variable. As we are assuming Ω is countable, the random variable X takes on at most countably many distinct values, so the range $X(\Omega)$ is at most countable. Let x_i denote a typical value. For each x_i, we can form the event $X(\omega) = x_i$; let us denote this event by A_i:

$$A_i \ = \ \{\omega \in \Omega : X(\omega) = x_i\} \subset \Omega. \tag{8.7}$$

Note that the A_i's are disjoint sets; if $\omega \in A_i \cap A_j$, then $X(\omega) = x_i$ as well as x_j. Further, $\cup_i A_i = \Omega$, because given any $\omega \in \Omega$, $X(\omega) = x_i$ for some i, hence ω is in some set A_i. The sets A_i form a **partition** of Ω (every $\omega \in \Omega$ is in one and only one A_i).

Remark 8.1.10 (Important). By the above, given an outcome space Ω with outcomes ω and a probability function p and a random variable X, we can form a new outcome space $\widetilde{\Omega}$ with outcomes x_i with probability function $\widetilde{p}$ given by

$$\widetilde{p}(x_i) \ = \ \sum_{\substack{\omega \in \Omega \\ X(\omega) = x_i}} p(\omega). \tag{8.8}$$

Remark 8.1.11 (Important). In a convenient abuse of notation, we often write

$$p(x_i) \ = \ p(X(\omega) = x_i) \ = \ \text{Prob}(\omega \in \Omega : X(\omega) = x_i). \tag{8.9}$$

We also call the random variable X an event, as the subsets of Ω corresponding to different values of X are events. Thus we can talk about the event "the value of the first roll," as the following example and Example 8.1.14 illustrate.

Example 8.1.12. *Let Ω be the set of all possible pairs of rolls of a fair die, and $X(\omega)$ equals the number of the first roll. We obtain events $A_1, \dots, A_6$. Let $Y(\omega)$ equal the number of the second roll, giving events $B_1, \dots, B_6$. If we consider the sum rolled, we have events $C_2, \dots, C_{12}$. For example, $C_7 = \{(1, 6), (2, 5), (3, 4), (4, 3), (5, 2), (6, 1)\}$. See Chapter 9 of [Sc] for a plethora of interesting problems on dice.*

Exercise 8.1.13. *Calculate the probabilities of the events $C_2, \ldots, C_{12}$ for Example 8.1.12.*

Example 8.1.14 (Characteristic or Indicator Functions). *We continue to reconcile our two notions of an event, namely a subset $A \subset \Omega$ and a random variable X. To any $A \subset \Omega$ we can associate a **characteristic** or **indicator random variable** 1_A as follows:*

$$1_A(\omega) = \begin{cases} 1 & \text{if } \omega \in A \\ 0 & \text{if } \omega \notin A. \end{cases} \tag{8.10}$$

Thus A is the set of ω where $1_A(\omega) = 1$.

Definition 8.1.15 (Complements). *The complement of a set $A \subset \Omega$ is the set of all $\omega \notin A$. We denote this by A^c:*

$$A^c = \{\omega : \omega \in \Omega, \omega \notin A\}. \tag{8.11}$$

Using complements, we can rewrite the definition of the indicator random variable X_A:

$$X_A(\omega) = \begin{cases} 1 & \text{if } \omega \in A \\ 0 & \text{if } \omega \in A^c. \end{cases} \tag{8.12}$$

Lemma 8.1.16. *Consider an outcome space Ω with outcomes ω and probability function p. Let $A \subset \Omega$ be an event. Then*

$$p(A) = 1 - p(A^c). \tag{8.13}$$

This simple observation is extremely useful for calculating many probabilities, as sometimes $p(A^c)$ is significantly easier to determine.

Exercise 8.1.17. *Prove the above lemma. Consider 100 tosses of a fair coin. What is the probability that at least three tosses are heads?*

Exercise[hr] **8.1.18.** *Consider 100 tosses of a fair coin. What is the probability that at least three consecutive tosses are heads? What about at least five consecutive tosses?*

Given an outcome space Ω with outcomes ω and random variable X, we can define a new random variable $Y = aX$, $a \in \mathbb{R}$, by $Y(\omega) = a \cdot X(\omega)$. This implies $p(Y(\omega) = ax_i) = p(X(\omega) = x_i)$. Thus if $X(\omega)$ takes on the values x_i with probabilities $p(x_i)$, $Y(\omega) = a \cdot X(\omega)$ takes on the values ax_i with probabilities $p(x_i)$.

Exercise 8.1.19. *Let X be a random variable on an outcome space Ω with probability function p. Fix a constant a and let $Y(\omega) = X(\omega) + a$. Determine the probability $Y(\omega) = y_i$.*

8.1.3 Conditional Probabilities

Consider two probability spaces Ω_1 and Ω_2 with outcomes ω_1 and ω_2. We can define a new outcome space

$$\Omega = \{\omega = (\omega_1, \omega_2) : \omega_1 \in \Omega_1 \text{ and } \omega_2 \in \Omega_2\}, \qquad (8.14)$$

with outcomes $\omega = (\omega_1, \omega_2)$. We need to define a probability function $p(\omega)$, i.e., we need to assign probabilities to these outcomes. One natural way is as follows: let p_i be the probability function for outcomes $\omega_i \in \Omega_i$. We define

$$p(\omega) = p_1(\omega_1) \cdot p_2(\omega_2) \text{ if } \omega = (\omega_1, \omega_2). \qquad (8.15)$$

Exercise 8.1.20. *Show the above defines a probability function.*

Of course, we could also define a probability function $p : \Omega \to \mathbb{R}$ directly. We again consider two tosses of a fair coin. We have outcomes $\omega = (\omega_1, \omega_2)$. Let us define $p(\omega) = \frac{1}{36}$, i.e., each of the 36 outcomes is equally likely. Let $X(\omega) = \omega_1$, the roll of the first die; similarly, set $Y(\omega) = \omega_2$, the roll of the second die.

Example 8.1.21. *What is* $\mathrm{Prob}(X(\omega) = 2)$*? There are 6 pairs with first roll 2:* $(2, 1), (2, 2), \ldots, (2, 6)$. *Each pair has probability* $\frac{1}{36}$. *Thus,* $\mathrm{Prob}(X(\omega) = 2) = \frac{6}{36} = \frac{1}{6}$.

More generally we have

$$\mathrm{Prob}\,(X(\omega) = x_i) = \sum_{\substack{\omega = (\omega_1, \omega_2) \\ X(\omega) = x_i}} p(\omega). \qquad (8.16)$$

The above is a simple recipe to find $\mathrm{Prob}\,(X(\omega) = a)$: it is the probability of all pairs (ω_1, ω_2) such that $X(\omega) = x_i$, ω_2 arbitrary.

Let us consider a third random variable, the sum of the two rolls. Thus let $Z(\omega) = \omega_1 + \omega_2$, each outcome $\omega = (\omega_1, \omega_2)$ occurs with probability $\frac{1}{36}$. We have just seen that, if we have no information about the second roll, the probability that the first roll is a 2 is $\frac{1}{6}$ (what we would expect). What if, however, we know the sum of the two rolls is 2, or 7 or 10? Now what is the probability that the first roll is a 2? We are looking for pairs (ω_1, ω_2) such that $\omega_1 = 2$ and $\omega_1 + \omega_2 = 2, 7$, or 10. A quick inspection shows there are no pairs with sum 2 or 10. For a sum of 7, only one pair works: $(2, 5)$.

This leads us to the concept of **conditional probability**: *what is the probability of an event A, given an event B has occurred?* For an event A we can write

$$\mathrm{Prob}(A) = \frac{\sum_{\omega \in A} p(w)}{\sum_{\omega \in \Omega} p(\omega)}. \qquad (8.17)$$

Note the denominator is 1. For conditional probabilities, we restrict to $\omega \in B$. Thus, we have

$$\mathrm{Prob}(A|B) = \frac{\sum_{\substack{w \in A \\ w \in B}} p(w)}{\sum_{w \in B} p(\omega)}. \qquad (8.18)$$

The numerator above may be regarded as the event $A \cap B$ (as both must happen, ω must be in A and B). $\mathrm{Prob}(A|B)$ is read *the probability of A, given B occurs* (or as the conditional probability of A given B). Thus,

Lemma 8.1.22. *If* $\mathrm{Prob}(B) \neq 0$,

$$\mathrm{Prob}(A|B) = \frac{\mathrm{Prob}(A \cap B)}{\mathrm{Prob}(B)}. \tag{8.19}$$

In the example above, let A be the event that the first roll is a 2 and B the event that the sum of the rolls is 7. As the die are fair, the probability of any pair (ω_1, ω_2) is $\frac{1}{36}$. Then

$$A = \{(2,1), (2,2), (2,3), (2,4), (2,5), (2,6)\}$$
$$B = \{(1,6), (2,5), (3,4), (4,3), (5,2), (6,1)\}$$
$$A \cap B = \{(2,5)\}$$
$$\mathrm{Prob}(A|B) = \frac{\mathrm{Prob}(A \cap B)}{\mathrm{Prob}(B)} = \frac{\frac{1}{36}}{6 \cdot \frac{1}{36}} = \frac{1}{6}. \tag{8.20}$$

Exercise 8.1.23. *Let* Ω *be the results of two rolls of two dice, where* ω_1 *is the number rolled first and* ω_2 *the number rolled second. For* $\omega = (\omega_1, \omega_2) \in \Omega$, *define the probabilities of the outcomes by*

$$p(\omega) = \begin{cases} \frac{1.5}{36} & \text{if } \omega_1 \text{ is even} \\ \frac{.5}{36} & \text{if } \omega_1 \text{ is odd.} \end{cases} \tag{8.21}$$

Show the above is a probability function of Ω. *Let* $X(\omega)$ *be the number of the first roll,* $Y(\omega)$ *the number of the second roll. For each* $k \in \{1, \ldots, 6\}$, *what is the probability that* $Y(\omega) = k$ *given* $X(\omega) = 2$? *Given* $X(\omega) = 1$?

Exercise 8.1.24. *Three players enter a room and a red or blue hat is placed on each person's head. The color of each hat is determined by a coin toss, with the outcome of one coin toss having no effect on the others. Each person can see the other players' hats but not their own. No communication of any sort is allowed, except for an initial strategy session before the game begins. Once they have had a chance to look at the other hats, the players must simultaneously guess the color of their own hats or pass. The group shares a $3 million prize if at least one player guesses correctly and no players guess incorrectly. One can easily find a strategy which gives them a 50% chance of winning; using conditional probability find one where they win 75% of the time! More generally find a strategy for a group of* n *players that maximizes their chances of winning. See [Ber, LS] for more details, as well as [CS, LS] for applications to error correcting codes.*

8.1.4 Independent Events

The concept of **independence** is one of the most important in probability. Simply put, two events are independent if knowledge of one gives no information about the other. Explicitly, the probability of A occurring given that B has occurred is the same as if we knew nothing about whether or not B occurred:

$$\mathrm{Prob}(A|B) = \frac{\mathrm{Prob}(A \cap B)}{\mathrm{Prob}(B)} = \mathrm{Prob}(A). \tag{8.22}$$

Knowing event B occurred gives no additional information on the probability that event A occurred.

Again, consider two rolls of a fair dice with outcome space Ω consisting of pairs of rolls $\omega = (\omega_1, \omega_2)$. Let $X(\omega) = \omega_1$ (the result of the first roll), $Y(\omega) = \omega_2$ (the result of the second roll) and $Z(\omega) = X(\omega) + Y(\omega) = \omega_1 + \omega_2$ (the sum of the two rolls). Let A be the event that the first roll is 2 and B the event that the sum of the two rolls is 7. We have shown

$$\text{Prob}(A|B) = \frac{1}{6} = \text{Prob}(A); \qquad (8.23)$$

thus, A and B are independent events. If, however, we had taken B to be the event that the sum of the two rolls is 2 (or 10), we would have found

$$\text{Prob}(A|B) = 0 \neq \text{Prob}(A); \qquad (8.24)$$

in this case, the two events are not independent.

We rewrite the definition of independence in a more useful manner. Since for two independent events A and B,

$$\text{Prob}(A|B) = \frac{\text{Prob}(A \cap B)}{\text{Prob}(B)} = \text{Prob}(A), \qquad (8.25)$$

we have

$$\text{Prob}(A \cap B) = \text{Prob}(A)\text{Prob}(B). \qquad (8.26)$$

Note the more symmetric form of the above. In general, events $A_1, \ldots, A_n$ are independent if for any subset $\{i_1, \ldots, i_k\}$ of $\{1, \ldots, n\}$ we have

$$\text{Prob}(A_{i_1} \cap A_{i_2} \cap \cdots \cap A_{i_k}) = \text{Prob}(A_{i_1})\text{Prob}(A_{i_2}) \cdots \text{Prob}(A_{i_l}). \qquad (8.27)$$

If events $A_1, \ldots, A_n$ are pairwise independent, it is possible that the events are not independent.

Exercise 8.1.25. *Consider two tosses of a fair coin, each pair occurs with probability $\frac{1}{4}$. Let A be the event that the first toss is a head, B the event that the second toss is a tail and C the event that the sum of the number of heads is odd. Prove the events are pairwise independent, but not independent.*

Example 8.1.26. *Consider a fair die. Let A be the event that the first roll equals a, B be the event that the second roll equals b and C be the event that the sum of the two rolls is c, $c \in \{2, \ldots, 12\}$. As each pair of rolls is equally likely, the probability that the first roll is a is $\frac{1}{6}$ (as six of the thirty-six pairs give a first roll of a). Thus, for any choices of a and b, the result of the first roll is independent of the second roll. We say that the two rolls (or the events A and B) are independent.*

Consider now event C, the sum of the two rolls. If the sum of the rolls is 7, then the probability that the first roll equals a is $\frac{1}{6}$ for all a; however, in general the conditional probabilities for the first roll will depend on the sum. For example, if the sum is 2 then the probability that the first roll is 1 is 1 and the probability that the first roll is 2 or more is 0. Thus, events A and C (the first roll and the sum of the rolls) are not independent.

Definition 8.1.27 (Independent Random Variables). *Let X and Y be two random variables. We can associate events $A_i = \{\omega \in \Omega : X(\omega) = x_i\}$ and $B_j = \{\omega \in \Omega : Y(\omega) = y_j\}$. If for all i and j the events A_i and B_j are independent, we say the random variables X and Y are independent:* knowledge of the value of Y yields no information about the value of X.

Exercise 8.1.28. *Again consider two tosses of a fair coin, with $X(\omega)$ the number of the first toss and $Y(\omega)$ the number of the second toss. Prove X and Y are independent. Let Z be the random variable which is the number of heads in two tosses. Prove X and Z are not independent.*

The above exercise appears throughout probability investigations. For example, if we choose a non-rational $\alpha \in (0, 1)$ "at random," we could let $X(\alpha)$ denote the value of the first decimal digit, and $Y(\alpha)$ denote the value of the second decimal digit. Are X and Y independent? The answer will depend on how we "randomly" choose α.

We give an example typical of the independence we will see in our later investigations. Let $\Omega_i = \{0, 1\}$ and for some finite N consider $\Omega = \Omega_1 \times \cdots \times \Omega_N$. For each i, define probability functions $p_i(1) = q_i$ and $p_i(0) = 1 - q_i$, $q_i \in [0, 1]$, and for $\omega = (\omega_1, \ldots, \omega_N) \in \Omega$, let $p(\omega) = \prod_i p_i(\omega_i)$. We may interpret this as follows: we toss N coins, where coin i has probability q_i of being heads. The outcome of each toss is independent of all the other tosses.

Exercise 8.1.29 (The Birthday Problem). *Assume each day of the year is equally likely to be someone's birthday, and no one is ever born on February 29th. How many people must there be in a room before there is at least a 50% chance that two share a birthday? How many other people must there be before at least one of them shares your birthday? Note the two questions have very different answers, because in the first we do not specify beforehand which is the shared day, while in the second we do. See also Exercise A.4.8.*

Exercise 8.1.30. *Redo the previous problem assuming that there are one-fourth as many people born on February 29th as on any other day.*

Exercise[hr] **8.1.31.** *Two players roll die with k sides, with each side equally likely of being rolled. Player one rolls m dice and player two rolls n dice. If player one's highest roll exceeds the highest roll of player two then player one wins, otherwise player two wins. Prove*

$$\text{Prob(Player one wins)} = \frac{1}{k^{m+n}} \sum_{a=2}^{k} [a^m - (a-1)^m] \cdot (a-1)^n, \quad (8.28)$$

which by the integral version of partial summation equals

$$\frac{1}{k^{m+n}} \left[k^m \cdot (k-1)^n - \int_1^k [u]^m \cdot n(u-1)^{n-1} du \right]. \quad (8.29)$$

If m, n and k are large and of approximately the same size, show

$$\text{Prob(Player one wins)} = \frac{m}{m+n} - \frac{m}{2(m+n-1)} \frac{n}{k}; \quad (8.30)$$

note if $m = n = k$ the probability is much less than 50%. See [Mil7] for more details.

8.1.5 Expectation

Definition 8.1.32 (Expected Value). *Consider an outcome space Ω with outcomes ω_i occurring with probabilities $p(\omega_i)$ and a random variable X. The expected value (or mean or average value) of the random variable X is defined by*

$$\overline{X} = \sum_i X(\omega_i)p(\omega_i). \tag{8.31}$$

We often write $\mathbb{E}[X]$, read as **the expected value** or **expectation of** X, for $\overline{X}$.

Exercise 8.1.33. *Show the mean of one roll of a fair dice is 3.5. Consider N tosses of a fair coin. Let $X(\omega)$ equal the number of heads in $\omega = (\omega_1, \ldots, \omega_N)$. Determine $\mathbb{E}[X]$.*

Remark 8.1.34. Remember we may regard random variables as events; thus it makes sense to talk about the mean value of such events, as the events are real numbers. If we considered an event not arising through a random variable, things would not be as clear. For example, consider $\Omega = \{HH, HT, TH, TT\}$, each with probability $\frac{1}{4}$. We cannot add a head and a tail; however, if we assign a 1 to a head and a 0 to the tail, we need only add numbers.

Exercise 8.1.35. *Consider all finite fair tosses of a coin where all but the last toss are tails (and the last toss is a head). We denote the outcome space by*

$$\Omega = \{H, TH, TTH, TTTH, \ldots\}. \tag{8.32}$$

Let X be the random variable equal to the number of the toss which is the head. For example, $X(TTH) = 3$. Calculate the probability that the first head is the ith toss. Calculate $\mathbb{E}[X]$.

Definition 8.1.36 (k^{th} Moment). *The kth moment of X is the expected value of x^k. If X is a random variable on an outcome space Ω with events ω_i, we write*

$$\mathbb{E}[X^k] = \sum_{\omega_i \in \Omega} X(\omega_i)^k \cdot p(\omega_i). \tag{8.33}$$

Note the first moment is the expected value of X, and the zeroth moment is always 1.

Definition 8.1.37 (Moments of Probability Distributions). *Let $\Omega \subset \mathbb{R}$; thus all events are real numbers, which we shall denote by $x \in \Omega$. Let p be a probability distribution on Ω so that the probability of x is just $p(x)$. We can consider a random variable X with $X(x) = x$; thus the probability that the random variable takes on the value x is $p(x)$. Equivalently we can consider p as a random variable (see Example 8.1.6). We define the kth moment of p by*

$$p_k = \mathbb{E}[X^k] = \sum_{x \in \Omega} x^k p(x). \tag{8.34}$$

Similar to how Taylor series coefficients can often determine a "nice" function, a sequence of moments often uniquely determines a probability distribution. We will use such a moment analysis in our Random Matrix Theory investigations in Part 5; see §15.3.2 for more details.

Exercise 8.1.38. *Prove the zeroth moment of any probability distribution is 1.*

Lemma 8.1.39 (Additivity of the Means). *If X and Y are two random variables on Ω with a probability function p, they induce a joint probability function P with*

$$P(x_i, y_j) := \text{Prob}(X(\omega) = x_i, Y(\omega) = y_j). \tag{8.35}$$

Consider the random variable Z, $Z = X + Y$. Then $\mathbb{E}[Z] = \mathbb{E}[X] + \mathbb{E}[Y]$.

Proof. First note

$$\text{Prob}(X(\omega) = x_i) = \sum_j \text{Prob}(X(\omega) = x_i, Y(\omega) = y_j) = \sum_j P(x_i, y_j). \tag{8.36}$$

Thus the expected value of the random variable X is

$$\mathbb{E}[X] = \sum_i x_i \sum_j P(x_i, y_j), \tag{8.37}$$

and similarly for the random variable Y. Therefore

$$\begin{aligned}
\mathbb{E}[X + Y] &= \sum_{(i,j)} (x_i + y_j) P(x_i, y_j) \\
&= \sum_i \sum_j x_i P(x_i, y_j) + \sum_i \sum_j y_j P(x_i, y_j) \\
&= \sum_i x_i \sum_j P(x_i, y_j) + \sum_j y_j \sum_i P(x_i, y_j) \\
&= \mathbb{E}[X] + \mathbb{E}[Y].
\end{aligned} \tag{8.38}$$

$\square$

The astute reader may notice that some care is needed to interchange the order of summations. If $\sum_i \sum_j |x_i + y_j| p(x_i, y_j) < \infty$, then Fubini's Theorem (Theorem A.2.8) is applicable and we may interchange the summations at will. For an example where the summations cannot be interchanged, see Exercise 11.4.12.

Lemma 8.1.40 (Expectation Is Linear). *Let X_1 through X_N be a finite collection of random variables. Let a_1 through a_N be real constants. Then*

$$\mathbb{E}[a_1 X_1 + \cdots + a_N X_N] = a_1 \mathbb{E}[X_1] + \cdots + a_N \mathbb{E}[X_N]. \tag{8.39}$$

See §10.5.2 for an application of the linearity of expected values to investigating digits of continued fractions.

Exercise 8.1.41. *Prove Lemma 8.1.40.*

Lemma 8.1.42. *Let X and Y be independent random variables. Then $\mathbb{E}[XY] = \mathbb{E}[X]\mathbb{E}[Y]$.*

Proof. From Definition 8.1.27, for all i and j the events $A_i = \{\omega : X(\omega) = x_i\}$ and $B_j = \{\omega : Y(\omega) = y_j\}$ are independent. This implies

$$\text{Prob}(A_i \cap B_j) = \text{Prob}(A_i)\text{Prob}(B_j) = p(x_i)q(y_j). \tag{8.40}$$

If $r(x_i, y_j)$ is the probability that the random variable X is x_i and the random variable Y is y_j, then independence implies $r(x_i, y_j) = p(x_i)q(y_j)$ for two probability functions p and q. Thus,

$$
\begin{aligned}
\mathbb{E}[XY] &= \sum_i \sum_j x_i y_j r(x_i, y_j) \\
&= \sum_i \sum_j x_i y_j p(x_i) q(y_j) \\
&= \sum_i x_i p(x_i) \cdot \sum_j y_j q(y_j) \\
&= \mathbb{E}[X] \cdot \mathbb{E}[Y].
\end{aligned}
\tag{8.41}
$$

$\square$

Exercise 8.1.43. *Find two random variables such that $\mathbb{E}[XY] \neq \mathbb{E}[X]\mathbb{E}[Y]$.*

Exercise 8.1.44 (Two Envelope Problem). *Consider two sealed envelopes; one has X dollars inside and the other has $2X$ dollars, $X > 0$. You are randomly given an envelope — you have an equal likelihood of receiving either. You calculate that you have a 50% chance of having the smaller (larger) amount. Let Y be the amount in your envelope. If you keep this envelope you expect to receive say Y dollars; if you switch your expected value is $.5 \cdot 2Y + .5 \cdot \frac{Y}{2}$, or $1.25Y$. But this is true without ever looking inside the envelope, so you should switch again! What is wrong with the above analysis?*

8.1.6 Variances

The **variance** σ_X^2 and its square root, the **standard deviation** σ_X measure how spread out the values taken on by a random variable are: the larger the variance, the more spread out the distribution.

Definition 8.1.45 (Variance). *Given an outcome space Ω with outcomes ω_i with probabilities $p(\omega_i)$ and a random variable $X : \Omega \to \mathbb{R}$, the variance σ_X^2 is*

$$
\sigma_X^2 = \sum_i \left(X(\omega_i) - \mathbb{E}[X] \right)^2 p(\omega_i) = \mathbb{E}\left[\left(X - \mathbb{E}[X] \right)^2 \right].
\tag{8.42}
$$

Exercise 8.1.46. *Let $\Omega_1 = \{0, 25, 50, 75, 100\}$ with probabilities $\{.2, .2, .2, .2, .2\}$, and let X be the random variable $X(\omega) = \omega$, $\omega \in \Omega_1$. Thus $X(0) = 0, X(25) = 25$, and so on. Let Ω_2 be the same outcome space but with probabilities $\{.1, .25, .3, .25, .1\}$, and define $Y(\omega) = \omega$, $\omega \in \Omega_2$. Calculate the means and the variances of X and Y.*

For computing variances, instead of (8.42) one often uses

Lemma 8.1.47. *For a random variable X we have $\sigma_X^2 = \mathbb{E}[X^2] - \mathbb{E}[X]^2$.*

Proof. Recall $\overline{X} = \mathbb{E}[X]$. Then

$$\sigma_X^2 = \sum_i (X_i(\omega) - \mathbb{E}[X])^2 \, p(\omega_i)$$

$$= \sum_i (X_i(\omega)^2 - 2X_i(\omega)\mathbb{E}[X] + \mathbb{E}[X]^2)p(\omega_i)$$

$$= \sum_i X_i(\omega)^2 p(\omega_i) - 2\mathbb{E}[X] \sum_i X_i(\omega)p(\omega_i) + \mathbb{E}[X]^2 \sum_i p(\omega_i)$$

$$= \mathbb{E}[X^2] - 2\mathbb{E}[X]^2 + \mathbb{E}[X]^2 = \mathbb{E}[X^2] - \mathbb{E}[X]^2. \qquad (8.43)$$

□

The main result on variances is

Lemma 8.1.48 (Variance of a Sum). *Let X and Y be two independent random variables on an outcome space Ω. Then $\sigma_{X+Y}^2 = \sigma_X^2 + \sigma_Y^2$.*

Proof. We use the fact that the expected value of a sum is the sum of expected values (Lemma 8.1.40).

$$\begin{aligned}
\sigma_{X+Y}^2 &= \mathbb{E}[(X+Y)^2] - \mathbb{E}[(X+Y)]^2 \\
&= \mathbb{E}[X^2 + 2XY + Y^2] - (\mathbb{E}[X] + \mathbb{E}[Y])^2 \\
&= \left(\mathbb{E}[X^2] + 2\mathbb{E}[XY] + \mathbb{E}[Y^2]\right) - \left(\mathbb{E}[X]^2 + 2\mathbb{E}[X]\mathbb{E}[Y] + \mathbb{E}[Y]^2\right) \\
&= \left(\mathbb{E}[X^2] - \mathbb{E}[X]^2\right) + \left(\mathbb{E}[Y^2] - \mathbb{E}[Y]^2\right) + 2\left(\mathbb{E}[XY] - \mathbb{E}[X]\mathbb{E}[Y]\right) \\
&= \sigma_X^2 + \sigma_Y^2 + 2\left(\mathbb{E}[XY] - \mathbb{E}[X]\mathbb{E}[Y]\right). \qquad (8.44)
\end{aligned}$$

By Lemma 8.1.42, as X and Y are independent, $\mathbb{E}[XY] = \mathbb{E}[X]\mathbb{E}[Y]$, completing the proof. □

Let Ω be an outcome space with outcomes ω and a random variable X. For $i \leq N$ let $\Omega_i = \Omega$ and let X_i be the same random variable as X except X_i lives on Ω_i. For example, we could have N rolls with X_i the outcome of the ith roll. We have seen in Lemma 8.1.40 that the mean of the random variable $X_1 + \cdots + X_N$ is $N\mathbb{E}[X]$. What is the variance?

Lemma 8.1.49. *Notation as above,*

$$\sigma_{X_1 + \cdots + X_N} = \sqrt{N}\sigma_X. \qquad (8.45)$$

Exercise 8.1.50. *Prove Lemma 8.1.49.*

Lemma 8.1.51. *Given an outcome space Ω with outcomes ω with probabilities $p(\omega)$ and a random variable X. Consider the new random variable $aX + b$. Then*

$$\sigma_{aX+b}^2 = a^2 \sigma_X^2. \qquad (8.46)$$

Exercise 8.1.52. *Prove 8.1.51.*

Note that if the random variable X has units of meters then the variance σ_X^2 has units of meters^2, and the standard deviation σ_X and the mean $\overline{X}$ have units meters.

Thus it is the standard deviation that gives a good measure of the deviations of X around its mean.

There are, of course, alternate measures one can use. For example, one could consider

$$\sum_i (x_i - \overline{X})p(x_i). \tag{8.47}$$

Unfortunately this is a signed quantity, and large positive deviations can cancel with large negatives. In fact, more is true.

Exercise 8.1.53. *Show* $\sum_i (x_i - \overline{X})p(x_i) = 0$.

This leads us to consider

$$\sum_i |x_i - \overline{X}|p(x_i). \tag{8.48}$$

While this has the advantage of avoiding cancellation of errors (as well as having the same units as the events), the absolute value function is not a good function analytically. For example, it is not differentiable. This is primarily why we consider the standard deviation (the square root of the variance).

Exercise 8.1.54 (Method of Least Squares). *Consider the following set of data: for* $i \in \{1, \ldots, n\}$, *given* t_i *one observes* y_i. *Believing that t and y are linearly related, find the best fit straight line. Namely, determine constants a and b that minimize the error (calculated via the variance)*

$$\sum_{i=1}^{n} (y_i - (at_i + b))^2 = \sum_{i=1}^{n} (\text{Observed}_i - \text{Predicted}_i)^2. \tag{8.49}$$

Hint: *Use multi-variable calculus to find linear equations for a and b, and then solve with linear algebra. If one requires that $a = 0$, show that the b leading to least error is $b = \overline{y} = \frac{1}{n}\sum_i y_i$.*

The method of proof generalizes to the case when one expects y is a linear combination of N fixed functions. The functions need not be linear; all that is required is that we have a linear combination, say $a_1 f_1(t) + \cdots + a_N f_N(t)$. One then determines the $a_1, \ldots, a_N$ that minimize the variance (the sum of squares of the errors) by calculus and linear algebra. If instead of measuring the total error by the squares of the individual error we used another measure (for example, using the absolute value), closed form expressions for the a_i become significantly harder, even in the simple case of fitting a line.

Exercise 8.1.55 (Chebyshev's Theorem). *Let X be a random variable with mean μ and finite variance σ^2. Prove Chebyshev's Theorem:*

$$\text{Prob}(|X - \mu| \geq k\sigma) \leq \frac{1}{k^2}, \tag{8.50}$$

where $\text{Prob}(|X - \mu| \geq a)$ is the probability that X takes on values at least a units from the mean. Chebyshev's theorem holds for all nice distributions, and provides bounds for being far away from the mean (where far is relative to the natural spacing, namely σ).

Exercise 8.1.56. *Use Chebyshev's Theorem to bound the probability of tossing a fair coin* 10000 *times and observing at least* 6000 *heads.*

If the probability distribution decays sufficiently rapidly we can use the Central Limit Theorem (Theorem 8.4.1) and obtain better estimates than those from Chebyshev's Theorem. See Exercise 8.4.3.

8.2 STANDARD DISTRIBUTIONS

We describe several common probability distributions. Consider the important case when the outcome space $\Omega \subset \mathbb{R}$ and is countable; thus the outcomes are real numbers. Let p be a probability function on Ω. For notational convenience we sometimes extend Ω to all of $\mathbb{R}$ and define the probabilities of the new outcomes as 0.

To each $x \in \mathbb{R}$ we have attached a non-negative number $p(x)$, which is zero except for at most countably many X. We let x_i denote a typical outcome where $p(x) \neq 0$. Similar to calculating the means, variances and higher moments of a random variable, we can compute these quantities for a probability distribution; see Definition 8.1.37. For example, for a discrete probability distribution p the mean is $\sum_i x_i p(x_i)$.

8.2.1 Bernoulli Distribution

Recall the binomial coefficient $\binom{N}{r} = \frac{N!}{r!(N-r)!}$ is the number of ways to choose r objects from N objects when order does not matter; see §A.1.3 for a review of binomial coefficients. Consider n independent repetitions of a process with only two possible outcomes. We typically call one outcome **success** and the other **failure**, the event a **Bernoulli trial**, and a collection of independent Bernoulli trials a **Bernoulli process**. In each Bernoulli trial let there be probability p of success and $q = 1 - p$ of failure. Often we represent a success with 1 and a failure with 0. In §8.2.4 we describe a Bernoulli trial to experimentally determine π!

Exercise 8.2.1. *Consider a Bernoulli trial with random variable X equal to 1 for a success and 0 for a failure. Show $\overline{X} = p$, $\sigma_X^2 = pq$, and $\sigma_X = \sqrt{pq}$. Note X is also an indicator random variable (see Exercise 8.1.14).*

Let Y_N be the number of successes in N trials. Clearly the possible values of Y_N are $\{0, 1, \ldots, N\}$. We analyze $p_N(k) = \text{Prob}(Y_N(\omega) = k)$. Here the sample space Ω is all possible sequences of N trials, and the random variable $Y_N : \Omega \to \mathbb{R}$ is given by $Y_N(\omega)$ equals the number of successes in ω.

If $k \in \{0, 1, \ldots, N\}$, we need k successes and $N - k$ failures. We do not care what order we have them (i.e., if $k = 4$ and $N = 6$ then $SSFSSF$ and $FSSSSF$ both contribute equally). Each such string of k successes and $N - k$ failures has probability of $p^k \cdot (1 - p)^{N-k}$. There are $\binom{N}{k}$ such strings, which implies $p_N(k) = \binom{N}{k} p^k \cdot (1 - p)^{N-k}$ if $k \in \{0, 1, \ldots, N\}$ and 0 otherwise.

By clever algebraic manipulations, one can directly evaluate the mean $\overline{Y_N}$ and the variance $\sigma^2_{Y_N}$; however, Lemmas 8.1.40 and 8.1.49 allow one to calculate both quantities immediately, once one knows the mean and variance for a single occurrence (see Exercise 8.2.1).

Lemma 8.2.2. *For a Bernoulli process with N trials, each having probability p of success, the expected number of successes is $\overline{Y_N} = Np$ and the variance is $\sigma^2_{Y_N} = Npq$.*

Lemma 8.2.2 states the expected number of successes is of size Np, and the fluctuations about Np are of size $\sigma^2_{Y_N} = \sqrt{Npq}$. Thus, if $p = \frac{1}{2}$ and $N = 10^6$, we expect 500,000 successes, with fluctuations on the order of 500. Note how much smaller the fluctuations about the mean are than the mean itself (the mean is of size N, the fluctuations of size $\sqrt{N}$). This is an example of a general phenomenon, which we describe in greater detail in §8.4.

Exercise 8.2.3. *Prove Lemma 8.2.2. Prove the variance is largest when $p = q = \frac{1}{2}$.*

Consider the following problem: Let $\Omega = \{S, FS, FFS, \dots\}$ and let Z be the number of trials before the first success. What is $\overline{Z}$ and σ^2_Z?

First we determine the **Bernoulli distribution** $p(k) = \text{Prob}(Z(\omega) = k)$, the probability that the first success occurs after k trials. Clearly this probability is non-zero only for k a positive integer, in which case the string of results must be $k - 1$ failures followed by 1 success. Therefore

$$p(k) = \begin{cases} (1 - p)^{k-1} \cdot p & \text{if } k \in \{1, 2, \dots\} \\ 0 & \text{otherwise.} \end{cases} \tag{8.51}$$

To determine the mean $\overline{Z}$ we must evaluate

$$\overline{Z} = \sum_{k=1}^{\infty} k(1 - p)^{k-1} p = p \sum_{k=1}^{\infty} kq^{k-1}, \quad 0 < q = 1 - p < 1. \tag{8.52}$$

Consider the geometric series

$$f(q) = \sum_{k=0}^{\infty} q^k = \frac{1}{1 - q}. \tag{8.53}$$

A careful analysis shows we can differentiate term by term if $-1 \le q < 1$. Then

$$f'(q) = \sum_{k=0}^{\infty} kq^{k-1} = \frac{1}{(1 - q)^2}. \tag{8.54}$$

Recalling $q = 1 - p$ and substituting yields

$$\overline{Z} = p \sum_{k=1}^{\infty} kq^{k-1} = \frac{p}{(1 - (1 - p))^2} = \frac{1}{p}. \tag{8.55}$$

Remark 8.2.4. Differentiating under the summation sign is a powerful tool in Probability Theory, and is a common technique for proving such identities. See [Mil4] for more on differentiating identities, where the expected number of alternations between heads and tails in n tosses of a coin with probability p of heads is derived, along with other combinatorial and probability results.

Exercise 8.2.5. *Calculate σ^2_Z. Hint: Differentiate $f(q)$ twice.*

8.2.2 Poisson Distribution

Divide the unit interval into N equal pieces. Consider N independent Bernoulli trials, one in each subinterval. If the probability of a success is $\frac{\lambda}{N}$, then by Lemma 8.2.2 the expected number of successes is $N \cdot \frac{\lambda}{N} = \lambda$. We consider the limit as $N \to \infty$. We still expect λ successes in each unit interval, but what is the probability of 3λ successes? How long do we expect to wait between successes?

We call this a **Poisson process with parameter** λ. For example, look at the midpoints of the N intervals. At each midpoint we have a Bernoulli trial with probability of success $\frac{\lambda}{N}$ and failure $1 - \frac{\lambda}{N}$. We determine the $N \to \infty$ limits. For fixed N, the probability of *exactly* k successes in a unit interval is

$$p_N(k) = \binom{N}{k} \left(\frac{\lambda}{N}\right)^k \left(1 - \frac{\lambda}{N}\right)^{N-k}$$

$$= \frac{N!}{k!(N-k)!} \frac{\lambda^k}{N^k} \left(1 - \frac{\lambda}{N}\right)^{N-k}$$

$$= \frac{N \cdot (N-1) \cdots (N-k+1)}{N \cdot N \cdots N} \frac{\lambda^k}{k!} \left(1 - \frac{\lambda}{N}\right)^N \left(1 - \frac{\lambda}{N}\right)^{-k}$$

$$= 1 \cdot \left(1 - \frac{1}{N}\right) \cdots \left(1 - \frac{k-1}{N}\right) \frac{\lambda^k}{k!} \left(1 - \frac{\lambda}{N}\right)^N \left(1 - \frac{\lambda}{N}\right)^{-k}. \quad (8.56)$$

For fixed, finite k and λ, as $N \to \infty$ the first k factors in $p_N(k)$ tend to 1, $\left(1 - \frac{\lambda}{N}\right)^N \to e^{-\lambda}$, and $\left(1 - \frac{\lambda}{N}\right)^{-k} \to 1$ (see §5.4 for a review of properties of e). Thus $p_N(k) \to \frac{\lambda^k}{k!} e^{-\lambda}$. We shall see similar calculations as these when we investigate the properties of $x_n = n^k \alpha \bmod 1$ in Chapter 12.

Using our investigations of Bernoulli trials as a motivation, we are led to the **Poisson Distribution**: Given a parameter λ (interpreted as the expected number of occurrences per unit interval), the probability of k occurrences in a unit interval is $p(k) = \frac{\lambda^k}{k!} e^{-\lambda}$ for $k \in \{0, 1, 2, \dots\}$. This is a discrete, integer valued process.

Exercise 8.2.6. *Check that $p(k)$ given above is a probability distribution. Namely, show $\sum_{k \geq 0} p(k) = 1$.*

Exercise[h] **8.2.7.** *Calculate the mean and variance for the Poisson Distribution.*

8.2.3 Continuous Distributions

Up to now we have only considered discrete probability distributions. We now study a continuous example. We consider a generalization of a Bernoulli process with λ successes in a unit interval. We divide the real line into subintervals of size $\frac{1}{N}$ and consider a Bernoulli trial at the midpoint of each subinterval with probability $\frac{\lambda}{N}$ of success. Start counting at 0, and let the first success be at X. How is X distributed as $N \to \infty$ (i.e., how long do we expect to wait before seeing the first success)? Denote this distribution by $p_S(x)$.

We have approximately $\frac{x-0}{1/N} = Nx$ midpoints from 0 to X (with N midpoints per unit interval). Let $\lceil y \rceil$ be the smallest integer greater than or equal to y. Then we have $\lceil Nx \rceil$ midpoints, where the results of the Bernoulli trials of the first $\lceil Nx \rceil - 1$ midpoints are all failures and the last is a success. Thus the probability of the first success occurring in an interval of length $\frac{1}{N}$ containing X (with N divisions per unit interval) is

$$p_{N,S}(x) = \left(1 - \frac{\lambda}{N}\right)^{\lceil Nx \rceil - 1} \cdot \left(\frac{\lambda}{N}\right)^1. \tag{8.57}$$

For N large the above is approximately $e^{-\lambda x} \frac{\lambda}{N}$.

Exercise 8.2.8. *For large N, calculate the size of $N\left(p_{N,s}(x) - e^{-\lambda x}\frac{\lambda}{N}\right)$. Show this difference tends to zero as N tends to infinity.*

Definition 8.2.9 (Continuous Probability Distribution). *We say $p(x)$ is a continuous probability distribution on $\mathbb{R}$ if*

1. *$p(x) \geq 0$ for all $x \in \mathbb{R}$.*

2. *$\int_{\mathbb{R}} p(x)dx = 1$.*

3. *$\text{Prob}(a \leq x \leq b) = \int_a^b p(x)dx$.*

We call $p(x)$ the probability density function or the density; $p(x)dx$ is interpreted as the probability of the interval $[x, x+dx]$.

In the previous example, as $N \to \infty$ we obtain the continuous probability density function

$$p_S(x) = \begin{cases} \lambda e^{-\lambda x} & \text{if } x \geq 0 \\ 0 & \text{if } x < 0; \end{cases} \tag{8.58}$$

note $\frac{1}{N}$ is like dx for N large. In the special case of $\lambda = 1$, we get the standard exponential decay, e^{-x}. We will see this distribution in Chapter 12 when we investigate the fractional parts of $n^k \alpha$ (k, α fixed, n varying).

For instance, let $\pi(M)$ be the number of primes that are at most M. The Prime Number Theorem states $\pi(M) = \frac{M}{\log M}$ plus lower order terms. Thus the average spacing between primes around M is about $\log M$. We can model the distribution of primes as a Poisson Process, with parameter $\lambda = \lambda_M = \frac{1}{\log M}$ (this is called the Cramér model). While possible locations of primes (obviously) is discrete (it must be an integer, and in fact the location of primes are not independent), a Poisson model often gives very good heuristics; see for example [Sch].

We often renormalize so that $\lambda = 1$. This is denoted **unit mean spacing**. For example, one can show the Mth prime p_M is about $M \log M$, and spacings between primes around p_M is about $\log M$. Then the normalized primes $q_M \approx \frac{p_M}{\log M}$ will have unit mean spacing and $\lambda = 1$.

Example 8.2.10 (**Uniform Distribution on** $[a, b]$). *Let $\Omega = \{x \in \mathbb{R} : a \leq x \leq b\}$. The uniform distribution has probability density function $p(x) = \frac{1}{b-a}$. Note for any $[c, d] \subset [a, b]$,*

$$\text{Prob}([c, d]) = \int_c^d p(x)dx = \frac{d-c}{b-a}. \tag{8.59}$$

The uniform distribution is one of the most common (and best understood!) continuous distributions; the probability of $x \in [c, d] \subset [a, b]$ depends only on the length of the subinterval $[c, d]$.

Example 8.2.11 (Gaussian Distribution). *For $x \in \mathbb{R}$, consider the probability density function $p(x) = \frac{1}{\sqrt{2\pi\sigma^2}} e^{-(x-\mu)^2/2\sigma^2}$. This is called the Gaussian (or normal or bell curve) distribution. By Exercise 8.2.12 it has mean μ and variance σ^2. If $\mu = 0$ and $\sigma^2 = 1$, it is called the standard normal or the standard Gaussian. See §8.4 for more details.*

We sketch the main idea in the proof that the above is a probability distribution. As it is clearly non-negative, we need only show it integrates to one. Consider

$$I = \int_{-\infty}^{\infty} e^{-x^2} dx. \tag{8.60}$$

Square I, and change from rectangular to polar coordinates, where $dxdy$ becomes $rdrd\theta$:

$$
\begin{aligned}
I^2 &= \int_{-\infty}^{\infty} e^{-x^2} dx \cdot \int_{-\infty}^{\infty} e^{-y^2} dy \\
&= \int_{-\infty}^{\infty} \int_{-\infty}^{\infty} e^{-x^2-y^2} dxdy \\
&= \int_{0}^{2\pi} d\theta \int_{0}^{\infty} e^{-r^2} rdr \\
&= 2\pi \cdot \left[-\frac{1}{2} e^{-r^2} \right]_{0}^{\infty} = \pi.
\end{aligned}
\tag{8.61}
$$

The reason the above works is that while $e^{-x^2} dx$ is hard to integrate, $re^{-r^2} dr$ is easy. Thus $I = \sqrt{\pi}$.

Exercise 8.2.12. *Let $p(x) = \frac{1}{\sqrt{2\pi\sigma^2}} e^{-(x-\mu)^2/2\sigma^2}$. Prove $\int_{-\infty}^{\infty} p(x)dx = 1$, $\int_{-\infty}^{\infty} xp(x)dx = \mu$ and $\int_{-\infty}^{\infty} (x-\mu)^2 p(x)dx = \sigma^2$. This justifies our claim that the Gaussian is a probability distribution with mean μ and variance σ^2.*

Example 8.2.13 (Cauchy Distribution). *Consider*

$$p(x) = \frac{1}{\pi} \frac{1}{1+x^2}. \tag{8.62}$$

This is a continuous distribution and is symmetric about zero. While we would like to say it therefore has mean zero, the problem is the integral $\int_{-\infty}^{\infty} xp(x)dx$ is not well defined as it depends on how we take the limit. For example,

$$\lim_{A\to\infty} \int_{-A}^{A} xp(x)dx = 0, \quad \lim_{A\to\infty} \int_{-A}^{2A} xp(x)dx = \infty. \tag{8.63}$$

Regardless, $p(x)$ has infinite variance. We shall see the Cauchy distribution again in Chapter 15; see also Exercises 3.3.28 and 3.3.29.

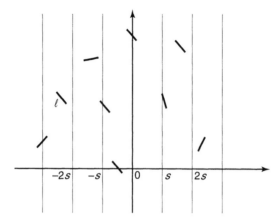

Figure 8.1 Buffon's needle

Exercise 8.2.14. *Prove the Cauchy distribution is a probability distribution by showing*

$$\int_{-\infty}^{\infty} \frac{1}{\pi} \frac{1}{1+x^2} dx \ = \ 1. \tag{8.64}$$

Show the variance is infinite. See also Exercise 3.3.29.

The Cauchy distribution shows that not all probability distributions have finite moments. When the moments do exist, however, they are a powerful tool for understanding the distribution. The moments play a similar role as coefficients in Taylor series expansions. We use moment arguments to investigate the properties of eigenvalues in Chapters 15 and 16; see in particular §15.3.2.

8.2.4 Buffon's Needle and π

We give a nice example of a continuous probability distribution in two dimensions. Consider a collection of infinitely long parallel lines in the plane, where the spacing between any two adjacent lines is s. Let the lines be located at $x = 0, \pm s, \pm 2s, \ldots$. Consider a rod of length ℓ where for convenience we assume $\ell < s$. If we were to randomly throw the rod on the plane, what is the probability it hits a line? See Figure 8.1. This question was first asked by Buffon in 1733. For a truly elegant solution which does not use calculus, see [AZ]; we present the proof below as it highlights many of the techniques for investigating probability problems in several variables.

Because of the vertical symmetry we may assume the center of the rod lies on the line $x = 0$, as shifting the rod (without rotating it) up or down will not alter the number of intersections. By the horizontal symmetry, we may assume $-\frac{s}{2} \le x < \frac{s}{2}$. We posit that all values of x are equally likely. As x is continuously distributed, we may add in $x = \frac{s}{2}$ without changing the probability. The probability density function of x is $\frac{dx}{s}$.

Let θ be the angle the rod makes with the x-axis. As each angle is equally likely, the probability density function of θ is $\frac{d\theta}{2\pi}$. We assume that x and θ are chosen independently. Thus the probability density for (x, θ) is $\frac{dx\,d\theta}{s \cdot 2\pi}$.

The projection of the rod (making an angle of θ with the x-axis) along the x-axis is $\ell \cdot |\cos \theta|$. If $|x| \le \ell \cdot |\cos \theta|$, then the rod hits exactly one vertical line exactly once; if $x > \ell \cdot |\cos \theta|$, the rod does not hit a vertical line. Note that if $\ell > s$, a rod could hit multiple lines, making the arguments more involved. Thus the probability a rod hits a line is

$$p = \int_{\theta=0}^{2\pi} \int_{x=-\ell \cdot |\cos \theta|}^{\ell \cdot |\cos \theta|} \frac{dx\,d\theta}{s \cdot 2\pi} = 2 \int_{\theta=0}^{2\pi} \frac{\ell \cdot |\cos \theta|}{s} \frac{d\theta}{2\pi} = \frac{2\ell}{\pi s}. \quad (8.65)$$

Exercise 8.2.15. *Show*

$$\frac{1}{2\pi} \int_0^{2\pi} |\cos \theta| d\theta = \frac{2}{\pi}. \quad (8.66)$$

Let A be the random variable which is the number of intersections of a rod of length ℓ thrown against parallel vertical lines separated by $s > \ell$ units. Then

$$A = \begin{cases} 1 & \text{with probability } \frac{2\ell}{\pi s} \\ 0 & \text{with probability } 1 - \frac{2\ell}{\pi s}. \end{cases} \quad (8.67)$$

If we were to throw N rods independently, since the expected value of a sum is the sum of the expected values (Lemma 8.1.40), we expect to observe $N \cdot \frac{2\ell}{\pi s}$ intersections.

Turning this around, let us throw N rods, and let I be the number of observed intersections of the rods with the vertical lines. Then

$$I \approx N \cdot \frac{2\ell}{\pi s} \quad \text{which implies} \quad \pi \approx \frac{N}{I} \cdot \frac{2\ell}{s}. \quad (8.68)$$

The above is an *experimental* formula for π!

Exercise 8.2.16. *Assume we are able to throw the rod randomly as described above, and the N throws are independent. We then have a Bernoulli process with N trials. We have calculated the expected number of successes; using the methods of §8.2.1, calculate the variance (and hence the size of the fluctuations in I). For each N, give the range of values we expect to observe for π.*

8.3 RANDOM SAMPLING

We introduce the notion of **random sampling**. Consider a countable set $\Omega \subset \mathbb{R}$ and a probability function p on Ω; we can extend p to all of $\mathbb{R}$ by setting $p(r) = 0$ if $r \notin \Omega$. Using the probability function p, we can choose elements from $\mathbb{R}$ **at random**. Explicitly, the probability that we choose $\omega \in \Omega$ is $p(\omega)$.

For example, let $\Omega = \{1, 2, 3, 4, 5, 6\}$ with each event having probability $\frac{1}{6}$ (the rolls of a fair die). If we were to roll a fair die N times (for N large), we observe a particular sequence of outcomes. It is natural to assume the rolls are independent of each other. Let X_i denote the outcome of the ith roll. The X_i's all have the same distribution (arising from p). We call the X_i **i.i.d.r.v.** (independent identically distributed random variables), and we say the X_i are a **sample** from the probability distribution p. We say we **randomly sample (with respect to** p) $\mathbb{R}$. Often we simply say we have **randomly chosen** N **numbers**.

A common problem is to sample some mathematical or physical process and use the observations to make inferences about the underlying system. For example, we may be given a coin without being told what its probabilities for heads and tails are. We can attempt to infer the probability p of a head by tossing the coin many times, and recoding the outcomes. Let X_i be the outcome of the ith toss (1 for head, 0 for tail). After N tosses we expect to see about Np heads; however, we observe some number, say S_N. Given that we observe S_N heads after N tosses, what is our best guess for p? By Lemma 8.1.40, we guess $p = \frac{S_N}{N}$. It is extremely unlikely that our guess is exactly right. This leads us to a related question: given that we observe S_N heads, can we give a small interval about our best guess where we are extremely confident the true value p lies? The solution is given by the Central Limit Theorem (see §8.4).

Exercise 8.3.1. *For the above example, if p is irrational show the best guess can never be correct.*

One can generalize the above to include the important case where p is a continuous distribution. For example, say we wish to investigate the digits of numbers in $[0, 1]$. It is natural to put the uniform distribution on this interval, and choose numbers at random relative to this distribution; we say we choose N numbers randomly with respect to the uniform distribution on $[0, 1]$, or simply we choose N numbers uniformly from $[0, 1]$. Two natural problems are to consider the nth digit in the base 10 expansion and the nth digit in the continued fraction expansion. By observing many choices, we hope to infer knowledge about how these digits are distributed. The first problem is theoretically straightforward. It is not hard to calculate the probability that the nth digit is d; it is just $\frac{1}{10}$. The probabilities of the digits of continued fractions are significantly harder (unlike decimal expansions, *any* positive integer can occur as a digit); see Chapter 10 for the answer.

Exercise 8.3.2 (Important for Computational Investigations). *For any continuous distribution p on $\mathbb{R}$, the probability we chose a number in $[a, b]$ is $\int_a^b p(x)dx$. If we were to choose N numbers, N large, then we expect approximately $N \int_a^b p(x)dx$ to be in $[a, b]$. Often computers have built in random number generators for certain continuous distributions, such as the standard Gaussian or the uniform, but not for less common ones. Show if one can randomly choose numbers from the uniform distribution, one can use this to randomly choose from any distribution. Hint: Use $C_p(x) = \int_{-\infty}^x p(x)dx$, the **Cumulative Distribution Function** of p (see also §15.3.2); it is the probability of observing a number at most x.*

Remark 8.3.3. The observant reader may notice a problem with sampling from a continuous distribution: the probability of choosing any particular real number is zero, but some number is chosen! One explanation is that, fundamentally, we cannot choose numbers from a continuous probability distribution. For example, if we use computers to choose our numbers, all computers can do is a finite number of manipulations of 0's and 1's; thus, they can only choose numbers from a countable (actually finite) set. The other interpretation of the probability of any $r \in \mathbb{R}$ is zero is that, while at each stage some number is chosen, no number is ever chosen twice. Thus, in some sense, any number we explicitly write down is "special." See also Exercise 8.1.44, where the resolution is that one cannot choose numbers uniformly on all of $(0, \infty)$.

For our investigations, we approximate continuous distributions by discrete distributions with many outcomes. From a practical point of view, this suffices for many experiments; however, one should note that while theoretically we can write statements such as "choose a real number uniformly from $[0, 1]$," we can never actually do this.

8.4 THE CENTRAL LIMIT THEOREM

We close our introduction to probability with a statement of *the* main theorem about the behavior of a sum of independent events. We give a proof in an important special case in §8.4.2 and sketch the proof in general in §11.5. For more details and weaker conditions, see [Bi, CaBe, Fe]. We discuss applications of the Central Limit Theorem to determining whether or not numerical experiments support a conjecture in Chapter 9.

8.4.1 Statement of the Central Limit Theorem

Let X_i ($i \in \{1, \ldots, N\}$) be independent identically distributed random variables (i.i.d.r.v.) as in §8.3, all sampled from the same probability distribution p with mean μ and variance σ^2; thus $\mathbb{E}[X_i] = \mu$ and $\sigma^2_{X_i} = \sigma^2$ for all i. Let $S_N = \sum_{i=1}^{N} X_i$. We are interested in the distribution of the random variable S_N as $N \to \infty$. As each X_i has expected value μ, by Lemma 8.1.40 $\mathbb{E}[S_N] = N\mu$. We now consider a more refined question: how is S_N distributed about $N\mu$? The Central Limit Theorem answers this, and tells us what the correct scale is to study the fluctuations about $N\mu$.

Theorem 8.4.1 (Central Limit Theorem). *For $i \in \{1, \ldots, N\}$, let X_i be i.i.d.r.v. with mean μ, finite variance σ^2 and finite third moment. Let $S_N = X_1 + \cdots + X_N$. As $N \to \infty$*

$$\mathrm{Prob}(S_N \in [\alpha, \beta]) \sim \frac{1}{\sqrt{2\pi\sigma^2 N}} \int_\alpha^\beta e^{-(t-\mu N)^2/2\sigma^2 N} dt. \qquad (8.69)$$

In other words, the distribution of S_N converges to a Gaussian with mean μN and variance $\sigma^2 N$. We may re-write this as

$$\lim_{N \to \infty} \mathrm{Prob}\left(\frac{S_N - \mu N}{\sqrt{\sigma^2 N}} \in [a, b]\right) = \frac{1}{\sqrt{2\pi}} \int_a^b e^{-t^2/2} dt. \qquad (8.70)$$

Here $Z_N = \frac{S_N - \mu N}{\sqrt{\sigma^2 N}}$ converges to a Gaussian with mean 0 and variance 1.

The probability density $\frac{1}{\sqrt{2\pi}} e^{-t^2/2}$ is the **standard Gaussian**. It is *the* universal curve of probability. Note how robust the Central Limit Theorem is: it does not depend on fine properties of the X_i, just that they all have the same distributions and finite variance (and a bit more). While this is true in most situations, it fails in some cases such as sampling from a Cauchy distribution (see Exercise 12.7.7 for another limit theorem which can handle such cases). Sometimes it is important to know how rapidly Z_N is converging to the Gaussian. The rate of convergence *does* depend on the higher moments; see §11.5 and [Fe].

Exercise 8.4.2. *The Central Limit Theorem gives us the correct scale to study fluctuations. For example, say we toss a fair coin N times (hence $\mu = \frac{1}{2}$ and $\sigma^2 = \frac{1}{4}$). We expect S_N to be about $\frac{N}{2}$. Find values of a and b such that the probability of $S_N - N\mu \in [a\sqrt{N}/2, b\sqrt{N}/2]$ converges to 95% (resp., 99%). For large N, show for any fixed $\delta > 0$ that the probability of $S_N - N\mu \in [aN^{\frac{1}{2}+\delta}/2, bN^{\frac{1}{2}+\delta}/2]$ tends to zero. Thus we expect to observe half of the tosses as heads, and we expect deviations from one-half to be of size $2/\sqrt{N}$.*

Exercise 8.4.3. *Redo Exercise 8.1.56 using the Central Limit Theorem and compare the two bounds.*

Exercise 8.4.4. *For $S_N = X_1 + \cdots + X_N$, calculate the variance of $Z_N = \frac{S_N - \mu N}{\sqrt{\sigma^2 N}}$; this shows $\sqrt{\sigma^2 N}$ is the correct scale to investigate fluctuations of S_N about μN.*

One common application of the Central Limit Theorem is to test whether or not we are sampling the X_i independently from a fixed probability distribution with mean μ and known standard deviation σ (if the standard deviation is not known, there are other tests which depend on methods to estimate σ). Choose N numbers randomly from what we expect has mean μ. We form S_N as before and investigate $\frac{S_N - \mu N}{\sqrt{\sigma^2 N}}$. As $S_N = \sum_{i=1}^{N} X_i$, we expect S_N to be of size N. If the X_i are not drawn from a distribution with mean μ, then $S_N - N\mu$ will also be of size N. Thus, $\frac{S_N - N\mu}{\sqrt{\sigma^2 N}}$ will be of size $\sqrt{N}$ if the X_i are not drawn from something with mean μ. If, however, the X_i are from sampling a distribution with mean μ, the Central Limit Theorem states that $\frac{S_N - N\mu}{\sqrt{\sigma^2 N}}$ will be of size 1. See Chapter 9 for more details and Exercise 12.7.7 for an alternate sampling statistic.

Finally, we note that the Central Limit Theorem is an example of the **Philosophy of Square Root Cancellation**: the sum is of size N, but the deviations are of size $\sqrt{N}$. We have already seen examples of such cancellation in Remark 3.3.1 and §4.4, and will see more in our investigations of writing integers as the sum of primes (see §13.3.2).

8.4.2 Proof for Bernoulli Processes

We sketch the proof of the Central Limit Theorem for Bernoulli Processes where the probability of success is $p = \frac{1}{2}$. Consider the random variable X that is 1 with probability $\frac{1}{2}$ and -1 with probability $\frac{1}{2}$ (for example, tosses of a fair coin; the advantage of making a tail -1 is that the mean is zero). Note the mean of X is $\overline{X} = 0$, the standard deviation is $\sigma_X = \frac{1}{2}$ and the variance is $\sigma_X^2 = \frac{1}{4}$.

Let $X_1, \ldots, X_{2N}$ be independent identically distributed random variables, distributed as X (it simplifies the expressions to consider an even number of tosses). Consider $S_{2N} = X_1 + \cdots + X_{2N}$. Its mean is zero and its variance is $2N/4$, and we expect fluctuations of size $\sqrt{2N/4}$. We show that for N large the distribution of S_{2N} is approximately normal. We need

Lemma 8.4.5 (Stirling's Formula). *For n large,*

$$n! \,=\, n^n e^{-n} \sqrt{2\pi n}\,(1 + O(1/n)). \tag{8.71}$$

For a proof, see [WW]. We show the above is a reasonable approximation.

$$\log n! \,=\, \sum_{k=1}^{n} \log k \,\approx\, \int_{1}^{n} \log t\, dt \,=\, (t\log t - t)|_1^n. \tag{8.72}$$

Thus $\log n! \approx n \log n - n$, or $n! \approx n^n e^{-n}$.

We now consider the distribution of S_{2N}. The probability that $S_{2N} = 2k$ is just $\binom{2N}{N+k}(\frac{1}{2})^{N+k}(\frac{1}{2})^{N-k}$. This is because for $S_{2N} = 2k$, we need $2k$ more 1's (heads) than -1's (tails), and the number of 1's and -1's add to $2N$. Thus we have $N + k$ heads (1's) and $N - k$ tails (-1's). There are 2^{2N} strings of 1's and -1's, $\binom{2N}{N+k}$ have exactly $N + k$ heads and $N - k$ tails, and the probability of each string is $(\frac{1}{2})^{2N}$. We have written $(\frac{1}{2})^{N+k}(\frac{1}{2})^{N-k}$ to show how to handle the more general case when there is a probability p of heads and $1 - p$ of tails.

We use Stirling's Formula to approximate $\binom{2N}{N+k}$. After elementary algebra we find

$$\binom{2N}{N+k} \approx \frac{2^{2N}}{(N+k)^{N+k}(N-k)^{N-k}} \sqrt{\frac{N}{\pi(N+k)(N-k)}}$$

$$= \frac{2^{2N}}{\sqrt{\pi N}} \frac{1}{(1 + \frac{k}{N})^{N+\frac{1}{2}+k}(1 - \frac{k}{N})^{N+\frac{1}{2}-k}}. \tag{8.73}$$

Using $\left(1 + \frac{w}{N}\right)^N \approx e^w$ from §5.4, after some more algebra we find

$$\binom{2N}{N+k} \approx \frac{2^{2N}}{\sqrt{\pi N}} e^{-2k^2/N}. \tag{8.74}$$

Thus

$$\mathrm{Prob}(S_{2N} = 2k) \,=\, \binom{2N}{N+k}\frac{1}{2^{2N}} \approx \frac{1}{\sqrt{2\pi \cdot (2N/4)}}\, e^{-k^2/2(2N/4)}. \tag{8.75}$$

The distribution of S_{2N} looks like a Gaussian with mean 0 and variance $2N/4$. While we can only observe an integer number of heads, for N enormous the Gaussian is very slowly varying and hence approximately constant from $2k$ to $2k + 2$.

Exercise 8.4.6. *Generalize the above arguments to handle the case when $p \neq \frac{1}{2}$.*

Exercise 8.4.7. *Use the integral test to bound the error in (8.72), and then use that to bound the error in the estimate of $n!$.*

Chapter Nine

Applications of Probability: Benford's Law and Hypothesis Testing

The Gauss-Kuzmin Theorem (Theorem 10.3.1) tells us that the probability that the millionth digit of a randomly chosen continued fraction expansion is k is approximately $q_k = \log_2\left(1 + \frac{1}{k(k+2)}\right)$. What if we choose N algebraic numbers, say the cube roots of N consecutive primes: how often do we expect to observe the millionth digit equal to k? If we believe that algebraic numbers other than rationals and quadratic irrationals satisfy the Gauss-Kuzmin Theorem, we expect to observe $q_k N$ digits equal to k, and probably fluctuations on the order of $\sqrt{N}$. If we observe M digits equal to k, how confident are we (as a function of M and N, of course) that the digits are distributed according to the Gauss-Kuzmin Theorem? This leads us to the subject of **hypothesis testing**: if we assume some process has probability p of success, and we observe M successes in N trials, does this provide support for or against the hypothesis that the probability of success is p?

We develop some of the theory of hypothesis testing by studying a concrete problem, the distribution of the first digit of certain sequences. In many problems (for example, 2^n base 10), the distribution of the first digit is given by Benford's Law, described below. We first investigate situations where we can easily prove the sequences are Benford, and then discuss how to analyze data in harder cases where the proofs are not as clear (such as the famous $3x + 1$ problem). The error analysis is, of course, the same as the one we would use to investigate whether or not the digits of the continued fraction expansions of algebraic numbers satisfy the Gauss-Kuzmin Theorem. In the process of investigating Benford's Law, we encounter equidistributed sequences (Chapter 12), logarithmic probabilities (similar to the Gauss-Kuzmin probabilities in Chapter 10), and Poisson Summation (Chapter 11), as well as many of the common problems in statistical testing (such as non-independent events and multiple comparisons).

9.1 BENFORD'S LAW

While looking through tables of logarithms in the late 1800s, Newcomb noticed a surprising fact: certain pages were significantly more worn out than others. People were looking up numbers whose logarithm started with 1 more frequently than other digits. In 1938 Benford [Ben] observed the same digit bias in a variety of phenomenon. See [Hi1, Rai] for a description and history, [Hi2, BBH, KonMi, LaSo,

MN] for recent results, [Knu] for connections between Benford's law and rounding errors in computer calculations and [Nig1, Nig2] for applications of Benford's Law by the IRS to detect corporate tax fraud!

A sequence of positive numbers $\{x_n\}$ is **Benford (base b)** if the probability of observing the first digit of x_n in base b is j is $\log_b\left(1 + \frac{1}{j}\right)$. More precisely,

$$\lim_{N\to\infty} \frac{\#\{n \le N : \text{first digit of } x_n \text{ in base } b \text{ is } j\}}{N} = \log_b\left(1 + \frac{1}{j}\right). \quad (9.1)$$

Note that $j \in \{1, \ldots, b-1\}$. This is a probability distribution as one of the $b-1$ events must occur, and the total probability is

$$\sum_{j=1}^{b-1} \log_b\left(1 + \frac{1}{j}\right) = \log_b \prod_{j=1}^{b-1}\left(1 + \frac{1}{j}\right) = \log_b \prod_{j=1}^{b-1} \frac{j+1}{j} = \log_b b = 1. \quad (9.2)$$

It is possible to be Benford to some bases but not others; we show the first digit of 2^n is Benford base 10, but clearly it is not Benford base 2 as the first digit is always 1. For many processes, we obtain a sequence of points, and the distribution of the first digits are Benford. For example, consider the **3x+1 problem**. Let a_0 be any positive integer, and consider the sequence where

$$a_{n+1} = \begin{cases} 3a_n + 1 & \text{if } a_n \text{ is odd} \\ a_n/2 & \text{if } a_n \text{ is even.} \end{cases} \quad (9.3)$$

For example, if $a_0 = 13$, we have

$$13 \longrightarrow 40 \longrightarrow 20 \longrightarrow 10 \longrightarrow 5 \longrightarrow 16 \longrightarrow 8 \longrightarrow 4 \longrightarrow 2 \longrightarrow 1$$
$$\longrightarrow 4 \longrightarrow 2 \longrightarrow 1 \longrightarrow 4 \longrightarrow 2 \longrightarrow 1 \cdots . \quad (9.4)$$

An alternate definition is to remove as many powers of two as possible in one step. Thus,

$$a_{n+1} = \frac{3a_n + 1}{2^k}, \quad (9.5)$$

where k is the largest power of 2 dividing $3a_n + 1$. It is conjectured that for *any* a_0, eventually the sequence becomes $4 \to 2 \to 1 \to 4 \cdots$ (or in the alternate definition $1 \to 1 \to 1 \cdots$). While this is known for all $a_0 \le 2^{60}$, the problem has resisted numerous attempts at proofs (Kakutani has described the problem as a conspiracy to slow down mathematical research because of all the time spent on it). See [Lag1, Lag2] for excellent surveys of the problem. How do the first digits behave for a_0 large? Do numerical simulations support the claim that this process is Benford? Does it matter which definition we use?

Exercise 9.1.1. *Show the Benford probabilities* $\log_{10}\left(1 + \frac{1}{j}\right)$ *for* $j \in \{1, \ldots, 9\}$ *are irrational. What if instead of base ten we work in base d for some integer d?*

9.2 BENFORD'S LAW AND EQUIDISTRIBUTED SEQUENCES

As we can write any positive x as b^u for some u, the following lemma shows that it suffices to investigate u mod 1:

Lemma 9.2.1. *The first digits of b^u and b^v are the same in base b if and only if $u \equiv v \bmod 1$.*

Proof. We prove one direction as the other is similar. If $u \equiv v \bmod 1$, we may write $v = u + m, m \in \mathbb{Z}$. If

$$b^u \;=\; u_k b^k + u_{k-1} b^{k-1} + \cdots + u_0 + u_{-1} b^{-1} + \cdots, \tag{9.6}$$

then

$$
\begin{aligned}
b^v &= b^{u+m}\\
&= b^u \cdot b^m\\
&= (u_k b^k + u_{k-1} b^{k-1} + \cdots + u_0 + u_{-1} b^{-1} + \cdots) b^m\\
&= u_k b^{k+m} + \cdots + u_0 b^m + u_{-1} b^{m-1} + \cdots . \tag{9.7}
\end{aligned}
$$

Thus the first digits of each are u_k, proving the claim. $\square$

Exercise 9.2.2. *Prove the other direction of the if and only if.*

Consider the unit interval $[0, 1)$. For $j \in \{1, \ldots, b\}$, define p_j by

$$b^{p_j} = j \quad \text{or equivalently} \quad p_j = \log_b j. \tag{9.8}$$

For $j \in \{1, \ldots, b-1\}$, let

$$I_j^{(b)} \;=\; [p_j, p_{j+1}) \subset [0, 1). \tag{9.9}$$

Lemma 9.2.3. *The first digit of b^y base b is j if and only if $y \bmod 1 \in I_j^{(b)}$.*

Proof. By Lemma 9.2.1 we may assume $y \in [0, 1)$. Then $y \in I_j^{(b)} = [p_j, p_{j+1})$ if and only if $b^{p_j} \le y < b^{p_{j+1}}$, which from the definition of p_j is equivalent to $j \le b^y < j + 1$, proving the claim. $\square$

The following theorem shows that the exponentials of equidistributed sequences (see Definition 12.1.4) are Benford.

Theorem 9.2.4. *If $y_n = \log_b x_n$ is equidistributed mod 1 then x_n is Benford (base b).*

Proof. By Lemma 9.2.3,

$$
\begin{aligned}
\{n \le N : y_n \bmod 1 \in [\log_b j, \log_b (j+1))\} \\
= \{n \le N : \text{first digit of } x_n \text{ in base } b \text{ is } j\}. \tag{9.10}
\end{aligned}
$$

Therefore

$$
\begin{aligned}
\lim_{N \to \infty} & \frac{\#\{n \le N : y_n \bmod 1 \in [\log_b j, \log_b (j+1))\}}{N}\\
&= \lim_{N \to \infty} \frac{\#\{n \le N : \text{first digit of } x_n \text{ in base } b \text{ is } j\}}{N}. \tag{9.11}
\end{aligned}
$$

If y_n is equidistributed, then the left side of (9.11) is $\log_b \left(1 + \frac{1}{j}\right)$ which implies x_n is Benford base b. $\square$

Remark 9.2.5. One can extend the definition of Benford's Law from statements concerning the distribution of the first digit to the distribution of the first k digits. With such an extension, Theorem 9.2.4 becomes $y_n = \log_b x_n \mod 1$ is equidistributed if and only if x_n is Benford base b. See [KonMi] for details.

Let $\{x\} = x - [x]$ denote the fractional part of x, where $[x]$ as always is the greatest integer at most x. In Theorem 12.3.2 we prove that for $\alpha \notin \mathbb{Q}$ the fractional parts of $n\alpha$ are equidistributed mod 1. From this and Theorem 9.2.4, it immediately follows that geometric series are Benford (modulo the irrationality condition):

Theorem 9.2.6. Let $x_n = ar^n$ with $\log_b r \notin \mathbb{Q}$. Then x_n is Benford (base b).

Proof. Let $y_n = \log_b x_n = n \log_b r + \log_b a$. As $\log_b r \notin \mathbb{Q}$, by Theorem 12.3.2 the fractional parts of y_n are equidistributed. Exponentiating by b, we obtain that x_n is Benford (base b) by Theorem 9.2.4. $\qquad\square$

Theorem 9.2.6 implies that 2^n is Benford base 10, but not surprisingly that it is not Benford base 2.

Exercise 9.2.7. *Do the first digits of e^n follow Benford's Law? What about $e^n + e^{-n}$?*

9.3 RECURRENCE RELATIONS AND BENFORD'S LAW

We show many sequences defined by recurrence relations are Benford. For more on recurrence relations, see Exercise 7.3.9. The interested reader should see [BrDu, NS] for more on the subject.

9.3.1 Recurrence Preliminaries

We consider recurrence relations of length k:

$$a_{n+k} = c_1 a_{n+k-1} + \cdots + c_k a_n, \tag{9.12}$$

where $c_1, \ldots, c_k$ are fixed real numbers. If the characteristic polynomial

$$r^k - c_1 r^{k-1} - c_2 r^{k-2} - \cdots - c_{k-1}r - c_k = 0 \tag{9.13}$$

has k distinct roots $\lambda_1, \ldots, \lambda_k$, there exist k numbers $u_1, \ldots, u_k$ such that

$$a_n = u_1 \lambda_1^n + \cdots + u_k \lambda_k^n, \tag{9.14}$$

where we have ordered the roots so that $|\lambda_1| \geq \cdots \geq |\lambda_k|$.

For the Fibonacci numbers $k = 2$, $c_1 = c_2 = 1$, $u_1 = -u_2 = \frac{1}{\sqrt{5}}$, and $\lambda_1 = \frac{1+\sqrt{5}}{2}$, $\lambda_2 = \frac{1-\sqrt{5}}{2}$ (see Exercise 7.3.11). If $|\lambda_1| = 1$, we do not expect the first digit of a_n to be Benford (base b). For example, if we consider

$$a_n = 2a_{n-1} - a_{n-2} \tag{9.15}$$

with initial values $a_0 = a_1 = 1$, every $a_n = 1$! If we instead take $a_0 = 0$, $a_1 = 1$, we get $a_n = n$. See [Kos] for many interesting occurrences of Fibonacci numbers and recurrence relations.

9.3.2 Recurrence Relations Are Benford

Theorem 9.3.1. *Let a_n satisfy a recurrence relation of length k with k distinct real roots. Assume $|\lambda_1| \neq 1$ with $|\lambda_1|$ the largest absolute value of the roots. Further, assume the initial conditions are such that the coefficient of λ_1 is non-zero. If $\log_b |\lambda_1| \notin \mathbb{Q}$, then a_n is Benford (base b).*

Proof. By assumption, $u_1 \neq 0$. For simplicity we assume $\lambda_1 > 0$, $\lambda_1 > |\lambda_2|$ and $u_1 > 0$. Again let $y_n = \log_b x_n$. By Theorem 9.2.4 it suffices to show y_n is equidistributed mod 1. We have

$$x_n = u_1 \lambda_1^n + \cdots + u_n \lambda_k^n$$
$$x_n = u_1 \lambda_1^n \left[1 + O\left(\frac{ku\lambda_2^n}{\lambda_1^n} \right) \right], \qquad (9.16)$$

where $u = \max_i |u_i| + 1$ (so $ku > 1$ and the big-Oh constant is 1). As $\lambda_1 > |\lambda_2|$, we "borrow" some of the growth from λ_1^n; this is a very useful technique. Choose a small ϵ and an n_0 such that

1. $|\lambda_2| < \lambda_1^{1-\epsilon}$;

2. for all $n > n_0$, $\frac{(ku)^{1/n}}{\lambda_1^\epsilon} < 1$, which then implies $\frac{ku}{\lambda_1^{n\epsilon}} = \left(\frac{(ku)^{1/n}}{\lambda_1^\epsilon} \right)^n$.

As $ku > 1$, $(ku)^{1/n}$ is decreasing to 1 as n tends to infinity. Note $\epsilon > 0$ if $\lambda_1 > 1$ and $\epsilon < 0$ if $\lambda_1 < 1$. Letting

$$\beta = \frac{(ku)^{1/n_0}}{\lambda_1^\epsilon} \frac{|\lambda_2|}{\lambda_1^{1-\epsilon}} < 1, \qquad (9.17)$$

we find that the error term above is bounded by β^n for $n > n_0$, which tends to 0. Therefore

$$y_n = \log_b x_n$$
$$= \log_b(u_1 \lambda_1^n) + O\left(\log_b(1 + \beta^n) \right)$$
$$= n \log_b \lambda_1 + \log_b u_1 + O(\beta^n), \qquad (9.18)$$

where the big-Oh constant is bounded by C say. As $\log_b \lambda_1 \notin \mathbb{Q}$, the fractional parts of $n \log_b \lambda_1$ are equidistributed mod 1, and hence so are the shifts obtained by adding the fixed constant $\log_b u_1$.

We need only show that the error term $O(\beta^n)$ is negligible. It is possible for the error term to change the first digit; for example, if we had 999999 (or 1000000), then if the error term contributes 2 (or -2), we would change the first digit base 10. However, for n sufficiently large, the error term will change a vanishingly small number of first digits. Say $n \log_b \lambda_1 + \log_b u_1$ exponentiates base b to first digit j, $j \in \{1, \ldots, b-1\}$. This means

$$n \log_b \lambda_1 + \log_b u_1 \in I_j^{(b)} = [p_{j-1}, p_j). \qquad (9.19)$$

The error term is at most $C\beta^n$ and y_n exponentiates to a different first digit than $n \log_b \lambda_1 + \log_b u_1$ only if one of the following holds:

1. $n \log_b \lambda_1 + \log_b u_1$ is within $C\beta^n$ of p_j, and adding the error term pushes us to or past p_j;

2. $n \log_b \lambda_1 + \log_b u_1$ is within $C\beta^n$ of p_{j-1}, and adding the error term pushes us before p_{j-1}.

The first set is contained in $[p_j - C\beta^n, p_j)$, of length $C\beta^n$. The second is contained in $[p_{j-1}, p_{j-1} + C\beta^n)$, also of length $C\beta^n$. Thus the length of the interval where $n \log_b \lambda_1 + \log_b u_1$ and y_n could exponentiate base b to different first digits is of size $2C\beta^n$. If we choose N sufficiently large then for all $n > N$ we can make these lengths arbitrarily small. As $n \log_b \lambda_1 + \log_b u_1$ is equidistributed mod 1, we can control the size of the subsets of $[0, 1)$ where $n \log_b \lambda_1 + \log_b u_1$ and y_n disagree. The Benford behavior (base b) of x_n now follows in the limit. □

Exercise 9.3.2. *Weaken the conditions of Theorem 9.3.1 as much as possible. What if several roots equal λ_1? What does a general solution to (9.12) look like now? What if λ_1 is negative? Can anything be said if there are complex roots?*

Exercise$^{(hr)}$ 9.3.3. *Consider the recurrence relation $a_{n+1} = 5a_n - 8a_{n-1} + 4a_{n-2}$. Show there is a choice of initial conditions such that the coefficient of λ_1 (a largest root of the characteristic polynomial) is non-zero but the sequence does not satisfy Benford's Law.*

Exercise$^{(hr)}$ 9.3.4. *Assume all the roots of the characteristic polynomial are distinct, and let λ_1 be the largest root in absolute value. Show for almost all initial conditions that the coefficient of λ_1 is non-zero, which implies that our assumption that $u_1 \neq 0$ is true most of the time.*

9.4 RANDOM WALKS AND BENFORD'S LAW

Consider the following (colorful) problem: A drunk starts off at time zero at a lamppost. Each minute he stumbles with probability p one unit to the right and with probability $q = 1 - p$ one unit to the left. Where do we expect the drunk to be after N tosses? This is known as a **Random Walk**. By the Central Limit Theorem (Theorem 8.4.1), his distribution after N tosses is well approximated by a Gaussian with mean $1 \cdot pN + (-1) \cdot (1 - p)N = (2p - 1)N$ and variance $p(1 - p)N$. For more details on Random Walks, see [Re].

For us, a **Geometric Brownian Motion** is a process such that its logarithm is a Random Walk (see [Hu] for complete statements and applications). We show below that the first digits of Geometric Brownian Motions are Benford. In [KonSi] the $3x + 1$ problem is shown to be an example of Geometric Brownian Motion. For heuristic purposes we use the first definition of the $3x + 1$ map, though the proof is for the alternate definition. We have two operators: T_3 and T_2, with $T_3(x) = 3x + 1$ and $T_2(x) = \frac{x}{2}$. If a_n is odd, $3a_n + 1$ is even, so T_3 must always be followed by T_2. Thus, we have really have two operators T_2 and $T_{3/2}$, with $T_{3/2}(x) = \frac{3x+1}{2}$. If we assume each operator is equally likely, half the time we go from $x \rightarrow \frac{3}{2}x + 1$, and half the time to $\frac{1}{2}x$.

If we take logarithms, $\log x$ goes to $\log \frac{3}{2}x = \log x + \log \frac{3}{2}$ half the time and $\log \frac{1}{2}x = \log x + \log \frac{1}{2}$ the other half. Hence on average we send $\log x \to \log x + \frac{1}{2}\log \frac{3}{4}$. As $\log \frac{3}{4} < 0$, on average our sequence is decreasing (which agrees with the conjecture that eventually we reach $4 \to 2 \to 1$). Thus we might expect our sequence to look like $\log x_k = \log x + \frac{k}{2}\log \frac{3}{4}$. As $\log \frac{3}{4} \notin \mathbb{Q}$, its multiples are equidistributed mod 1, and thus when we exponentiate we expect to see Benford behavior. Note, of course, that this is simply a heuristic, suggesting we might see Benford's Law.

While we can consider Random Walks or Brownian Motion with non-zero means, for simplicity below we assume the means are zero. Thus, in the example above, $p = \frac{1}{2}$.

Exercise 9.4.1. *Give a better heuristic for the Geometric Brownian Motion of the $3x + 1$ map by considering the alternate definition: $a_{n+1} = \frac{3a_n + 1}{2^k}$.*

9.4.1 Needed Gaussian Integral

Consider a sequence of Gaussians G_σ with mean 0 and variance σ^2, with $\sigma^2 \to \infty$. The following lemma shows that for any $\delta > 0$ as $\sigma \to \infty$ almost all of the probability is in the interval $[-\sigma^{1+\delta}, \sigma^{1+\delta}]$. We will use this lemma to show that it is enough to investigate Gaussians in the range $[-\sigma^{1+\delta}, \sigma^{1+\delta}]$.

Lemma 9.4.2.

$$\frac{2}{\sqrt{2\pi\sigma^2}} \int_{\sigma^{1+\delta}}^{\infty} e^{-x^2/2\sigma^2} dx \ll e^{-\sigma^{2\delta}/2}. \tag{9.20}$$

Proof. Change the variable of integration to $w = \frac{x}{\sigma\sqrt{2}}$. Denoting the above integral by I, we find

$$I = \frac{2}{\sqrt{2\pi\sigma^2}} \int_{\sigma^\delta/\sqrt{2}}^{\infty} e^{-w^2} \cdot \sigma\sqrt{2}dw = \frac{2}{\sqrt{\pi}} \int_{\sigma^\delta/\sqrt{2}}^{\infty} e^{-w^2} dw. \tag{9.21}$$

The integrand is monotonically decreasing. For $w \in \left[\frac{\sigma^\delta}{\sqrt{2}}, \frac{\sigma^\delta}{\sqrt{2}} + 1\right]$, the integrand is bounded by substituting in the left endpoint, and the region of integration is of length 1. Thus,

$$I < 1 \cdot \frac{2}{\sqrt{\pi}}e^{-\sigma^{2\delta}/2} + \frac{2}{\sqrt{\pi}} \int_{\frac{\sigma^\delta}{\sqrt{2}}+1}^{\infty} e^{-w^2} dw$$

$$= \frac{2}{\sqrt{\pi}}e^{-\sigma^{2\delta}/2} + \frac{2}{\sqrt{\pi}} \int_{\frac{\sigma^\delta}{\sqrt{2}}}^{\infty} e^{-(u+1)^2} du$$

$$= \frac{2}{\sqrt{\pi}}e^{-\sigma^{2\delta}/2} + \frac{2}{\sqrt{\pi}} \int_{\frac{\sigma^\delta}{\sqrt{2}}}^{\infty} e^{-u^2} e^{-2u} e^{-1} du$$

$$< \frac{2}{\sqrt{\pi}}e^{-\sigma^{2\delta}/2} + \frac{2}{e\sqrt{\pi}}e^{-\sigma^{2\delta}/2} \int_{\frac{\sigma^\delta}{\sqrt{2}}}^{\infty} e^{-2u} du$$

$$< \frac{2(e+1)}{\sqrt{\pi}}e^{-\sigma^{2\delta}/2}$$

$$< 4e^{-\sigma^{2\delta}/2}. \tag{9.22}$$

□

Exercise 9.4.3. *Prove a similar result for intervals of the form* $[-\sigma g(\sigma), \sigma g(\sigma)]$ *where* $g(\sigma)$ *is a positive increasing function and* $\lim_{\sigma \to \infty} g(\sigma) = +\infty$.

9.4.2 Geometric Brownian Motions Are Benford

We investigate the distribution of digits of processes that are Geometric Brownian Motions. By Theorem 9.2.4 it suffices to show that the Geometric Brownian Motion converges to being equidistributed mod 1. Explicitly, we have the following: after N iterations, by the Central Limit Theorem the expected value converges to a Gaussian with mean 0 and variance proportional to $\sqrt{N}$. We must show that the Gaussian with growing variance is equidistributed mod 1.

For convenience we assume the mean is 0 and the variance is $N/2\pi$. This corresponds to a fair coin where for each head (resp., tail) we move $\frac{1}{\sqrt{4\pi}}$ units to the right (resp., left). By the Central Limit Theorem the probability of being x units to the right of the origin after N tosses is asymptotic to

$$p_N(x) = \frac{e^{-\pi x^2/N}}{\sqrt{N}}. \tag{9.23}$$

For ease of exposition, we assume that rather than being asymptotic to a Gaussian, the distribution is a Gaussian. For our example of flipping a coin, this cannot be true. If every minute we flip a coin and record the outcome, after N minutes there are 2^N possible outcomes, a finite number. To each of these we attach a number equal to the excess of heads to tails. There are technical difficulties in working with discrete probability distributions; thus we study instead continuous processes such that at time N the probability of observing x is given by a Gaussian with mean 0 and variance $N/2\pi$. For complete details see [KonMi].

Theorem 9.4.4. *As* $N \to \infty$, $p_N(x) = \frac{e^{-\pi x^2/N}}{\sqrt{N}}$ *becomes equidistributed mod 1.*

Proof. For each N we calculate the probability that for $x \in \mathbb{R}$, x mod $1 \in [a, b] \subset [0, 1)$. This is

$$\int_{\substack{x=-\infty \\ x \text{ mod } 1 \in [a,b]}}^{\infty} p_N(x)dx = \frac{1}{\sqrt{N}} \sum_{n \in \mathbb{Z}} \int_{x=a}^{b} e^{-\pi(x+n)^2/N} dx. \tag{9.24}$$

We need to show the above converges to $b - a$ as $N \to \infty$. For $x \in [a, b]$, standard calculus (Taylor series expansions, see §A.2.3) gives

$$e^{-\pi(x+n)^2/N} = e^{-\pi n^2/N} + O\left(\frac{\max(1, |n|)}{N} e^{-n^2/N}\right). \tag{9.25}$$

We claim that in (9.24) it is sufficient to restrict the summation to $|n| \leq N^{5/4}$. The proof is immediate from Lemma 9.4.2: we increase the integration by expanding to $x \in [0, 1]$, and then trivially estimate. Thus, up to negligible terms, all the contribution is from $|n| \leq N^{5/4}$.

In §11.4.2 we prove the Poisson Summation formula, which in this case yields

$$\frac{1}{\sqrt{N}} \sum_{n \in \mathbb{Z}} e^{-\pi n^2/N} = \sum_{n \in \mathbb{Z}} e^{-\pi n^2 N}. \tag{9.26}$$

The beauty of Poisson Summation is that it converts one infinite sum with *slow* decay to another sum with *rapid* decay; because of this, Poisson Summation is an extremely useful technique for a variety of problems. The exponential terms on the left of (9.26) are all of size 1 for $n \leq \sqrt{N}$, and do not become small until $n \gg \sqrt{N}$ (for instance, once $n > \sqrt{N} \log N$, the exponential terms are small for large N); however, almost all of the contribution on the right comes from $n = 0$. The power of Poisson Summation is it often allows us to approximate well long sums with short sums. We therefore have

$$\frac{1}{\sqrt{N}} \sum_{|n| \leq N^{5/4}} \int_{x=a}^{b} e^{-\pi(x+n)^2/N} dx$$

$$= \frac{1}{\sqrt{N}} \sum_{|n| \leq N^{5/4}} \int_{x=a}^{b} \left[e^{-\pi n^2/N} + O\left(\frac{\max(1, |n|)}{N} e^{-n^2/N} \right) \right] dx$$

$$= \frac{b-a}{\sqrt{N}} \sum_{|n| \leq N^{5/4}} e^{-\pi n^2/N} + O\left(\frac{1}{N} \sum_{n=0}^{N^{5/4}} \frac{n+1}{\sqrt{N}} e^{-\pi(n/\sqrt{N})^2} \right)$$

$$= \frac{b-a}{\sqrt{N}} \sum_{|n| \leq N^{5/4}} e^{-\pi n^2/N} + O\left(\frac{1}{N} \int_{w=0}^{N^{3/4}} (w+1) e^{-\pi w^2} \sqrt{N} dw \right)$$

$$= \frac{b-a}{\sqrt{N}} \sum_{|n| \leq N^{5/4}} e^{-\pi n^2/N} + O\left(N^{-1/2} \right). \tag{9.27}$$

By Lemma 9.4.2 we can extend all sums to $n \in \mathbb{Z}$ in (9.27) with negligible error. We now apply Poisson Summation and find that up to lower order terms,

$$\frac{1}{\sqrt{N}} \sum_{n \in \mathbb{Z}} \int_{x=a}^{b} e^{-\pi(x+n)^2/N} dx \approx (b-a) \cdot \sum_{n \in \mathbb{Z}} e^{-\pi n^2 N}. \tag{9.28}$$

For $n = 0$ the right hand side of (9.28) is $b - a$. For all other n, we trivially estimate the sum:

$$\sum_{n \neq 0} e^{-\pi n^2 N} \leq 2 \sum_{n \geq 1} e^{-\pi n N} \leq \frac{2 e^{-\pi N}}{1 - e^{-\pi N}}, \tag{9.29}$$

which is less than $4 e^{-\pi N}$ for N sufficiently large. $\qquad\square$

We can interpret the above arguments as follows: for each N, consider a Gaussian $p_N(x)$ with mean 0 and variance $N/2\pi$. As $N \to \infty$ for each x (which occurs with probability $p_N(x)$) the first digit of 10^x converges to the Benford base 10 probabilities.

Remark 9.4.5. The above framework is very general and applicable to a variety of problems. In [KonMi] it is shown that these arguments can be used to prove Benford behavior in discrete systems such as the $3x + 1$ problem as well as continuous systems such as the absolute values of the Riemann zeta function (and any "good" L-function) near the critical line! For these number theory results, the crucial ingredients are Selberg's result (near the critical line, $\log |\zeta(s + it)|$ for $t \in [T, 2T]$ converges to a Gaussian with variance tending to infinity in T) and estimates by Hejhal on the rate of convergence. For the $3x + 1$ problem the key ingredients are the structure theorem (see [KonSi]) and the approximation exponent of Definition 5.5.1; see [LaSo] for additional results on Benford behavior of the $3x + 1$ problem.

9.5 STATISTICAL INFERENCE

Often we have reason to believe that some process occurs with probability p of success and $q = 1 - p$ of failure. For example, consider the $3x + 1$ problem. Choose a large a_0 and look at the first digit of the a_n's. There is reason to believe the distribution of the first digits is given by Benford's Law for most a_0 as $a_0 \to \infty$. We describe how to test this and similar hypotheses. We content ourselves with describing one simple test; the interested reader should consult a statistics textbook (for example, [BD, CaBe, LF, MoMc]) for the general theory and additional applications.

9.5.1 Null and Alternative Hypotheses

Suppose we think some population has a parameter with a certain value. If the population is small, it is possible to investigate every element; in general this is not possible.

For example, say the parameter is how often the millionth decimal or continued fraction digit is 1 in two populations: all rational numbers in $[0, 1)$ with denominator at most 5, and all real numbers in $[0, 1)$. In the first, there are only 10 numbers, and it is easy to check them all. In the second, as there are infinitely many numbers, it is impossible to numerically investigate each. What we do in practice is we sample a large number of elements (say N elements) in $[0, 1)$, and calculate the average value of the parameter for this sample.

We thus have two **populations**, the **underlying population** (in the second case, all numbers in $[0, 1)$), and the **sample population** (in this case, the N sampled elements).

Our goal is to test whether or not the underlying population's parameter has a given value, say p. To this end, we want to compare the sample population's value to p. The **null hypothesis**, denoted H_0, is the claim that there is no difference between the sample population's value and the underlying population's value; the **alternative hypothesis**, denoted H_a, is the claim that there is a difference between the sample population's value and the underlying population's value.

When we analyze the data from the sample population, either we reject the null hypothesis, or we fail to reject the null hypothesis. It is important to note that we *never* prove the null or alternative hypothesis is true or false. We are always rejecting or failing to reject the null hypothesis, we are never accepting it. If we flip a coin 100 times and observe all heads, this does not mean the coin is not fair:

it is possible the coin is fair but we had a very unusual sample (though, of course, it is extremely unlikely).

We now discuss how to test the null hypothesis. Our main tool is the Central Limit Theorem. This is just one of many possible inference tests; we refer the reader to [BD, CaBe, LF, MoMc] for more details.

9.5.2 Bernoulli Trials and the Central Limit Theorem

Assume we have some process where we expect a probability p of observing a given value. For example, if we choose numbers uniformly in $[0, 1)$ and look at the millionth decimal digit, we believe that the probability this digit is 1 is $\frac{1}{10}$. If we look at the continued fraction expansion, by Theorem 10.3.1 the probability that the millionth digit is 1 is approximately $\log_2 \frac{4}{3}$. What if we restrict to algebraic numbers? What is the probability the millionth digit (decimal or continued fraction expansion) equals 1?

In general, once we formalize our conjecture we test it by choosing N elements from the population independently at random (see §8.3). Consider the claim that a process has probability p of success. We have N independent Bernoulli trials (see §8.2.1). The null hypothesis is the claim that p percent of the sample are a success. Let S_N be the number of successes; if the null hypothesis is correct, by the Central Limit Theorem (see §8.4) we expect S_N to have a Gaussian distribution with mean pN and variance pqN (see Exercise 8.2.1 for the calculations of the mean and variance of a Bernoulli process). This means that if we were to look at many samples with N elements, on average each sample would have $pN \pm O(\sqrt{pqN})$ successes. We calculate the probability of observing a difference $|S_N - pN|$ as large or larger than a. This is given by the area under the Gaussian with mean pN and variance pqN:

$$\frac{1}{\sqrt{2\pi pqN}} \int_{|s-pN|\geq a} e^{-(s-pN)^2/2pqN} ds. \tag{9.30}$$

If this integral is small, it is extremely unlikely that we choose N independent trials from a process with probability p of success and we reject the null hypothesis; if the integral is large, we do not reject the null hypothesis, and we have support for our claim that the underlying process does have probability p of success.

Unfortunately, the Gaussian is a difficult function to integrate, and we would need to tabulate these integrals for *every* different pair of mean and variance. It is easier, therefore, to renormalize and look at a new statistic which should also be Gaussian, but with mean 0 and variance 1. The advantage is that we need only tabulate *one* special Gaussian, the standard normal.

Let $Z = \frac{S_N - pN}{\sqrt{pqN}}$. This is known as the **z-statistic**. If S_N's distribution is a Gaussian with mean pN and variance pqN, note Z will be a Gaussian with mean 0 and variance 1.

Exercise 9.5.1. *Prove the above statement about the distribution of z.*

Let

$$I(a) = \frac{1}{\sqrt{2\pi}} \int_{|z|\geq a} e^{-z^2/2} dz, \tag{9.31}$$

the area under the standard normal (mean 0, standard deviation 1) that is at least a units from the mean. We consider different **confidence intervals**. If we were to randomly choose a number z from such a Gaussian, what is the probability (as a function of a) that z is at most a units from the mean? Approximately 68% of the time $|z| \leq 1$ ($I(1) \approx .32$), approximately 95% of the time $z \leq 1.96$ ($I(1.96) \approx .05$), and approximately 99% of the time $|z| \leq 2.57$ ($I(2.57) = .01$). In other words, there is only about a 1% probability of observing $|z| \geq 2.57$. If $|z| \geq 2.57$, we have strong evidence against the hypothesis that the process occurs with probability p, and we would be reasonably confident in rejecting the null hypothesis; of course, it is possible we were unlucky and obtained an unrepresentative set of data (but it is extremely unlikely that this occurred; in fact, the probability is at most 1%).

Remark 9.5.2. For a Gaussian with mean μ and standard deviation σ, the probability that $|X - \mu| \leq \sigma$ is approximately .68. Thus if X is drawn from a normal with mean μ and standard deviation σ, then approximately 68% of the time $\mu \in [x - \sigma, x + \sigma]$ (where x is the observed value of the random variable X).

To test the claim that some process occurs with probability p, we observe N independent trials, calculate the z-statistic, and see how likely it is to observe $|Z|$ that large or larger. We give two examples below.

9.5.3 Digits of the $3x + 1$ Problem

Consider again the $3x + 1$ problem. Choose a large integer a_0, and look at the iterates: $a_1, a_2, a_3, \ldots$. We study how often the first digit of terms in the sequence equal $d \in \{1, \ldots, 9\}$. We can regard the first digit of a term as a Bernoulli trial with a success (or 1) if the first digit is d and a failure (or 0) otherwise. If the distribution of digits is governed by Benford's Law, the theoretical prediction is that the percentage of the first digits that equal d is $p = \log_{10}(\frac{d+1}{d})$. Assume there are N terms in our sequence (before we hit the pattern $4 \rightarrow 2 \rightarrow 1 \rightarrow 4 \cdots$), and say M of them have first digit d. For what M does this experiment provide support that the digits follow Benford's Law?

Exercise 9.5.3. *The terms in the sequence generated by a_0 are not independent, as a_{n+1} is determined by a_n. Show that if the first digit of a_n is 2 then the first digit of a_{n+1} cannot be a 2.*

The above exercise shows that the first digit of the terms *cannot* be considered independent Bernoulli trials. As the sequence is completely determined by the first term, this is not surprising. If we look at an enormous number of terms, however, these effects "should" average out. Another possible experiment is to look at the first digit of the millionth term for N different a_0's.

Let $a_0 = 333 \ldots 333$ be the integer that is 10,000 threes. There are 177,857 terms in the sequence before we hit $4 \rightarrow 2 \rightarrow 1$. The following data comparing the number of first digits equal to d to the Benford predictions are from [Min]:

digit	observed	predicted	variance	z-statistic	$I(z)$
1	53425	53540	193.45	−0.596	0.45
2	31256	31310	160.64	−0.393	0.31
3	22257	22220	139.45	0.257	0.21
4	17294	17230	124.76	0.464	0.36
5	14187	14080	113.88	0.914	0.63
6	11957	11900	105.40	0.475	0.36
7	10267	10310	98.57	−0.480	0.37
8	9117	9090	92.91	0.206	0.16
9	8097	8130	88.12	−0.469	0.36

As the values of the z-statistics are all small (well below 1.96 and 2.57), the above table provides evidence that the first digits in the $3x + 1$ problem follow Benford's Law, and we would not reject the null hypothesis for any of the digits. If we had obtained large z-statistics, say 4, we would reject the null hypothesis and doubt that the distribution of digits follow Benford's Law.

Remark 9.5.4 (Important). One must be very careful when analyzing all the digits. Once we know how many digits are in $\{1, \ldots, 8\}$, then the number of 9's is forced: these are not nine independent tests, and a different statistical test (a chi-square test with eight degrees of freedom) should be done. Our point here is not to write a treatise on statistical inference, but merely highlight some of the tools and concepts. See [BD, CaBe, LF, MoMc] for more details, and [Mil5] for an amusing analysis of a baseball problem involving chi-square tests.

Additionally, if we have many different experiments, then "unlikely" events should happen. For example, if we have 100 different experiments we would not be surprised to see an outcome which only has a 1% chance of occurring (see Exercise 9.5.5). Thus, if there are many experiments, the confidence intervals need to be adjusted. One common method is the Bonferroni adjustment method for multiple comparisons. See [BD, MoMc].

Exercise 9.5.5. *Assume for each trial there is a 95% chance of observing the percentage of first digits equal to 1 is in $[\log_{10} 2 - 1.96\sigma, \log_{10} 2 + 1.96\sigma]$ (for some σ). If we have 10 independent trials, what is the probability that all the observed percentages are in this interval? If we have 14 independent trials?*

Remark 9.5.6. How does one calculate with $10,000$ digit numbers? Such large numbers are greater than the standard number classes (int, long, double) of many computer programming languages. The solution is to represent numbers as arrays. To go from a_n to $3a_n + 1$, we multiply the array by 3, carrying as needed, and then add 1; we leave space-holding zeros at the start of the array. For example,

$$3 \cdot [0, \ldots, 0, 0, 5, 6, 7] = [0, \ldots, 0, 1, 7, 0, 1]. \tag{9.32}$$

We need only do simple operations on the array. For example, $3 \cdot 7 = 21$, so the first entry of the product array is 1 and we carry the 2 for the next multiplication. We must also compute $a_n/2$ if a_n is even. Note this is the same as $5a_n$ divided by 10. The advantage of this approach is that it is easy to calculate $5a_n$, and as a_n is even, the last digit of $5a_n$ is zero, hence array division by 10 is trivial.

Exercise 9.5.7. *Consider the first digits of the $3x + 1$ problem in base 6. Choose a large integer a_0, and look at the iterates $a_1, a_2, a_3, \ldots$. Show that as $a_0 \to \infty$ the distribution of digits is Benford base 6.*

Exercise 9.5.8 (Recommended). *Here is another variant of the $3x + 1$ problem:*

$$a_{n+1} = \begin{cases} 3a_n + 1 & \text{if } a_n \text{ is odd} \\ a_n/2^k & \text{if } a_n \text{ is even and } 2^k || a_n; \end{cases} \tag{9.33}$$

$2^k || a_n$ means 2^k divides a_n, but 2^{k+1} does not. Consider the distribution of first digits of this sequence for various a_0. What is the null hypothesis? Do the data support the null hypothesis, or the alternative hypothesis? Do you think these numbers also satisfy Benford's Law? What if instead we define

$$a_{n+1} = \frac{3a_n + 1}{2^k}, \quad 2^k || a_n. \tag{9.34}$$

9.5.4 Digits of Continued Fractions

Let us test the hypothesis that the digits of algebraic numbers are given by the Gauss-Kuzmin Theorem (Theorem 10.3.1). Let us look at how often the 1000th digit equals 1. By the Gauss-Kuzmin Theorem this should be approximately $\log_2 \frac{4}{3}$. Let p_n be the nth prime. In the continued fraction expansions of $\sqrt[3]{p_n}$ for $n \in \{100000, 199999\}$, exactly 41565 have the 1000th digit equal to 1. Assuming we have a Bernoulli process with probability of success (a digit of 1) of $p = \log_2 \frac{4}{3}$, the z-statistic is .393. As the z-statistic is small (95% of the time we expect to observe $|z| \leq 1.96$), we do not reject the null hypothesis, and we have obtained evidence supporting the claim that the probability that the 1000th digit is 1 is given by the Gauss-Kuzmin Theorem. See Chapter 10 for more detailed experiments on algebraic numbers and the Gauss-Kuzmin Theorem.

9.6 SUMMARY

We have chosen to motivate our presentation of statistical inference by investigating the first digits of the $3x + 1$ problem, but of course the methods apply to a variety of problems. Our main tool is the Central Limit Theorem: if we have a process with probability p (resp., $q = 1 - p$) of success (resp., failure), then in N independent trials we expect about pN successes, with fluctuations of size $\sqrt{pqN}$. To test whether or not the underlying probability is p we formed the z-statistic: $\frac{S_N - pN}{\sqrt{pqN}}$, where S_N is the number of successes observed in the N trials.

If the process really does have probability p of success, then by the Central Limit Theorem the distribution of S_N is approximately a Gaussian with mean pN and standard deviation $\sqrt{pqN}$, and we then expect the z-statistic to be of size 1. If, however, the underlying process occurs not with probability p but p', then we expect S_N to be approximately a Gaussian with mean $p'N$ and standard deviation $\sqrt{p'q'N}$. We now expect the z-statistic to be of size $\frac{(p'-p)N}{\sqrt{p'q'N}}$. This is of size $\sqrt{N}$, much larger than 1.

We see the z-statistic is very sensitive to $p' - p$: if p' is differs from p, for large N we quickly observe large values of z. Note, of course, that statistical tests can only provide compelling evidence in favor or against a hypothesis, never a proof.

Chapter Ten

Distribution of Digits of Continued Fractions

Given $\alpha \in \mathbb{R}$, we can calculate its continued fraction expansion and investigate the distribution of its digits. Without loss of generality we assume $\alpha \in (0, 1)$, as this shift changes only the zeroth digit. Thus

$$\alpha = [0, a_1, a_2, a_3, a_4, \dots]. \tag{10.1}$$

Given any sequence of positive integers a_i, we can construct a number α with these as its digits. However, for a generic α chosen uniformly in $(0, 1)$, how often do we expect to observe the nth digit in the continued fraction expansion equal to 1? To 2? To 3? And so on.

If $\alpha \in \mathbb{Q}$ then by Theorem 7.3.20 α has a finite continued fraction expansion; if α is a quadratic irrational then by Theorem 7.6.1 its continued fraction expansion is periodic. In both of these cases there are really only finitely many digits; however, if we stay away from rationals and quadratic irrationals, then α will have a bona fide infinite continued fraction expansion, and it makes sense to ask the above questions.

For the decimal expansion of a generic $\alpha \in (0, 1)$, we expect each digit to take the values 0 through 9 with equal probability; as there are infinitely many values for the digits of a continued fraction, each value cannot be equally likely. We will see, however, that as $n \to \infty$ the probability of the nth digit equalling k converges to $\log_2\left(1 + \frac{1}{k(k+2)}\right)$. An excellent source is [Kh], whose presentation strongly influenced our exposition.

In this chapter we assume the reader is familiar with the basic properties of continued fractions (Chapter 7), probability theory (Chapter 8) and elementary notions of length or measure of sets (Appendix A.5). For the numerical experiments we assume knowledge of statistical inference and hypothesis testing at the level of Chapter 9.

10.1 SIMPLE RESULTS ON DISTRIBUTION OF DIGITS

We recall some notation: we can truncate the continued fraction expansion of $\alpha \in (0, 1)$ after the nth digit, obtaining an approximation

$$\frac{p_n}{q_n} = [0, a_1, \dots, a_n] \tag{10.2}$$

to α, with each $a_i \in \mathbb{N}$. Further there exists $r_{n+1} \in \mathbb{R}$, $r_{n+1} \in [1, \infty)$ such that

$$\alpha = [0, a_1, \dots, a_n, r_{n+1}]. \tag{10.3}$$

We begin with a simple question: What is the measure of $\alpha \in (0,1)$ such that $a_1(\alpha) = k$? Such α must have the following expansion:

$$\alpha = \cfrac{1}{k + \cfrac{1}{a_2 + \cfrac{1}{\ddots}}}. \tag{10.4}$$

Clearly, if $\alpha \le \frac{1}{k+1}$ then $a_1 \ge k+1$. A simple calculation shows that if $\frac{1}{k+1} < \alpha \le \frac{1}{k}$, then $a_1(\alpha) = k$ because $a_1 = \left[\frac{1}{\alpha}\right]$. Remember that $[x]$ is the greatest integer less than or equal to x. Therefore the measure (or length of the interval) of $\alpha \in [0,1)$ such that $a_1 = k$ is $\frac{1}{k} - \frac{1}{k+1} = \frac{1}{k(k+1)}$, which is approximately $\frac{1}{k^2}$.

Exercise 10.1.1. *Show for k large, $\frac{1}{k(k+1)} \approx \log_2\left(1 + \frac{1}{k(k+2)}\right)$. Thus the claimed answer for the distribution of digits is "reasonable."*

We start counting with the zeroth digit; as $\alpha \in [0,1)$, the zeroth digit is always zero. For notational convenience, we adopt the following convention. Let $A_{1,\ldots,n}(a_1,\ldots,a_n)$ be the event that $\alpha \in [0,1)$ has its continued fraction expansion $\alpha = [0, a_1, \ldots, a_n, \ldots]$. Similarly $A_{n_1,\ldots,n_k}(a_{n_1},\ldots,a_{n_k})$ is the event where the zeroth digit is 0, digit n_1 is a_{n_1}, ..., and digit n_k is a_{n_k}, and $A_n(k)$ is the event that the zeroth digit is 0 and the nth digit is k.

Exercise 10.1.2. *Show for n finite, $A_{1,\ldots,n}(a_1,\ldots,a_n)$ is a subinterval of $[0,1)$. Show $A_{n_1,\ldots,n_k}(a_{n_1},\ldots,a_{n_k})$ is an at most countable (see §5.2) union of subintervals of $[0,1)$. When is it a finite union?*

Remark 10.1.3. By Theorem 7.4.6, the continued fraction expansion for an α is unique if α has infinitely many digits. As only the rationals have finite continued fractions, and they have length zero (see §A.5.1), *we assume all α are irrational in all future investigations*. Note this will not change the lengths of any subintervals.

The length of the subinterval $A_{1,\ldots,n}(a_1,\ldots,a_n)$ is the probability that $\alpha \in [0,1)$ has its first digits $a_1,\ldots,a_n$; we denote this by

$$\text{Prob}(A_{1,\ldots,n}(a_1,\ldots,a_n)) = |A_{1,\ldots,n}(a_1,\ldots,a_n)|. \tag{10.5}$$

For $\alpha \in A_{1,\ldots,n}(a_1,\ldots,a_n)$, one can ask: given that α has these first n digits, what is the probability that the next digit is k? How does this depend on n and a_1 through a_n? This is a conditional probability. By Lemma 8.1.22 the answer is

$$\text{Prob}(a_{n+1}(\alpha) = k | A_{1,\ldots,n}(a_1,\ldots,a_n)) = \frac{|A_{1,\ldots,n,n+1}(a_1,\ldots,a_n,k)|}{|A_{1,\ldots,n}(a_1,\ldots,a_n)|}. \tag{10.6}$$

Equivalently, this is the conditional probability of observing $a_{n+1}(\alpha) = k$, given the first n digits have the prescribed values $a_1,\ldots,a_n$.

Lemma 10.1.4.

$$|A_{1,\ldots,n}(a_1,\ldots,a_n)| = \frac{1}{q_n^2\left(1 + \frac{q_{n-1}}{q_n}\right)}. \tag{10.7}$$

Proof. We want the length of the subinterval of $[0, 1)$ that corresponds to $\alpha \in [0, 1)$ whose continued fraction begins with $[0, a_1, \ldots, a_n]$. We have

$$\frac{p_n}{q_n} = [0, a_1, \ldots, a_n]$$

$$\alpha = [0, a_1, \ldots, a_n, r_{n+1}], \quad 1 < r_{n+1} < \infty. \tag{10.8}$$

Remark 10.1.5. If $\alpha \notin \mathbb{Q}$ then the continued fraction expansion of α has infinitely many digits; thus, $r_{n+1} \neq 1$.

The recursion relations for the p_n's and q_n's (see Theorem 7.3.5 and Lemma 7.3.13) are

$$p_{n+1} = a_{n+1}p_n + p_{n-1}$$
$$q_{n+1} = a_{n+1}q_n + q_{n-1}$$
$$p_n q_{n-1} - p_{n-1}q_n = (-1)^{n-1}. \tag{10.9}$$

By considering what p_{n+1} and q_{n+1} are when $r_{n+1} = 1$ and infinity, we find the subinterval is

$$\left[\frac{p_n}{q_n}, \frac{p_n + p_{n-1}}{q_n + q_{n-1}}\right). \tag{10.10}$$

A straightforward calculation (using the recursion relations) shows this has length

$$\frac{1}{q_n^2\left(1 + \frac{q_{n-1}}{q_n}\right)}, \tag{10.11}$$

which completes the proof. $\square$

Remark 10.1.6 (Notation). We wrote the interval as $\left[\frac{p_n}{q_n}, \frac{p_n + p_{n-1}}{q_n + q_{n-1}}\right)$; however, it is possible that it should have been written $\left(\frac{p_n + p_{n-1}}{q_n + q_{n-1}}, \frac{p_n}{q_n}\right]$. For definiteness we always write the intervals open on the right.

Exercise 10.1.7. *Prove* (10.11).

Lemma 10.1.8. *We have*

$$|A_{1,\ldots,n,n+1}(a_1, \ldots, a_n, k)| = \frac{1}{q_n^2 k^2}\frac{1}{\left(1 + \frac{q_{n-1}}{kq_n}\right)\left(1 + \frac{1}{k} + \frac{q_{n-1}}{kq_n}\right)}. \tag{10.12}$$

Proof. For such α, we have

$$\frac{p_n}{q_n} = [0, a_1, \ldots, a_n]$$

$$\alpha = [0, a_1, \ldots, a_n, r_{n+1}], \quad k \leq r_{n+1} < k+1. \tag{10.13}$$

Using the recursion relations again and $q_m = a_{m+1}q_m + q_{m-1}$ (and similarly for p_m), we find the interval is

$$\left[\frac{p_n k + p_{n-1}}{q_n k + q_{n-1}}, \frac{p_n(k+1) + p_{n-1}}{q_n(k+1) + q_{n-1}}\right). \tag{10.14}$$

We can see this by considering the endpoints, which occur when $r_{n+1} = k$ and $k + 1$. A straightforward calculation shows the length equals

$$\frac{1}{q_n^2 k^2 \left(1 + \frac{q_{n-1}}{kq_n}\right)\left(1 + \frac{1}{k} + \frac{q_{n-1}}{kq_n}\right)}, \tag{10.15}$$

which completes the proof. $\square$

Lemma 10.1.9. *We have*

$$\frac{1}{3k^2} \leq \frac{|A_{1,\ldots,n,n+1}(a_1,\ldots,a_n,k)|}{|A_{1,\ldots,n}(a_1,\ldots,a_n)|} \leq \frac{2}{k^2}, \tag{10.16}$$

which implies that the conditional probability is proportional to $\frac{1}{k^2}$.

Proof. Using Lemmas 10.1.4 and 10.1.8, by (10.6) we find the conditional probability that $a_{n+1}(\alpha) = k$, given that $a_i(\alpha) = a_i$ for $i \leq n$, is (after simple algebra)

$$\frac{1}{k^2} \cdot \frac{1 + \frac{q_{n-1}}{q_n}}{\left(1 + \frac{q_{n-1}}{kq_n}\right)\left(1 + \frac{1}{k} + \frac{q_{n-1}}{kq_n}\right)}. \tag{10.17}$$

We bound the second factor from above and below, independently of k. As $q_{n-1} \leq q_n$, the second factor is at most 2 (as the denominator is greater than 1 and the numerator is at most 2). For the lower bound, note the second factor is monotonically increasing with k. Thus its smallest value is attained when $k = 1$, which is at least $\frac{1}{3}$. $\square$

This implies the conditional probability is proportional to $\frac{1}{k^2}$. While the exact answer will almost surely depend on k and a_1 through a_n, we have found bounds *that depend only on k*; this will be useful for many applications. One application is to estimate the probability that a given digit equals k. We can pass from knowledge of conditional probabilities by summing over all n-tuples:

Theorem 10.1.10. *We have*

$$\frac{1}{3k^2} \leq |A_{n+1}(k)| \leq \frac{2}{k^2}. \tag{10.18}$$

Proof. We find the conditional probability that $a_{n+1}(\alpha) = k$ for all possible choices of $a_1(\alpha),\ldots,a_n(\alpha)$; the result then follows by summing over all appropriately weighted choices.

$$|A_{n+1}(k)| = \sum_{(a_1,\ldots,a_n)\in\mathbb{N}^n} \frac{|A_{1,\ldots,n,n+1}(a_1,\ldots,a_n,k)|}{|A_{1,\ldots,n}(a_1,\ldots,a_n)|} \cdot |A_{1,\ldots,n}(a_1,\ldots,a_n)|.$$

$$\tag{10.19}$$

As each conditional probability is at least $\frac{1}{3k^2}$ and at most $\frac{2}{k^2}$, and

$$\sum_{(a_1,\ldots,a_n)\in\mathbb{N}^n} |A_{1,\ldots,n}(a_1,\ldots,a_n)| = 1 \tag{10.20}$$

(because every $\alpha \in [0, 1)$ is in exactly one of the above intervals, and the intervals are disjoint), we have

$$\frac{1}{3k^2} \leq |A_{n+1}(k)| \leq \frac{2}{k^2}, \tag{10.21}$$

which completes the proof.   □

In the arguments above, we multiplied by one in the form $\frac{|A_{1,\ldots,n}(a_1,\ldots,a_n)|}{|A_{1,\ldots,n}(a_1,\ldots,a_n)|}$; this is a very common technique, and allows us to incorporate our knowledge of the conditional probabilities (which are easy to bound). Note the above bounds are independent of $n + 1$, and for large k are of the same order of magnitude as $\log_2\left(1 + \frac{1}{k(k+2)}\right)$. Later we derive a better estimate for $|A_{n+1}(k)|$.

Corollary 10.1.11. *There exist constants $0 < C_1 < 1 < C_2 < \infty$, independent of K, such that*

$$\frac{C_1}{K} < \text{Prob}(a_{n+1}(\alpha) \geq K) < \frac{C_2}{K}. \tag{10.22}$$

Proof. $\text{Prob}(a_{n+1}(\alpha) \geq K) = \sum_{k=K}^{\infty} |A_{n+1}(k)|$, and $\frac{1}{K^2} + \frac{1}{(K+1)^2} + \cdots$ is of size $\frac{1}{K}$.   □

Exercise 10.1.12. *Estimate C_1 and C_2 in Corollary 10.1.11.*

10.2 MEASURE OF α WITH SPECIFIED DIGITS

Another immediate application of Theorem 10.1.10 is that the length of $\alpha \in [0, 1)$ with every digit less than K is zero. Explicitly,

Theorem 10.2.1. *Consider all $\alpha \in [0, 1)$ such that for all n, $a_n(\alpha) < K$ for some fixed constant K. The set of such α has length 0. In other words,*

$$\text{Prob}(\forall n, a_n(\alpha) < K) = 0. \tag{10.23}$$

Proof. We would like to argue that by Corollary 10.1.11, the probability of the nth digit being less than K is at most $\beta = 1 - \frac{C_1}{K} < 1$. Thus the probability that the first N digits are less than K would be β^N, which tends to zero as $N \to \infty$. Unfortunately, these are not necessarily independent events, and hence the probabilities are not multiplicative. We see in §10.4 that knowledge of some digits does give knowledge about other digits.

This defect is easily remedied by using Lemma 10.1.9. We proceed by induction. By Corollary 10.1.11 $\text{Prob}(a_1(\alpha) \leq K) < \beta^1$. Assume we have shown that $\text{Prob}(a_i(\alpha) < K : 1 \leq i \leq N) \leq \beta^N$; the theorem follows by showing the above holds for $N + 1$. We have

$$\text{Prob}(a_i(\alpha) < K : 1 \leq i \leq N + 1)$$

$$= \sum_{k<K} \sum_{\substack{(a_1,\ldots,a_N)\in\mathbb{N}^N \\ a_i < K}} \frac{|A_{1,\ldots,n,n+1}(a_1,\ldots,a_n,k)|}{|A_{1,\ldots,n}(a_1,\ldots,a_n)|} \cdot |A_{1,\ldots,n}(a_1,\ldots,a_n)|$$

$$= \sum_{\substack{(a_1,\ldots,a_N)\in\mathbb{N}^N \\ a_i < K}} |A_{1,\ldots,n}(a_1,\ldots,a_n)| \sum_{k<K} \frac{|A_{1,\ldots,n,n+1}(a_1,\ldots,a_n,k)|}{|A_{1,\ldots,n}(a_1,\ldots,a_n)|}.$$

$$\tag{10.24}$$

By Lemma 10.1.9 and arguments similar to those in Corollary 10.1.11, for each fixed n-tuple $(a_1, \ldots, a_n)$ we have

$$\sum_{k < K} \frac{|A_{1,\ldots,n,n+1}(a_1, \ldots, a_n, k)|}{|A_{1,\ldots,n}(a_1, \ldots, a_n)|} < 1 - \frac{C_1}{K} = \beta. \qquad (10.25)$$

By the inductive assumption

$$\sum_{\substack{(a_1, \ldots, a_N) \in \mathbb{N}^N \\ a_i < K}} |A_{1,\ldots,n}(a_1, \ldots, a_n)| < \beta^N. \qquad (10.26)$$

Combining the pieces gives $\mathrm{Prob}(a_i(\alpha) \leq K : 1 \leq i \leq N+1) < \beta^{N+1}$, as claimed. $\qquad\square$

Remark 10.2.2. If α has $a_{n+1}(\alpha) > N$, then by Exercise 7.5.2 $\left| \alpha - \frac{p_n}{q_n} \right| \leq \frac{1}{N q_n^2}$. Letting $\epsilon = \frac{1}{N}$ we see we can approximate α to within $\frac{\epsilon}{q_n^2}$. As almost all α have infinitely many i with $a_i(\alpha) > N$, given any $\epsilon > 0$ for almost all α we can find infinitely many $\frac{p_n}{q_n}$ such that the approximation is at least as good as $\frac{\epsilon}{q_n^2}$.

Exercise 10.2.3. *What is the measure of the subset of $[0, 1)$ such that all decimal digits are at most d, $d \in \{0, \ldots, 9\}$? What is the measure of the subset of $[0,1)$ such that the digit d is not used in the decimal expansion? State and prove similar results for binary expansions.*

Theorem 10.2.4. *If $\{k_n\}$ is a sequence of positive integers such that $\sum_{n=1}^{\infty} \frac{1}{k_n}$ converges, the set $\{\alpha \in [0, 1) : a_n(\alpha) < k_n\}$ has positive measure or length.*

Proof. We sketch the proof. Let C_2 be as in Corollary 10.1.11. For convenience, assume $C_2 < k_n$ for all n; as the k_n-sum converges there can be only finitely many n where $C_2 \geq k_n$. For fixed N, consider the sets $\{a_n(\alpha) < k_n : 1 \leq i \leq N\}$. Arguing similarly as in Theorem 10.2.1, we find this set has measure at least $\prod_{n=1}^{N} \left(1 - \frac{C_2}{k_n}\right)$. To see this, we proceed by induction. $N = 1$ is clear, and if the result is true for N, then at the next stage we keep at least $1 - \frac{C_2}{k_{N+1}}$ of the interval. If we can show the product converges to a non-zero value, we are done. It is often easier to work with sums than products. By taking the logarithm we reduce the proof to showing a sum is not $-\infty$. For $|x|$ small, Taylor expanding gives

$$\ln(1 - x) = -\left(x + \frac{x^2}{2} + \cdots \right) = -x - O(x^2); \qquad (10.27)$$

i.e., $\log(1 - x) = -x$ plus an error of size x^2, and the error is negative. Therefore

$$\log \prod_{n=1}^{N} \left(1 - \frac{C_2}{k_n}\right) = \sum_{n=1}^{N} \log \left(1 - \frac{C_2}{k_n}\right) = -\sum_{n=1}^{N} \frac{1}{k_n} - O\left(\sum_{n=1}^{N} \frac{1}{k_n^2}\right). \qquad (10.28)$$

As $\sum \frac{1}{k_n}$ converges, the right hand side is bounded away from $-\infty$; exponentiating both sides gives the product is greater than 0. $\qquad\square$

Note the right hand side of (10.28) is negative, which implies the product is less than 1 (as one would expect since this is a probability). This is a good consistency check.

Exercise 10.2.5. *Prove the above theorem when C_2 is greater than some of the k_n.*

10.3 THE GAUSS-KUZMIN THEOREM

10.3.1 Statement of the Gauss-Kuzmin Theorem

We have shown that, independent of n, the probability that the nth digit is k is at least $\frac{1}{3k^2}$ and at most $\frac{2}{k^2}$. Gauss conjectured that as $n \to \infty$ the probability that the nth digit equals k converges to $\log_2\left(1 + \frac{1}{k(k+2)}\right)$. In 1928, Kuzmin [Ku] proved Gauss' conjecture, with an explicit error term:

Theorem 10.3.1 (Kuzmin). *There exist positive constants A and B such that*

$$\left| A_n(k) - \log_2\left(1 + \frac{1}{k(k+2)}\right) \right| \leq \frac{A}{k(k+1)} e^{-B\sqrt{n-1}}. \qquad (10.29)$$

This is clearly compatible with Gauss' conjecture, as for $B > 0$ the expression $e^{-B\sqrt{n-1}}$ tends to zero when n approaches $+\infty$. The error term has been improved by Lévy [Le] to Ae^{-Cn}, and then further by Wirsling [Wir].

We closely follow Khinchine's outstanding exposition (see [Kh]); we strongly recommend his book to anyone interested in continued fractions. As we are interested in the limit as $n \to \infty$, it is sufficient to investigate the properties of the digits for *irrational* $\alpha \in (0,1)$. The general formulation is

$$\begin{aligned}
\alpha &= [0, a_1, a_2, \ldots, a_{n-1}, a_n, \ldots] \in (0,1) \\
&= [0, a_1, a_2, \ldots, a_{n-1}, r_n(\alpha)] \\
r_n(\alpha) &= [a_n, a_{n+1}, \ldots] \in (1, \infty) \\
z_n(\alpha) &= r_n(\alpha) - a_n \in (0,1) \\
m_n(x) &= |\{\alpha \in (0,1) : z_n(\alpha) < x\}|.
\end{aligned} \qquad (10.30)$$

Thus $m_n(x)$ is the measure (or length, proportion or probability) of $\alpha \in (0,1)$ with $z_n(\alpha) = [0, a_{n+1}, \ldots] < x$. Also observe that we can safely assume that all of our continued fractions are infinite, as finite continued fractions will have no contribution to lengths of intervals.

Remark 10.3.2. Note $a_n(\alpha) = k$ is equivalent to $\frac{1}{k+1} < z_{n-1}(\alpha) \leq \frac{1}{k}$. Thus to calculate $A_n(k)$ (the probability that $a_n(\alpha) = k$), it is sufficient to evaluate $m_n(\frac{1}{k}) - m_n(\frac{1}{k+1})$. Thus we begin to see why these functions are natural to consider.

10.3.2 Proof of the Gauss-Kuzmin Theorem

The Gauss-Kuzmin Theorem follows from the following:

Theorem 10.3.3. *Let $f_0(x)$, $f_1(x), \ldots$ be a sequence of real-valued functions on $(0,1)$ such that for all $n \geq 0$,*

$$f_{n+1}(x) = \sum_{k=1}^{\infty} \frac{1}{(k+x)^2} f_n\left(\frac{1}{k+x}\right). \qquad (10.31)$$

If $0 < f_0(x) < M$ and $|f_0'(x)| < \mu$ for some M and μ, then for all x

$$\left| f_n(x) - \frac{a}{1+x} \right| \le A e^{-B\sqrt{n}}, \quad a = \frac{1}{\ln 2} \int_0^1 f_0(x) dx \qquad (10.32)$$

with $A = A(M, \mu) > 0$ and $B > 0$ absolute positive constants.

We first show how Theorem 10.3.3 implies the Gauss-Kuzmin Theorem. We have

$$r_{n+1}(\alpha) = [a_{n+1}, a_{n+2}, \dots]$$

$$\frac{1}{r_{n+1}(\alpha)} = [0, a_{n+1}, \dots]$$

$$\frac{1}{a_{n+1} + z_{n+1}(\alpha)} = [0, a_{n+1}, \dots] = z_n(\alpha). \qquad (10.33)$$

For $x \in [0, 1)$ the above relation implies that $z_{n+1}(\alpha) < x$ if and only if for some k, $\frac{1}{k+x} < z_n(\alpha) \le \frac{1}{k}$; here k is the value of a_{n+1} and the bounds arise from taking $z_{n+1}(\alpha) = 0$ and x. If $a_{n+1} = k$ the corresponding set has measure $m_n(\frac{1}{k}) - m_n(\frac{1}{k+x})$. Summing over k yields

$$m_{n+1}(x) = \sum_{k=1}^{\infty} \left[m_n\left(\frac{1}{k}\right) - m_n\left(\frac{1}{k+x}\right) \right]. \qquad (10.34)$$

Differentiating term by term gives

$$m_{n+1}'(x) = \sum_{k=1}^{\infty} \frac{1}{(k+x)^2} m_n'\left(\frac{1}{k+x}\right). \qquad (10.35)$$

Exercise 10.3.4. *Justify the term by term differentiation. Note $m_0(x) = x$ and if $m_n(x)$ is continuous and bounded, Taylor expanding for $x \in [0, 1)$ yields a uniformly convergent series.*

Remark 10.3.5. We begin to see why Theorem 10.3.3 is a natural approach to the Gauss-Kuzmin Theorem. By Remark 10.3.2, we related $A_n(k)$ to $z_{n-1}(\alpha)$ to $m_n(\frac{1}{k}) - m_n(\frac{1}{k+1})$, and we have just shown summing this over k is related to $m_{n+1}(x)$.

The functions $\{m_n'(x)\}$ satisfy the conditions of the Theorem 10.3.3 for any $M > 1$ and any $\mu > 0$. As $m_0(x) = x$, $a = \frac{1}{\ln 2}$. Therefore

$$\left| m_n'(x) - \frac{1}{\ln 2} \frac{1}{1+x} \right| \le A e^{-B\sqrt{n}}. \qquad (10.36)$$

Integrating over $(0, 1)$ (which has length 1) gives

$$\left| m_n(x) - \frac{\ln(1+x)}{\ln 2} \right| \le A e^{-B\sqrt{n}} \cdot 1. \qquad (10.37)$$

To have $a_n(\alpha) = k$, we must have

$$\frac{1}{k+1} < z_{n-1}(\alpha) \le \frac{1}{k}. \qquad (10.38)$$

Thus

$$|\{\alpha \in (0,1) : a_n(\alpha) = k\}| = m_{n-1}\left(\frac{1}{k}\right) - m_{n-1}\left(\frac{1}{k+1}\right)$$

$$= \int_{\frac{1}{k+1}}^{\frac{1}{k}} m'_{n-1}(x)dx. \tag{10.39}$$

By (10.37), the integral is

$$\frac{\ln(1+x)}{\ln 2}\bigg|_{\frac{1}{k+1}}^{\frac{1}{k}} = \frac{\ln\left(1 + \frac{1}{k(k+2)}\right)}{\ln 2} = \log_2\left(1 + \frac{1}{k(k+2)}\right), \tag{10.40}$$

plus an error at most $Ae^{-B\sqrt{n-1}}$, which is what we set out to prove.

Exercise 10.3.6. *Let $\{f_n\}$ be a sequence of functions satisfying (10.31). If $f_0(x) > 0$ for all $x \in [0,1]$, show $f_n(x) > 0$ for all $x \in [0,1]$.*

10.3.3 Preliminary Lemmas

We now state some preliminary results which are needed in the proof of Theorem 10.3.3. We write

$$\frac{p_n}{q_n} = [0, a_1, \ldots, a_n]; \tag{10.41}$$

we should really write $\frac{p_n(\alpha)}{q_n(\alpha)}$, but for convenience we suppress α below. We use the recursion relations (10.9) for the p's and q's and the notation from Remark 10.1.6.

For each n-tuple $(a_1, \ldots, a_n) \in \mathbb{N}^n$, we look at all $\alpha \in [0,1)$ whose continued fraction expansion begins with $[0, a_1, \ldots, a_n]$. Numbers beginning with $[0, a_1, \ldots, a_n]$ are given by the following union of intervals:

$$\bigcup_{k=1}^{\infty} \left[\frac{p_n k + p_{n-1}}{q_n k + q_{n-1}}, \frac{p_n(k+1) + p_{n-1}}{q_n(k+1) + q_{n-1}}\right) = \left[\frac{p_n}{q_n}, \frac{p_n + p_{n-1}}{q_n + q_{n-1}}\right), \tag{10.42}$$

as the extremes are given by $k=1$ and $k=\infty$. We have thus shown

Lemma 10.3.7. *The set of $\alpha \in [0,1)$ such that the first n digits are $[0, a_1, \ldots, a_n]$ is the interval $\left[\frac{p_n}{q_n}, \frac{p_n+p_{n-1}}{q_n+q_{n-1}}\right)$ of length $\left|\frac{p_n}{q_n} - \frac{p_n+p_{n-1}}{q_n+q_{n-1}}\right|$. Note each $\frac{p_n}{q_n} = [0, a_1, \ldots, a_n]$ corresponds to a unique subinterval. As each $[0, a_1, \ldots, a_n]$ leads to a different subinterval, and the subintervals are disjoint, we have*

$$[0,1) = \bigcup_{(a_1,\ldots,a_n)\in\mathbb{N}^n} \left[\frac{p_n}{q_n}, \frac{p_n+p_{n-1}}{q_n+q_{n-1}}\right)$$

$$1 = \sum_{(a_1,\ldots,a_n)\in\mathbb{N}^n} \left|\frac{p_n}{q_n} - \frac{p_n+p_{n-1}}{q_n+q_{n-1}}\right|. \tag{10.43}$$

We now derive useful properties of functions satisfying the conditions of Theorem 10.3.3.

Lemma 10.3.8. *Assume the functions $f_n(x)$ satisfy the conditions of Theorem 10.3.3. For all $n \geq 0$,*

$$f_n(x) = \sum_{(a_1,\ldots,a_n)\in\mathbb{N}^n} f_0\left(\frac{p_n + xp_{n-1}}{q_n + xq_{n-1}}\right) \cdot \frac{1}{(q_n + xq_{n-1})^2}. \tag{10.44}$$

Note $(a_1, \ldots, a_n) \in \mathbb{N}^n$ means consider all n-tuples where each a_i is a positive integer. Summing over such n-tuples is the same as summing over all possible length n beginnings for continued fractions.

Exercise 10.3.9. *Prove the above lemma by induction. For $n = 0$, remember we have $p_0 = 0$, $q_0 = 1$, $p_{-1} = 1$ and $q_{-1} = 0$. Assuming the result is true for $f_n(x)$, substituting in (10.31) yields the claim for $f_{n+1}(x)$.*

Lemma 10.3.10. *Assume the functions $f_n(x)$ satisfy the conditions of Theorem 10.3.3. For all $n \geq 0$,*

$$|f_n'(x)| < \frac{\mu}{2^{n-3}} + 4M. \tag{10.45}$$

Proof. Differentiating term by term (justify this), using $p_n q_{n-1} - p_{n-1}q_n = (-1)^{n-1}$, and letting $u_n = \frac{p_n + xp_{n-1}}{q_n + xq_{n-1}}$ yields

$$f_n'(x) = \sum_{(a_1,\ldots,a_n)\in\mathbb{N}^n} f_0'(u_n)\frac{(-1)^{n-1}}{(q_n + xq_{n-1})^4}$$

$$- 2\sum_{(a_1,\ldots,a_n)\in\mathbb{N}^n} f_0(u_n)\frac{q_{n-1}}{(q_n + xq_{n-1})^3}. \tag{10.46}$$

Using

1. $\frac{1}{(q_n+xq_{n-1})^2} < \frac{2}{q_n(q_n+q_{n-1})}$,

2. $q_n(q_n + q_{n-1}) > q_n^2 > 2^{n-1}$ (see Theorem 7.4.4),

3. $\sum_{(a_1,\ldots,a_n)\in\mathbb{N}^n} \frac{1}{q_n(q_n+q_{n-1})} = \sum_{(a_1,\ldots,a_n)\in\mathbb{N}^n} \left|\frac{p_n}{q_n} - \frac{p_n+p_{n-1}}{q_n+q_{n-1}}\right| = 1$

completes the proof (the last result follows from Lemma 10.3.7). $\square$

Lemma 10.3.11. *Assume the functions $f_n(x)$ satisfy the conditions of Theorem 10.3.3. For $x \in (0, 1)$, if*

$$\frac{t}{1 + x} < f_n(x) < \frac{T}{1 + x} \tag{10.47}$$

then

$$\frac{t}{1 + x} < f_{n+1}(x) < \frac{T}{1 + x}. \tag{10.48}$$

Exercise 10.3.12. *Prove the above. Substitute the assumed bounds for $f_n(x)$ in (10.31).*

Lemma 10.3.13. *Assume the functions $f_n(x)$ satisfy the conditions of Theorem 10.3.3. For all $n \geq 0$,*

$$\int_0^1 f_n(x)dx = \int_0^1 f_0(x)dx. \tag{10.49}$$

Exercise 10.3.14. *Prove the above lemma by induction, using (10.31).*

10.3.4 Proof of Theorem 10.3.3

In this technical section we prove Theorem 10.3.3, which in §10.3.2 we showed implies the Gauss-Kuzmin Theorem. We highlight the main arguments, following Khinchine's outstanding exposition (see [Kh]). We show that if $h_n(x) = C\ln(1+x)$, then the $h_n'(x)$ satisfy the conditions of Theorem 10.3.3, (10.31), and the conclusions of the theorem are immediate for $h_n'(x)$. For any functions $f_n(x)$ satisfying the conditions of Theorem 10.3.3, we bound the difference between $f_n(x)$ and $h_n(x)$ for an appropriate C, and show the difference tends to zero as $N \to \infty$ (which completes the proof).

We now supply the details. Choose m such that for all $x \in [0,1]$, $m \le f_0(x) < M$. For our applications to continued fractions, we will take $f_0(x) = m_0'(x)$. As $m_0(x) = x$ for the bounds on $f_0(x)$ we may take $m = 1$ and M any number greater than 1. Letting $g = \frac{m}{2}$ and $G = 2m$, we immediately see that

$$\frac{g}{1+x} < f_0(x) < \frac{G}{1+x}, \quad x \in [0,1]. \tag{10.50}$$

For each $n \ge 0$, let

$$\phi_n(x) = f_n(x) - \frac{g}{1+x}. \tag{10.51}$$

Exercise 10.3.15. Show $\phi_n(x) > 0$ for all $x \in [0,1]$. Hint: *The functions f_n satisfy (10.31). See also Exercise 10.3.6.*

Lemma 10.3.16. *For any C, $h(x) = C\ln(1+x)$ satisfies*

$$h(x) = \sum_{k=1}^{\infty} \left[h\left(\frac{1}{k}\right) - h\left(\frac{1}{k+x}\right) \right]. \tag{10.52}$$

Proof. The series converges as for fixed x, eventually the summands are of size $\frac{x}{k^2}$. We show

$$\lim_{K\to\infty} \sum_{k\le K} \left[h\left(\frac{1}{k}\right) - h\left(\frac{1}{k+x}\right) \right] = C\ln(1+x) = h(x). \tag{10.53}$$

By the definition of h, each summand is

$$h\left(\frac{1}{k}\right) - h\left(\frac{1}{k+x}\right) = C\ln\left(\frac{k+x}{k+x+1}\frac{k+1}{k}\right). \tag{10.54}$$

By standard properties of the logarithm,

$$\sum_{k\le K} \left[h\left(\frac{1}{k}\right) - h\left(\frac{1}{k+x}\right) \right] = C\ln\left(\prod_{k=1}^{K} \frac{k+x}{k+x+1}\frac{k+1}{k}\right)$$

$$= C\ln\left(\frac{1+x}{K+x+1}\frac{K+1}{1}\right). \tag{10.55}$$

For fixed x, as $K \to \infty$, $\frac{K+1}{K+x+1} \to 1$. Therefore the above converges to $C\ln(1+x)$. $\square$

Exercise 10.3.17. *Is the convergence in Lemma 10.3.16 uniform?*

Lemma 10.3.18. *For h as above,*

$$h'(x) = \sum_{k=1}^{\infty} \frac{1}{(k+x)^2} h'\left(\frac{1}{k+x}\right). \tag{10.56}$$

Proof. Differentiating (10.52) term by term (using the Chain Rule from calculus) gives

$$h'(x) = \sum_{k=1}^{\infty} h'\left(\frac{1}{k+x}\right) \cdot \frac{1}{(k+x)^2}. \tag{10.57}$$

$\square$

Exercise 10.3.19. *Justify the term by term differentiation above.*

Let $h_n(x) = h(x)$ for all n; then the family $\{h_n\}$ satisfies (10.31); by assumption, so does the family $\{f_n\}$. Therefore the family $\{\phi_n\}$, where $\phi_n(x) = f_n(x) - gh(x)$, also satisfies (10.31). In particular, all consequences that we derived for families satisfying (10.31) hold for the family $\{\phi_n\}$.

For notational convenience let $u_n = \frac{p_n + xp_{n-1}}{q_n + xq_{n-1}}$. Then

$$\phi_n(x) = \sum_{(a_1,\ldots,a_n)\in\mathbb{N}^n} \phi_0(u_n)\frac{1}{(q_n + xq_{n-1})^2}. \tag{10.58}$$

As the denominator above is less than $2q_n(q_n + q_{n-1})$, we obtain the lower bound

$$\phi_n(x) > \frac{1}{2} \sum_{(a_1,\ldots,a_n)\in\mathbb{N}^n} \phi_0(u_n)\frac{1}{q_n(q_n + q_{n-1})}. \tag{10.59}$$

For fixed $(a_1,\ldots,a_n) \in \mathbb{N}^n$, by Lemma 10.3.7 we are working with the subinterval $I_{(a_1,\ldots,a_n)} = \left[\frac{p_n}{q_n}, \frac{p_n+p_{n-1}}{q_n+q_{n-1}}\right)$. Summing over all n-tuples gives us $[0,1)$, and each subinterval has length $\frac{1}{q_n(q_n+q_{n-1})}$. Thus on each subinterval, by the Mean Value Theorem (Theorem A.2.2) we have

$$\int_{\frac{p_n}{q_n}}^{\frac{p_n+p_{n-1}}{q_n+q_{n-1}}} \phi_0(z)dz = \phi_0(u_n')\frac{1}{q_n(q_n + q_{n-1})}, \quad u_n' \in \left[\frac{p_n}{q_n}, \frac{p_n+p_{n-1}}{q_n+q_{n-1}}\right]. \tag{10.60}$$

Summing over all n-tuples and multiplying by $\frac{1}{2}$ yields

$$\frac{1}{2}\int_0^1 \phi_0(z)dz = \frac{1}{2} \sum_{(a_1,\ldots,a_n)\in\mathbb{N}^n} \phi_0(u_n')\frac{1}{q_n(q_n + q_{n-1})}. \tag{10.61}$$

As the ϕ_n are positive (Exercise 10.3.15), subtracting (10.61) from (10.59) gives

$$\phi_n(x) - \frac{1}{2}\int_0^1 \phi_0(z)dz > \frac{1}{2} \sum_{(a_1,\ldots,a_n)\in\mathbb{N}^n} [\phi_0(u_n) - \phi_0(u_n')]\frac{1}{q_n(q_n + q_{n-1})}. \tag{10.62}$$

Because f_0 satisfies the conditions of Theorem 10.3.3, the derivative of $\phi_0(x) = f_0(x) - \frac{g}{1+x}$ is less than $\mu + g$. Hence

$$|\phi_0(u_n) - \phi_0(u_n')| < (\mu + g)|u_n - u_n'|. \tag{10.63}$$

As $u_n, u'_n \in I_{(a_1,\ldots,a_n)}$, $|u_n - u'_n| < \frac{1}{q_n(q_n + q_{n-1})}$. As $q_n^2 > 2^{n-1}$ (Theorem 7.4.4), $|\phi_0(u_n) - \phi_0(u'_n)| < \frac{\mu + g}{2^{n-1}}$. Substituting into (10.62) gives

$$\phi_n(x) > \frac{1}{2}\int_0^1 \phi_0(z)dz - \frac{\mu + g}{2^n} = L - \frac{\mu + g}{2^n}, \qquad (10.64)$$

where $L = \frac{1}{2}$. Recalling the definition of $\phi_n(x)$ as $f_n(x) - \frac{g}{1+x}$ gives

$$f_n(x) > \frac{g}{1+x} + L - \frac{\mu + g}{2^n} > \frac{g + L - 2^{-n+1}(\mu + g)}{1+x} = \frac{g_1}{1+x}. \quad (10.65)$$

A similar argument applied to $\psi_n(x) = \frac{G}{1+x} - f_n(x)$ would yield

$$f_n(x) < \frac{G - L' + 2^{-n+1}(\mu + G)}{1+x} = \frac{G_1}{1+x}, \qquad (10.66)$$

with $L' = \frac{1}{2}\int_0^1 \psi_0(z)dz$. Note $L + L' = \frac{(G-g)\ln 2}{2}$. As $L, L' > 0$, for n large

$$g < g_1 < G_1 < G \qquad (10.67)$$

and

$$G_1 - g_1 < G - g - (L + L') + 2^{-n+2}(\mu + G). \qquad (10.68)$$

Letting $\delta = 1 - \frac{\ln 2}{2} < 1$ and using $L + L' = \frac{(G-g)\ln 2}{2}$ we find

$$G_1 - g_1 < (G - g)\delta + 2^{-n+2}(\mu + G). \qquad (10.69)$$

We have proved

Lemma 10.3.20. *Assume*

1. $\frac{g}{1+x} < f_0(x) < \frac{G}{1+x}$.

2. $|f'_0(x)| < \mu$.

Then for n sufficiently large,

1. $\frac{g_1}{1+x} < f_n(x) < \frac{G_1}{1+x}$.

2. $g < g_1 < G_1 < G$, $G_1 - g_1 < (G - g)\delta + 2^{-n+2}(\mu + G)$.

We inductively apply the above lemma to $f_{rn}(x)$ instead of $f_0(x)$. We need to have $|f'_{rn}(x)| < \mu_r$; by Lemma 10.3.10 we may take $\mu_r = \frac{\mu}{2^{rn-3}} + 4M$, and for n large, $\mu_r < 5M$. We obtain for n large

$$\frac{g_r}{1+x} < f_{rn}(x) < \frac{G_r}{1+x}, \qquad (10.70)$$

where for $r > 0$

$$g_{r-1} < g_r < G_r < G_{r-1}, \quad G_r - g_r < (G_{r-1} - g_{r-1})\delta + 2^{-n+2}(\mu_{r-1} + G_{r-1}). \qquad (10.71)$$

As $\delta < 1$ is an absolute constant, successive applications of (10.71) (apply for r up to n) give

$$G_n - g_n < (G - g)\delta^n + 2^{-n+2}\left[(\mu + 2M)\delta^{n-1} + 7M\delta^{n-2} + \cdots + 7M\delta + 7M\right]$$

$$< C(M, \mu)e^{-Bn}, \qquad (10.72)$$

where $B > 0$ is an absolute constant and $C = C(M, \mu) > 0$ depends only on M and μ. Therefore G_n and g_n have the same limit, say a. As $r = n$ we are working with the function $f_{n \cdot n}(x) = f_{n^2}(x)$.

We have shown

$$\left| f_{n^2}(x) - \frac{a}{1+x} \right| < Ce^{-Bn}.$$
(10.73)

Therefore as $n \to \infty$

$$\int_0^1 f_{n^2}(z)dz \longrightarrow a \ln 2.$$
(10.74)

By Lemma 10.3.13, all the integrals $\int_0^1 f_m(z)dz$ are equal to each other, which yields

$$a = \frac{1}{\ln 2} \int_0^1 f_0(z)dz.$$
(10.75)

All that is left is to bound the difference between $f_N(x)$ and $\frac{a}{1+x}$. Choose $n \in \mathbb{N}$ such that $n^2 \leq N < (n+1)^2$. From (10.73),

$$\frac{a - 2Ce^{-Bn}}{1+x} < f_{n^2}(x) < \frac{a + 2Ce^{-Bn}}{1+x}.$$
(10.76)

By Lemma 10.3.11, by incrementing n^2 by 1 a total of $N - n^2$ times, we obtain

$$\frac{a - 2Ce^{-Bn}}{1+x} < f_N(x) < \frac{a + 2Ce^{-Bn}}{1+x}.$$
(10.77)

Thus, as $\sqrt{N} < n + 1$,

$$\left| f_N(x) - \frac{a}{1+x} \right| < 2Ce^{-Bn}$$

$$= 2Ce^1 \cdot e^{-B(n+1)}$$

$$< Ae^{-B\sqrt{N}},$$
(10.78)

(finally) completing the proof of Theorem 10.3.3.

Exercise 10.3.21. *Prove* (10.66). *See also Project 10.6.5.*

10.4 DEPENDENCIES OF DIGITS

Are the digits of a continued fraction independent, or does knowledge that $a_n(\alpha) = k_1$ provide some information as to the probability that $a_{n+1}(\alpha) = k_2$? We show, unlike the digits of a decimal expansion, the digits of a continued fraction are *not*, in general, independent.

10.4.1 Simple Example

For $\alpha \in [0, 1)$ let us consider

$$\text{Prob}(a_2(\alpha) = 1 | a_1(\alpha) = 1) = \frac{|A_{1,2}(1,1)|}{|A_1(1)|}$$

$$\text{Prob}(a_2(\alpha) = 1 | a_1(\alpha) = 2) = \frac{|A_{1,2}(2,1)|}{|A_1(2)|}. \tag{10.79}$$

We need to determine the length of the subintervals of $[0, 1)$ corresponding to these events. In the first case, $a_1(\alpha) = 1$ means $\alpha \in (\frac{1}{2}, 1]$. If $a_1(\alpha) = a_2(\alpha) = 1$, then

$$\alpha = 0 + \cfrac{1}{1 + \cfrac{1}{1 + \cfrac{1}{r_3(\alpha)}}}, \quad r_3(\alpha) \in [1, \infty). \tag{10.80}$$

We obtain the extreme values of such α when $r_3(\alpha) = 1$ and $r_3(\alpha) = \infty$. We find that $a_1(\alpha) = a_2(\alpha) = 1$ corresponds to $A_{1,1}(1, 1) = (\frac{1}{2}, \frac{2}{3}]$. As $A_1(1) = (\frac{1}{2}, 1]$, we have

$$\text{Prob}(a_2(\alpha) = 1 | a_1(\alpha) = 1) = \frac{|A_{1,2}(1,1)|}{|A_1(1)|} = \frac{\frac{2}{3} - \frac{1}{2}}{1 - \frac{1}{2}} = \frac{1}{3}. \tag{10.81}$$

Similarly, in the second case $a_1(\alpha) = 2$ corresponds to the interval $(\frac{1}{3}, \frac{1}{2}]$, and $a_1(\alpha) = 1$, $a_2(\alpha) = 1$ corresponds to

$$\alpha = 0 + \cfrac{1}{2 + \cfrac{1}{1 + \cfrac{1}{r_3(\alpha)}}}, \quad r_3(\alpha) \in [1, \infty). \tag{10.82}$$

We obtain the extreme values for α when $r_3(\alpha) = 1$ and ∞, respectively yielding the boundary values of $\frac{2}{5}$ and $\frac{1}{3}$. Thus $a_1(\alpha) = 2$, $a_2(\alpha) = 1$ corresponds to $A_{1,2}(2, 1) = (\frac{1}{3}, \frac{2}{5}]$. As $A_1(2) = (\frac{1}{3}, \frac{1}{2}]$, we have

$$\text{Prob}(a_2(\alpha) = 1 | a_1(\alpha) = 2) = \frac{|A_{1,2}(2,1)|}{|A_1(1)|} = \frac{\frac{2}{5} - \frac{1}{3}}{\frac{1}{2} - \frac{1}{3}} = \frac{2}{5}. \tag{10.83}$$

We have shown

$$\text{Prob}(a_2(\alpha) = 1 | a_1(\alpha) = 1) = \frac{1}{3}$$

$$\text{Prob}(a_2(\alpha) = 1 | a_1(\alpha) = 2) = \frac{2}{5}. \tag{10.84}$$

Exercise 10.4.1. *Generalize the above argument to calculate the probability that* $a_2(\alpha) = k_2$ *given* $a_1(\alpha) = k_1$.

We observe a significant difference in the probability of the second digit being a 1, given the first digit is 1 or 2. Is this a misleading observation, the result of working with the first few digits, or is there a similar discrepancy for $\text{Prob}(a_{n+1}(\alpha) =$

$k_2|a_n(\alpha) = k_1)$ as $n \to \infty$? See Remark 7.10.2 for an example where small data sets can be misleading.

For example, the Gauss-Kuzmin Theorem tells us that to three significant figures

$$\lim_{n \to \infty} \text{Prob}(a_n(\alpha) = 1) = .415$$

$$\lim_{n \to \infty} \text{Prob}(a_n(\alpha) = 2) = .170. \tag{10.85}$$

Direct calculation gives (to three significant digits)

$$\text{Prob}(a_1(\alpha) = 1) = \left| \left(\frac{1}{2}, 1 \right] \right| = .500$$

$$\text{Prob}(a_1(\alpha) = 2) = \left| \left(\frac{1}{3}, \frac{1}{2} \right] \right| = .167. \tag{10.86}$$

Thus the probability that the first digit equals 1 is significantly different than the probability that the nth digit is 1 (as $n \to \infty$). Note, however, that already the probability that the first digit is a 2 is very close to the probability that the nth digit is a 2 (as $n \to \infty$).

In §10.4.2 we show the probability that the $(n+1)^{st}$ digit is k_2 does depend on what the nth digit is; the digits are *not* independent, and the difference in probabilities does remain significant.

10.4.2 Dependencies of Digits

Theorem 10.4.2. *We have*

$$\lim_{n \to \infty} \text{Prob}(a_{n+1}(\alpha) = k_2|a_n(\alpha) = k_1)$$

$$= \frac{\log \left(1 + \dfrac{1}{[(k_2 + 1)k_1 + 1] [(k_1 + 1)k_2 + 1]} \right)}{\log \left(1 + \dfrac{1}{k_1(k_1 + 2)} \right)}. \tag{10.87}$$

Sketch of proof. We have

$$\text{Prob}(a_{n+1}(\alpha) = k_2|a_n(\alpha) = k_1) = \frac{|A_{n,n+1}(k_1, k_2)|}{|A_n(k_1)|}. \tag{10.88}$$

By the Gauss-Kuzmin Theorem, up to a small error for n large

$$|A_n(k_1)| = \log_2 \left(1 + \frac{1}{k_1(k_1 + 2)} \right) = \frac{1}{\log 2} \int_{\frac{1}{k_1+1}}^{\frac{1}{k_1}} \frac{dx}{1 + x}. \tag{10.89}$$

We disregard the first $n-1$ digits, as well as digits $n+2$ and onward. If $a_n(\alpha) = k_1$, we have $\frac{1}{k_1+1} \le r_{n+1}(\alpha) \le \frac{1}{k_1}$. Integrating the density function $\frac{1}{\log 2} \frac{1}{1+x}$ over the interval $\left(\frac{1}{k_1+1}, \frac{1}{k_1} \right]$ gives $\log_2 \left(1 + \frac{1}{k_1(k_1+2)} \right)$, which is the probability of the nth digit being equal to k_1. We calculate what subinterval of $\left(\frac{1}{k_1+1}, \frac{1}{k_1} \right]$ gives $a_{n+1}(\alpha) = k_2$, and integrate the density $\frac{1}{\log 2} \frac{1}{1+x}$ over this subinterval.

The subinterval of

$$\frac{1}{k_1 + 1} < \cfrac{1}{k_1 + \cfrac{1}{r_{n+1}(\alpha)}} \leq \frac{1}{k_1} \qquad (10.90)$$

with $(n + 1)^{\text{st}}$ digit k_2 is

$$\cfrac{1}{k_1 + \cfrac{1}{k_2}} < \cfrac{1}{k_1 + \cfrac{1}{k_2 + \cfrac{1}{r_{n+2}(\alpha)}}} \leq \cfrac{1}{k_1 + \cfrac{1}{k_2 + 1}}. \qquad (10.91)$$

The length of the subinterval is

$$\left| \left(\frac{1}{k_1 + \frac{1}{k_2}}, \frac{1}{k_1 + \frac{1}{k_2+1}} \right] \right|$$

$$= \frac{1}{\log 2} \int_{\frac{1}{k_1 + \frac{1}{k_2}}}^{\frac{1}{k_1 + \frac{1}{k_2+1}}} \frac{dx}{1 + x}$$

$$= \frac{1}{\log 2} \cdot \log \left(\frac{1 + \frac{1}{k_1 + \frac{1}{k_2+1}}}{1 + \frac{1}{k_1 + \frac{1}{k_2}}} \right) \qquad (10.92)$$

$$= \frac{1}{\log 2} \cdot \log \left(1 + \frac{1}{[(k_2 + 1)k_1 + 1] \, [(k_1 + 1)k_2 + 1]} \right),$$

where the last line follows from straightforward algebra. The theorem follows from dividing by $\text{Prob}(a_n(\alpha) = k_1) = |A_n(k_1)|$. □

Note a more enlightening (but not as useful for calculations) way to write the desired probability is by not evaluating the integrals, giving

$$\lim_{n \to \infty} \text{Prob}(a_{n+1}(\alpha) = k_2 | a_n(\alpha) = k_1) = \cfrac{\frac{1}{\log 2} \int_{\frac{1}{k_1 + \frac{1}{k_2}}}^{\frac{1}{k_1 + \frac{1}{k_2+1}}} \frac{dx}{1+x}}{\frac{1}{\log 2} \int_{\frac{1}{k_1+1}}^{\frac{1}{k_1}} \frac{dx}{1+x}}. \qquad (10.93)$$

By the Gauss-Kuzmin Theorem, as $n \to \infty$ to four significant figures we have

$$\begin{aligned}
\text{Prob}(a_n(\alpha) = 1) &= A_n(1) \longrightarrow .4150 \\
\text{Prob}(a_n(\alpha) = 2) &= A_n(2) \longrightarrow .1699 \\
\text{Prob}(a_n(\alpha) = 3) &= A_n(3) \longrightarrow .0931.
\end{aligned} \qquad (10.94)$$

Compare the above to

$$\begin{aligned}
\text{Prob}(a_{n+1}(\alpha) = 1 | a_n(\alpha) = 1) &\longrightarrow .3662 \\
\text{Prob}(a_{n+1}(\alpha) = 2 | a_n(\alpha) = 1) &\longrightarrow .1696 \\
\text{Prob}(a_{n+1}(\alpha) = 3 | a_n(\alpha) = 1) &\longrightarrow .0979,
\end{aligned} \qquad (10.95)$$

and

$$\text{Prob}(a_{n+1}(\alpha) = 1 | a_n(\alpha) = 2) \longrightarrow .4142$$
$$\text{Prob}(a_{n+1}(\alpha) = 2 | a_n(\alpha) = 2) \longrightarrow .1715$$
$$\text{Prob}(a_{n+1}(\alpha) = 3 | a_n(\alpha) = 2) \longrightarrow .0938, \tag{10.96}$$

and

$$\text{Prob}(a_{n+1}(\alpha) = 1 | a_n(\alpha) = 3) \longrightarrow .4364$$
$$\text{Prob}(a_{n+1}(\alpha) = 2 | a_n(\alpha) = 3) \longrightarrow .1712$$
$$\text{Prob}(a_{n+1}(\alpha) = 3 | a_n(\alpha) = 3) \longrightarrow .0914. \tag{10.97}$$

As the above are the limiting values, numerical experiments for small n could mis-leadingly suggest that the digits are independent.

Finally, arguing along similar lines, we can calculate the probabilities of events such as $A_{n+1,\dots,n+m}(k_1, \dots, k_m)$, as well as more complicated events such as $a_m(\alpha) = k_m$ for $m \in N_1$ given that $a_n(\alpha) = k_n$ for $n \in N_2$, where N_1, N_2 are disjoint subsets of positive integers.

Exercise 10.4.3. *What is the probability that* $a_{n+2}(\alpha) = k_2$, *given that* $a_n(\alpha) = k_1$ *(i.e.,* $a_{n+1}(\alpha)$ *is arbitrary?*

Exercise 10.4.4. *Find the smallest integer* K_{50} *(resp.,* K_{99}) *such that as* $n \to \infty$ *there is as close to 50% (resp., 99%) as possible chance of* $a_n(\alpha) \le K$.

Exercise 10.4.5. *For fixed* k_2 *calculate*

$$\lim_{k_1 \to \infty} \left[\lim_{n \to \infty} \text{Prob}(a_{n+1}(\alpha) = k_2 | a_n(\alpha) = k_1) \right]. \tag{10.98}$$

Exercise 10.4.6. *If they exist, calculate the mean (§8.1.5) and the variance (§8.1.5) of the Gauss-Kuzmin probability distribution* $p_{GK}(x) = \log_2 \left(1 + \frac{1}{k(k+2)} \right)$.

Exercise 10.4.7 (Hard). *For a typical* $\alpha \in [0,1]$, *as* $N \to \infty$ *how large do you expect* $\prod_{n=1}^{N} a_n(\alpha)$ *to be? For almost all* $\alpha \in [0,1)$, *the geometric mean (see §A.6) of the first* n *digits,* $\sqrt[n]{a_1 \cdots a_n}$, *converges to Khinchine's constant*

$$\prod_{r=1}^{\infty} \left(1 + \frac{1}{r(r+2)} \right)^{\log_2 r} \approx 2.68545. \tag{10.99}$$

It is not known if this number is irrational. See [Kh] for details.

10.5 GAUSS-KUZMIN EXPERIMENTS

The Gauss-Kuzmin Theorem states that

$$|A_n(k)| = \log_2 \left(1 + \frac{1}{k(k+2)} \right) \tag{10.100}$$

plus an error at most $\frac{A}{k(k+1)} \cdot e^{-B\sqrt{n-1}}$ for some constants A and B (the error has been improved, see [Le, Wir]). Thus for n large the probability that the nth digit is k is approximately independent of n. Fix a range of digits, say n runs from M to $N-1$. What is the expected number of digits equal to k? Is it approximately $(N-M)\log_2(1+\frac{1}{k(k+2)})$? For algorithms to compute continued fractions, as well as reports on additional investigations, see §7.7.

Exercise 10.5.1. *Prove the Gauss-Kuzmin probabilities are always irrational. What does this imply about numerical experiments?*

10.5.1 Direct Solution

We give a computationally difficult, but direct, solution. We have a range of digits that we are investigating. For definiteness, let us say we are looking at digits 1 to 1000. Let S_{1000} be the set of all strings of 1000 numbers where each number is a positive integer.

Given a number $x \in [0, 1)$, there is a unique $s = s(x) \in S_{1000}$ such that the first 1000 digits in the continued fraction expansion of x are the same (and in the same order) as those from s. Given x, let $\text{num}_{1000;k}(x)$ be the number of k's in the first 1000 digits of its continued fraction expansion, and similarly for $\text{num}_{1000;k}(s)$. We say $x \equiv s$ if the first 1000 digits of the continued fraction of x is s. Thus the expected number of k's in the first 1000 digits of continued fraction expansions of numbers in $(0, 1)$ is

$$\sum_{s \in S_{1000}} \text{num}_{1000;k}(s) \cdot \text{Length}\left(\{x : x \equiv s\}\right). \qquad (10.101)$$

To solve the original question using this method, it is necessary to know what is the length or measure of the sets $\{x : x \equiv s\}$ for each $s \in S_{1000}$. This is a very complicated question. It can be computed by brute force, but it is quite involved, and the Gauss-Kuzmin Theorem is not applicable. Obviously there is nothing special about the first 1000 digits; we could investigate the number of occurrences of k from digits M to $N-1$.

We now show a better method which replaces the lengths of sets such as $\{x : x \equiv s\}$ with sets where the Gauss-Kuzmin Theorem is applicable.

10.5.2 Solution via Linearity of Expected Values

Let $a_n(x)$ be the nth digit of the continued fraction expansion of x. We define the following **binary indicator variables**

$$I_{n,k}(x) = \begin{cases} 1 & \text{if } a_n(x) = k \\ 0 & \text{if } a_n(x) \neq k. \end{cases} \qquad (10.102)$$

Thus

$$\text{num}_{M,N;k}(x) = \sum_{n=M}^{N-1} I_{n,k}(x) \qquad (10.103)$$

is the number of digits $a_n(x)$ which equal k with $M \leq n < N$.

Let $q_k = \log_2\left(1 + \frac{1}{k(k+2)}\right)$. The Gauss-Kuzmin Theorem states that up to a small error for n large, the probability that $a_n(x)$ equals k is q_k. Thus each

$I_{n,k}(x) = 1$ with probability approximately q_k and 0 with probability approximately $1 - q_k$. Hence the expected value of $I_{n,k}(x)$ is

$$\mathbb{E}[I_{n,k}(x)] \approx 1 \cdot q_k + 0 \cdot (1 - q_k) \approx q_k. \tag{10.104}$$

The power of the Gauss-Kuzmin Theorem is that the above expectation is, up to a very small error, independent of n. In other words, the main term of these expectations is q_k.

Exercise 10.5.2. *For M and N large, bound the error in $\mathbb{E}[I_{n,k}(x)]$.*

We now use the linearity of the expected value: the expected value of a sum is the sum of the expected values (Lemma 8.1.40). This simple observation allows us to avoid having to calculate the measure of sets $\{x : x \equiv_{M,N-1} s\}$ for each $s \in S_{M,N-1}$, where the notation is an obvious generalization of before.
We have

$$
\begin{aligned}
\mathbb{E}\left[\mathrm{num}_{M,N;k}(x)\right] &= \mathbb{E}\left[\sum_{n=M}^{N-1} I_{n,k}(x)\right] \\
&= \sum_{n=M}^{N-1} \mathbb{E}\left[I_{n,k}(x)\right] \\
&= \sum_{n=M}^{N-1} (q_k + \text{small error}) \\
&= (N - M)q_k + (\text{small error}) \cdot (N - M). \tag{10.105}
\end{aligned}
$$

Therefore, if we look at $N - M$ digits, we expect to see $(N - M)q_k$ occurrences of the digit k. While the proof above is for *consecutive* digits, a similar proof works for *any* set of digits.

Very concretely, if we were to look at digits 50001 to 100000, the number of 1's we expect to see would be

$$50000 \cdot \log_2\left(1 + \frac{1}{1(1+2)}\right) \approx 20752. \tag{10.106}$$

Exercise 10.5.3. *Determine the error in (10.105), and calculate the expected variance. Are the $I_{n,k}$ independent? If not, find an interpretation of $I_{n_1,k}(x) \cdot I_{n_2,k}(x)$.*

10.5.3 Generalization

Fix k_1, k_2, M and $N - 1$. What is the expected number of occurrences of the pair (k_1, k_2) among digits M to $N - 1$? For example, let $(k_1, k_2) = (1, 1)$. If we had a string of digits

$$1, 1, 1, 2, 3, 453, 1, 17, 5, 4, 1, 2, 10, 2, 1, 1, 19 \tag{10.107}$$

in the continued fraction expansion of x, this would contribute three pairs of $(1, 1)$ as $1, 1, 1$ counts as two pairs.

Using the methods of §10.4, one can show for n large that the probability that $a_n(x) = k_1$ and $a_{n+1}(x) = k_2$ is q_{k_1, k_2} plus significantly smaller corrections (see Theorem 10.4.2). Similar to §10.5.2, where we studied just one digit at a time, we have approximate n-independence in the probabilities q_{k_1,k_2}. Note that $q_{k_1,k_2} \neq q_{k_1} q_{k_2}$.

We again define binary indicator variables:

$$I_{n,k_1,k_2}(x) = \begin{cases} 1 & \text{if } a_n(x) = k_1 \text{ and } a_{n+1}(x) = k_2 \\ 0 & \text{otherwise.} \end{cases} \qquad (10.108)$$

Similar to before, we have the expected value of $I_{n,k_1,k_2}(x)$ is q_{k_1,k_2} plus small corrections. We define $\text{num}_{M,N,k_1,k_2}(x)$ to equal the number of $n \in [M, N-1]$ such that $a_n(x) = k_1$ and $a_{n+1}(x) = k_2$. We have

$$\mathbb{E}\left[\text{num}_{M,N,k_1,k_2}(x)\right] = \mathbb{E}\left[\sum_{n=M}^{N-1} I_{n,k_1,k_2}(x)\right]$$

$$= \sum_{n=M}^{N-1} \mathbb{E}\left[I_{n,k_1,k_2}(x)\right]$$

$$= \sum_{n=M}^{N-1} (q_{k_1,k_2} + \text{small error})$$

$$= (N-M)q_{k_1,k_2} + (\text{small error}) \cdot (N-M)$$

$$\approx (N-M)q_{k_1,k_2}. \qquad (10.109)$$

Exercise 10.5.4. *Using the Gauss-Kuzmin Theorem, calculate the probability that* $a_n(x) = k_1$ *and* $a_{n+1}(x) = k_2$ *for* n *large; show this equals* q_{k_1,k_2} *plus a very rapidly decaying error term in* n. *Show* $q_{k_1,k_2} \neq q_{k_1} q_{k_2}$.

Exercise 10.5.5. *Determine the error in* (10.109), *and calculate the expected variance.*

In the case of base 10 expansions, a number x is called **normal** if given any string of n numbers the percentage of strings of n consecutive digits of x that equal the given string tends to $\frac{1}{10^n}$. Explicitly, let s be a string of n decimal digits; there are 10^n such strings. If we look at the first N digits of x, if $p_N(s)$ is the number of n-tuples $x_{k+1} \ldots x_{k+n}$ that equal this string (with $k+1, \ldots, k+n \leq N$), then x is normal if $\lim_{N\to\infty} \frac{p_N(s)}{N} = \frac{1}{10^n}$.

Exercise 10.5.6. *Show Champernowne's constant [Ch]*

$$.01234567891011121314\ldots 99100101102103\ldots 999100010011002\ldots$$

is normal. Calculate its continued fraction; its continued fraction expansion is very interesting.

Similarly, one can define the concept of a normal continued fraction expansion; in this case, all strings of digits will *not* be equally likely. Normal continued fractions are known to exist, though there is no easy example like Champernowne's constant; however, in the sense of cardinality, there are many non-normal continued fractions (see [Ols]).

Exercise 10.5.7. *Construct an $\alpha \in (0, 1)$ such that as $N \to \infty$ the percentage of the first N digits that are k tends to $\log_2\left(1 + \frac{1}{k(k+2)}\right)$ (the Gauss-Kuzmin probabilities).*

Exercise 10.5.8. *Building on the previous exercise, can you find a $\beta \in (0, 1)$ such that not only do the limiting percentages of each digit agree with the Gauss-Kuzmin probabilities, but so do the percentages of consecutive pairs (k_1, k_2) for all positive integers k_1 and k_2. Can you generalize this construction to find a normal continued fraction?*

10.5.4 General Comments

Using binary indicator variables is an extremely useful and standard technique. Note that the digits in a continued fraction expansion are not independent of each other. Fortunately, such a fact is not needed to calculate the expected number of occurrences of $a_n(x) = k$ or $a_n(x) = k_1$ and $a_{n+1}(x) = k_2$ for $n \in [M, N-1]$.

We cannot avoid having to do some averaging over all $x \in [0, 1)$, but we do not want to have to deal with sets such as *the set of $x \in [0, 1)$ such that there are exactly* 237 *ones in the first* 1000 *digits of the continued fraction expansion*. It would be very difficult to determine the measure of each set of this form (for all the sets we would need).

Using binary indicator variables, we can avoid such a subdivision. Instead, we study significantly simpler sets such as *the set of $x \in [0, 1)$ such that $a_n(x) = k$*. This is much easier, and in fact the measure of such sets for n large is given by the Gauss-Kuzmin Theorem.

10.6 RESEARCH PROJECTS

There are many open questions concerning the digits of a generic continued fraction expansion. We know the digits in the continued fraction expansions of rationals and quadratic irrationals do not satisfy the Gauss-Kuzmin densities in the limits; in the first case there are only finitely many digits, while in the second the expansion is periodic. What can one say about the structure of the set of $\alpha \in [0, 1)$ whose distribution of digits satisfy the Gauss-Kuzmin probabilities? We know such a set has measure 1, but what numbers are in this set?

The set of algebraic numbers is countable, hence of measure zero. Thus it is possible for the digits of every algebraic number to violate the Gauss-Kuzmin law. Computer experimentation, however, indicates that the digits of algebraic numbers do seem to follow the Gauss-Kuzmin probabilities (except for quadratic irrationals, of course). The following subsets of real algebraic numbers were extensively tested by students at Princeton (where the number of digits with given values were compared with the predictions from the Gauss-Kuzmin Theorem, and in some cases pairs and triples were also compared) and shown to have excellent agreement with predictions: $\sqrt[n]{p}$ for p prime and $n \leq 5$ ([Ka, Law1, Mic1]) and roots of polynomials with different Galois groups ([AB]). To analyze the data from such experiments, one should perform basic hypothesis testing (see §9.5). For some results on numbers whose digits violate the Gauss-Kuzmin Law, see [Mic2].

Research Project 10.6.1. Investigate the digits of other families of algebraic numbers, for example, the positive real roots of $x^n - p = 0$ (see the mentioned student reports for more details and suggestions). Alternatively, for a fixed real algebraic number α, one can investigate its powers or rational multiples. There are two different types of experiments one can perform. First, one can fix a digit, say the millionth digit, and examine its value as we vary the algebraic number. Second, one could look at the same large block of digits for an algebraic number, and then vary the algebraic number.

While similar, there are different features in the two experiments. In the first we are checking digit by digit. For a fixed number, its nth digit is either k or not; thus, the only probabilities we see are 0 or 1. To have a chance of observing the Gauss-Kuzmin probabilities, we need to perform some averaging (which is accomplished by looking at roots of many different polynomials).

For the second, since we are looking at a large block of digits there is already a chance of observing probabilities close to the Gauss-Kuzmin predictions. For each root and each value (or pairs of values and so on), we obtain a probability in $[0, 1]$. One possibility is to perform a *second* level of averaging by averaging these numbers over roots of different polynomials. Another possibility is to construct a histogram plot of the probabilities for each value. This allows us to investigate more refined questions. For example, are the probabilities as likely to undershoot the predicted values as overshoot? How does that depend on the value? How are the observed probabilities for the different values for each root distributed about the predictions: does it look like a uniform distribution or a normal distribution?

Remark 10.6.2. If one studies say $x^3 - p = 0$, as we vary p the first few digits of the continued fraction expansions of $\sqrt[3]{p}$ are often similar. For example,

$$\sqrt[3]{1000000087} = [1000, 34482, 1, 3, 6, 4, \ldots]$$
$$\sqrt[3]{1000000093} = [1000, 32258, 15, 3, 1, 3, 1, \ldots]$$
$$\sqrt[3]{1000000097} = [1000, 30927, 1, 5, 10, 19, \ldots]. \quad (10.110)$$

The zeroth digit is 1000, which isn't surprising as these cube roots are all approximately 10^3. Note the first digit in the continued fraction expansions is about 30000 for each. Hence if we know the continued fraction expansion for $\sqrt[3]{p}$ for one prime p around 10^9, then we have some idea of the first few digits of $\sqrt[3]{q}$ for primes q near p. Thus if we were to look at the first digit of the cube roots of ten thousand consecutive primes near 10^9, we would not expect to see the Gauss-Kuzmin probabilities.

Consider a large number n_0. Primes near it can be written as $n_0 + x$ for x small.

Then

$$
\begin{aligned}
(n_0 + x)^{1/3} &= n_0^{1/3} \cdot \left(1 + \frac{x}{n_0}\right)^{1/3} \\
&\approx n_0^{1/3} \cdot \left(1 + \frac{1}{3}\frac{x}{n_0}\right) \\
&= n_0^{1/3} + \frac{x}{3n_0^{2/3}}.
\end{aligned}
\tag{10.111}
$$

If n_0 is a perfect cube, then for small x relative to n_0 we see these numbers should have a large first digit. Thus, if we want investigate cube roots of lots of primes p that are approximately the same size, the first few digits are not independent as we vary p. In many of the experiments digits 50,000 to 1,000,000 were investigated: for cube roots of primes of size 10^9, this was sufficient to see independent behavior (though ideally one should look at autocorrelations to verify this claim). Also, the Gauss-Kuzmin Theorem describes the behavior for n large; thus, it is worthwhile to throw away the first few digits so we only study regions where the error term is small.

There are many special functions in number theory. If we evaluate countably many special functions at countably many points, we again obtain a countable set of measure 0. Thus, all these numbers' digits could violate the Gauss-Kuzmin probabilities. Experiments have shown, however, that special values of $\Gamma(s)$ at rational arguments ([Ta])) and the Riemann zeta function $\zeta(s)$ at positive integers ([Kua]) seem to follow the Gauss-Kuzmin probabilities.

Research Project 10.6.3. Consider the non-trivial zeros of $\zeta(s)$, or, more generally, the zeros of any L-function; see Chapter 3 for definitions. Do the digits follow the Gauss-Kuzmin distribution? For the Fourier coefficients of an elliptic curve, $a_p = 2\cos(\theta_p)$; how are the digits of θ_p distributed? How are the digits of $\log n$ distributed? How are the digits of $2^{\sqrt{n}}$ distributed for n square-free?

We know quadratic irrationals are periodic, and hence cannot follow Kuzmin's Law. Only finitely many numbers occur in the continued fraction expansion. Thus, only finitely many numbers have a positive probability of occurring in the expansion, but the Gauss-Kuzmin probabilities are positive for all positive integers.

Research Project 10.6.4. What if we consider a family of quadratic irrationals with growing period? As the size of the period grows, does the distribution of digits tend to the Gauss-Kuzmin probabilities? See the warnings in §7.10 as well as Project 7.10.3 for more details.

Research Project 10.6.5. Assign explicit values to the constants A and B in the Gauss-Kuzmin Theorem, or find A_0, B_0, N_0 such that for all $n \geq N_0$, one may take $A = A_0$ and $B = B_0$.

Chapter Eleven

Introduction to Fourier Analysis

Everyone has experience representing quantities in terms of a basis set. Many cities are laid out in a grid, and locations are represented by x blocks east and y blocks north. We represent this mathematically by writing any point in the plane $\mathbb{R}^2$ as $x \binom{1}{0} + y \binom{0}{1}$, with $x, y \in \mathbb{R}$. Here $\binom{1}{0}$ is the unit vector in the east direction, $\binom{0}{1}$ is the unit vector in the north direction, and our "basis" is the set of the two unit vectors.

Another example is Taylor series expansions (see §A.2.3). For "nice" functions (although see Exercise A.2.7)

$$f(x) = a_0 \cdot 1 + a_1 \cdot x + a_2 \cdot x^2 + \cdots, \qquad (11.1)$$

at least for x sufficiently close to 0. We are expanding $f(x)$ in terms of $\{1, x, x^2, x^3, x^4, \dots\}$. Unlike our example in the plane, here we have an infinite dimensional space. Because there are infinitely many directions, we will have to work to show how to deal with issues such as convergence (does any such series converge to a nice function?) and representation (can any nice function be written as such a series?).

Fourier Analysis is concerned with expanding periodic functions (often with period 1) in terms of the Fourier basis $e_n(x) = e^{2\pi i n x}$, $n \in \mathbb{Z}$. Recall a function f is **periodic** of period a if $f(x + a) = f(x)$ for all x. This basis turns out to be extremely useful for a variety of problems; we will see several instances below, ranging from the equidistribution of sequences in Chapter 12 (if $\alpha \notin \mathbb{Q}$ then $n\alpha \bmod 1$ is equidistributed in $[0, 1]$) to Goldbach-type problems in Chapter 13 (representing numbers as the sum of primes). While the investigations in Chapter 12 require many of the technical results on convergence proved below, in Chapter 13 all we really need is the notation and results of §11.1. We will also prove Poisson Summation, one of the most useful techniques in number theory with applications ranging from proving the functional equation of $\zeta(s)$ (Theorem 3.1.19) to investigating digit bias (see §9.4.2). Finally, when we investigate the distribution of zeros of L-functions in Chapter 18, we shall use Fourier analysis to derive formulas connecting sums over zeros to sums over primes.

It is a deep problem to determine what functions are given by an expansion in the Fourier basis. We prove for many "nice" periodic functions that f is equal to its expansion in the Fourier basis. This is but one of many applications of Fourier analysis; others include solving the heat and wave equations (how systems evolve with time), the isoperimetric inequality (of all smooth closed curves in the plane, for a given perimeter the circle encloses the greatest area), the uncertainty principle (one cannot localize arbitrarily well a function and its Fourier transform), computation of special values of $\zeta(s)$ and L-functions (Chapter 3, and is related to a proof that there are infinitely many primes), the Central Limit Theorem (Chapter 8) and Poissonian behavior of $n^d\alpha$ mod 1 (Chapter 12), to name a few. We sketch some of these applications in §11.6. For a comprehensive treatment of Fourier analysis, see [Be, SS1, Zy].

11.1 INNER PRODUCT OF FUNCTIONS

For $x \in \mathbb{C}$ we define the exponential function by means of the series

$$e^x = \sum_{n=0}^{\infty} \frac{x^n}{n!}, \tag{11.2}$$

which converges everywhere; see §5.4 for more properties of e. Given the Taylor series expansion of $\sin x$ and $\cos x$, we can verify the identity

$$e^{ix} = \cos x + i \sin x. \tag{11.3}$$

Exercise 11.1.1. *Prove e^x converges for all $x \in \mathbb{R}$ (even better, for all $x \in \mathbb{C}$). Show the series for e^x also equals*

$$\lim_{n \to \infty} \left(1 + \frac{x}{n}\right)^n, \tag{11.4}$$

which you may remember from compound interest problems.

Exercise 11.1.2. *Prove, using the series definition, that $e^{x+y} = e^x e^y$ and calculate the derivative of e^x.*

Recall the definition of the **inner** or **dot product**: for two complex-valued vectors $\vec{v} = (v_1, \cdots, v_n)$, $\vec{w} = (w_1, \cdots, w_n)$, we define the inner product $\vec{v} \cdot \vec{w}$ (also denoted $\langle \vec{v}, \vec{w} \rangle$) by

$$\vec{v} \cdot \vec{w} = \langle \vec{v}, \vec{w} \rangle = \sum_{i=1}^{n} v_i \overline{w}_i, \tag{11.5}$$

where $\overline{z}$ is the complex conjugate of z (if $z = x + iy$, $\overline{z} = x - iy$). The length of a vector $\vec{v}$ is

$$|\vec{v}| = \sqrt{\langle \vec{v}, \vec{v} \rangle}. \tag{11.6}$$

We generalize the dot product to functions. Let $f : \mathbb{R} \to \mathbb{C}$, say $f(x) = u(x) + iv(x)$ with u, v real valued functions. We define

$$\int_a^b f(x)dx = \int_a^b u(x)dx + i \int_a^b v(x)dx. \tag{11.7}$$

For a complex number z, $|z|^2 = z\overline{z}$. We will see the generalization of length to a function will be $\int f(x)\overline{f}(x)dx$ (while $\int f(x)^2 dx$ can be zero (or even negative) for complex valued f, the first integral is always non-negative).

For definiteness, assume f and g are functions from $[0,1]$ to $\mathbb{C}$. Divide the interval $[0,1]$ into n equal pieces. Then we can associate to f a vector in $\mathbb{R}^n$ by

$$f(x) \longmapsto \left(f(0), f\left(\frac{1}{n}\right), \ldots, f\left(\frac{n-1}{n}\right)\right), \tag{11.8}$$

and similarly for g. Call these vectors f_n and g_n. As before, we consider

$$\langle f_n, g_n \rangle = \sum_{j=0}^{n-1} f\left(\frac{j}{n}\right) \cdot \overline{g}\left(\frac{j}{n}\right); \tag{11.9}$$

note if $f = g$ this sum is real and non-negative. In general, as we continue to divide the interval ($n \to \infty$), the above sum diverges. For example, if f and g are identically 1, the above sum is n. We expect that the inner product of the constant function on the unit interval with itself (its length) should be 1.

There is a natural re-scaling: we multiply each term in the sum by $\frac{1}{n}$, the size of the subinterval. Note for the constant function the sum is now independent of n. Thus, for good f and g, we are led to define $\langle f, g \rangle$ by a Riemann integral

$$\langle f, g \rangle = \lim_{n \to \infty} \sum_{j=0}^{n-1} f\left(\frac{j}{n}\right) \cdot \overline{g}\left(\frac{j}{n}\right) \frac{1}{n} = \int_0^1 f(x)\overline{g(x)}dx. \tag{11.10}$$

Taking the complex conjugate of g ensures that $\langle f, f \rangle$ is non-negative. Here, "good" means any class of functions such that the Riemann integral converges (for example, continuous or piecewise continuous functions).

Exercise 11.1.3. *Let f, g and h be continuous functions on $[0, 1]$, and $a, b \in \mathbb{C}$. Prove*

1. *$\langle f, f \rangle \geq 0$, and equals 0 if and only if f is identically zero;*

2. *$\langle f, g \rangle = \overline{\langle g, f \rangle}$;*

3. *$\langle af + bg, h \rangle = a\langle f, h \rangle + b\langle g, h \rangle$.*

Exercise 11.1.4. *Find a vector $\vec{v} = \begin{pmatrix} v_1 \\ v_2 \end{pmatrix} \in \mathbb{C}^2$ such that $v_1^2 + v_2^2 = 0$, but $\langle \vec{v}, \vec{v} \rangle \neq 0$.*

Definition 11.1.5 (Orthogonal). *Two continuous functions on $[0, 1]$ are orthogonal (or perpendicular) if their inner product equals zero.*

Exercise 11.1.6. *Prove x^n and x^m are not perpendicular on $[0, 1]$. Find a $c \in \mathbb{R}$ such that $x^n - cx^m$ is perpendicular to x^m; c is related to the projection of x^n in the direction of x^m.*

We will see that the exponential function behaves very nicely under the inner product. For $n \in \mathbb{Z}$, define

$$e_n(x) = e^{2\pi i n x}. \tag{11.11}$$

Exercise 11.1.7 (Important). *Show for $m, n \in \mathbb{Z}$ that*

$$\langle e_m(x), e_n(x) \rangle = \begin{cases} 1 & \text{if } m = n \\ 0 & \text{otherwise.} \end{cases} \tag{11.12}$$

Thus $\{\ldots, e_{-1}(x), e_0(x), e_1(x), \ldots\}$ is a set of mutually perpendicular, unit-length functions. By an **orthogonal set** we mean a set of vectors or functions which are mutually perpendicular; if additionally each vector or function has unit length, we say the set is **orthonormal**. Thus the Fourier basis is an orthonormal set.

Much care is needed, however, in expanding a general periodic function in terms of the $e_n(x)$'s. First, we have issues arising because we have infinitely many basis functions – we must show the infinite series converge, and further that it converges to the initial function (and this is, sadly, not always the case). This is very different from the case of the plane, when we had just two basis vectors, $\binom{1}{0}$ and $\binom{0}{1}$. Second, we have questions of completeness. Is the above list of the $e_n(x)$'s complete? Do these infinitely many functions capture everything, or do we need to add more functions which are orthogonal to each $e_n(x)$?

Definition 11.1.8 (Periodic). *A function $f(x)$ is periodic with period a if for all $x \in \mathbb{R}$, $f(x + a) = f(x)$.*

Note the exponential functions $e_n(x)$ are periodic with period 1. Thus, if f is periodic with period 1, it makes no difference if we study f on $[0, 1]$ or $[-\frac{1}{2}, \frac{1}{2}]$ or, more generally, on *any* interval of length one.

Exercise 11.1.9. *Let f and g be periodic functions with period a. Prove $\alpha f(x) + \beta g(x)$ is periodic with period a.*

Definition 11.1.10 (Even, Odd). *A function $f(x)$ is even (resp., odd) if $f(x) = f(-x)$ (if $f(x) = -f(-x)$).*

Exercise 11.1.11. *Prove any function can be written as the sum of an even and an odd function.*

11.2 FOURIER SERIES

11.2.1 Introduction

Let f be continuous and periodic on $\mathbb{R}$ with period one. Define the nth **Fourier coefficient** $\widehat{f}(n)$ of f to be

$$\widehat{f}(n) = \langle f(x), e_n(x) \rangle = \int_0^1 f(x) e^{-2\pi i n x} dx. \qquad (11.13)$$

Returning to the intuition of $\mathbb{R}^m$, we can think of the $e_n(x)$'s as an infinite set of perpendicular unit directions. The above is simply the projection of f in the direction of $e_n(x)$. Often one writes a_n for $\widehat{f}(n)$.

Exercise 11.2.1. *Show*

$$\langle f(x) - \widehat{f}(n) e_n(x), e_n(x) \rangle = 0. \qquad (11.14)$$

This agrees with our intuition: after removing the projection in a certain direction, what is left is perpendicular to that direction.

The Nth **partial Fourier series** of f is

$$S_N(x) = \sum_{n=-N}^{N} \widehat{f}(n)e_n(x). \qquad (11.15)$$

Exercise 11.2.2. *Prove*

1. $\langle f(x) - S_N(x), e_n(x) \rangle = 0$ *if* $|n| \leq N$.

2. $|\widehat{f}(n)| \leq \int_0^1 |f(x)| dx$.

3. *Bessel's Inequality: if* $\langle f, f \rangle < \infty$ *then* $\sum_{n=-\infty}^{\infty} |\widehat{f}(n)|^2 \leq \langle f, f \rangle$.

4. *Riemann-Lebesgue Lemma: if* $\langle f, f \rangle < \infty$ *then* $\lim_{|n| \to \infty} \widehat{f}(n) = 0$ *(this holds for more general f; it suffices that* $\int_0^1 |f(x)| dx < \infty$).

5. *Assume f is differentiable k times; integrating by parts, show* $|\widehat{f}(n)| \ll \frac{1}{n^k}$ *and the constant depends only on f and its first k derivatives.*

As $\langle f(x) - S_N(x), e_n(x) \rangle = 0$ if $|n| \leq N$, we might think that we just have to let N tend to infinity to obtain a series $S_\infty(x)$ such that

$$\langle f(x) - S_\infty(x), e_n(x) \rangle = 0. \qquad (11.16)$$

Assume that a periodic function $g(x)$ is orthogonal to $e_n(x)$ for every n if and only if $g(x)$ is identically zero. Then $f(x) - S_\infty(x) = 0$, and hence $f(x) = S_\infty(x)$: we have expressed $f(x)$ as a sum of exponentials! We must be very careful. We have just glossed over the two central issues – completeness (are the $e_n(x)$'s all the "directions"?) and, even worse, convergence (do the sums agree with f for all x?). For many f, the Fourier series does converge pointwise, but much care is required to prove such results. By looking at modified Fourier series, we will easily give examples of finite approximations to f with good pointwise convergence.

Exercise 11.2.3. *Let $h(x) = f(x) + g(x)$. Does $\widehat{h}(n) = \widehat{f}(n) + \widehat{g}(n)$? Let $k(x) = f(x)g(x)$. Does $\widehat{k}(n) = \widehat{f}(n)\widehat{g}(n)$?*

Remark 11.2.4. In many of our theorems below we assume that $\langle f, f \rangle = \int_0^1 |f(x)|^2 dx < \infty$. This is a natural condition, as the Cauchy-Schwarz inequality (Appendix A.6) implies that if f and g are two such functions, $\langle f, g \rangle < \infty$ and $\int_0^1 |f(x)| dx < \infty$; the second statement is false if $f, g : \mathbb{R} \to \mathbb{C}$ and not $[0,1] \to \mathbb{C}$. If $\int_0^1 |f(x)|^r dx < \infty$, one often writes $f \in L^r([0,1])$.

Exercise 11.2.5. *Remark 11.2.4 shows that if $\langle f, f \rangle, \langle g, g \rangle < \infty$ then the dot product of f and g exists: $\langle f, g \rangle < \infty$. Do there exist $f, g : [0,1] \to \mathbb{C}$ such that $\int_0^1 |f(x)| dx, \int_0^1 |g(x)| dx < \infty$ but $\int_0^1 f(x)\overline{g}(x) dx = \infty$? Is $f \in L^2([0,1])$ a stronger or an equivalent assumption as $f \in L^1([0,1])$?*

11.2.2 Approximations to the Identity

We assume the reader is familiar with the basics of probability functions (see Chapter 8, especially §8.2.3). A sequence $A_1(x), A_2(x), A_3(x), \ldots$ of functions is an **approximation to the identity on** $[0, 1]$ if

1. for all x and N, $A_N(x) \geq 0$;

2. for all N, $\int_0^1 A_N(x)dx = 1$;

3. for all δ, $0 < \delta < \frac{1}{2}$, $\lim_{N \to \infty} \int_\delta^{1-\delta} A_N(x)dx = 0$.

Similar definitions hold with $[0, 1]$ replaced by other intervals; it is often more convenient to work on $[-\frac{1}{2}, \frac{1}{2}]$, replacing the third condition with

$$\lim_{N \to \infty} \int_{|x| > \delta} A_N(x)dx = 0 \quad \text{if} \quad 0 < \delta < \frac{1}{2}. \tag{11.17}$$

We could also replace $[0, 1]$ with $\mathbb{R}$, which would make the third condition

$$\lim_{N \to \infty} \int_{-\infty}^{-\delta} + \int_\delta^\infty A_N(x)dx = 0 \quad \text{if} \quad \delta > 0. \tag{11.18}$$

There is a natural interpretation of these conditions. The first two (non-negativity and integral equals 1) allow us to think of each $A_N(x)$ as a probability distribution. The third condition states that, as $N \to \infty$, all of the probability is concentrated arbitrarily close to one point. Physically, we may regard the $A_n(x)$'s as densities for a unit mass of smaller and smaller width. In the limit, we obtain a **unit point mass**; it will have finite mass, but infinite density at one point, and zero density elsewhere.

Exercise 11.2.6. *Define*

$$A_N(x) = \begin{cases} N & \text{for } |x| \leq \frac{1}{N} \\ 0 & \text{otherwise.} \end{cases} \tag{11.19}$$

Prove A_N is an approximation to the identity on $[-\frac{1}{2}, \frac{1}{2}]$. If f is continuously differentiable and periodic with period 1, calculate

$$\lim_{N \to \infty} \int_{-\frac{1}{2}}^{\frac{1}{2}} f(x)A_N(x)dx. \tag{11.20}$$

Exercise 11.2.7. *Let $A(x)$ be a non-negative function with $\int_\mathbb{R} A(x)dx = 1$. Prove $A_N(x) = N \cdot A(Nx)$ is an approximation to the identity on $\mathbb{R}$.*

Exercise 11.2.8 (Important). *Let $A_N(x)$ be an approximation to the identity on $[-\frac{1}{2}, \frac{1}{2}]$. Let $f(x)$ be a continuous function on $[-\frac{1}{2}, \frac{1}{2}]$. Prove*

$$\lim_{N \to \infty} \int_{-\frac{1}{2}}^{\frac{1}{2}} f(x)A_N(x)dx = f(0). \tag{11.21}$$

By Exercise 11.2.8, in the limit the functions $A_N(x)$ are acting like unit point masses at the origin.

Definition 11.2.9 (Dirac Delta Functional). *We define a map from continuous complex valued functions to the complex numbers by $\delta(f) = f(0)$. We often write this in the more suggestive notation*

$$\int f(x)\delta(x)dx = f(0), \tag{11.22}$$

where the integration will usually be over $[0,1]$, $[-\frac{1}{2}, \frac{1}{2}]$ or $\mathbb{R}$.

By a standard abuse of notation, we often call $\delta(x)$ the delta function. We can consider the probability densities $A_n(x)dx$ and $\delta(x)dx$. For $A_N(x)dx$, as $N \to \infty$ almost all the probability (mass) is concentrated in a narrower and narrower band about the origin; $\delta(x)dx$ is the limit with all the mass at one point. It is a discrete (as opposed to continuous) probability measure, with infinite density but finite mass. Note that $\delta(x-a)$ acts like a unit point mass; however, instead of having its mass concentrated at the origin, it is now concentrated at a.

11.2.3 Dirichlet and Fejér Kernels

We define two functions which will be useful in investigating convergence of Fourier series. Set

$$D_N(x) := \sum_{n=-N}^{N} e_n(x) = \frac{\sin((2N+1)\pi x)}{\sin \pi x}$$

$$F_N(x) := \frac{1}{N}\sum_{n=0}^{N-1} D_n(x) = \frac{\sin^2(N\pi x)}{N\sin^2 \pi x}. \tag{11.23}$$

Exercise 11.2.10. *Prove the two formulas above. The geometric series formula will be helpful:*

$$\sum_{n=N}^{M} r^n = \frac{r^N - r^{M+1}}{1-r}. \tag{11.24}$$

Here F stands for Fejér, D for Dirichlet. $F_N(x)$ and $D_N(x)$ are two important examples of (integral) **kernels**. By integrating a function against a kernel, we obtain a new function related to the original. We will study integrals of the form

$$g(x) = \int_0^1 f(x)K(x-y)dy. \tag{11.25}$$

Such an integral is called the **convolution** of f and K. The Fejér and Dirichlet kernels yield new functions related to the Fourier expansion of $f(x)$.

Theorem 11.2.11. *The Fejér kernels $F_1(x), F_2(x), F_3(x), \ldots$ are an approximation to the identity on $[0,1]$.*

Proof. The first property is immediate. The second follows from the observation that $F_N(x)$ can be written as

$$F_N(x) = e_0(x) + \frac{N-1}{N}(e_{-1}(x) + e_1(x)) + \cdots, \qquad (11.26)$$

and all integrals are zero but the first, which is 1. To prove the third property, note that $F_N(x) \le \frac{1}{N\sin^2 \pi \delta}$ for $\delta \le x \le 1 - \delta$. $\qquad\square$

Exercise 11.2.12. *Show that the Dirichlet kernels are* not *an approximation to the identity. How large are $\int_0^1 |D_N(x)|dx$ and $\int_0^1 D_N(x)^2 dx$?*

Let f be a continuous, periodic function on $\mathbb{R}$ with period one. We may consider f as a function on just $[0,1]$, with $f(0) = f(1)$. Define

$$T_N(x) = \int_0^1 f(y)F_N(x-y)dy. \qquad (11.27)$$

In other words, $T_N(x)$ is the integral transform of $f(x)$ with respect to the Fejér kernel. We show below that, for many f, $T_N(x)$ has good convergence properties to $f(x)$. To do so requires some basic facts from analysis, which are recalled in Appendix A.3.

11.3 CONVERGENCE OF FOURIER SERIES

We investigate when the Fourier series converges to the original function. For continuous functions, a related series always converges. An important application is that instead of proving results for "general" f, it often suffices to prove results for Fourier series (see Chapter 12).

11.3.1 Convergence of Fejér Series to f

Theorem 11.3.1 (Fejér). *Let $f(x)$ be a continuous, periodic function on $[0,1]$. Given $\epsilon > 0$ there exists an N_0 such that for all $N > N_0$,*

$$|f(x) - T_N(x)| \le \epsilon \qquad (11.28)$$

for every $x \in [0,1]$. Hence as $N \to \infty$, $T_N f(x) \to f(x)$.

Proof. The starting point of the proof is multiplying by 1 in a clever way, a very powerful technique. We have

$$f(x) = f(x)\int_0^1 F_N(y)dy = \int_0^1 f(x)F_N(y)dy; \qquad (11.29)$$

this is true as $F_N(y)$ is an approximation to the identity and thus integrates to 1.

For any positive N and $\delta \in (0, 1/2)$,

$$
\begin{aligned}
T_N(x) - f(x) &= \int_0^1 f(x - y) F_N(y) dy - f(x) \cdot 1 \\
&= \int_0^1 f(x - y) F_N(y) dy - \int_0^1 f(x) F_N(y) dy \\
&\qquad \text{(by property 2 of } F_N) \\
&= \int_0^\delta (f(x - y) - f(x)) F_N(y) dy \\
&\quad + \int_\delta^{1-\delta} (f(x - y) - f(x)) F_N(y) dy \\
&\quad + \int_{1-\delta}^1 (f(x - y) - f(x)) F_N(y) dy.
\end{aligned}
\tag{11.30}
$$

As the $F_N(x)$'s are an approximation to the identity, we find

$$
\left| \int_\delta^{1-\delta} (f(x - y) - f(x)) F_N(y) dy \right| \le 2 \max |f(x)| \cdot \int_\delta^{1-\delta} F_N(y) dy. \tag{11.31}
$$

By Theorem A.3.13, $f(x)$ is bounded, so there exists a B such that $\max |f(x)| \le B$. Since

$$
\lim_{N \to \infty} \int_\delta^{1-\delta} F_N(y) dy = 0, \tag{11.32}
$$

we obtain

$$
\lim_{N \to \infty} \int_\delta^{1-\delta} (f(x - y) - f(x)) F_N(y) dy = 0. \tag{11.33}
$$

Thus, by choosing N large enough (where large depends on δ), we can ensure that this piece is at most $\frac{\epsilon}{3}$.

It remains to estimate what happens near zero, *and this is where we use f is periodic*. Since f is continuous and $[0, 1]$ is a finite closed interval, f is uniformly continuous (Theorem A.3.7). Thus we can choose δ small enough that $|f(x - y) - f(x)| < \frac{\epsilon}{3}$ for any x and any positive $y < \delta$. Then

$$
\left| \int_0^\delta (f(x - y) - f(x)) F_N(y) dy \right| \le \int_0^\delta \frac{\epsilon}{3} F_N(y) dy \le \frac{\epsilon}{3} \int_0^1 F_N(y) dy \le \frac{\epsilon}{3}.
\tag{11.34}
$$

Similarly

$$
\left| \int_{1-\delta}^1 (f(x - y) - f(x)) F_N(y) dy \right| \le \frac{\epsilon}{3}. \tag{11.35}
$$

Therefore

$$
|T_N(x) - f(x)| \le \epsilon \tag{11.36}
$$

for all N sufficiently large. $\qquad \square$

Remark 11.3.2. Where did we use f periodic? Recall we had expressions such as $f(x - y) - f(x)$. For example, if $x = .001$ and $y = .002$, we have $f(-.001) - f(.001)$. The periodicity of f allows us to extend f to a continuous function on $\mathbb{R}$.

One often uses the interval $[-\frac{1}{2}, \frac{1}{2}]$ instead of $[0, 1]$; the proof follows analogously. Proofs of this nature are often called *three epsilon proofs*; splitting the interval as above is a common technique for analyzing such functions.

Definition 11.3.3 (Trigonometric Polynomials). *Any finite linear combination of the functions $e_n(x)$ is called a trigonometric polynomial.*

From Fejér's Theorem (Theorem 11.3.1) we immediately obtain the

Theorem 11.3.4 (Weierstrass Approximation Theorem). *Any continuous periodic function can be uniformly approximated by trigonometric polynomials.*

Remark 11.3.5. Weierstrass proved (many years before Fejér) that if f is continuous on $[a, b]$, then for any $\epsilon > 0$ there is a polynomial $p(x)$ such that $|f(x) - p(x)| < \epsilon$ for all $x \in [a, b]$. This important theorem has been extended numerous times (see, for example, the Stone-Weierstrass Theorem in [Rud]).

Exercise 11.3.6. *Prove the Weierstrass Approximation Theorem implies the original version of Weierstrass' Theorem (see Remark 11.3.5).*

We have shown the following: if f is a continuous, periodic function, given any $\epsilon > 0$ we can find an N_0 such that for $N > N_0$, $T_N(x)$ is within ϵ of $f(x)$. As ϵ was arbitrary, as $N \to \infty$, $T_N(x) \to f(x)$.

11.3.2 Pointwise Convergence of Fourier Series

Theorem 11.3.1 shows that given a continuous, periodic f, the Fejér series $T_N(x)$ converges pointwise to $f(x)$. The Fejér series is a weighted Fourier series, though; what can be said about pointwise convergence of the initial Fourier series to $f(x)$?

Recall $\widehat{f}(n)$ is the nth Fourier coefficient of $f(x)$. Consider the Fourier series

$$S_N(x) = \sum_{n=-N}^{N} \widehat{f}(n) e^{2\pi i n x}. \tag{11.37}$$

Exercise 11.3.7. *Show*

$$S_N(x_0) = \int_{-\frac{1}{2}}^{\frac{1}{2}} f(x) D_N(x - x_0) dx = \int_{-\frac{1}{2}}^{\frac{1}{2}} f(x_0 - x) D_N(x) dx. \tag{11.38}$$

Theorem 11.3.8 (Dirichlet). *Suppose*

1. *$f(x)$ is periodic with period 1;*

2. *$|f(x)|$ is bounded;*

3. *$f(x)$ is differentiable at x_0.*

Then $\lim_{N\to\infty} S_N(x_0) = f(x_0)$.

Proof. Let $D_N(x)$ be the Dirichlet kernel. Previously we have shown that $D_N(x) = \frac{\sin((2N+1)\pi x)}{\sin(\pi x)}$ and $\int_{-\frac{1}{2}}^{\frac{1}{2}} D_N(x)dx = 1$. Thus

$$f(x_0) - S_N(x_0) = f(x_0)\int_{-\frac{1}{2}}^{\frac{1}{2}} D_N(x)dx - \int_{-\frac{1}{2}}^{\frac{1}{2}} f(x_0 - x)D_N(x)dx$$

$$= \int_{-\frac{1}{2}}^{\frac{1}{2}} [f(x_0) - f(x_0 - x)] D_N(x)dx$$

$$= \int_{-\frac{1}{2}}^{\frac{1}{2}} \frac{f(x_0) - f(x_0 - x)}{\sin(\pi x)} \cdot \sin((2N+1)\pi x)dx$$

$$= \int_{-\frac{1}{2}}^{\frac{1}{2}} g_{x_0}(x) \sin((2N+1)\pi x)dx. \tag{11.39}$$

We claim $g_{x_0}(x) = \frac{f(x_0)-f(x_0-x)}{\sin(\pi x)}$ is bounded. As f is bounded, the numerator is bounded. The denominator is only troublesome near $x = 0$; however, as f is differentiable at x_0,

$$\lim_{x\to 0} \frac{f(x_0 + x) - f(x_0)}{x} = f'(x_0). \tag{11.40}$$

Multiplying by 1 in a clever way (one of the most useful proof techniques) gives

$$\lim_{x\to 0} \frac{f(x_0 + x) - f(x_0)}{\sin(\pi x)} = \lim_{x\to 0} \frac{f(x_0 + x) - f(x_0)}{\pi x} \cdot \frac{\pi x}{\sin(\pi x)} = \frac{f'(x_0)}{\pi}. \tag{11.41}$$

Therefore $g_{x_0}(x)$ is bounded everywhere, say by B. We have

$$|f(x_0) - S_N(x_0)| \le \int_{-\frac{1}{2}}^{\frac{1}{2}} B|\sin((2N+1)\pi x)|dx. \tag{11.42}$$

The above integral is bounded by $\frac{B}{(2N+1)\pi}$. To see this, break $[-\frac{1}{2}, \frac{1}{2}]$ into regions where $\sin((2N+1)\pi x)$ is always positive (resp., negative). Then we can handle the absolute value, and the anti-derivative will be $\frac{B\cos((2N+1)\pi x)}{(2N+1)\pi}$, evaluated at the endpoints of the interval. Thus for each interval where $\sin((2N+1)\pi x)$ is always positive (resp., negative), the integral is bounded by the length of that interval times $\frac{B}{(2N+1)\pi}$. As the intervals' lengths sum to 1, this proves the claim. We have shown

$$|f(x_0) - S_N(x_0)| \le \frac{B}{(2N+1)\pi}. \tag{11.43}$$

Hence as $N \to \infty$, $S_N(x_0)$ converges (pointwise) to $f(x_0)$. □

Remark 11.3.9. If f is twice differentiable, by Exercise 11.2.2 $\hat{f}(n) \ll \frac{1}{n^2}$ and the series $S_N(x)$ has good convergence properties.

What can be said about pointwise convergence for general functions? It is possible for the Fourier series of a continuous function to diverge at a point (see §2.2

of [SS1]). Kolmogorov (1926) constructed a function such that $\int_0^1 |f(x)|dx$ is finite and the Fourier series diverges everywhere; however, if $\int_0^1 |f(x)|^2 dx < \infty$, the story is completely different. For such f, Carleson proved that for almost all $x \in [0, 1]$ the Fourier series converges to the original function (see [Ca, Fef]).

Exercise 11.3.10. *Let* $\widehat{f}(n) = \frac{1}{2^{|n|}}$. *Does* $\sum_{-\infty}^{\infty} \widehat{f}(n)e_n(x)$ *converge to a continuous, differentiable function? If so, is there a simple expression for that function?*

11.3.3 Parseval's Identity

Theorem 11.3.11 (Parseval's Identity). *Assume* $\int_0^1 |f(x)|^2 dx < \infty$. *Then*

$$\sum_{n=-\infty}^{\infty} |\widehat{f}(n)|^2 = \int_0^1 |f(x)|^2 dx. \tag{11.44}$$

In other words, Bessel's Inequality (Exercise 11.2.2) is an equality for such f.

We sketch the proof for continuous functions. From the definition of $S_N(x)$ we obtain Bessel's Inequality:

$$0 \le \int_0^1 (f(x) - S_N(x))\overline{(f(x) - S_N(x))}dx = \int_0^1 |f(x)|^2 - \sum_{|n|\le N} |\widehat{f}(n)|^2. \tag{11.45}$$

Rearranging yields

$$\int_0^1 |f(x)|^2 dx = \int_0^1 |f(x) - S_N(x)|^2 dx + \sum_{|n|\le N} |\widehat{f}(n)|^2. \tag{11.46}$$

To complete the proof, we need only show that as $N \to \infty$, $\int_0^1 |f(x)-S_N(x)|^2 dx \to 0$. Note Bessel's Inequality, (11.46), immediately implies

$$\sum_{n=-\infty}^{\infty} |\widehat{f}(n)|^2 < \int_0^1 |f(x)|^2 dx < \infty; \tag{11.47}$$

therefore the sum converges and given any ϵ, for N sufficiently large

$$\sum_{|n|>\sqrt{N}} |\widehat{f}(n)|^2 < \epsilon. \tag{11.48}$$

In the proof of Theorem 11.3.8 we multiplied by 1; we now do another common trick: adding zero in a clever way. By Theorem 11.3.1, given any $\epsilon > 0$ there exists an N_0 such that for all $N > N_0$, $|f(x) - T_N(x)| < \epsilon$. We apply the inequality $|a + b|^2 \le 4|a|^2 + 4|b|^2$ to $|f(x) - S_N(x)|$ with $a = f(x) - T_N(x)$ and $b = T_N(x) - S_N(x)$. We have

$$\int_0^1 |f(x)-S_N(x)|^2 dx \le 4\int_0^1 |f(x)-T_N(x)|^2 dx + 4\int_0^1 |T_N(x)-S_N(x)|^2 dx. \tag{11.49}$$

The first term on the right is at most $4\epsilon^2$. To handle the second integral, note

$$T_N(x) = \sum_{n=-N}^{N} \frac{N-|n|}{N} \widehat{f}(n) e^{2\pi i n x}, \tag{11.50}$$

which implies

$$\int_0^1 |T_N(x) - S_N(x)|^2 dx = \sum_{n=-N}^{N} \frac{|n|^2}{N^2} |\widehat{f}(n)|^2. \tag{11.51}$$

Since f is continuous, f is bounded (Theorem A.3.13), hence by Exercise 11.2.2 $\widehat{f}(n)$ is bounded, say by B. The sum in (11.51) can be made arbitrarily small (the terms with $|n| \le \sqrt{N}$ contribute at most $\frac{2B^2}{N}$, and the remaining contributes at most ϵ by (11.48)).

Exercise 11.3.12. *Fill in the details for the above proof. Prove the result for all f satisfying $\int_0^1 |f(x)|^2 dx < \infty$.*

Exercise 11.3.13. *If $\int_0^1 |f(x)|^2 dx < \infty$, show Bessel's Inequality implies there exists a B such that $|\widehat{f}(n)| \le B$ for all n.*

Exercise 11.3.14. *Though we used $|a+b|^2 \le 4|a|^2 + 4|b|^2$, any bound of the form $c|a|^2 + c|b|^2$ would suffice. What is the smallest c that works for all $a, b \in \mathbb{C}$?*

11.3.4 Sums of Series

One common application of pointwise convergence and Parseval's identity is to evaluate infinite sums. For example, if we know at some point x_0 that $S_N(x_0) \to f(x_0)$, we obtain

$$\sum_{n=-\infty}^{\infty} \widehat{f}(n) e^{2\pi i n x_0} = f(x_0). \tag{11.52}$$

Additionally, if $\int_0^1 |f(x)|^2 dx < \infty$ we obtain

$$\sum_{n=-\infty}^{\infty} |\widehat{f}(n)|^2 = \int_0^1 |f(x)|^2 dx. \tag{11.53}$$

Thus, if the terms in a series correspond to Fourier coefficients of a "nice" function, we can evaluate the series.

Exercise 11.3.15. *Let $f(x) = \frac{1}{2} - |x|$ on $[-\frac{1}{2}, \frac{1}{2}]$. Calculate $\sum_{n=0}^{\infty} \frac{1}{(2n+1)^2}$. Use this to deduce the value of $\sum_{n=1}^{\infty} \frac{1}{n^2}$. This is often denoted $\zeta(2)$ (see Exercise 3.1.7). For connections with continued fractions, see [BP].*

Exercise 11.3.16. *Let $f(x) = x$ on $[0,1]$. Evaluate $\sum_{n=1}^{\infty} \frac{1}{n^2}$.*

Exercise 11.3.17. *Let $f(x) = x$ on $[-\frac{1}{2}, \frac{1}{2}]$. Prove $\frac{\pi}{4} = \sum_{n=1}^{\infty} \frac{(-1)^{n+1}}{(2n-1)^2}$. See also Exercise 3.3.29; see Chapter 11 of [BB] or [Sc] for a history of calculations of π.*

Exercise 11.3.18. *Find a function to determine $\sum_{n=1}^{\infty} \frac{1}{n^4}$; compare your answer with Exercise 3.1.25.*

11.4 APPLICATIONS OF THE FOURIER TRANSFORM

To each periodic function (say with period 1) we associate its Fourier coefficients

$$\widehat{f}(n) = \int_0^1 f(x)e^{-2\pi inx}. \tag{11.54}$$

Note the integral is well defined for all n if $\int_0^1 |f(x)|dx < \infty$. If f is continuous and differentiable, we have seen that we can recover f from its Fourier coefficients $\widehat{f}(n)$. We briefly discuss the generalization to non-periodic functions on all of $\mathbb{R}$, the Fourier transform; see [SS1] for complete details.

We give two applications. The first is Poisson Summation, which relates sums of f at integers to sums of its Fourier transform at integers. Often this converts a long, slowly decaying sum to a short, rapidly decaying one (see for example Theorem 3.1.19 on the functional equation of $\zeta(s)$, as well as Theorem 9.4.2 which shows there is digit bias in Geometric Brownian Motions). Poisson Summation is one of the most important tools in a number theorist's arsenal. As a second application we sketch the proof of the Central Limit Theorem (Theorem 8.4.1).

The Fourier transform also appears in Chapter 18 when we investigate the zeros of L-functions. When we derive formulas relating sums of a function f at zeros of an L-function to sums of the product of the coefficients of the L-function times the Fourier transform of f at primes, we shall use some properties of the Fourier transform to show that the sums converge. Relations like this are the starting point of many investigations of properties of zeros of L-functions, primarily because often we can evaluate the sums of the Fourier transform at primes and then use that knowledge to glean information about the zeros.

11.4.1 Fourier Transform

We define the **Fourier transform** by

$$\widehat{f}(y) = \int_{-\infty}^{\infty} f(x)e^{-2\pi ixy}dx; \tag{11.55}$$

(sometimes the Fourier transform is defined with e^{-ixy} instead of $e^{-2\pi ixy}$). Instead of countably many Fourier coefficients, we now have one for each $y \in \mathbb{R}$. While $\widehat{f}(y)$ is well defined whenever $\int_{-\infty}^{\infty} |f(x)|dx < \infty$, much more is true for functions with $\int_{-\infty}^{\infty} |f(x)|^2 dx < \infty$.

The **Schwartz Space** $\mathcal{S}(\mathbb{R})$ is the space of all infinitely differentiable functions whose derivatives are rapidly decreasing. Explicitly,

$$\forall j, k \geq 0, \quad \sup_{x \in \mathbb{R}}(|x| + 1)^j |f^{(k)}(x)| < \infty. \tag{11.56}$$

Thus as $|x| \to \infty$, f and all its derivatives decay faster than any polynomial. One can show the Fourier transform of a Schwartz function is a Schwartz function, and

Theorem 11.4.1 (Fourier Inversion Formula). *For $f \in \mathcal{S}(\mathbb{R})$,*

$$f(x) = \int_{-\infty}^{\infty} \widehat{f}(y)e^{2\pi ixy}dy. \tag{11.57}$$

In fact, for any $f \in \mathcal{S}(\mathbb{R})$,

$$\int_{-\infty}^{\infty} |f(x)|^2 dx = \int_{-\infty}^{\infty} |\widehat{f}(y)|^2 dy. \qquad (11.58)$$

Definition 11.4.2 (Compact Support). *A function $f : \mathbb{R} \to \mathbb{C}$ has compact support if there is a finite closed interval $[a, b]$ such that for all $x \notin [a, b]$, $f(x) = 0$.*

Remark 11.4.3 (Advanced). Schwartz functions with compact support are extremely useful in many arguments. It can be shown that given any continuous function g on a finite closed interval $[a, b]$, there is a Schwartz function f with compact support arbitrarily close to g; i.e., for all $x \in [a, b]$, $|f(x) - g(x)| < \epsilon$. Similarly, given any such continuous function g, one can find a sum of step functions of intervals arbitrarily close to g (in the same sense as above). Often, to prove a result for step functions it suffices to prove the result for continuous functions, which is the same as proving the result for Schwartz functions. Schwartz functions are infinitely differentiable and as the Fourier Inversion formula holds, we can pass to the Fourier transform space, which is sometimes easier to study.

Exercise 11.4.4. *Show the Gaussian $f(x) = \frac{1}{\sqrt{2\pi\sigma^2}} e^{-(x-\mu)^2/2\sigma^2}$ is in $\mathcal{S}(\mathbb{R})$ for any $\mu, \sigma \in \mathbb{R}$.*

Exercise 11.4.5. *Let $f(x)$ be a Schwartz function with compact support contained in $[-\sigma, \sigma]$ and denote its Fourier transform by $\widehat{f}(y)$. Prove for any integer $A > 0$ that $|\widehat{f}(y)| \le c_f y^{-A}$, where the constant c_f depends only on f, its derivatives and σ. As such a bound is useless at $y = 0$, one often derives bounds of the form $|\widehat{f}(y)| \le \frac{\widetilde{c}_f}{(1+|y|)^A}$.*

11.4.2 Poisson Summation

We say a function $f(x)$ decays like x^{-a} if there are constants x_0 and C such that for all $|x| > x_0$, $|f(x)| \le \frac{C}{x^a}$.

Theorem 11.4.6 (Poisson Summation). *Assume f is twice continuously differentiable and that f, f' and f'' decay like $x^{-(1+\eta)}$ for some $\eta > 0$. Then*

$$\sum_{n=-\infty}^{\infty} f(n) = \sum_{n=-\infty}^{\infty} \widehat{f}(n), \qquad (11.59)$$

where $\widehat{f}$ is the Fourier transform of f.

The theorem is true for more general f. We confine ourselves to this useful case, whose conditions are often met in applications. See, for example, Theorem 9.4.2, where Poisson Summation allowed us to replace a long slowly decaying sum with just one term (plus a negligible error), as well as Theorem 3.1.19, where we proved the functional equation of $\zeta(s)$.

It is natural to study $F(x) = \sum_{n\in\mathbb{Z}} f(x + n)$; the theorem follows from understanding $F(0)$. As $F(x)$ is periodic with period 1, it is natural to try to apply our results on Fourier series to approximate F. To use the results from §11.3.2, we need F to be continuously differentiable on $[0,1]$; however, before we show F is differentiable, we must first show F is continuous and well defined! For example, consider $f(x)$ to be narrow spikes at the integers, say of height n and width $\frac{1}{n^4}$ centered around $x = n$. Note the sum $F(0)$ does not exist.

Exercise 11.4.7. *Consider*

$$f(x) = \begin{cases} n^5\left(\frac{1}{n^4} - |n - x|\right) & \text{if } |x - n| \le \frac{1}{n^4} \text{ for some } n \in \mathbb{Z} \\ 0 & \text{otherwise.} \end{cases} \quad (11.60)$$

Show $f(x)$ is continuous but $F(0)$ is undefined. Show $F(x)$ converges and is well defined for any $x \notin \mathbb{Z}$.

Lemma 11.4.8. *If $g(x)$ decays like $x^{-(1+\eta)}$ for some $\eta > 0$, then $G(x) = \sum_{n\in\mathbb{Z}} g(x + n)$ converges for all x, and is continuous.*

Exercise 11.4.9. *Prove Lemma 11.4.8.*

Lemma 11.4.10. *If f, f', f'' decay like $x^{-(1+\eta)}$, then $F'(x)$ is continuously differentiable.*

Proof. We must show that for any $\epsilon > 0$ there exists a $\delta > 0$ such that

$$\left| \frac{F(x + \delta) - F(x)}{\delta} - F'(x) \right| < \epsilon. \quad (11.61)$$

The natural candidate for $F'(x)$ is

$$F'(x) = \sum_{n\in\mathbb{Z}} f'(x + n), \quad (11.62)$$

which converges and is continuous by Lemma 11.4.8 (applied to f'). Let $\delta < 1$. We may write

$$\frac{F(x + \delta) - F(x)}{\delta} - F'(x) = \sum_{|n|\le N} \left[\frac{f(x + n + \delta) - f(x + n)}{\delta} - f'(x + n) \right]$$
$$+ \sum_{|n|>N} \left[\frac{f(x + n + \delta) - f(x + n)}{\delta} - f'(x + n) \right]. \quad (11.63)$$

By the Mean Value Theorem (Theorem A.2.2),

$$f(x + n + \delta) - f(x + n) = \delta \cdot f'(x + n + c_n), \quad c_n \in [0, \delta], \quad (11.64)$$

and similarly with f replaced by f'. Therefore

$$\frac{F(x + \delta) - F(x)}{\delta} - F'(x) = \sum_{|n|\le N} [f'(x + n + c_n) - f'(x + n)]$$
$$+ \sum_{|n|>N} [f'(x + n + c_n) - f'(x + n)]$$
$$= \sum_{|n|\le N} [f'(x + n + c_n) - f'(x + n)]$$
$$+ \sum_{|n|>N} f''(x + n + d_n) \quad (11.65)$$

with $d_n \in [0, \delta]$. Since f'' is of rapid decay, x is fixed and $\delta < 1$, by taking N sufficiently large we can make

$$\sum_{|n|>N} f''(x+n+d_n) \leq \sum_{|n|>N} \frac{C}{n^{1+\eta}} \leq \frac{2C}{\eta(N-1)^\eta} < \frac{\epsilon}{2}. \qquad (11.66)$$

Since f' is continuous at $x + n$, we can find $\delta_n < 1$ such that if $|c_n| < \delta_n$ then $|f'(x+n+c_n) - f'(x+n)| < \frac{\epsilon}{2(2N+1)}$. Letting $\delta = \min_{|n| \geq N} \delta_n$, we find

$$\sum_{|n| \leq N} [f'(x+n+c_n) - f'(x+n)] < \sum_{|n| \leq N} \frac{\epsilon}{2(2N+1)} < \frac{\epsilon}{2}, \qquad (11.67)$$

completing the proof.   □

Exercise 11.4.11. *For what weaker assumptions on f, f', f'' is the conclusion of Lemma 11.4.10 still true?*

We have shown that the assumptions in Theorem 11.4.6 imply that F, F' exist and are continuous, and clearly F is periodic of period 1. Let $\widehat{F}(m)$ be the mth Fourier coefficient of F:

$$\widehat{F}(m) = \int_0^1 F(x)e^{-2\pi imx}dx. \qquad (11.68)$$

Because F is continuously differentiable for all x, by Theorem 11.3.8

$$F(x) = \sum_{m=-\infty}^{\infty} \widehat{F}(m)e^{2\pi imx}. \qquad (11.69)$$

In particular,

$$F(0) = \sum_{m=-\infty}^{\infty} \widehat{F}(m). \qquad (11.70)$$

As $F(0) = \sum_n f(n)$, from (11.70) it suffices to show $\widehat{F}(m) = \widehat{f}(m)$. While we use the same notation for the Fourier Transform and the Fourier coefficients, the Fourier Transform is for a function defined on $\mathbb{R}$ (such as f) and the Fourier coefficients are for periodic functions (such as F). We have

$$\widehat{F}(m) = \int_0^1 \sum_{n=-\infty}^{\infty} f(x+n)e^{-2i\pi mx}dx. \qquad (11.71)$$

By Fubini's Theorem (Theorem A.2.8), if we take absolute values of the integrand and either $\int \sum |*|$ or $\sum \int |*|$ exists, then we can interchange order of summation and integration. One must be careful, as it is not always possible to interchange orders. See Exercise 11.4.12.

Hence in the integral-sum for $\widehat{F}(m)$, we find

$$\sum_{n=-\infty}^{\infty} \int_0^1 |f(x+n)e^{-2i\pi mx}| = \sum_{n=-\infty}^{\infty} \int_0^1 |f(x+n)e^{-2i\pi mx}| dx$$

$$= \sum_{n=-\infty}^{\infty} \int_0^1 |f(x+n)e^{-2i\pi m(x+n)}e^{2i\pi mn}| dx$$

$$= \int_{-\infty}^{\infty} |f(x)e^{-2i\pi mx}| dx$$

$$= \int_{-\infty}^{\infty} |f(x)| dx < \infty, \qquad (11.72)$$

as f decays like $x^{-(1+\eta)}$. We can therefore interchange the order of integration and summation. Removing the absolute values above gives $\hat{f}(m)$, the Fourier Transform of f evaluated at m. We have shown $\widehat{F}(m) = \hat{f}(m)$, and substituting into (11.70) completes the proof of Theorem 11.4.6.   □

While the following exercise is not needed for the investigations above, it indicates how dangerous interchanging orders of summation and integration can be. The reader is advised to study and remember this example!

Exercise 11.4.12. *One cannot always interchange orders of integration. For simplicity, we give a sequence a_{mn} such that $\sum_m (\sum_n a_{m,n}) \neq \sum_n (\sum_m a_{m,n})$. For $m, n \geq 0$ let*

$$a_{m,n} = \begin{cases} 1 & \text{if } n = m \\ -1 & \text{if } n = m+1 \\ 0 & \text{otherwise.} \end{cases} \qquad (11.73)$$

Show that the two different orders of summation yield different answers.

11.4.3 Convolutions and Probability Theory

An important property of both Fourier series and Fourier transforms is that they behave nicely under convolution. We denote the convolution of two functions f and g by $h = f * g$, where

$$h(y) = \int_I f(x)g(y-x)dx = \int_I f(x-y)g(x)dx \qquad (11.74)$$

and $I = [0, 1]$ if f, g are periodic of period 1 and $I = \mathbb{R}$ if $f, g : \mathbb{R} \to \mathbb{C}$. We assume the reader is familiar with the Cauchy-Schwarz inequality (see Appendix A.6). Recall $\langle f, g \rangle = \int_I f(x)\overline{g}(x)dx$, with I as above.

Exercise 11.4.13. *Let f, g be continuous functions on $I = [0, 1]$ or $I = \mathbb{R}$. Show if $\langle f, f \rangle, \langle g, g \rangle < \infty$ then $h = f * g$ exists. Hint: Use the Cauchy-Schwarz inequality. Show further that $\widehat{h}(n) = \hat{f}(n)\hat{g}(n)$ if $I = [0, 1]$ or if $I = \mathbb{R}$. Thus the Fourier transform converts convolution to multiplication.*

We can now return to the proof of the Central Limit Theorem, Theorem 8.4.1. We assume the reader is familiar with the notations from Chapter 8. The following example is the starting point to the proof of the Central Limit Theorem. Let p be a probability density on $\mathbb{R}$ such that $\langle p, p \rangle < \infty$. Let X_1 and X_2 be two random variables chosen independently with probability density p. Thus the probability of $X_i \in [x, x + \Delta x]$ is $\int_x^{x+\Delta x} p(t)dt$, which is approximately $p(x)\Delta x$. The probability that $X_1 + X_2 \in [x, x + \Delta x]$ is just

$$\int_{x_1=-\infty}^{\infty} \int_{x_2=x-x_1}^{x+\Delta x - x_1} p(x_1)p(x_2)dx_2dx_1. \tag{11.75}$$

As $\Delta x \to 0$ we obtain the convolution of p with itself, and find

$$\text{Prob}(X_1 + X_2 \in [a, b]) = \int_a^b (p * p)(z)dz. \tag{11.76}$$

We must justify our use of the word "probability" in (11.76); namely, we must show $p * p$ is a probability density. Clearly $(p * p)(z) \geq 0$, and for any two f, g with $\langle f, f \rangle, \langle g, g \rangle < \infty$,

$$\int_{-\infty}^{\infty} (f * g)(x)dx = \int_{-\infty}^{\infty} \int_{-\infty}^{\infty} f(x - y)g(y)dydx$$

$$= \int_{-\infty}^{\infty} \int_{-\infty}^{\infty} f(x - y)g(y)dxdy$$

$$= \int_{-\infty}^{\infty} g(y) \left(\int_{-\infty}^{\infty} f(x - y)dx \right) dy$$

$$= \int_{-\infty}^{\infty} g(y) \left(\int_{-\infty}^{\infty} f(t)dt \right) dy. \tag{11.77}$$

If we take $f = g = p$, the last integrals are 1. We used general f and g as the above arguments yield

Lemma 11.4.14. *The convolution of two "nice" probability densities is a probability density.*

Exercise 11.4.15. *Prove (11.76).*

Exercise 11.4.16 (Important). *If for all $i = 1, 2, \ldots$ we have $\langle f_i, f_i \rangle < \infty$, prove for all i and j that $\langle f_i * f_j, f_i * f_j \rangle < \infty$. What about $f_1 * (f_2 * f_3)$ (and so on)? Prove $f_1 * (f_2 * f_3) = (f_1 * f_2) * f_3$. Therefore convolution is associative, and we may write $f_1 * \cdots * f_N$ for the convolution of N functions.*

Exercise 11.4.17. *Suppose $X_1, \ldots, X_N$ are i.i.d.r.v. from a probability distribution p on $\mathbb{R}$. Determine the probability that $X_1 + \cdots + X_N \in [a, b]$. What must be assumed about p for the integrals to converge?*

11.5 CENTRAL LIMIT THEOREM

As another application of Fourier analysis, we sketch the proof of the Central Limit Theorem (Theorem 8.4.1). We highlight the key steps, but we do not provide detailed justifications (which would require several standard lemmas about the Fourier transform; see for example [SS1]).

For simplicity, we consider the case where we have a probability density p on $\mathbb{R}$ that has mean zero, variance one, finite third moment and is of sufficiently rapid decay so that all convolution integrals that arise converge; see Exercise 15.1.6. Specifically, let p be an infinitely differentiable function satisfying

$$\int_{-\infty}^{\infty} xp(x)dx = 0, \quad \int_{-\infty}^{\infty} x^2 p(x)dx = 1, \quad \int_{-\infty}^{\infty} |x|^3 p(x)dx < \infty. \quad (11.78)$$

Assume $X_1, X_2, \ldots$ are independent identically distributed random variables (i.i.d.r.v.) drawn from p; thus, $\mathrm{Prob}(X_i \in [a, b]) = \int_a^b p(x)dx$. Define $S_N = \sum_{i=1}^{N} X_i$. Recall the standard Gaussian (mean zero, variance one) is $\frac{1}{\sqrt{2\pi}}e^{-x^2/2}$.

Theorem 11.5.1 (Central Limit Theorem). *Let X_i, S_N be as above and assume the third moment of each X_i is finite. Then $\frac{S_N}{\sqrt{N}}$ converges in probability to the standard Gaussian:*

$$\lim_{N\to\infty} \mathrm{Prob}\left(\frac{S_N}{\sqrt{N}} \in [a, b]\right) = \frac{1}{\sqrt{2\pi}} \int_a^b e^{-\frac{x^2}{2}} dx. \quad (11.79)$$

We sketch the proof. The Fourier transform of p is

$$\widehat{p}(y) = \int_{-\infty}^{\infty} p(x)e^{-2\pi i xy} dx. \quad (11.80)$$

Clearly, $|\widehat{p}(y)| \le \int_{-\infty}^{\infty} p(x)dx = 1$, and $\widehat{p}(0) = \int_{-\infty}^{\infty} p(x)dx = 1$.

Exercise 11.5.2. *One useful property of the Fourier transform is that the derivative of $\widehat{g}$ is the Fourier transform of $2\pi i x g(x)$; thus, differentiation (hard) is converted to multiplication (easy). Explicitly, show*

$$\widehat{g}'(y) = \int_{-\infty}^{\infty} 2\pi i x \cdot g(x)e^{-2\pi i xy} dx. \quad (11.81)$$

If g is a probability density, note $\widehat{g}'(0) = 2\pi i \mathbb{E}[x]$ and $\widehat{g}''(0) = -4\pi^2 \mathbb{E}[x^2]$.

The above exercise shows why it is, at least potentially, natural to use the Fourier transform to analyze probability distributions. The mean and variance (and the higher moments) are simple multiples of the derivatives of $\widehat{p}$ at zero. By Exercise 11.5.2, as p has mean zero and variance one, $\widehat{p}'(0) = 0, \widehat{p}''(0) = -4\pi^2$. We Taylor expand $\widehat{p}$ (we do not justify that such an expansion exists and converges; however, in most problems of interest this can be checked directly, and this is the reason we need technical conditions about the higher moments of p), and find near the origin that

$$\widehat{p}(y) = 1 + \frac{p''(0)}{2}y^2 + \cdots = 1 - 2\pi^2 y^2 + O(y^3). \quad (11.82)$$

Near the origin, the above shows $\widehat{p}$ looks like a concave down parabola.

By Exercises 11.4.13, 11.4.16 and 11.4.17, we have the following:

1. The probability that $X_1 + \cdots + X_N \in [a, b]$ is $\int_a^b (p * \cdots * p)(z)dz$.

2. The Fourier transform converts convolution to multiplication. If $FT[f](y)$ denotes the Fourier transform of f evaluated at y, then we have

$$FT[p * \cdots * p](y) = \widehat{p}(y) \cdots \widehat{p}(y). \qquad (11.83)$$

However, we do not want to study the distribution of $X_1 + \cdots + X_N = x$, but rather the distribution of $S_N = \frac{X_1 + \cdots + X_N}{\sqrt{N}} = x$.

Exercise 11.5.3. *If $B(x) = A(cx)$ for some fixed $c \neq 0$, show $\widehat{B}(y) = \frac{1}{c}\widehat{A}\left(\frac{y}{c}\right)$.*

Exercise 11.5.4. *Show that if the probability density of $X_1 + \cdots + X_N = x$ is $(p * \cdots * p)(x)$ (i.e., the distribution of the sum is given by $p * \cdots * p$), then the probability density of $\frac{X_1 + \cdots + X_N}{\sqrt{N}} = x$ is $(\sqrt{N}p * \cdots * \sqrt{N}p)(x\sqrt{N})$. By Exercise 11.5.3, show*

$$FT\left[(\sqrt{N}p * \cdots * \sqrt{N}p)(x\sqrt{N})\right](y) = \left[\widehat{p}\left(\frac{y}{\sqrt{N}}\right)\right]^N. \qquad (11.84)$$

The previous exercises allow us to determine the Fourier transform of the distribution of S_N. It is just $\left[\widehat{p}\left(\frac{y}{\sqrt{N}}\right)\right]^N$. We take the limit as $N \to \infty$ for **fixed** y. From (11.82), $\widehat{p}(y) = 1 - 2\pi^2 y^2 + O(y^3)$. Thus we have to study

$$\left[1 - \frac{2\pi^2 y^2}{N} + O\left(\frac{y^3}{N^{3/2}}\right)\right]^N. \qquad (11.85)$$

Exercise 11.5.5. *Show for any fixed y that*

$$\lim_{N \to \infty} \left[1 - \frac{2\pi^2 y^2}{N} + O\left(\frac{y^3}{N^{3/2}}\right)\right]^N = e^{-2\pi y^2}. \qquad (11.86)$$

Exercise 11.5.6. *Show that the Fourier transform of $e^{-2\pi y^2}$ at x is $\frac{1}{\sqrt{2\pi}}e^{-x^2/2}$. Hint: This problem requires contour integration from complex analysis.*

We would like to conclude that as the Fourier transform of the distribution of S_N converges to $e^{-2\pi y^2}$ and the Fourier transform of $e^{-2\pi y^2}$ is $\frac{1}{\sqrt{2\pi}}e^{-x^2/2}$, then the distribution of S_N equalling x converges to $\frac{1}{\sqrt{2\pi}}e^{-x^2/2}$. Justifying these statements requires some results from complex analysis. We refer the reader to [Fe] for details.

The key point in the proof is that we used Fourier Analysis to study the sum of independent identically distributed random variables, as Fourier transforms convert convolution to multiplication. The universality is due to the fact that *only* terms up to the second order contribute in the Taylor expansions. Explicitly, for "nice" p the distribution of S_N converges to the standard Gaussian, independent of the fine structure of p. The fact that p has mean zero and variance one is really just a normalization to study all probability distributions on a similar scale; see Exercise 15.1.6.

The higher order terms are important in determining the *rate* of convergence in the Central Limit Theorem (see [Fe] for details and [KonMi] for an application to Benford's Law).

Exercise 11.5.7. *Modify the proof to deal with the case of p having mean μ and variance σ^2.*

Exercise 11.5.8. *For reasonable assumptions on p, estimate the rate of convergence to the Gaussian.*

Exercise 11.5.9. *Let p_1, p_2 be two probability densities satisfying (11.78). Consider $S_N = X_1 + \cdots + X_N$, where for each i, X_1 is equally likely to be drawn randomly from p_1 or p_2. Show the Central Limit Theorem is still true in this case. What if we instead had a fixed, finite number of such distributions $p_1, \ldots, p_k$, and for each i we draw X_i from p_j with probability q_j (of course, $q_1 + \cdots + q_k = 1$)?*

11.6 ADVANCED TOPICS

Below we briefly highlight additional applications of Fourier Series and the Fourier Transform. The first problem complements Dirichlet's Theorem (Theorem 11.3.8) by describing what can go wrong at points where the function is discontinuous. We then give an example of a continuous function that is nowhere differentiable, followed by a proof that of all smooth curves with a given perimeter a circle encloses the most area. We then end with some applications to differential equations. Several of these problems require more mathematical pre-requisites. For more details, see for example [Be, SS1].

Exercise 11.6.1 (Gibbs Phenomenon). *Define a periodic with period 1 function by*

$$f(x) = \begin{cases} -1 & \text{if } -\frac{1}{2} \le x < 0 \\ 1 & \text{if } 0 \le x < \frac{1}{2}. \end{cases} \tag{11.87}$$

Prove that the Fourier coefficients are

$$\widehat{f}(n) = \begin{cases} 0 & \text{if } n \text{ is even} \\ \frac{4}{n\pi i} & \text{if } n \text{ is odd}. \end{cases} \tag{11.88}$$

Show that the Nth partial Fourier series $S_N(x)$ converges pointwise to $f(x)$ wherever f is continuous, but overshoots and undershoots for x near 0. Hint: Express the series expansion for $S_N(x)$ as a sum of sines. Note $\frac{\sin(2m\pi x)}{2m\pi} = \int_0^x \cos(2m\pi t)dt$. Express this as the real part of a geometric series of complex exponentials, and use the geometric series formula. This will lead to

$$S_{2N-1}(x) = 8\int_0^x \Re\left(\frac{1}{2i}\frac{e^{4n\pi it}-1}{\sin(2\pi t)}\right)dt = 4\int_0^x \frac{\sin(4n\pi t)}{\sin(2\pi t)}dt, \tag{11.89}$$

which is about 1.179 (or an overshoot of about 18%) when $x = \frac{1}{4n\pi}$. What can you say about the Fejér series $T_N(x)$ for x near 0?

Exercise 11.6.2 (Nowhere Differentiable Function). *Weierstrass constructed a continuous but nowhere differentiable function! We give a modified example and sketch the proof. Consider*

$$f(x) = \sum_{n=0}^{\infty} a^n \cos(2^n \cdot 2\pi x), \quad \frac{1}{2} < a < 1. \tag{11.90}$$

Show f is continuous but nowhere differentiable. Hint: *First show $|a| < 1$ implies f is continuous. Our claim on f follows from: if a periodic continuous function g is differentiable at x_0 and $\widehat{g}(n) = 0$ unless $n = \pm 2^m$, then there exists C such that for all n, $|\widehat{g}(n)| \leq Cn2^{-n}$. To see this, show it suffices to consider $x_0 = 0$ and $g(0) = 0$. Our assumptions imply that $(g, e_m) = 0$ if $2^{n-1} < m < 2^{n+1}$ and $m \neq 2^n$. We have $\widehat{g}(2^n) = (g, e_{2^n} F_{2^{n-1}}(x))$ where F_N is the Fejér kernel. The claim follows from bounding the integral $(g, e_{2^n} F_{2^{n-1}}(x))$. In fact, more is true: Baire showed that, in a certain sense, "most" continuous functions are nowhere differentiable! See, for example, [Fol].*

Exercise 11.6.3 (Isoperimetric Inequality). *Let $\gamma(t) = (x(t), y(t))$ be a smooth closed curve in the plane; we may assume it is parametrized by arc length and has length 1. Prove the enclosed area A is largest when $\gamma(t)$ is a circle.* Hint: *By Green's Theorem (Theorem A.2.9),*

$$\oint_\gamma x\,dy - y\,dx = 2\text{Area}(A). \tag{11.91}$$

The assumptions on $\gamma(t)$ imply $x(t), y(t)$ are periodic functions with Fourier series expansions and $\left(\frac{dx}{dt}\right)^2 + \left(\frac{dy}{dt}\right)^2 = 1$. Integrate this equality from $t = 0$ to $t = 1$ to obtain a relation among the Fourier coefficients of $\frac{dx}{dt}$ and $\frac{dx}{dt}$ (which are related to those of $x(t)$ and $y(t)$); (11.91) gives another relation among the Fourier coefficients. These relations imply $4\pi\text{Area}(A) \leq 1$ with strict inequality unless the Fourier coefficients vanish for $|n| > 1$. After some algebra, one finds this implies we have a strict inequality unless γ is a circle.

Exercise 11.6.4 (Applications to Differential Equations). *One reason for the introduction of Fourier series was to solve differential equations. Consider the vibrating string problem: a unit string with endpoints fixed is stretched into some initial position and then released; describe its motion as time passes. Let $u(x, t)$ denote the vertical displacement from the rest position x units from the left endpoint at time t. For all t we have $u(0, t) = u(1, t) = 0$ as the endpoints are fixed. Ignoring gravity and friction, for small displacements Newton's laws imply*

$$\frac{\partial^2 u(x, t)}{\partial x^2} = c^2 \frac{\partial^2 u(x, t)}{\partial t^2}, \tag{11.92}$$

where c depends on the tension and density of the string. Guessing a solution of the form

$$u(x, t) = \sum_{n=1}^{\infty} a_n(t) \sin(n\pi x), \tag{11.93}$$

solve for $a_n(t)$.

One can also study problems on $\mathbb{R}$ by using the Fourier Transform. Its use stems from the fact that it converts multiplication to differentiation, and vice versa: if $g(x) = f'(x)$ and $h(x) = xf(x)$, prove that $\widehat{g}(y) = iy\widehat{f}(y)$ and $\frac{d\widehat{f}(y)}{dy} = -i\widehat{h}(y)$. This and Fourier Inversion allow us to solve problems such as the heat equation

$$\frac{\partial u(x, t)}{\partial t} = \frac{\partial^2 u(x, t)}{\partial x^2}, \quad x \in \mathbb{R}, \ t > 0 \tag{11.94}$$

with initial conditions $u(x, 0) = f(x)$.

Chapter Twelve

$\{n^k\alpha\}$ and Poissonian Behavior

For $x \in \mathbb{R}$, let $\{x\}$ denote the fractional part of x and $[x]$ denote the greatest integer less than or equal to x, so $x = [x] + \{x\}$. Given a sequence z_n, let $x_n = \{z_n\}$. Thus $\{x_n\}_{n=1}^{\infty}$ is a sequence of numbers in $[0, 1]$, and we can investigate its properties. We ask three questions of increasing difficulty about $x_n = \{n^k\alpha\}$ (where k and α are fixed): is the sequence dense, is it equidistributed, and what are the spacing statistics between ordered adjacent spacings. In many cases it is conjectured that the answers for these sequences should be the same as what we would observe if we chose the x_n uniformly in $[0, 1]$. This is known as **Poissonian behavior**, and arises in many mathematical and physical systems. For example, numerical investigations of spacings between adjacent primes support the conjecture that the primes exhibit Poissonian behavior; see for example [Sch, Weir]; we encounter another example in Project 17.4.3. We assume the reader is familiar with probability theory and elementary Fourier analysis at the level of Chapters 8 and 11. See [Py] for a general survey on spacing results.

12.1 DEFINITIONS AND PROBLEMS

In this chapter we fix a positive integer k and an $\alpha \in \mathbb{R}$, and investigate the sequence $\{x_n\}_{n=1}^{\infty}$, where $x_n = \{n^k\alpha\}$. The first natural question (and the easiest) to ask about such sequences is whether or not the sequence gets arbitrarily close to every point:

Definition 12.1.1 (Dense). *A sequence* $\{x_n\}_{n=1}^{\infty}$, $x_n \in [0, 1]$, *is dense in* $[0, 1]$ *if for all* $x \in [0, 1]$ *there is a subsequence* $\{x_{n_k}\}_{k=1}^{\infty}$ *such that* $x_{n_k} \to x$.

Question 12.1.2. *Are the fractional parts* $\{n^k\alpha\}$ *dense in* $[0, 1]$? *How does the answer depend on* k *and* α?

We show in §12.2 that if $\alpha \notin \mathbb{Q}$ then the fractional parts $\{n^k\alpha\}$ are dense. We prove this only for $k = 1$, and sketch the arguments for larger k. For a dense sequence, the next natural question to ask concerns how often the sequence is near a given point:

Definition 12.1.3 (Characteristic Function).

$$\chi_{(a,b)}(x) = \begin{cases} 1 & \text{if } x \in [a, b] \\ 0 & \text{otherwise.} \end{cases} \tag{12.1}$$

We call $\chi_{(a,b)}$ the characteristic (or indicator) function of the interval $[a, b]$.

Definition 12.1.4 (Equidistributed). *A sequence $\{x_n\}_{n=-\infty}^{\infty}$, $x_n \in [0, 1]$, is equidistributed in $[0, 1]$ if*

$$\lim_{N \to \infty} \frac{\#\{n : |n| \le N, x_n \in [a, b]\}}{2N + 1} = \lim_{N \to \infty} \frac{\sum_{n=-N}^{N} \chi_{(a,b)}(x_n)}{2N + 1} = b - a$$

(12.2)

for all $(a, b) \subset [0, 1]$. A similar definition holds for $\{x_n\}_{n=0}^{\infty}$.

Question 12.1.5. *Assume the fractional parts $\{n^k \alpha\}$ are dense; are they equidistributed? How does the answer depend on k and α?*

We show in Theorem 12.3.2 that if $\alpha \notin \mathbb{Q}$ then the fractional parts $\{n\alpha\}$ are equidistributed. Equivalently, $n\alpha \bmod 1$ is equidistributed. For $k > 1$, $\{n^k \alpha\}$ is also equidistributed, and we sketch the proof in §12.3.2.

We have satisfactory answers to the first two questions. The last natural question is still very much open. Given a sequence of numbers $\{x_n\}_{n=1}^{N}$ in $[0, 1)$, we arrange the terms in increasing order, say

$$0 \le y_1(= x_{n_1}) \le y_2(= x_{n_2}) \cdots \le y_N(= x_{n_N}) < 1. \qquad (12.3)$$

The y_i's are called order statistics; see [DN] for more information.

Definition 12.1.6 (Wrapped unit interval). *We call $[0, 1)$, when all arithmetic operations are done mod 1, the wrapped unit interval. The distance between $x, y \in \mathbb{R}$ is given by*

$$||x - y|| = \min_{n \in \mathbb{Z}} |x - y - n|. \qquad (12.4)$$

For example, $||8.45 - .41|| = .04$, and $||.999 - .001|| = .002$.

Exercise 12.1.7. *Show $||x - y|| \le \frac{1}{2}$ and $||x - y|| \le |x - y|$. Is $||x - z|| \le ||x - y|| + ||y - z||$?*

If the N elements $y_n \in [0, 1)$ are distinct, every element has a unique element to the left and to the right, except for y_1 (no element to the left) and y_N (no element to the right). If we consider the wrapped unit interval, this technicality vanishes, as y_1 and y_N are neighbors. We have N spacings between neighbors: $||y_2 - y_1||, \dots, ||y_N - y_{N-1}||, ||y_1 - y_N||$.

Exercise 12.1.8. *Consider the spacings $||y_2 - y_1||, \dots, ||y_N - y_{N-1}||, ||y_1 - y_N||$ where all the y_n's are distinct. Show the average spacing is $\frac{1}{N}$.*

How are the y_n spaced? How likely is it that two adjacent y_n's are very close or far apart? Since the average spacing is about $\frac{1}{N}$, it becomes very unlikely that two adjacent terms are far apart on an absolute scale. Explicitly, as $N \to \infty$, it is unlikely that two adjacent y_n's are separated by $\frac{1}{2}$ or more. This is not the natural question; the natural questions concern the normalized spacings; for example:

1. How often are two y_n's less than half their average spacing apart?

2. How often are two y_n's more than twice their average spacing apart?

3. How often do two adjacent y_n's differ by $\frac{c}{N}$? How does this depend on c?

Question 12.1.9. *Let $x_n = \{n^k \alpha\}$ for $1 \leq n \leq N$. Let $\{y_n\}_{n=1}^N$ be the x_n's arranged in increasing order. What rules govern the spacings between the y_n's? How does this depend on k and α and N?*

We answer such questions below for some choices of sequences x_n, and describe interesting conjectural results for $\{n^k \alpha\}$, $\alpha \notin \mathbb{Q}$. Of course, it is possible to study sequences other than $n^k \alpha$, and in fact much is known about the fractional parts of $g(n)\alpha$ for certain $g(n)$.

For example, let $g(n)$ be a **lacunary** sequence of integers. This means that $\liminf \frac{g(n+1)}{g(n)} > 1$, which implies that there are large gaps between adjacent values. A typical example is to take $g(n) = b^n$ for any integer $b \geq 2$. In [RZ2] it is shown that for any lacunary sequence $g(n)$ and almost all α, the fractional parts of $g(n)\alpha$ are equidistributed and exhibit Poissonian behavior. We refer the reader to [RZ1, RZ2] for complete details; we have chosen to concentrate on $n^k \alpha$ for historical reasons as well as the number of open problems.

Exercise 12.1.10. *Show $g(n) = n!$ is lacunary but $g(n) = n^k$ is not. Is $g(n) = \binom{2n}{1}$ or $g(n) = \binom{2n}{n}$ lacunary?*

Exercise 12.1.11. *For any integer b, show α is normal base b if and only if the fractional parts $b^n \alpha$ are equidistributed (see §10.5.3 for a review of normal numbers). As it can be shown that for almost all α the fractional parts $b^n \alpha$ are equidistributed, this implies almost all numbers are normal base b.*

12.2 DENSENESS OF $\{n^k \alpha\}$

We tackle Question 12.1.2, namely, when is $\{n^k \alpha\}$ dense in $[0, 1]$?

Exercise 12.2.1. *For $\alpha \in \mathbb{Q}$, prove $\{n^k \alpha\}$ is never dense.*

Theorem 12.2.2 (Kronecker). *For $k = 1$ and $\alpha \notin \mathbb{Q}$, $x_n = \{n\alpha\}$ is dense in $[0, 1]$.*

The idea of the proof is as follows: by Dirichlet's Pigeon-Hole Principle, we can find a multiple of α that is "close" to zero. By taking sufficiently many copies of this multiple, we can move near any x.

Proof. We must show that, for any $x \in [0, 1]$, there is a subsequence $x_{n_j} \to x$. It suffices to show that for each $\epsilon > 0$ we can find an x_{n_j} such that $\|x_{n_j} - x\| < \epsilon$ (remember, .001 is close to .999). By Dirichlet's Pigeon-Hole Principle (see Theorem 5.5.4 or 7.9.4), as α is irrational there are *infinitely many relatively prime* p *and* q such that $|\alpha - \frac{p}{q}| < \frac{1}{q^2}$. Thus, $\|q\alpha\| \leq |q\alpha - p| < \frac{1}{q}$. Choose q large enough so that $\frac{1}{q} < \epsilon$.

Either $0 < q\alpha - p < \frac{1}{q}$ or $-\frac{1}{q} < q\alpha - p < 0$. As the two cases are handled similarly, we only consider the first. Given $x \in [0, 1]$, choose j so that $j(q\alpha - p)$ is within $\frac{1}{q}$ of x. This is always possible. Each time j increases by one, we increase

$j(q\alpha - p)$ by a fixed amount independent of j and at most $\frac{1}{q}$. As $||j(q\alpha - p)|| = ||jq\alpha||$, we have shown that $x_{n_j} = \{jq\alpha\}$ is within ϵ of x.   □

Exercise 12.2.3. *In the above argument, show that $q\alpha - p \neq 0$.*

Exercise 12.2.4. *Handle the second case, and show the above argument does generate a sequence $x_{n_j} \to x$ for any $x \in [0, 1]$.*

Remark 12.2.5 (Kronecker's Theorem). Kronecker's Theorem is more general than what is stated above. Let $\alpha_1, \alpha_2, \dots, \alpha_k \in \mathbb{R}$ have the property that if

$$c_0 1 + c_1 \alpha_1 + \cdots + c_n \alpha_k = 0 \qquad (12.5)$$

with all $c_i \in \mathbb{Q}$, then $c_i = 0$ for all i. The standard terminology is to say the α_i are linearly independent (see Definition B.1.7) over $\mathbb{Q}$. Then Kronecker's theorem asserts that the sequence of points $(v_n)_{n \in \mathbb{N}}$ given by

$$v_n = (\{n\alpha_1\}, \dots, \{n\alpha_k\}) \qquad (12.6)$$

is dense in $[0, 1]^k$; we proved the case $k = 1$.

Exercise 12.2.6. *Show it is not enough to assume just $c_1 \alpha_1 + \cdots + c_k \alpha_k = 0$ implies all the $c_i = 0$.*

Exercise 12.2.7. *Prove Kronecker's Theorem in full generality (a proof is given in [HW], Chapter XXIII).*

Theorem 12.2.8. *For k a positive integer greater than 1 and $\alpha \notin \mathbb{Q}$, $x_n = \{n^k \alpha\}$ is dense in $[0, 1]$.*

If we can show $\{n^k \alpha\}$ is equidistributed, then the above theorem is an immediate corollary. As we sketch the proof of equidistribution later (see §12.3), we confine ourselves to proving the $k = 2$ case for very special α, which highlight how the algebraic structure of α *can* enter proofs as well as the different notions of sizes of sets. There are other methods to prove this theorem which can handle larger classes of α; we have chosen the method below because of the techniques it introduces (especially in showing how the approximation exponent can enter).

Proof. Assume α has approximation exponent $4 + \eta$ for *some* $\eta > 0$ (see §5.5). Then there are infinitely many solutions to $|\alpha - \frac{p}{q}| < \frac{1}{q^4}$ with p, q relatively prime. We must show that, given x and $\epsilon > 0$, there is an x_{n_j} within ϵ of x. Choose p, q so large that $\frac{1}{q} < \frac{\epsilon}{100}$ and $|\alpha - \frac{p}{q}| < \frac{1}{q^4}$. Thus there exists a δ such that

$$\alpha - \frac{p}{q} = \frac{\delta}{q^4}. \qquad (12.7)$$

We assume $\delta > 0$ (the other case is handled similarly), and clearly $\delta < 1$. We have

$$q^2 m^2 \alpha - pqm^2 = (\delta m^2) \cdot \frac{1}{q^2}. \qquad (12.8)$$

We claim we may choose m so that $\delta m^2 \in [1, 4]$. If $\delta m_0^2 < 1$ then $m_0 < \frac{1}{\sqrt{\delta}}$ and

$$\delta(m_0 + 1)^2 - \delta m_0^2 = \delta(2m_0 + 1) < \frac{2\delta}{\sqrt{\delta}} + \delta < 3. \qquad (12.9)$$

As we move at most 3 units each time we increment m_0 by 1, we can find an m such that $\delta m^2 \in [1,4]$. Later we will see why it is necessary to have upper bound for δm^2.

Given $x \in [0,1]$, as $\delta m^2 \geq 1$ we have $\frac{x}{\delta m^2} \in [0,1]$. We can find n such that $\frac{n}{q}$ is within $\frac{1}{q}$ of $\sqrt{\frac{x}{\delta m^2}}$ and $\frac{n}{q} < \sqrt{\frac{x}{\delta m^2}}$. Further we may take $\frac{n}{q} \in [0,1]$ so $n \leq q$. For some $\theta \in [0,1]$ we have

$$\frac{n}{q} + \frac{\theta}{q} = \sqrt{\frac{x}{\delta m^2}}, \qquad \frac{n^2}{q^2} + \frac{2n\theta + \theta^2}{q^2} = \frac{x}{\delta m^2}. \tag{12.10}$$

The point is that $\delta m^2 \frac{n^2}{q^2}$ is very close to x. We want to write x as almost a square times α (minus an integer), as we are working with numbers of the form $\{b^2 \alpha\}$. Multiplying (12.8) by n^2 yields

$$q^2 m^2 n^2 \alpha - pqm^2 n^2 = (\delta m^2) \cdot \frac{n^2}{q^2}$$

$$= x - \frac{(\delta m^2) \cdot (2n\theta + \theta^2)}{q^2}. \tag{12.11}$$

As $\delta m^2 \leq 4$, $n \leq q$, $\theta \leq 1 \leq q$ and $\frac{1}{q} < \frac{\epsilon}{100}$,

$$\frac{(\delta m^2) \cdot (2n\theta + \theta^2)}{q^2} < \frac{12}{q} < \epsilon, \tag{12.12}$$

which shows that $(qmn)^2 \alpha$ minus an integer is within ϵ of x. $\square$

Remark 12.2.9. It was essential that δm^2 was bounded from above, as the final error used $\delta m^2 \leq 4$. Of course, we could replace 4 with any finite number.

Remark 12.2.10. We do not need α to have order of approximation $4 + \eta$; all we need is that there are infinitely many solutions to $|\alpha - \frac{p}{q}| < \frac{1}{q^4}$.

Remark 12.2.11. Similar to our investigations on the Cantor set (§A.5.2 and Remark A.5.10), we have a set of numbers that is "small" in the sense of measure, but "large" in the sense of cardinality. By Theorem 5.5.9 the measure of $\alpha \in [0,1]$ with infinitely many solutions to $|\alpha - \frac{p}{q}| < \frac{1}{q^4}$ is zero; on the other hand, every Liouville number (see §5.6.2) has infinitely many solutions to this equation, and in Exercise 5.6.5 we showed there are uncountably many Liouville numbers.

Exercise 12.2.12. *Prove that if α has order of approximation $4 + \eta$ for some $\eta > 0$, then there are infinitely many solutions to $|\alpha - \frac{p}{q}| < \frac{1}{q^4}$ with p, q relatively prime.*

Exercise 12.2.13. *To show $\{n^k \alpha\}$ is dense (using the same arguments), what must we assume about the order of approximation of α?*

Remark 12.2.14. We can weaken the assumptions on the approximability of α by rationals and the argument follows similarly. Let $f(q)$ be any monotone increasing function tending to infinity. If $|\alpha - \frac{p}{q}| < \frac{1}{q^2 f(q)}$ has infinitely many solutions, then $\{n^2 \alpha\}$ is dense. We argue as before to find m so that $\delta m^2 \in [1,4]$, and then find n so that $\frac{n^2}{f(q)}$ is close to $\frac{x}{\delta m^2}$. Does the set of such α still have measure zero (see Theorem 5.5.9)? Does the answer depend on f?

Exercise 12.2.15. *Prove the claim in Remark 12.2.14.*

Exercise 12.2.16. *Let* $x_n = \sin n$; *is this sequence dense in* $[-1, 1]$? *We are of course measuring angles in radians and not degrees.*

Exercise 12.2.17. *In Exercise 5.5.6 we showed* $\sum_{n=1}^{\infty} (\cos n)^n$ *diverge by showing* $|(\cos n)^n|$ *is close to 1 infinitely often. Is* $x_n = (\cos n)^n$ *dense in* $[-1, 1]$?

Exercise 12.2.18. *Let* α *be an irrational number, and as usual set* $x_n = \{n\alpha\}$. *Is the sequence* $\{x_{n^2}\}_{n \in \mathbb{N}}$ *dense? How about* $\{x_{p_n}\}$? *Here* $\{p_1, p_2, p_3, \dots\}$ *is the sequence of prime numbers.*

12.3 EQUIDISTRIBUTION OF $\{n^k \alpha\}$

We now turn to Question 12.1.5. We prove that $\{n^k \alpha\}$ is equidistributed for $k = 1$, and sketch the proof for $k > 1$. We will use the following functions in our proof:

Definition 12.3.1 $(e(x), e_m(x))$. *We set*

$$e(x) = e^{2\pi i x}, \quad e_m(x) = e^{2\pi i m x}. \tag{12.13}$$

Theorem 12.3.2 (Weyl). *Let* α *be an irrational number in* $[0, 1]$, *and let* k *be a fixed positive integer. Let* $x_n = \{n^k \alpha\}$. *Then* $\{x_n\}_{n=1}^{\infty}$ *is equidistributed.*

Contrast the above with

Exercise 12.3.3. *Let* $\alpha \in \mathbb{Q}$. *If* $x_n = \{n^k \alpha\}$ *prove* $\{x_n\}_{n=1}^{\infty}$ *is not equidistributed.*

We give complete details for the $k = 1$ case, and provide a sketch in the next subsection for general k. According to [HW], note on Chapter XXIII, the theorem for $k = 1$ was discovered independently by Bohl, Sierpiński and Weyl at about the same time. We follow Weyl's proof which, according to [HW], is undoubtedly "the best proof" of the theorem. There are other proofs, however; one can be found in [HW], Chapter XXIII. A more modern, useful reference is [KN], Chapter 1.

The following argument is very common in analysis. Often we want to prove a limit involving characteristic functions of intervals (and such functions are not continuous); here the characteristic function is the indicator function for the interval $[a, b]$. We first show that to each such characteristic function there is a continuous function "close" to it, reducing the original problem to a related one for continuous functions. We then show that, given any continuous function, there is a trigonometric polynomial that is "close" to the continuous function, reducing the problem to one involving trigonometric polynomials (see §11.3.4).

Why is it advantageous to recast the problem in terms of trigonometric polynomials rather than characteristic functions of intervals? Since $x_n = \{n\alpha\} = n\alpha - [n\alpha]$ and $e_m(x) = e_m(x + h)$ for every integer h,

$$e_m(x_n) = e^{2\pi i m \{n\alpha\}} = e^{2\pi i m n \alpha}. \tag{12.14}$$

The complications of looking at $n\alpha \bmod 1$ vanish, and it suffices to evaluate exponential functions at the simpler sequence $n\alpha$.

Remark 12.3.4. Here our "data" is discrete, namely we have a sequence of points $\{a_n\}$. See [MN] for another technique which is useful in investigating equidistribution of continuously distributed data (where one studies values $f(t)$ for $t \leq T$, which corresponds to taking terms a_n with $n \leq N$).

12.3.1 Weyl's Theorem when $k = 1$

Theorem 12.3.5 (Weyl Theorem, $k = 1$). *Let α be an irrational number in $[0, 1]$. Let $x_n = \{n\alpha\}$. Then $\{x_n\}_{n=1}^{\infty}$ is equidistributed.*

Proof. For $\chi_{a,b}$ as in Definition 12.1.3 we must show

$$\lim_{N \to \infty} \frac{1}{2N+1} \sum_{n=-N}^{N} \chi_{(a,b)}(x_n) = b - a. \tag{12.15}$$

We have

$$\frac{1}{2N+1} \sum_{n=-N}^{N} e_m(x_n) = \frac{1}{2N+1} \sum_{n=-N}^{N} e_m(n\alpha)$$

$$= \frac{1}{2N+1} \sum_{n=-N}^{N} (e^{2\pi i m \alpha})^n \tag{12.16}$$

$$= \begin{cases} 1 & \text{if } m = 0 \\ \frac{1}{2N+1} \frac{e_m(-N\alpha) - e_m((N+1)\alpha)}{1 - e_m(\alpha)} & \text{if } m > 0, \end{cases}$$

where the last follows from the geometric series formula. For a fixed irrational α, $|1 - e_m(\alpha)| > 0$; *this is where we use* $\alpha \notin \mathbb{Q}$. Therefore if $m \neq 0$,

$$\lim_{N \to \infty} \frac{1}{2N+1} \frac{e_m(-N\alpha) - e_m((N+1)\alpha)}{1 - e_m(\alpha)} = 0. \tag{12.17}$$

Let $P(x) = \sum_m a_m e_m(x)$ be a finite sum (i.e., $P(x)$ is a trigonometric polynomial). By possibly adding some zero coefficients, we can write $P(x)$ as a sum over a symmetric range: $P(x) = \sum_{m=-M}^{M} a_m e_m(x)$.

Exercise 12.3.6. *Show $\int_0^1 P(x)dx = a_0$.*

We have shown that for any finite trigonometric polynomial $P(x)$:

$$\lim_{N \to \infty} \frac{1}{2N+1} \sum_{n=-N}^{N} P(x_n) \longrightarrow a_0 = \int_0^1 P(x)dx. \tag{12.18}$$

Consider two continuous approximations to the characteristic function $\chi_{(a,b)}$:

1. A_{1j}: $A_{1j}(x) = 1$ if $a + \frac{1}{j} \leq x \leq b - \frac{1}{j}$, drops linearly to 0 at a and b, and is zero elsewhere (see Figure 12.1).

2. A_{2j}: $A_{1j}(x) = 1$ if $a \leq x \leq b$, drops linearly to 0 at $a - \frac{1}{j}$ and $b + \frac{1}{j}$, and is zero elsewhere.

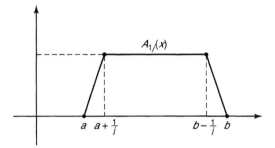

Figure 12.1 Plot of $A_{1j}(x)$

Note there are trivial modifications if $a = 0$ or $b = 1$. Clearly

$$A_{1j}(x) \le \chi_{(a,b)}(x) \le A_{2j}(x). \tag{12.19}$$

Therefore

$$\frac{1}{2N+1} \sum_{n=-N}^{N} A_{1j}(x_n) \le \frac{1}{2N+1} \sum_{n=-N}^{N} \chi_{(a,b)}(x_n) \le \frac{1}{2N+1} \sum_{n=-N}^{N} A_{2j}(x_n). \tag{12.20}$$

By Theorem 11.3.1, for each j, given $\epsilon > 0$ we can find symmetric trigonometric polynomials $P_{1j}(x)$ and $P_{2j}(x)$ such that $|P_{1j}(x) - A_{1j}(x)| < \epsilon$ and $|P_{2j}(x) - A_{2j}(x)| < \epsilon$. As A_{1j} and A_{2j} are continuous functions, we can replace

$$\frac{1}{2N+1} \sum_{n=-N}^{N} A_{ij}(x_n) \quad \text{with} \quad \frac{1}{2N+1} \sum_{n=-N}^{N} P_{ij}(x_n) \tag{12.21}$$

at a cost of at most ϵ. As $N \to \infty$,

$$\frac{1}{2N+1} \sum_{n=-N}^{N} P_{ij}(x_n) \longrightarrow \int_0^1 P_{ij}(x)dx. \tag{12.22}$$

But $\int_0^1 P_{1j}(x)dx = (b-a) - \frac{1}{j}$ and $\int_0^1 P_{2j}(x)dx = (b-a) + \frac{1}{j}$. Therefore, given j and ϵ, we can choose N large enough so that

$$(b-a) - \frac{1}{j} - \epsilon \le \frac{1}{2N+1} \sum_{n=-N}^{N} \chi_{(a,b)}(x_n) \le (b-a) + \frac{1}{j} + \epsilon. \tag{12.23}$$

Letting j tend to ∞ and ϵ tend to 0, we see $\frac{1}{2N+1} \sum_{n=-N}^{N} \chi_{(a,b)}(x_n) \to b - a$, completing the proof. $\qquad\square$

Exercise 12.3.7. *Rigorously do the necessary book-keeping to prove the previous theorem.*

Exercise 12.3.8. *Prove for k a positive integer, if $\alpha \in \mathbb{Q}$ then $\{n^k\alpha\}$ is periodic while if $\alpha \notin \mathbb{Q}$ then no two $\{n^k\alpha\}$ are equal.*

12.3.2 Weyl's Theorem for $k > 1$

We sketch the proof of the equidistribution of $\{n^k \alpha\}$. Recall that by Theorem 12.3.17 a sequence $\{\xi_n\}_{n=1}^{\infty}$ is equidistributed mod 1 if and only if for every integer $m \neq 0$,

$$\lim_{N \to \infty} \frac{1}{2N+1} \sum_{n=-N}^{N} e_m(z_n) = 0. \tag{12.24}$$

We follow the presentation in [Ca]. We need the following technical lemma:

Lemma 12.3.9. *Let* $u_1, u_2, \ldots, u_N \in \mathbb{C}$ *and let* $1 \leq H \leq N$. *Then*

$$H^2 \left| \sum_{1 \leq n \leq N} u_n \right|^2 \leq H(H+N-1) \sum_{1 \leq n \leq N} |u_n|^2$$
$$+ 2(H+N-1) \sum_{0 < h < H} (H-h) \sum_{1 \leq n \leq N-h} \overline{u}_n u_{n+h}. \tag{12.25}$$

For a proof, see [Ca], page 71. From the above lemma we can conclude

Corollary 12.3.10. *For a sequence* $\{z_n\}_{n=1}^{\infty}$, *suppose that for each* $h > 0$ *and integer* $m \neq 0$

$$\lim_{N \to \infty} \frac{1}{2N+1} \sum_{n=-N}^{N} e_m(z_{n+h} - z_n) = 0. \tag{12.26}$$

Then

$$\lim_{N \to \infty} \frac{1}{2N+1} \sum_{n=-N}^{N} e_m(z_n) = 0. \tag{12.27}$$

In particular if $\{z_{n+h} - z_n\}_{n=1}^{\infty}$ *is equidistributed for all* $h \in \mathbb{N}$ *then* $\{z_n\}_{n=1}^{\infty}$ *is equidistributed.*

Exercise 12.3.11. *Prove the corollary follows from the lemma.*

Exercise[h] **12.3.12.** *Prove* $\{n\alpha\}$ *is equidistributed using the results of this subsection.*

Exercise[h] **12.3.13.** *Prove* $\{n^2\alpha\}$ *is equidistributed.*

Exercise[h] **12.3.14.** *Prove* $\{n^k\alpha\}$ *is equidistributed for* $k \geq 1$.

Exercise 12.3.15. *Let* $\theta = 0.123456789101112\ldots$ *be the transcendental number from Exercise 10.5.6. By Theorem 12.3.5,* $\{n\theta\}$ *is equidistributed as* $\theta \notin \mathbb{Q}$. *Show directly that* $\{10^n\theta\}$ *is also equidistributed.*

Exercise 12.3.16. *Show that the sequence* $\{n!e\}$ *is not equidistributed. In fact, the only limit point of this sequence is 0.*

12.3.3 Weyl's Criterion

We generalize the methods of §12.3.1 to provide a useful test for equidistribution.

Theorem 12.3.17 (Weyl's Criterion). *A sequence $\{\xi_n\}$ is equidistributed (modulo 1) in $[0,1)$ if and only if*

$$\forall h \in \mathbb{Z} - \{0\}, \quad \lim_{N \to \infty} \frac{1}{N} \sum_{n=1}^{N} e_h(\xi_n) = 0. \tag{12.28}$$

Proof. We sketch the proof of the sufficiency of (12.28); for necessity see Exercise 12.3.21. Suppose that we are given that (12.28) is true, and we wish to prove that $\{\xi_n\}$, considered modulo 1, is equidistributed in $[0,1)$.

Claim 12.3.18. *Let f be any continuous periodic function with period 1. If (12.28) holds, then*

$$\lim_{N \to \infty} \frac{1}{N} \sum_{n=1}^{N} f(\xi_n) = \int_0^1 f(x)\,dx. \tag{12.29}$$

Proof. First note that (12.29) holds for $f(x) = e(hx)$ for all $h \in \mathbb{Z}$ (for $h = 0$ trivially and for $h \neq 0$ by assumption). Thus (12.29) holds for all trigonometric polynomials. Now suppose f is any continuous periodic function with period 1. As in §12.3.1, fix an $\epsilon > 0$ and choose a trigonometric polynomial P satisfying

$$\sup_{x \in \mathbb{R}} |f(x) - P(x)| < \epsilon. \tag{12.30}$$

Choose N large enough so that

$$\left| \frac{1}{N} \sum_{n=1}^{N} P(\xi_n) - \int_0^1 P(x)\,dx \right| < \epsilon. \tag{12.31}$$

By the triangle inequality (Exercise 12.3.19) we have

$$\left| \frac{1}{N} \sum_{n=1}^{N} f(\xi_n) - \int_0^1 f(x)\,dx \right| \leq \frac{1}{N} \sum_{n=1}^{N} |f(\xi_n) - P(\xi_n)|$$

$$+ \left| \frac{1}{N} \sum_{n=1}^{N} P(\xi_n) - \int_0^1 P(x)\,dx \right|$$

$$+ \int_0^1 |P(x) - f(x)|\,dx < 3\epsilon, \tag{12.32}$$

which proves the claim. □

The above argument uses several standard techniques: we *approximate our function with a trigonometric polynomial, add zero* and then do a *three epsilon proof*.

The rest of the proof of Theorem 12.3.17 mirrors the argument in §12.3.1. We choose functions $A_{1j}(x)$ and $A_{2k}(x)$ as before. We leave the rest of the proof to the reader. For a proof of the necessity of (12.28), see Exercise 12.3.21. □

Exercise 12.3.19 (Triangle Inequality). *Prove for $a, b \in \mathbb{C}$ that $|a + b| \leq |a| + |b|$.*

Exercise 12.3.20. *Complete the proof of Theorem 12.3.17.*

Exercise 12.3.21. *Prove the following statements:*

1. *If* $0 \leq x - y < \frac{\pi}{2}$, *then* $|e^{ix} - e^{iy}| \leq |x - y|$. Hint: *If* $0 < x < \frac{\pi}{2}$, *then* $\sin x \leq x$.

2. *Show that if the sequence* $\{\xi_n\}$ *is equidistributed then* (12.28) *holds for all* h. Hint: *One can proceed as follows:*

    (a) *Define* $I_j = [\frac{j}{L}, \frac{j+1}{L})$ *for each* $j \in \{0, 1, \ldots, L-1\}$ *and let* $A_{j,N} = \{n : 1 \leq n \leq N, \xi_n \in I_j\}$. *Show that there exists* N_0 *such that* $\forall N \geq N_0$,

    $$\frac{1}{L} - \frac{1}{L^2} \leq \frac{|A_{j,N}|}{N} \leq \frac{1}{L} + \frac{1}{L^2} \quad \forall j \in \{0, 1, \ldots, L-1\}.$$

    (b) *Using the triangle inequality and part (a) of this exercise show that*

    $$\left| \frac{1}{N} \sum_{n=1}^{N} e_h(\xi_n) \right| \leq \frac{2\pi h}{L} + \left| \frac{1}{N} \sum_{j=0}^{L-1} \sum_{n \in A_{j,N}} e_h\left(\frac{j}{L}\right) \right|. \tag{12.33}$$

    (c) *Show that for* $N \geq N_0$ *we have*

    $$\left| \frac{1}{N} \sum_{j=0}^{L-1} \sum_{n \in A_{j,N}} e_h\left(\frac{j}{L}\right) \right| \leq \frac{1}{L}. \tag{12.34}$$

    (d) *Use equation* (12.33) *and equation* (12.34) *to prove the desired statement.*

Exercise[h] 12.3.22. *Is the sequence* $\{\log_b n\}_{n=1}^{\infty}$ *equidistributed for* $b = e$ *or* $b = 10$?

12.4 SPACING PRELIMINARIES

For many α and k we have shown $\{n^k \alpha\}$ is equidistributed in $[0, 1)$. We now ask finer questions about the sequences. Consider the uniform distribution on $[0, 1)$. If we were to choose N points from this distribution, we expect for N large that the points should look equidistributed. This *suggests* the possibility that these two processes (looking at the fractional parts $\{n^k \alpha\}$ and choosing N points uniformly in $[0, 1)$) could share other behavior as well. We see this is the case for some triples (N, k, α) and violently false for others. The answer often depends on the structure of α and its approximations by continued fractions.

We call the behavior of points chosen uniformly in $[0, 1)$ **Poissonian behavior**, and conjecture that often $\{x_n\}_{n=1}^{N}$ ($x_n = \{n^k \alpha\}$) exhibits Poissonian behavior. One must be very careful, of course, in making such conjectures. The Prime Number Theorem implies that for primes of size x, the average spacing between primes is like $\frac{1}{\log x}$. One natural model for the distribution of primes (the **Cramér model**) is that a number near x is prime with probability $\frac{1}{\log x}$, and each number's "primality" is determined independent of its neighbors. For many statistics (for example the number of primes, number of twin primes, spacings between primes; see for example [Sch, Weir]) this gives answers that are reasonably fit by the actual data; however, in [MS] a statistic was investigated where the actual data disagrees with the Cramér model. It agrees with another model, Random Matrix Theory; see Chapter 15 for an introduction to Random Matrix Theory.

12.5 POINT MASSES AND INDUCED PROBABILITY MEASURES

In §11.2.2 we introduced the Dirac delta functional $\delta(x - a)$, which we interpreted as a unit point mass at a. Recall $\int f(x)\delta(x - a)dx = f(a)$. Given N point masses located at $x_1, x_2, \ldots, x_N$, we can form a probability measure

$$\mu_N(x)dx = \frac{1}{N} \sum_{n=1}^{N} \delta(x - x_n)dx. \tag{12.35}$$

We say probability measure and not probability density as we do not have a nice, continuous density – our density "function" involves the Dirac functional. Note $\int \mu_N(x)dx = 1$, and if $f(x)$ is continuous

$$\int f(x)\mu_N(x)dx = \frac{1}{N} \sum_{n=1}^{N} f(x_n). \tag{12.36}$$

Exercise 12.5.1. *Prove* (12.36) *for continuous* $f(x)$.

Note the right hand side of (12.36) looks like a Riemann sum. If the x_n's are equidistributed, we expect this to converge to $\int f(x)dx$. In general the x_n's might not be equidistributed; for example, they may be drawn from a fixed probability distribution $p(x)$. We *sketch* what should happen.

For simplicity, assume $x_n \in [0, 1]$ and assume for any interval $[a, b] \subset [0, 1]$, as $N \to \infty$ the fraction of x_n's ($1 \le n \le N$) in $[a, b]$ tends to $\int_a^b p(x)dx$ for some continuous function $p(x)$:

$$\lim_{N \to \infty} \frac{\#\{n : 1 \le n \le N \text{ and } x_n \in [a, b]\}}{N} = \int_a^b p(x)dx. \tag{12.37}$$

Assume f', p' are bounded. We want to compare $\int f(x)p(x)dx$ with $\int f(x)\mu_N(x)dx$. Then

$$\int_0^1 f(x)\mu_N(x)dx = \frac{1}{N} \sum_{n=1}^{N} f(x_n)$$

$$\approx \sum_{k=0}^{M-1} f\left(\frac{k}{M}\right) \frac{\#\{n : 1 \le n \le N \text{ and } x_n \in \left[\frac{k}{M}, \frac{k+1}{M}\right]\}}{N}$$

$$\approx \sum_{k=0}^{M-1} f\left(\frac{k}{M}\right) \int_{\frac{k}{M}}^{\frac{k+1}{M}} p(x)dx$$

$$\approx \int_0^1 f(x)p(x)dx. \tag{12.38}$$

Exercise 12.5.2 (Monte Carlo Method). *Make the above argument rigorous when f', p' are bounded on $[0, 1]$. Explicitly, show given an $\epsilon > 0$ there exists an N_0 such that for all $N > N_0$, $|\int f(x)\mu_N(x)dx - \int f(x)p(x)dx| < \epsilon$. This gives a numerical method to evaluate integrals, known as the* **Monte Carlo Method.** *See [Met, MU] for more details and references.*

Monte Carlo methods are extremely popular and useful for determining multidimensional integrals with nice boundary conditions. For example, say we wish to determine the n-dimensional volume of a set $S \subset \mathbb{R}^n$ that is contained in some finite n-dimensional box B. If for each point $x \in B$ we can easily determine whether or not $x \in S$, we generate points uniformly in B and see what fraction lie in S. For "nice" regions S the percent of points in B that are also in S will converge to $\frac{\text{vol}(S)}{\text{vol}(B)}$. Monte Carlo integration is one of the most frequently used methods to numerically approximate such solutions.

Definition 12.5.3 (Convergence to $p(x)$). *If the sequence of points x_n satisfies (12.37) for some nice function $p(x)$, we say the probability measures $\mu_N(x)dx$ converge to $p(x)dx$.*

12.6 NEIGHBOR SPACINGS

Let $\{\alpha_n\}_{n=1}^N$ be a collection of points in $[0, 1)$. We arrange them in increasing order:

$$0 \leq \alpha_{n_1} \leq \alpha_{n_2} \leq \cdots \leq \alpha_{n_N}. \tag{12.39}$$

For notational convenience, let $\beta_j = \alpha_{n_j}$. We investigate how the differences $\beta_{j+1} - \beta_j$ are distributed. Remember as we are working on the wrapped unit interval, the distance is $||\beta_{j+1} - \beta_j||$ (see §12.1). In looking at spacings between the β_j's, we have $N - 1$ pairs of neighbors:

$$(\beta_2, \beta_1), \ (\beta_3, \beta_2), \ \ldots, \ (\beta_N, \beta_{N-1}). \tag{12.40}$$

These pairs give rise to spacings $\beta_{j+1} - \beta_j \in [0, 1)$. We can also consider the pair (β_1, β_N). This gives rise to the spacing $\beta_1 - \beta_N \in [-1, 0)$; however, as we are studying this sequence mod 1, this is equivalent to $\beta_1 - \beta_N + 1 \in [0, 1)$.

Definition 12.6.1 (Neighbor Spacings). *Given a sequence of numbers α_n in $[0, 1)$, fix an N and arrange the numbers α_n $(n \leq N)$ in increasing order. Label the new sequence β_j; note the ordering will depend on N. It is convenient to periodically extend our sequence, letting $\beta_{N+j} = \beta_j$ for all j.*

1. The nearest neighbor spacings are the numbers $||\beta_{j+1} - \beta_j||$, $j = 1$ to N.

2. The mth neighbor spacings are the numbers $||\beta_{j+m} - \beta_j||$, $j = 1$ to N.

Exercise 12.6.2 (Surprising). *Let $\alpha = \sqrt{2}$, and let $\alpha_n = \{n\alpha\}$ or $\{n^2\alpha\}$. Calculate the nearest neighbor and the next nearest neighbor spacings in each case for $N = 10$ and $N = 20$. Note the different behavior for $n\alpha$ and $n^2\alpha$. See also Exercise 12.6.3.*

Exercise 12.6.3. *Prove that if* $\alpha \notin \mathbb{Q}$ *then for each* N *there are at most three different values for the nearest neighbor spacings. Bound the number of possible values for the* m^{th} *neighbor spacings. Is the bound sharp? For more on such spacings, see [Bl1, Bl2].*

Remark 12.6.4. While we have concentrated on the distribution of the differences of independent uniformly distributed random variables, there are many other interesting questions we could ask. For example, how are the leading digits of these differences distributed? See [MN] for an interesting application of Poisson summation to this problem, which shows the distribution of the leading digits of these differences almost obey Benford's Law.

12.7 POISSONIAN BEHAVIOR

Before investigating Question 12.1.9 (concerning the spacings of $\{n^k \alpha\}$), we consider a simpler case. Fix N and consider N independent random variables x_n. Each random variable is chosen from the uniform distribution on $[0, 1)$; thus the probability that $x_n \in [a, b]$ is $b - a$. Let $\{y_n\}_{n=1}^N$ be the x_n's arranged in increasing order. How do the neighbor spacings behave?

We first need to decide what is the correct scale to use for our investigations. As we have N objects on the wrapped unit interval, we have N nearest neighbor spacings and we expect the average spacing to be $\frac{1}{N}$.

Definition 12.7.1 (Unfolding). *Let* $z_n = N y_n$. *The numbers* $z_n = N y_n$ *have unit mean spacing. While we expect the average spacing between adjacent* y_n's *to be* $\frac{1}{N}$ *units for typical sequences, we expect the average spacing between adjacent* z_n's *to be* 1 *unit.*

The probability of observing a spacing as large as $\frac{1}{2}$ between adjacent y_n's becomes negligible as $N \to \infty$. What we should ask is what is the probability of observing a nearest neighbor spacing of adjacent y_n's that is *half* the average spacing. In terms of the z_n's, this corresponds to a spacing between adjacent z_n's of $\frac{1}{2}$ a unit.

For the rest of the section, x_n, y_n *and* z_n *are as defined above.*

12.7.1 Nearest Neighbor Spacings

Let the x_n's in increasing order be labeled $y_1 \le y_2 \le \cdots \le y_N$, $y_n = x_{n_j}$. As the average distance between adjacent y_n's is $\frac{1}{N}$, it is natural to look at nearest neighbor spacings of size $\frac{t}{N}$. In particular, we study the probability of observing a nearest neighbor spacing between $\frac{t}{N}$ and $\frac{t+\Delta t}{N}$. By symmetry, on the wrapped unit interval the expected nearest neighbor spacing is independent of j. Explicitly, we expect $y_{l+1} - y_l$ to have the same distribution as $y_{i+1} - y_i$. *Remember, the* y_n's *are ordered and the* x_n's *are unordered.* We assume familiarity with elementary properties of binomial coefficients (see §1.2.4 and §A.1.3).

We give a simple argument first which highlights the ideas, although it is slightly wrong (though only in lower order terms). As the x_n's are chosen independently, there are $\binom{N-1}{1}$ choices of subscript n such that x_n is the *first neighbor to the right* of x_1. This can also be seen by symmetry, as each x_n is equally likely to be the first to the *right* of x_1 (where, of course, .001 is just a little to the right of .999), and we have $N-1$ choices left for x_n. As x_n was chosen uniformly in $[0,1)$, the probability that $x_n \in \left[\frac{t}{N}, \frac{t+\Delta t}{N}\right]$ is $\frac{\Delta t}{N}$.

For the remaining $N-2$ of the x_n's, each must be further than $\frac{t+\Delta t}{N}$ to the right x_n. They must *all* lie in an interval (or possibly two intervals if we wrap around) of length $1 - \frac{t+\Delta t}{N}$. The probability that they all lie in this region is $\left(1 - \frac{t+\Delta t}{N}\right)^{N-2}$.

Thus, if $x_1 = y_l$, we want to calculate the probability that $\|y_{l+1} - y_l\| \in \left[\frac{t}{N}, \frac{t+\Delta t}{N}\right]$. This is

$$\text{Prob}\left(\|y_{l+1} - y_l\| \in \left[\frac{t}{N}, \frac{t+\Delta t}{N}\right]\right) = \binom{N-1}{1} \cdot \frac{\Delta t}{N} \cdot \left(1 - \frac{t+\Delta t}{N}\right)^{N-2}$$

$$= \left(1 - \frac{1}{N}\right) \cdot \left(1 - \frac{t+\Delta t}{N}\right)^{N-2} \Delta t. \tag{12.41}$$

For N enormous and Δt small,

$$\left(1 - \frac{1}{N}\right) \approx 1$$

$$\left(1 - \frac{t+\Delta t}{N}\right)^{N-2} \approx e^{-(t+\Delta t)} \approx e^{-t}(1 + O(\Delta t)); \tag{12.42}$$

See §5.4 for a review of the needed properties of e. Thus

$$\text{Prob}\left(\|y_{l+1} - y_l\| \in \left[\frac{t}{N}, \frac{t+\Delta t}{N}\right]\right) \xrightarrow[N \to \infty]{} e^{-t} \Delta t. \tag{12.43}$$

In terms of the z_n's, which have mean spacing 1, this yields

$$\text{Prob}\left(\|z_{l+1} - z_l\| \in [t, t+\Delta t]\right) \xrightarrow[N \to \infty]{} e^{-t} \Delta t. \tag{12.44}$$

Remark 12.7.2. The above argument is infinitesimally wrong. Once we have located y_{l+1}, the remaining x_n's do not need to be more than $\frac{t+\Delta t}{N}$ units to the right of $x_1 = y_l$; they only need to be further to the right than y_{l+1}. As the incremental gain in probabilities for the locations of the remaining x_n's is of order Δt, these contributions will not influence the large N, small Δt limits, and we may safely ignore these effects.

To rigorously derive the limiting behavior of the nearest neighbor spacings using the above arguments, one would integrate over x_m ranging from $\frac{t}{N}$ to $\frac{t+\Delta t}{N}$, and the remaining events x_n would be in the a segment of length $1 - x_m$. As

$$\left|(1 - x_m) - \left(1 - \frac{t+\Delta t}{N}\right)\right| \leq \frac{\Delta t}{N}, \tag{12.45}$$

this will lead to corrections of higher order in Δt, hence negligible.

We can rigorously avoid this complication by instead considering the following:

1. Calculate the probability that *all* $N - 1$ of the other x_n's are at least $\frac{t}{N}$ units to the right of x_1. This is

$$p_N(t) = \left(1 - \frac{t}{N}\right)^{N-1}, \qquad \lim_{N \to \infty} p_N(t) = e^{-t}. \qquad (12.46)$$

2. Calculate the probability that *all* $N - 1$ of the other x_n's are at least $\frac{t + \Delta t}{N}$ units to the right of x_1. This is

$$p_N(t + \Delta t) = \left(1 - \frac{t + \Delta t}{N}\right)^{N-1}, \qquad \lim_{N \to \infty} p_N(t + \Delta t) = e^{-(t+\Delta t)}. \qquad (12.47)$$

3. The probability that no x_n's are within $\frac{t}{N}$ units to the right of x_1 but at least one x_n is between $\frac{t}{N}$ and $\frac{t + \Delta t}{N}$ units to the right is $p_N(t) - p_N(t + \Delta t)$, and

$$\lim_{N \to \infty} (p_N(t) - p_N(t + \Delta t)) = e^{-t} - e^{-(t + \Delta t)}$$

$$= e^{-t}\left(1 - e^{-\Delta t}\right)$$

$$= e^{-t}\left(1 - 1 + \Delta t + O\left((\Delta t)^2\right)\right)$$

$$= e^{-t}\Delta t + O\left((\Delta t)^2\right). \qquad (12.48)$$

We have shown

Theorem 12.7.3 (Nearest Neighbor Spacings). *As $N \to \infty$ and then $\Delta t \to 0$,*

$$\mathrm{Prob}\left(\|y_{l+1} - y_l\| \in \left[\frac{t}{N}, \frac{t + \Delta t}{N}\right]\right) \longrightarrow e^{-t}\Delta t$$

$$\mathrm{Prob}\left(\|z_{l+1} - z_l\| \in [t, t + \Delta t]\right) \longrightarrow e^{-t}\Delta t. \qquad (12.49)$$

Exercise[h] 12.7.4. *Generalize the above arguments to analyze the nearest neighbor spacings when $x_1, \ldots, x_N$ are independently drawn from a "nice" probability distribution. Explicitly, for any $\delta \in (0, 1)$, show that as $N \to \infty$ the normalized nearest neighbor spacings for any N^δ consecutive x_i's tend to independent standard exponentials. See Appendix A of [MN] for a proof.*

12.7.2 m^{th} Neighbor Spacings

Similarly, one can easily analyze the distribution of the mth nearest neighbor spacings (for fixed m) when each x_n is chosen independently from the uniform distribution on $[0, 1)$.

1. We first calculate the probability that exactly $m - 1$ of the other x_n's are at most $\frac{t}{N}$ units to the right of x_1, and the remaining $(N - 1) - (m - 1)$ of the x_n's are at least $\frac{t}{N}$ units to the right of x_1. There are $\binom{N-1}{m-1}$ ways to choose $m - 1$ of the x_n's to be at most $\frac{t}{N}$ units to the right of x_1. As m is fixed, for N large, m is significantly smaller than N. The probability is

$$p_N(t) = \binom{N-1}{m-1}\left(\frac{t}{N}\right)^{m-1}\left(1 - \frac{t}{N}\right)^{(N-1)-(m-1)}$$

$$\lim_{N \to \infty} p_N(t) = \frac{t^{m-1}}{(m-1)!}e^{-t}, \qquad (12.50)$$

because

$$\lim_{N \to \infty} \binom{N-1}{m-1} \frac{1}{N!} = \frac{1}{(m-1)!}. \tag{12.51}$$

It is not surprising that the limiting arguments break down for m of comparable size to N; for example, if $m = N$, the Nth nearest neighbor spacing is always 1.

2. We calculate the probability that exactly $m - 1$ of the other x_n's are at most $\frac{t}{N}$ units to the right of x_1, and the remaining $(N-1) - (m-1)$ of the x_n's are at least $\frac{t+\Delta t}{N}$ units to the right of x_1. Similar to the above, this gives

$$p_N(t + \Delta t) = \binom{N-1}{m-1} \left(\frac{t}{N}\right)^{m-1} \left(1 - \frac{t + \Delta t}{N}\right)^{(N-1)-(m-1)}$$

$$\lim_{N \to \infty} p_N(t + \Delta t) = \frac{t^{m-1}}{(m-1)!} e^{-(t+\Delta t)}. \tag{12.52}$$

3. The probability that exactly $m - 1$ of the x_n's are within $\frac{t}{N}$ units to the right of x_1 and at least one x_n is between $\frac{t}{N}$ and $\frac{t+\Delta t}{N}$ units to the right is $p_N(t) - p_N(t + \Delta t)$, and

$$\lim_{N \to \infty} (p_N(t) - p_N(t + \Delta t)) = \frac{t^{m-1}}{(m-1)!} e^{-t} - \frac{t^{m-1}}{(m-1)!} e^{-(t+\Delta t)}$$

$$= \frac{t^{m-1}}{(m-1)!} e^{-t} \left(1 - e^{-\Delta t}\right)$$

$$= \frac{t^{m-1}}{(m-1)!} e^{-t} \Delta t + O((\Delta t)^2). \tag{12.53}$$

Note that when $m = 1$, we recover the nearest neighbor spacings. We have shown

Theorem 12.7.5 (Neighbor Spacings). *For fixed m, as $N \to \infty$ and then $\Delta t \to 0$,*

$$\text{Prob}\left(\|y_{l+m} - y_l\| \in \left[\frac{t}{N}, \frac{t + \Delta t}{N}\right]\right) \longrightarrow \frac{t^{m-1}}{(m-1)!} e^{-t} \Delta t$$

$$\text{Prob}\left(\|z_{l+1} - z_l\| \in [t, t + \Delta t]\right) \longrightarrow \frac{t^{m-1}}{(m-1)!} e^{-t} \Delta t. \tag{12.54}$$

Exercise 12.7.6. *Keep track of the lower order terms in N and Δt in (12.50) and (12.52) to prove Theorem 12.7.5.*

Exercise 12.7.7 (Median). *We give another application of order statistics. Let $X_1, \ldots, X_n$ be independent random variables chosen from a probability distribution p. The median of p is defined as the number $\tilde{\mu}$ such that $\int_{-\infty}^{\tilde{\mu}} p(x)dx = \frac{1}{2}$. If $n = 2m + 1$, let the random variable $\tilde{X}$ be the median of $X_1, \ldots, X_n$. Show the density function of $\tilde{X}$ is*

$$f(\tilde{x}) = \frac{(2m+1)!}{m!m!} \left[\int_{-\infty}^{\tilde{x}} p(x)dx\right]^m \cdot p(\tilde{x}) \cdot \left[\int_{\tilde{x}}^{\infty} p(x)dx\right]^m. \tag{12.55}$$

It can be shown that if p is continuous and $p(\widetilde{\mu}) \neq 0$ then for large m, $f(\widetilde{x})$ is approximately a normal distribution with mean $\widetilde{\mu}$ and variance $\frac{1}{8p(\widetilde{\mu})^2 m}$. For distributions that are symmetric about the mean μ, the mean μ equals the median $\widetilde{\mu}$ and the above is another way to estimate the mean. This is useful in situations where the distribution has infinite variance and the Central Limit Theorem is not applicable, for example, if p is a translated Cauchy distribution:

$$p(x) \;=\; \frac{1}{\pi} \frac{1}{1 + (x - \widetilde{\mu})^2}. \tag{12.56}$$

In fact, while the above is symmetric about μ, the expected value does not exist yet the Median Theorem can still be used to estimate the parameter $\widetilde{\mu}$. Another advantage of the median over the mean is that changing one or two data points will not change the median much but could greatly change the mean. Thus, in situations where there is a good chance of recording the wrong value of an observation, the median is often a better statistic to study.

Exercise$^{(hr)}$ 12.7.8. *Prove the Median Theorem from the previous problem if p is sufficiently nice.*

12.7.3 Induced Probability Measures

We reinterpret our results in terms of probability measures.

Theorem 12.7.9. *Consider N independent random variables x_n chosen from the uniform distribution on the wrapped unit interval $[0, 1)$. For fixed N, arrange the x_n's in increasing order, labeled $y_1 \leq y_2 \leq \cdots \leq y_N$. Fix m and form the probability measure from the mth nearest neighbor spacings. Then as $N \to \infty$*

$$\mu_{N,m;y}(t)dt \;=\; \frac{1}{N} \sum_{n=1}^{N} \delta\left(t - N(y_n - y_{n-m})\right) dt \;\longrightarrow\; \frac{t^{m-1}}{(m-1)!} e^{-t} dt. \tag{12.57}$$

Equivalently, using $z_n = N y_n$ gives

$$\mu_{N,m;z}(t)dt \;=\; \frac{1}{N} \sum_{n=1}^{N} \delta\left(t - (z_n - z_{n-m})\right) dt \;\longrightarrow\; \frac{t^{m-1}}{(m-1)!} e^{-t} dt. \tag{12.58}$$

Definition 12.7.10 (Poissonian Behavior). *Let $\{x_n\}_{n=1}^{\infty}$ be a sequence of points in $[0, 1)$. For each N, let $\{y_n\}_{n=1}^{N}$ be $x_1, \ldots, x_N$ arranged in increasing order. We say x_n has Poissonian behavior if in the limit as $N \to \infty$, for each m the induced probability measure $\mu_{N,m;y}(t)dt$ converges to $\frac{t^{m-1}}{(m-1)!} e^{-t} dt$.*

Exercise 12.7.11. *Let $\alpha \in \mathbb{Q}$ and define $\alpha_n = \{n^m \alpha\}$ for some positive integer m. Show the sequence of points α_n does not have Poissonian Behavior.*

Exercise$^{(h)}$ 12.7.12. *Let $\alpha \notin \mathbb{Q}$ and define $\alpha_n = \{n\alpha\}$. Show the sequence of points α_n does not have Poissonian behavior.*

12.8 NEIGHBOR SPACINGS OF $\{n^k \alpha\}$

We now come to Question 12.1.9. Note there are three pieces of data: k, α and N.

Conjecture 12.8.1. *For any integer $k \geq 2$, for most irrational α (in the sense of measure), if $\alpha_n = \{n^k \alpha\}$ then the sequence $\{\alpha_n\}_{n=1}^N$ is Poissonian as $N \to \infty$.*

Many results towards this conjecture are proved in n [RS2, RSZ]. We merely state some of them; see [RS2, RSZ] for complete statements and proofs. We concentrate on the case $k = 2$.

Let $\alpha \notin \mathbb{Q}$ be such that there are infinitely many rationals b_j and q_j (with q_j prime) such that $|\alpha - \frac{b_j}{q_j}| < \frac{1}{q_j^3}$. Then there is a sequence $\{N_j\}_{j=1}^\infty$ such that $\{n^2 \alpha\}$ exhibits Poissonian behavior along this subsequence. Explicitly, $\{\{n^2 \alpha\}\}_{n=1}^N$ is Poissonian as $N \to \infty$ *if* we restrict N to the sequence $\{N_j\}$. If α has slightly better approximations by rationals, one can construct a subsequence where the behavior is *not* Poissonian. For $k = 2$ and α that are well approximated by rationals, we show there is a sequence of N_j's such that the triples (k, α, N_j) do not exhibit Poissonian behavior (see Theorem 12.8.5).

Thus, depending on how N tends to infinity, wildly different behavior can be seen. This leads to many open problems, which we summarize at the end of the chapter. For more information, see [Li, Mi1].

12.8.1 Poissonian Behavior

Consider first the case when $\alpha \in \mathbb{Q}$. If $\alpha = \frac{p}{q}$, the fractional parts $\{n^2 \frac{p}{q}\}$ are periodic with period N; contrast this with $\alpha \notin \mathbb{Q}$, where the fractional parts are distinct. *Assume q is prime.* Davenport [Da3, Da4] investigated the neighbor spacings of $\{\{n^2 \frac{p}{q}\}\}_{n=1}^N$ for $N = q$ as both tend to infinity; [RSZ] investigate the above for smaller N. As the sequence is periodic with period q, it suffices to study $1 \leq N \leq q$. For simplicity, let $p = 1$. If $N \leq \sqrt{q}$, the sequence is already in increasing order, and the adjacent spacings are given by $\frac{2n+1}{q}$. Clearly, for such small N, one does not expect the sequence $\{n^2 \alpha\}$ to behave like N points chosen uniformly in $[0, 1)$. Once N is larger than $\sqrt{q}$, say $N > q^{\frac{1}{2}+\epsilon}$, we have q^ϵ blocks of $\sqrt{q}$ terms. These will now be intermingled in $[0, 1)$, and there is now the possibility of seeing "randomness." Similarly, we must avoid N being too close to q, as if $N = q$ then the normalized spacings are integers. In [RSZ], evidence towards Poissonian behavior is proved for $N \in [q^{\frac{1}{2}+\epsilon}, \frac{q}{\log q}]$.

For $\alpha \notin \mathbb{Q}$, [RSZ] note the presence of large square factors in the convergents to α (the q_n in the continued fraction expansion) prevent Poissonian behavior. While it is possible to construct numbers that always have large square factors in the convergents (if q_{m-1} and q_m are squares, letting $a_{m+1} = q_m + 2\sqrt{q_{m-1}}$ yields $q_{m+1} = a_{m+1}q_m + q_{m-1} = (q_m + \sqrt{q_{m-1}})^2$), for most α in the sense of measure this will not occur, and one obtains subsequences N_j where the behavior is Poissonian. In particular, their techniques can prove

Theorem 12.8.2 (RSZ). *Let $\alpha \notin \mathbb{Q}$ have infinitely many rational approximations satisfying $|\alpha - \frac{p_j}{q_j}| < \frac{1}{q_j^3}$ and $\lim_{j\to\infty} \frac{\log \widetilde{q}_j}{\log q_j} = 1$, where $\widetilde{q}_j$ is the square-free part of*

q_j. *Then there is a subsequence $N_j \to \infty$ with $\frac{\log N_j}{\log q_j} \to 1$ such that the fractional parts $\{n^2\alpha\}$ are Poissonian along this subsequence.*

For data on rates of convergence to Poissonian behavior, see [Li, Mi1].

Exercise 12.8.3. *Construct an $\alpha \notin \mathbb{Q}$ such that the denominators in the convergents are perfect cubes (or, more generally, perfect k^{th} powers).*

One useful statistic to investigate whether or not a sequence exhibits Poissonian behavior is the n-**level correlation**. We content ourselves with discussing the **2-level** or **pair correlation** (see [RS2, RSZ, RZ1] for more details). Given a sequence of increasing numbers x_n in $[0, 1)$, let

$$R_2([-s, s], x_n, N) = \frac{1}{N}\#\{1 \le j \ne k \le N : ||x_j - x_k|| \le s/N\}. \quad (12.59)$$

As the expected difference between such x's is $\frac{1}{N}$, this statistic measures the distribution of spacings between x's on the order of the average spacing. The adjacent neighbor spacing distribution looks at $||x_{j+1} - x_j||$; here we allow x's to be between x_k and x_j. One can show that knowing *all* the n-level correlations allows one to determine all the neighbor spacings; see for example [Meh2].

Exercise 12.8.4 (Poissonian Behavior). *For $j \le N$ let the θ_j be chosen independently and uniformly in $[0, 1)$. Prove as $N \to \infty$ that $R_2([-s, s], \theta_j, N) \to 2s$.*

In [RS2] it is shown that the 2-level correlation for almost all α (in the sense of measure) converges to $2s$, providing support for the conjectured Poissonian behavior. The proof proceeds in three steps. First, they show the mean over α is $2s$. This is not enough to conclude that most α have 2-level correlations close to $2s$. For example, consider a sequence that half the time is 0 and the other half of the time is 2. The mean is 1, but clearly no elements are close to the mean. If, however, we knew the variance were small, *then* we could conclude that most elements are close to the mean. The second step in [RS2] is to show the variance is small, and then standard bounds on exponential sums and basic probability arguments allow them to pass to almost all α satisfy the claim.

Numerous systems exhibit Poissonian behavior. We see another example in Project 17.4.3, where we study the spacings between eigenvalues of real symmetric Toeplitz matrices (these matrices are constant along diagonals). See [GK, KR] for additional systems.

12.8.2 Non-Poissonian Behavior

While we give the details of a sequence that is non-Poissonian, remember that the *typical* α is expected to exhibit Poissonian behavior. See also [RZ1, RZ2] for Poissonian behavior of other (lacunary) sequences.

Theorem 12.8.5 ([RSZ]). *Let $\alpha \notin \mathbb{Q}$ be such that $\left|\alpha - \frac{p_j}{q_j}\right| < \frac{a_j}{q_j^3}$ holds infinitely often, with $a_j \to 0$. Then there exist integers $N_j \to \infty$ such that $\mu_{N_j,1}(t)$ does not converge to $e^{-t}dt$. In particular, along this subsequence the behavior is non-Poissonian.*

The idea of the proof is as follows: since α is well approximated by rationals, when we write the $\{n^2\alpha\}$ in increasing order and look at the normalized nearest neighbor differences (the z_n's in Theorem 12.7.3), the differences will be arbitrarily close to integers. This cannot be Poissonian behavior, as the limiting distribution there is $e^{-t}dt$, which is clearly not concentrated near integers.

As $a_j \to 0$, eventually $a_j < \frac{1}{10}$ for all j large. Let $N_j = q_j$, where $\frac{p_j}{q_j}$ is a good rational approximation to α:

$$\left| \alpha - \frac{p_j}{q_j} \right| < \frac{a_j}{q_j^3}. \tag{12.60}$$

We look at $\alpha_n = \{n^2\alpha\}$, $1 \le n \le N_j = q_j$. Let the β_n's be the α_n's arranged in increasing order, and let the γ_n's be the numbers $\{n^2\frac{p_j}{q_j}\}$ arranged in increasing order:

$$\beta_1 \le \beta_2 \le \cdots \le \beta_{N_j}$$
$$\gamma_1 \le \gamma_2 \le \cdots \le \gamma_{N_j}. \tag{12.61}$$

Lemma 12.8.6. *If $\beta_l = \alpha_n = \{n^2\alpha\}$, then $\gamma_l = \{n^2\frac{p_j}{q_j}\}$. Thus the same permutation orders both the α_n's and the γ_n's.*

Proof. Multiplying both sides of (12.60) by $n^2 \le q_j^2$ yields for j large

$$\left| n^2\alpha - n^2\frac{p_j}{q_j} \right| < n^2\frac{a_j}{q_j^3} \le \frac{a_j}{q_j} < \frac{1}{10q_j}, \tag{12.62}$$

and $n^2\alpha$ and $n^2\frac{p_j}{q_j}$ differ by at most $\frac{1}{10q_j}$. Therefore

$$\left\| \{n^2\alpha\} - \left\{ n^2\frac{p_j}{q_j} \right\} \right\| < \frac{1}{10q_j}. \tag{12.63}$$

As p_j and q_j are relatively prime, and all numbers $\{m^2\frac{p_j}{q_j}\}$ have denominator at most q_j, two such numbers cannot be closer than $\frac{1}{q_j}$ unless they are equal. For example, if q_j is a perfect square, $m_1 = \sqrt{q_j}$ and $m_2 = 2\sqrt{q_j}$ give the same number; the presence of large square factors of q_j has important consequences (see [RSZ]).

Thus $\{n^2\frac{p_j}{q_j}\}$ is the closest (or tied for being the closest) of the $\{m^2\frac{p_j}{q_j}\}$ to $\{n^2\alpha\}$. This implies that if $\beta_l = \{n^2\alpha\}$, then $\gamma_l = \{n^2\frac{p_j}{q_j}\}$, completing the proof. $\square$

Exercise 12.8.7. *Prove two of the $\{m^2\frac{p_j}{q_j}\}$ are either equal, or at least $\frac{1}{q_j}$ apart.*

Exercise 12.8.8. *Assume $\|a-b\|, \|c-d\| < \frac{1}{10}$. Show*

$$\|(a-b) - (c-d)\| < \|a-b\| + \|c-d\|. \tag{12.64}$$

We now prove Theorem 12.8.5: We have shown

$$\|\beta_l - \gamma_l\| < \frac{a_j}{q_j}. \tag{12.65}$$

As $N_j = q_j$,

$$\|N_j(\beta_l - \gamma_l)\| < a_j, \tag{12.66}$$

and the same result holds with l replaced by $l - 1$. By Exercise 12.8.8,

$$\|N_j(\beta_l - \gamma_l) - N_j(\beta_{l-1} - \gamma_{l-1})\| < 2a_j. \tag{12.67}$$

Rearranging gives

$$\|N_j(\beta_l - \beta_{l-1}) - N_j(\gamma_l - \gamma_{l-1})\| < 2a_j. \tag{12.68}$$

As $a_j \to 0$, this implies that the difference between $\|N_j(\beta_l - \beta_{l-1})\|$ and $\|N_j(\gamma_l - \gamma_{l-1})\|$ tends to zero.

The above distance calculations were done modulo 1. The actual differences will differ by an integer. Therefore

$$\mu_{N_j,1;\alpha}(t)dt = \frac{1}{N_j} \sum_{l=1}^{N_j} \delta\left(t - N_j(\beta_l - \beta_{l-1})\right) \tag{12.69}$$

and

$$\mu_{N_j,1;\frac{p_j}{q_j}}(t)dt = \frac{1}{N_j} \sum_{l=1}^{N_j} \delta\left(t - N_j(\gamma_l - \gamma_{l-1})\right) \tag{12.70}$$

are in some sense extremely close to one another: each point mass from the difference between adjacent β_l's is within $k + a_j$ units of a point mass from the difference between adjacent γ_l's for some integer k, and a_j tends to zero. Note, however, that if $\gamma_l = \{n^2 \frac{p_j}{q_j}\}$, then

$$N_j \cdot \gamma_l = q_j \left\{ n^2 \frac{p_j}{q_j} \right\} \in \mathbb{N}. \tag{12.71}$$

Thus the induced probability measure $\mu_{N_j,1;\frac{p_j}{q_j}}(t)dt$ formed from the γ_l's is supported on the integers! It is therefore impossible for $\mu_{N_j,1;\frac{p_j}{q_j}}(t)dt$ to converge to $e^{-t}dt$.

As $\mu_{N_j,1;\alpha}(t)dt$, modulo some possible integer shifts, is arbitrarily close to $\mu_{N_j,1;\frac{p_j}{q_j}}(t)dt$, the sequence $\{n^2\alpha\}$ is *not* Poissonian along the subsequence of Ns given by N_j, where $N_j = q_j$, q_j is a denominator in a good rational approximation to α.

12.9 RESEARCH PROJECTS

We have shown that the fractional parts $\{\{n^k\alpha\}\}_{n=1}^{N}$ (for $\alpha \notin \mathbb{Q}$) are dense and equidistributed for all positive integers k as $N \to \infty$; however, the finer question of neighbor spacings depends greatly on α and how N tends to infinity.

Research Project 12.9.1. Much is known about the fractional parts $\{n^2\alpha\}$ if $\alpha = \frac{p}{q}$ with q prime. What if q is the product of two primes (of comparable size or of wildly different size)? Is the same behavior true for $\{n^k \frac{p}{q}\}$?

Research Project 12.9.2. For certain α there is a subsequence N_j such that, along this subsequence, the behavior of $\{\{n^2\alpha\}\}_{n\leq N_j}$ is non-Poissonian. What about the behavior of $\{n^2\alpha\}$ away from such bad N_j's? Is the behavior Poissonian between N_j and N_{j+1}, and if so, how far must one move? See [Mi1] for some observations and results.

Research Project 12.9.3. For $\alpha \notin \mathbb{Q}$, the fractional parts $x_n = \{n\alpha\}$ are equidistributed but not Poissonian. In particular, for each N there are only three possible values for the nearest neighbor differences (the values depend on N and α); for the mth neighbor spacings, there are also only finitely many possible differences. How are the differences distributed among the possible values? Is each value equally likely? Does the answer depend on α, and if so, on what properties of α? For more details, see [Bl1, Bl2, Mar].

Research Project 12.9.4. Generalize the results of [RSZ] for the fractional parts $\{n^k\alpha\}$, $k \geq 3$. It is likely that the obstruction to non-Poissonian behavior will be the presence of convergents to α with q_n a large k^{th} power.

Research Project 12.9.5 (Primes). Similar to investigating the Poissonian behavior of $\{n^k\alpha\}$, one can study primes. For $\{n^k\alpha\}$ we looked at $n \leq N$. The average spacing was $\frac{1}{N}$ and we then took the limit as $N \to \infty$. For primes we can investigate $p \in [x, x + y]$. If we choose x large and y large on an absolute scale but small relative to x, then there will be a lot of primes and the average spacing will be approximately constant. For example, the Prime Number Theorem states the number of primes at most x is about $\frac{x}{\log x}$. If $x = 10^{14}$ and $y = 10^6$, then there should be a lot of primes in this interval and the average spacing should be fairly constant (about $\log 10^{14}$). Are the spacings between primes Poissonian? What about primes in arithmetic progression, or twin primes or generalized twin primes or prime tuples (see the introduction to Chapter 14 for more details about such primes). See [Sch, Weir] for numerical investigations along these lines. The Circle Method (Chapters 13 and 14) provides estimates as to the number of such primes. Another sequence to investigate are Carmichael numbers; see Project 1.4.25.

PART 4
The Circle Method

Chapter Thirteen

Introduction to the Circle Method

The Circle Method is a beautiful idea for studying many problems in additive number theory. It originated in investigations by Hardy and Ramanujan [HR] on the partition function $P(n)$, the number of ways to write n as a sum of positive integers. Since then it has been used to study problems in additive number theory ranging from writing numbers as sums of primes or k^{th} powers (for fixed k) to trying to count how many twin primes there are less than x. In some sense primes were created with multiplicative properties; by unique factorization every number can be written as a product of prime powers. While asking additive questions about primes might at first seem unnatural, we hope to show below that these are reasonable questions. Further the techniques we develop to study sums of primes are also useful in counting the number of twin primes, and we have seen throughout the book that one of the most natural questions to ask about a set is its density.

We start our study of the Circle Method in §13.1 by reviewing the basic properties of $P(n)$ via generating functions, and then we explore the generating functions of a variety of problems. In §13.2 we state the main ideas of the Circle Method, and then in §13.3 we sketch its applications to writing numbers as a sum of primes. We then give a detailed analysis of the Circle Method and handle most of the technicalities for a special application (Germain primes) in Chapter 14.

Our goal is to describe the key features of the Circle Method *without* handling all of the technical complications that arise in its use; we refer the reader to the excellent books [EE, Na] for complete details (the more advanced reader is encouraged to read Chapters 19 and 20 of [IK]), as well as Chapter 15 of [Sc] for more on partitions and generating functions. We highlight the main ideas and needed ingredients for its application, and describe the types of problems it either solves or predicts the answer.

13.1 ORIGINS

In this section we study various problems of additive number theory that motivated the development of the Circle Method. For example, consider the problem of writing n as a sum of s perfect k-powers. If $k = 1$ we have seen a combinatorial solution (see §1.2.4 and Lemma 1.2.26): the number of ways of writing n as a sum of s non-negative integers is $\binom{n+s-1}{s-1}$. Unfortunately, this argument does not generalize to higher k (it is easy to partition a set into s subsets; it is not clear how to partition it into s subsets where the number of elements in each subset is a perfect square). There is another method, an analytical approach, which solves the $k = 1$

case and can be generalized.

For $|x| < 1$ define the **generating function**

$$f(x) = \sum_{m=0}^{\infty} x^m = \frac{1}{1-x}. \qquad (13.1)$$

Let $r_{1,s}(n)$ denote the number of solutions to $m_1 + \cdots + m_s = n$ where each m_i is a non-negative integer. We claim

$$f(x)^s = \left(\sum_{m_1=0}^{\infty} x^{m_1} \right) \cdots \left(\sum_{m_s=0}^{\infty} x^{m_s} \right) = \sum_{n=0}^{\infty} r_{1,s}(n) x^n. \qquad (13.2)$$

This follows by expanding the product in (13.1). We have terms such as $x^{m_1} \cdots x^{m_s}$, which is $x^{m_1 + \cdots + m_s} = x^n$ for some n. Assuming everything converges, when we expand the product we obtain x^n many times, once for each choice of $m_1, \ldots, m_s$ that adds to n. Thus the coefficient of x^n in the expansion is $r_{1,s}(m)$. On the other hand, we have

$$f(x)^s = \left(\frac{1}{1-x} \right)^s = \frac{1}{(s-1)!} \frac{d^{s-1}}{dx^{s-1}} \frac{1}{1-x}. \qquad (13.3)$$

Substituting the geometric series expansion for $\frac{1}{1-x}$ gives

$$f(x)^s = \frac{1}{(s-1)!} \frac{d^{s-1}}{dx^{s-1}} \sum_{n=0}^{\infty} x^n = \sum_{n=0}^{\infty} \binom{n+s-1}{s-1} x^n, \qquad (13.4)$$

which yields $r_{1,s}(n) = \binom{n+s-1}{s-1}$. It is this second method of proof that we generalize. Below we describe a variety of problems and show how to find their generating functions. In most cases, exact formulas such as (13.3) are unavailable; we develop sufficient machinery to analyze the generating functions in a more general setting.

Exercise 13.1.1. *Justify the arguments above. Show all series converge, and prove* (13.3) *and* (13.4).

13.1.1 Partitions

We describe several problems where we can identify the generating functions. For $n \in \mathbb{N}$, $P(n)$ is the **partition function**, the number of ways of writing n as a sum of positive integers where we do not distinguish re-orderings. For example, if $n = 4$ then

$$\begin{aligned}
4 &= 4 \\
  &= 3 + 1 \\
  &= 2 + 2 \\
  &= 2 + 1 + 1 \\
  &= 1 + 1 + 1 + 1, \qquad (13.5)
\end{aligned}$$

and $P(4) = 5$. Note we do not count both $3+1$ and $1+3$. If we add the requirement that no two parts can be equal, there are only two ways to partition 4: 4 and $3 + 1$.

Proposition 13.1.2 (Euler). *We have as an identity of formal power series*

$$F(x) = \frac{1}{(1-x)(1-x^2)(1-x^3)\cdots} = 1 + \sum_{n=1}^{\infty} P(n)x^n. \tag{13.6}$$

An **identity of formal power series** means that, without worrying about convergence, the two sides have the same coefficients of x^n for all n. For this example, if we use the geometric series expansion on each $\frac{1}{(1-x^k)}$ and then collect terms with the same power of x, we would have the series on the right; however, we do not know that the series on the right is finite for any x.

Exercise[h] 13.1.3. *Prove the above proposition. Do the product or series converge for any $x > 0$?*

$F(x)$ is called the generating function of the partition function. If $f(n)$ is an arithmetic function (see Chapter 2), we can associate a generating function to f through a power series:

$$F_f(x) = 1 + \sum_{n=1}^{\infty} f(n)x^n. \tag{13.7}$$

Exercise 13.1.4. *1. Fix $m \in \mathbb{N}$. For each n, let $p_m(n)$ be the number of partitions of n into numbers less than or equal to the given number m. Show that*

$$\frac{1}{(1-x)(1-x^2)\cdots(1-x^m)} = 1 + \sum_{n=1}^{\infty} p_m(n)x^n. \tag{13.8}$$

Does the series converge for any $x > 0$?

2. Show that

$$(1+x)(1+x^2)(1+x^3)\cdots = 1 + \sum_{n=1}^{\infty} q(n)x^n. \tag{13.9}$$

where $q(n)$ is the number of partitions of n into non-equal parts. Does this series converge for any $x > 0$? For any $x < 0$?

3. Give similar interpretations for

$$\frac{1}{(1-x)(1-x^3)(1-x^5)\cdots} \tag{13.10}$$

and

$$(1+x^2)(1+x^4)(1+x^6)\cdots. \tag{13.11}$$

Do these products converge for any $x > 0$?

One can use generating functions to obtain interesting properties of the partition functions:

Proposition 13.1.5. *Let $n \in \mathbb{N}$. The number of partitions of n into unequal parts is equal to the number of partitions of n into odd numbers.*

Exercise[(h)] **13.1.6.** *Prove Proposition 13.1.5.*

For more examples of this nature, see Chapter XIX of [HW].

So far, we have studied power series expansions where the coefficients are related to the function we want to study. We now consider more quantitative questions. Is there a simple formula for $P(n)$? How rapidly does $P(n)$ grow as $n \to \infty$? Using the Circle Method, Hardy and Ramanujan showed that

$$P(n) \sim \frac{e^{\pi\sqrt{2n/3}}}{4n\sqrt{3}}. \tag{13.12}$$

We prove similar results for other additive problems.

13.1.2 Waring's Problem

It is useful to think of the partition problem in §13.1.1 as the study of the number of ways that a given number n can be written as a sum

$$\sum_i n_i^k \tag{13.13}$$

for $k = 1$, with the number of terms ranging from n (when each $n_i = 1$) to 1 (when $n_1 = n$). See also §1.2.4. We can now formulate the following question:

Question 13.1.7. *Let $k \in \mathbb{N}$. Let $P_k(n)$ be the number of ways that n can be written as the sum of perfect k^{th} powers. Can one calculate $P_k(n)$, or estimate its size for large n?*

It is clear that for all n, $P_n(n)$ is non-zero as n can be written as the sum of n ones. There is a striking difference between this case and the problem of $P(n)$ in §13.1.1. The difference is that if n is a natural number and $m < n$, then one can easily write a partition of n into m numbers. For higher powers, however, this is false; in fact not true even for $k = 2$. For example, 3 cannot be written as the sum of two squares. Hence we ask the following questions:

Question 13.1.8 (Waring's Problem). *Let $k \in \mathbb{N}$. What is the smallest number s such that every natural number can be written as the sum of at most s perfect k^{th} powers? Does such an s exist? If s exists, how does s depend on k?*

These questions can easily be translated to questions involving appropriate generating functions, as we now explain. For Question 13.1.7, we easily see that

$$1 + \sum_{n=1}^{\infty} P_k(n)x^n = \frac{1}{(1 - x^{1^k})(1 - x^{2^k})(1 - x^{3^k})\cdots}; \tag{13.14}$$

however, this expansion is only useful if we can use it to calculate the $P_k(n)$'s. For Question 13.1.8, consider the auxiliary function

$$Q_k(x) = \sum_{n=0}^{\infty} x^{n^k}. \tag{13.15}$$

As an identity of formal power series, we have

$$Q_k(x)^s = 1 + \sum_{n=1}^{\infty} a(n; k, s)x^n, \tag{13.16}$$

where $a(n; k, s)$ is the number of ways to write n as a sum of exactly s perfect k^{th} powers.

Exercise 13.1.9. *Prove* (13.16).

Remark 13.1.10 (Important). So far, all we have done is to use generating functions to find an equivalent formulation for the original problem. We must find a good way to determine $a(n; k, s)$.

If we could show that given a k there exists an s such that for all n, $a(n; k, s) \neq 0$, then we would have proved every number is the sum of s perfect k^{th} powers. The smallest such s, if it exists, is usually denoted by $g(k)$. In 1770 Waring stated without proof that every natural number is the sum of at most nine positive perfect cubes, also the sum of at most 19 perfect fourth powers, and so on. It was already known that every number is a sum of at most four squares. It is usually assumed that Waring believed that for all k, $g(k)$ exists. Hilbert [Hil] proved Waring's conjecture in 1909, though his method yielded poor bounds for the true value of $g(k)$.

Exercise 13.1.11. *Show that no number of the form $4k + 3$ can be the sum of two squares. Show that no number of the form $4^a(8k + 7)$ is the sum of three squares. This exercise shows that we cannot write all sufficiently large numbers as the sum of three squares.*

Exercise 13.1.12. *Let $n_k = 2^k \left[\left(\frac{3}{2}\right)^k \right] - 1$. How many perfect k^{th} powers are needed to represent n_k as a sum of kth powers? Conclude that $g(k) \geq 2^k + \left[\left(\frac{3}{2}\right)^k \right] - 2$. This gives $g(2) \geq 4$, $g(3) \geq 9$, $g(4) \geq 19, \ldots$.*

Exercise 13.1.13. *Using density arguments, we can often prove certain problems have no solutions. Show there are not enough perfect squares to write all large numbers as the sum of two squares. Use this method to determine a lower bound for how many perfect k^{th} powers are needed for each k.*

Let us concentrate on $g(2) = 4$. As we now know that infinitely many numbers cannot be the sum of three squares, we need to show that every natural number can be written as a sum of four squares. There are many proofs of this important fact, the first of which is due to Lagrange (though it is believed that Diophantus was familiar with the theorem). One proof uses geometric considerations based on Minkowski's theorem (see [StTa]). We refer the reader to Chapter XX of [HW] for three interesting proofs of the theorem, as well as [Na]. We are particularly interested in the proof in §20.11 and §20.12 of [HW] which uses generating functions. We set

$$\theta(x) = \sum_{m=-\infty}^{\infty} x^{m^2}. \tag{13.17}$$

If $r(n)$ is defined by

$$\theta(x)^4 = 1 + \sum_{n=1}^{\infty} r(n)x^n, \tag{13.18}$$

then $r(n)$ is equal to the number of representations of n as the sum of four squares:

$$r(n) = \#\{(m_1, m_2, m_3, m_4) : m_i \in \mathbb{Z}, \ n = m_1^2 + m_2^2 + m_3^2 + m_4^2\}. \tag{13.19}$$

Here the m_i's are integers, and different permutations of the m_i's are counted as distinct. One can show (Theorem 386 of [HW], for example) that

$$\theta(x)^4 = 1 + 8 \sum_{n=1}^{\infty} c_n x^n \tag{13.20}$$

where

$$c_n = \sum_{m|n, 4 \nmid m} m. \tag{13.21}$$

Thus, $r(n) = 8c(n)$. As $c_n > 0$, this implies that every integer is the sum of four squares.

The above is a common feature of such proofs: we show the *existence* of at least one solution by showing there are many. We proved n is the sum of four squares by actually finding out how many different ways n is the sum of four squares. In our investigations of other problems, we will argue similarly.

13.1.3 Goldbach's Conjecture

Previously we considered the question of determining the smallest number of perfect k^{th} powers needed to represent all natural numbers as a sum of kth powers. One can consider the analogous question for other sets of numbers. Namely, given a set A, is there a number s_A such that every natural number can be written as a sum of at most s_A elements of A? A set of natural arithmetic interest is the set P of all prime numbers. Goldbach, in a letter to Euler (June 7, 1742), conjectured that every integer is the sum of three primes. Euler reformulated this conjecture to every even integer is the sum of two primes.

Exercise 13.1.14. *Prove that if every integer is the sum of at most three primes, then every even number must be the sum of at most two primes. Conversely show if every even integer is the sum of at most two primes then every integer is the sum of at most three primes.*

To date, Goldbach's conjecture has been verified for all even numbers up to $2 \cdot 10^{16}$ (see [Ol]). There are deep unconditional results in the direction of Goldbach's conjecture:

1. Shnirel'man proved $s_P < \infty$. The proof is based on an ingenious density argument (see [Na], Chapter 7).

2. Estermann [Est1] proved that almost every even number is the sum of two primes.

3. Vinogradov showed every sufficiently large enough odd number is the sum of three primes. We discuss the proof of Vinogradov's theorem later. Vinogradov proved his theorem in [Vin1, Vin2], where he reformulated the Circle Method from the language of complex analysis to that of Fourier series. [ChWa] has shown that sufficiently large may be taken to be $e^{e^{11.503}}$.

4. Chen proved every even number is the sum of a prime and a number that is the product of at most two primes. Chen's theorem is based on a sieve argument (see [Na], Chapters 9 and 10).

In the next section we describe the key ideas of the Circle Method. This will allow us to approximate quantities such as $a(n; k, s)$ (see (13.16)). We return to generating function approaches to Goldbach's conjecture in §13.3.

13.2 THE CIRCLE METHOD

We explain the key features of the Circle Method. We reinterpret some of the problems discussed in §13.1 in this new language.

13.2.1 Problems

The Circle Method was devised to deal with additive problems of the following nature:

Problem 13.2.1. *Given some subset $A \subset \mathbb{N}$ and a positive integer s, what natural numbers can be written as a sum of s elements of A, and in how many ways? Explicitly, what is*

$$\{a_1 + \cdots + a_s : a_i \in A\} \cap \mathbb{N}. \tag{13.22}$$

More generally, one has

Problem 13.2.2. *Fix a collection of subsets $A_1, \ldots, A_s \subset \mathbb{N}$ and study*

$$\{a_1 + \cdots + a_s : a_i \in A_i\} \cap \mathbb{N}. \tag{13.23}$$

We give several problems where the Circle Method is useful. We confine ourselves to two common choices for A. The first choice is P, the set of primes: $P = \{2, 3, 5, 7, 11, \ldots\}$. We denote elements of P by p. The second choice is K, the set of kth powers of non-negative integers; $K = \{0, 1, 2^k, 3^k, 4^k, \ldots\}$. We denote elements of K by n^k.

1. Consider $A = P$ and $s = 2$. Thus we are investigating

$$\{p_1 + p_2 : p_i \text{ prime}\} \cap \mathbb{N}. \tag{13.24}$$

   This is Goldbach's conjecture for even numbers.

2. Again let $A = P$ but now let $s = 3$. Thus we are investigating

$$\{p_1 + p_2 + p_3 : p_i \text{ prime}\} \cap \mathbb{N}. \tag{13.25}$$

   Vinogradov's theorem asserts that every large enough odd number is included in the intersection.

3. Let $A = K$ and fix a positive integer s. We are studying

$$\{n_1^k + \cdots + n_s^k : n_i \in \mathbb{N}\} \cap \mathbb{N}. \tag{13.26}$$

This is Waring's problem.

4. Let $-P = \{-2, -3, -5, \ldots\}$. If we consider $P - P$, we have

$$\{p_1 - p_2\} \cap \mathbb{N}. \tag{13.27}$$

This tells us which numbers are the differences between primes. A related question is to study how many pairs (p_1, p_2) satisfy $p_1 - p_2 = n$. If we take $n = 2$, p_1 and p_2 are called **twin primes**.

In the following paragraphs we sketch the main ideas of the Circle Method, first without worrying about convergence issues, then highlighting where the technicalities lie. In Chapter 14 we work through all but one of these technicalities for a specific problem; the remaining technicality for this problem has resisted analysis to this day. We have chosen to describe an open problem rather than a problem where all the difficulties can be handled for several reasons. The first is that to handle these technicalities for one of the standard problems would take us too far afield, and there are several excellent expositions for those desiring complete details (see [Da2, EE, Na]). Further, there are numerous open problems where the Circle Method provides powerful heuristics that agree with experimental investigations; after working through the problem in Chapter 14 the reader will have no trouble deriving such estimates for additional problems.

13.2.2 Setup

Let us consider Problem 13.2.1. As before, we consider a generating function

$$F_A(x) = \sum_{a \in A} x^a. \tag{13.28}$$

Next, we write

$$F_A(x)^s = \sum_{n=1}^{\infty} r(n; s, A)x^n. \tag{13.29}$$

Exercise 13.2.3. *Prove $r(n; s, A)$ is the number of ways of writing n as a sum of s elements of A.*

An equivalent formulation of Problem 13.2.1 is the following:

Problem 13.2.4. *Determine $r(n; s, A)$.*

In order to extract individual coefficients from a power series we have the following standard fact from complex analysis:

Proposition 13.2.5. *1. Let γ be the unit circle oriented counter-clockwise. Then*

$$\frac{1}{2\pi i} \int_\gamma z^n \, dz = \begin{cases} 1 & \text{if } n = -1, \\ 0 & \text{otherwise.} \end{cases} \tag{13.30}$$

2. *Let $P(z) = \sum_{k=0}^{\infty} a_k z^k$ be a power series with radius of convergence larger than one. Then*

$$\frac{1}{2\pi i} \int_\gamma P(z) z^{-n-1} \, dz = a_n. \tag{13.31}$$

See §3.2.2 for a sketch of the proof, or any book on complex analysis (for example, [Al, La6, SS2]). Consequently, ignoring convergence problems yields

$$r(n; s, A) = \frac{1}{2\pi i} \int_\gamma F_A(z)^s z^{-n-1} \, dz. \tag{13.32}$$

Definition 13.2.6 $(e(x))$. *We set*

$$e(x) = e^{2\pi i x}. \tag{13.33}$$

Exercise 13.2.7. *Let $m, n \in \mathbb{Z}$. Prove*

$$\int_0^1 e(nx)e(-mx)dx = \begin{cases} 1 & \text{if } n = m, \\ 0 & \text{otherwise.} \end{cases} \tag{13.34}$$

An alternative, but equivalent, formulation is to consider a different generating function for A:

$$f_A(x) = \sum_{a \in A} e(ax). \tag{13.35}$$

Again, ignoring convergence problems,

$$\int_0^1 f_A(x)^s e(-nx)dx = r(n; s, A). \tag{13.36}$$

If we can evaluate the above integral, not only will we know which n can be written as the sum of s elements of A, but we will know in how many ways.

Exercise 13.2.8. *Using Exercise 13.2.7, prove* (13.36).

13.2.3 Convergence Issues

The additive problem considered in Problem 13.2.1 is interesting only if A is infinite; otherwise, we can just enumerate $a_1 + \cdots + a_s$ in a finite number of steps. If A is infinite, the defining sum for the generating function $f_A(x)$ need not converge. For each N, define

$$A_N = \{a \in A : a \leq N\} = A \cap \{0, 1, \ldots, N\}. \tag{13.37}$$

Note the A_N's are an increasing sequence of subsets

$$A_N \subset A_{N+1}, \tag{13.38}$$

and

$$\lim_{N \to \infty} A_N = A. \tag{13.39}$$

For each N, we consider the truncated generating function attached to A_N:

$$f_N(x) = \sum_{a \in A_N} e(ax). \tag{13.40}$$

As $f_N(x)$ is a finite sum, all the convergence issues vanish. A similar argument as before yields

$$f_N(x)^s = \sum_{n \leq sN} r_N(n; s, A)e(nx), \qquad (13.41)$$

except now we have $r_N(n; s, A)$, which is the number of ways of writing n as the sum of s elements of A with each element at most N. If $n \leq N$, then $r_N(n; s, A) = r(n; s, A)$, the number of ways of writing n as the sum of s elements of A; note $f_N(x)^s$ is the generating function for the sum of s elements (at most N) of A

For example, if $A = P$ (the set of primes), $N = 10$ and $s = 2$, then $A_{10} = P_{10} = \{2, 3, 5, 7\}$. An easy calculation gives $r_{10}(8; 2, P) = r(8; 2, P) = 2$. However, $r_{10}(14; 2, P) = 1$ (from $7 + 7$) but $r(14; 2, P) = 3$ (from $7 + 7$, $3 + 11$, and $11 + 3$).

We have shown the following, which is the key re-formulation of these additive problems:

Lemma 13.2.9. *If $n \leq N$ then*

$$r(n; s, A) = r_N(n; s, A) = \int_0^1 f_N(x)^s e(-nx)dx. \qquad (13.42)$$

It is a common technique to use a generating function to solve additive problems; however, having an integral expression for $r_N(n; s, A)$ is not enough. We must be able to *evaluate* the integral, either exactly or at least bound it away from zero. Note $f_N(x)$ has $|A_N|$ terms, each term of absolute value 1. In many problems, for most $x \in [0, 1]$ the size of $f_N(x)$ is about $\sqrt{|A_N|}$, while for special $x \in [0, 1]$ one has $f_N(x)$ is of size $|A_N|$. The main contribution to the integral is expected to come from x where $f_N(x)$ is large, and often this integration can be performed. If we can show that the contribution of the remaining x is smaller, we will have bounded $r_N(n; s, A)$ away from zero.

13.2.4 Major and Minor Arcs

The difficultly is evaluating the integral in Lemma 13.2.9. Many successful applications of the Circle Method proceed in the following manner:

1. Given a set A, we construct a generating function $f_N(x)$ for A_N. As $f_N(x)$ is a sum of complex exponentials of size 1, by the Philosophy of Square Root Cancellation we expect there will often be significant cancellation. See the comments after Theorem 4.4.19 and the index for other examples of similar cancellation in number theory.

2. Split $[0, 1]$ into two disjoint pieces, called the **Major arcs** $\mathcal{M}$ and the **Minor arcs** m. For $m < N$

$$r(m; s, A) = r_N(m; s, A) = \int_{\mathcal{M}} f_N^s(x)e(-mx)dx + \int_m f_N^s(x)e(-mx)dx. \qquad (13.43)$$

The construction of $\mathcal{M}$ and m depends on N and the problem being studied.

3. On the Major arcs $\mathcal{M}$ we find a function which, up to lower order terms, agrees with $f_N^s(x)$ and is easily integrated. We then perform the integration and are left with a contribution over the Major arcs which is bounded away from zero and is large.

4. One shows that as $N \to \infty$ the Minor arcs' contribution is of lower order than the Major arcs' contribution. This implies that for m large, $r_N(m; s, A)$ > 0, which proves that large m can be represented as a sum of s elements of A.

The last is the most difficult step. It is often highly non-trivial to obtain the required cancellation over the Minor arcs. For the problems mentioned, we are able to obtain the needed cancellation for $A = P$ and $s = 3$ (every large odd number is the sum of three primes), but not $A = P$ and $s = 2$; we give some heuristics in §13.3.7 as to why $s = 2$ is so much harder than $s = 3$. For $A = K$ (the set of kth powers of integers), we can obtain the desired cancellation for $s = s(k)$ sufficiently large. Hardy and Littlewood proved we may take $s(k) = 2^k + 1$. Wooley and others have improved this result; however, in general we expect the result to hold for smaller s than the best results to date. See [VW, Wo] for more details.

13.2.5 Historical Remark

We briefly comment on the nomenclature: we have been talking about the Circle Method and arcs, yet there are no circles anywhere in sight! Let us consider an example. Recall from Proposition 13.1.2 that the generating function for the partition problem is

$$F(x) = \frac{1}{(1-x)(1-x^2)(1-x^3)\cdots} = 1 + \sum_{n=1}^{\infty} P(n)x^n. \qquad (13.44)$$

By (13.32), and ignoring convergence issues, we need to consider

$$P(n) = \frac{1}{2\pi i} \int_\gamma F(z)z^{-n-1}\, dz. \qquad (13.45)$$

The integrand is not defined at any point of the form $e(\frac{a}{q})$. The idea is to consider a small arc around each point $e(\frac{a}{q})$. This is where $|F(z)|$ is large. At least intuitively one expects that the integral of $F(z)$ along these arcs should be the major part of the integral. Thus, we break the unit circle into two disjoint pieces, the Major arcs (where we expect the generating function to be large), and the Minor arcs (where we expect the function to be small). While many problems proceed through generating functions that are sums of exponentials, as well as integrating over $[0, 1]$ instead of a circle, we keep the original terminology.

13.2.6 Needed Number Theory Results

In our applications of the Circle Method we need several results concerning prime numbers. These will be used to analyze the size of the generating function on the

Major arcs. As we have seen in §2.3.4 and §3.2.2, it is often easier to weight primes by $\log p$ in sums, and then remove these weights through partial summation. We use the following statements freely (see, for example, [Da2] for proofs). We will also use partial summation numerous times; the reader is advised to review the material in §2.2.2.

Theorem 13.2.10 (Prime Number Theorem). *Let $\pi(x)$ denote the number of primes at most x. Then there is a constant $c < 1$ such that*

$$\sum_{p \leq x} \log p = x + O\left(x \exp(-c\sqrt{\log x})\right). \tag{13.46}$$

Equivalently, by partial summation we have

$$\pi(x) = \sum_{p \leq x} 1 = \mathrm{Li}(x) + O\left(x \exp\left(-\frac{c}{2}\sqrt{\log x}\right)\right), \tag{13.47}$$

*where $\mathrm{Li}(x)$ is the **logarithmic integral**, which for any fixed positive integer k has the Taylor expansion*

$$\mathrm{Li}(x) = \int_2^x \frac{dt}{\log t} = \frac{x}{\log x} + \frac{1!x}{\log^2 x} + \cdots + \frac{(k-1)!x}{\log^k x} + O\left(\frac{x}{\log^{k+1} x}\right). \tag{13.48}$$

The above is the original version of the Prime Number Theorem. The error term has been strengthened by Korobov and Vinogradov to $O\left(x \exp(-c_\theta \sqrt[\theta]{\log x})\right)$ for any $\theta < \frac{3}{5}$. All we will need is

$$\pi(x) = \frac{x}{\log x} + o\left(\frac{x}{\log x}\right), \qquad \sum_{p \leq x} \log p = x + o(x). \tag{13.49}$$

Exercise 13.2.11. *Using partial summation, deduce a good estimate for $\pi(x)$ from* (13.46).

Exercise 13.2.12. *Prove* (13.48).

Theorem 13.2.13 (Siegel-Walfisz). *Let $C, B > 0$ and let a and q be relatively prime. Then*

$$\sum_{\substack{p \leq x \\ p \equiv a(q)}} \log p = \frac{x}{\phi(q)} + O\left(\frac{x}{\log^C x}\right) \tag{13.50}$$

for $q \leq \log^B x$, and the constant above does not depend on x, q or a (i.e., it only depends on C and B).

One may interpret the Siegel-Walfisz Theorem as saying each residue class has, to first order, the same number of primes. Explicitly, for a fixed q there are $\phi(q)$ numbers a relatively prime to q. Up to lower order terms each residue class has $\frac{\pi(x)}{\phi(q)}$ primes (see §2.3.2, §3.3.3). *Note the main term is larger than the error term if we choose C sufficiently large.* If we were to take q as large as x^δ for some $\delta > 0$, then the error term would exceed the main term; we want to apply this theorem when q is much smaller than x. The choice of the Major arcs is crucially influenced by the error term in the Siegel-Walfisz Theorem.

Remark 13.2.14. If the Generalized Riemann Hypothesis is true for Dirichlet L-functions, for any $\delta, \epsilon > 0$ we could take $q \leq x^{\frac{1}{2}-\delta}$ and have an error of size $x^{\frac{1}{2}+\epsilon}$. We use the Siegel-Walfisz Theorem in order to have unconditional results, though for heuristic purposes one can assume GRH.

13.3 GOLDBACH'S CONJECTURE REVISITED

While we discuss the complications from estimating the integral over the Minor arcs below, we do not give details on actually bounding these integrals; the interested reader should consult [Da2, EE, Est2, Na]. It is our intention to only *introduce* the reader to the broad brush strokes of this elegant theory.

Unfortunately such an approach means that at the end of the day we have not solved the original problem. We have chosen this approach for several reasons. While the technical details can be formidable, for many problems these details are beautifully presented in the above (and many other) sources. Further, there are many applications of the Circle Method where the needed estimates on the Minor arcs are not known, even assuming powerful conjectures such as the Generalized Riemann Hypothesis. In these cases, while it is often reasonable to assume that the contribution from the Major arcs is the main term, one cannot prove such statements. Thus, the techniques we develop are sufficient to allow the reader to *predict* the answer for a variety of open problems; these answers can often be tested numerically.

For these reasons, we describe the ideas of the Circle Method for Goldbach's problem: What are the Major and Minor arcs? Why do we obtain the necessary cancellation when $s = 3$ but not when $s = 2$? These examples are well known in the literature, and we content ourselves with a very brief introduction. In Chapter 14 we give a very thorough treatment of another Circle Method problem, Germain primes, which has applications to cryptography. The techniques for this problem suffice to estimate the Major arc contributions in many other problems, for example, how many twin primes are there less than x.

We do not always explicitly compute the error terms below, often confining ourselves to writing the main term and remarking the correction terms are smaller. As an exercise, the reader is encouraged to keep track of these errors.

13.3.1 Setup

The Circle Method begins with a choice of a generating function specific to the problem, in this case, Goldbach's conjecture. For analytical reasons (see Remark 3.2.18 and §13.2.6), it is often convenient to analyze the weighted generating function

$$F_N(x) = \sum_{p \leq N} \log p \cdot e(px) \tag{13.51}$$

instead of $f_N(x)$, and pass to the unweighted function by partial summation. One could worked only with $f_N(x)$ (see [Est2], Chapter 3); however, we prefer to use

$F_N(x)$ as the weights are easily removed and simplify several formulas. Working analogously as before, to write m as a sum of s primes leads us to

$$R_{N,s}(m) = \int_0^1 F_N^s(x)e(-mx)\,dx, \qquad (13.52)$$

where now

$$R_{N,s}(m) = \sum_{\substack{p_1+\cdots+p_s=m \\ p_i \leq N}} \log p_1 \cdots \log p_s. \qquad (13.53)$$

Exercise 13.3.1. *Relate $R_{N,s}(m)$ and $r_N(m; s, P)$. For details, see §14.7.*

Thus, if we can show $R_{N,s}(m)$ is positive for N and m sufficiently large, then $r(m; s, P)$ is also positive.

13.3.2 Average Value of $|F_N(x)|^2$

We use the little-Oh notation (see Definition 2.2.3). Thus, $N + o(N)$ means the answer is N plus lower order terms. Recall

$$F_N(x) = \sum_{p \leq N} \log p \cdot e(px). \qquad (13.54)$$

Lemma 13.3.2. $|F_N(x)| \leq N + o(N)$.

Proof. By the Prime Number Theorem, (13.46), we have

$$|F_N(x)| = \left| \sum_{p \leq N} \log p \cdot e(px) \right| \leq \sum_{p \leq N} \log p = N + o(N). \qquad (13.55)$$

$\square$

Lemma 13.3.3. $F_N(0) = F_N(1) = N + o(N)$, and $F_N(\frac{1}{2}) = -N + o(N)$.

Proof. $F_N(0)$ and $F_N(1)$ are immediate, as $e(p \cdot 1) = 1$ for all p. For $F_N(\frac{1}{2})$, note

$$e\left(p \cdot \frac{1}{2}\right) = e^{\pi i p} = \begin{cases} -1 & \text{if } p \text{ is odd}, \\ +1 & \text{if } p \text{ is even}. \end{cases} \qquad (13.56)$$

As there is only one even prime,

$$F_N\left(\frac{1}{2}\right) = \log 2 - \sum_{3 \leq p \leq N} \log p, \qquad (13.57)$$

and the argument proceeds as before. $\square$

Exercise 13.3.4. *How large are $F_N(\frac{1}{4})$ and $F_N(\frac{3}{4})$? How big can $o(N)$ be in Lemma 13.3.3?*

Thus $F_N(x)$ is occasionally as large as N; in §13.3.5 we describe the x where $F_N(x)$ is large. We can, however, show that the average square of $F_N(x)$ is significantly smaller:

Lemma 13.3.5. *The average value of $|F_N(x)|^2$ is $N \log N + o(N \log N)$.*

Proof. The following trivial observation will be extremely useful in our arguments, and is a common technique to analyze average values. Let $g(x)$ be a complex-valued function, and let $\bar{g}(x)$ be its complex conjugate. Then $|g(x)|^2 = g(x)\bar{g}(x)$. In our case, as $\overline{F_N(x)} = F_N(-x)$ we have

$$\int_0^1 |F_N(x)|^2 dx = \int_0^1 F_N(x) F_N(-x) dx$$

$$= \int_0^1 \sum_{p \leq N} \log p \cdot e(px) \sum_{q \leq N} \log q \cdot e(-qx) dx$$

$$= \sum_{p \leq N} \sum_{q \leq N} \log p \log q \int_0^1 e\left((p-q)x\right) dx. \qquad (13.58)$$

By Exercise 13.2.7, the integral is 1 if $p = q$ and 0 otherwise. Therefore the only pairs (p, q) that contribute are when $p = q$, and we have

$$\int_0^1 |F_N(x)|^2 dx = \sum_{p \leq N} \log^2 p. \qquad (13.59)$$

Using partial summation (see Exercise 13.3.9), we can show

$$\sum_{p \leq N} \log^2 p = N \log N + o(N \log N). \qquad (13.60)$$

Thus

$$\int_0^1 |F_N(x)|^2 dx = N \log N + o(N \log N). \qquad (13.61)$$

$\square$

Remark 13.3.6. The above argument is very common. The absolute value function is not easy to work with; however, $g(x)\overline{g(x)}$ is very tractable (see also §8.1.6). In many problems, it is a lot easier to study $\int |g(x)|^2$ than $\int |g(x)|$ or $\int |g(x)|^3$.

Remark 13.3.7 (Philosophy of Square Root Cancellation). The average value of $|F_N(x)|^2$ is about $N \log N$, significantly smaller than the maximum possible value of N^2. Thus we have almost square root cancellation on average. In general, if one adds a "random" set of N numbers of absolute value 1 then the sum could be as large as N, but often is about $\sqrt{N}$. For more details and examples, see §4.4 and §8.4.2.

Exercise[h] 13.3.8. *Investigate the size of $\sum_{x=0}^{p-1} e^{2\pi i x^2 / p}$ for p prime.*

Exercise 13.3.9. *Using the Prime Number Theorem and Partial Summation, prove*

$$\sum_{p \leq N} \log^2 p = N \log N + o(N \log N). \qquad (13.62)$$

13.3.3 Large Values of $F_N(x)$

For a fixed B, let $Q = \log^B N$. Fix a $q \leq Q$ and an $a \leq q$ with a and q relatively prime. We evaluate $F_N\left(\frac{a}{q}\right)$. While on average $F_N(x)$ is of size $\sqrt{N \log N}$, for x near such $\frac{a}{q}$ we shall see that $F_N(x)$ is large.

$$F_N\left(\frac{a}{q}\right) = \sum_{p \leq N} \log p \cdot e\left(p\frac{a}{q}\right). \qquad (13.63)$$

The summands on the right hand side depend weakly on p. Specifically, the exponential terms only depend on $p \bmod q$, which allows us to rewrite $F_N\left(\frac{a}{q}\right)$ as a sum over congruence classes:

$$F_N\left(\frac{a}{q}\right) = \sum_{r=1}^{q} \sum_{\substack{p \equiv r(q) \\ p \leq N}} \log p \cdot e\left(\frac{ap}{q}\right)$$

$$= \sum_{r=1}^{q} \sum_{\substack{p \equiv r(q) \\ p \leq N}} \log p \cdot e\left(\frac{ar}{q}\right)$$

$$= \sum_{r=1}^{q} e\left(\frac{ar}{q}\right) \sum_{\substack{p \equiv r(q) \\ p \leq N}} \log p. \qquad (13.64)$$

We use the Siegel-Walfisz Theorem to evaluate the sum over $p \equiv r \bmod q$. We first remark that we may assume r and q are relatively prime (see Exercise 13.3.10). Briefly, if $p \equiv r \bmod q$, this means $p = \alpha q + r$ for some $\alpha \in \mathbb{N}$. If r and q have a common factor, there can be at most one prime p (namely r) such that $p \equiv r \bmod q$, and this can easily be shown to give a negligible contribution. For any $C > 0$, by the Siegel-Walfisz Theorem

$$\sum_{\substack{p \equiv r(q) \\ p \leq N}} \log p = \frac{N}{\phi(q)} + O\left(\frac{N}{\log^C N}\right). \qquad (13.65)$$

As $\phi(q)$ is at most q which is at most $\log^B N$, we see that if we take $C > B$ then the main term is significantly greater than the error term. Note the Siegel-Walfisz Theorem would be useless if q were large, say $q \approx N^\delta$. Then the main term would be like $N^{1-\delta}$, which would be smaller than the error term. Thus we find

$$F_N\left(\frac{a}{q}\right) = \sum_{\substack{r=1 \\ (r,q)=1}}^{q} e\left(\frac{ar}{q}\right) \frac{N}{\phi(q)} + O\left(\frac{qN}{\log^C N}\right)$$

$$= \frac{N}{\phi(q)} \sum_{\substack{r=1 \\ (r,q)=1}}^{q} e\left(\frac{ar}{q}\right) + O\left(\frac{N}{\log^{C-B} N}\right). \qquad (13.66)$$

If the sum over r in (13.66) is not too small, then $F_N\left(\frac{a}{q}\right)$ is "approximately" of size $\frac{N}{\phi(q)}$, with an error of size $\frac{N}{\log^{C-B} N}$. If $C > 2B$, the main term is significantly larger than the error term, and $F_N\left(\frac{a}{q}\right)$ is large.

The Siegel-Walfisz Theorem is our main tools for evaluating the necessary prime sums, and it is useful only when the error term is less than the main term. Our investigations of the (potential) size of $F_N(x)$ lead us to the proper definitions for the Major and Minor arcs in §13.3.4.

Exercise 13.3.10. *Show the terms with r and q not relatively prime in (13.64) contribute lower order terms.*

13.3.4 Definition of the Major and Minor Arcs

We split $[0, 1]$ into two disjoint parts, the Major and the Minor arcs. As $|F_N(x)|^2$ is of size $N \log N$ on average, there is significant cancellation in $F_N(x)$ most of the time. The Major arcs will be a union of very small intervals centered at rationals with small denominator relative to N. Near these rationals we can approximate $F_N(x)$ very well, and $F_N(x)$ will be large (of size N). The Minor arcs will be the rest of $[0, 1]$, and here we expect $F_N(x)$ to be significantly smaller than N. Obtaining such cancellation in the series expansion is *not* easy – this is the hardest part of the problem. In many cases we are unable to prove the integral over the Minor arcs is smaller than the contribution from the Major arcs, though we often believe this is the case, and numerical investigations support such claims.

13.3.4.1 Major Arcs

The choice of the Major arcs depend on the problem being investigated. In problems where the Siegel-Walfisz Theorem is used, the results from §13.3.3 suggest the following choice. Let $B > 0$, and let $Q = \log^B N \ll N$. For each $q \in \{1, 2, \dots, Q\}$ and $a \in \{1, 2, \dots, q\}$ with a and q relatively prime, consider the set

$$\mathcal{M}_{a,q} = \left\{ x \in [0, 1] : \left| x - \frac{a}{q} \right| < \frac{Q}{N} \right\}. \tag{13.67}$$

We also add in one interval centered at either 0 or 1, i.e., the interval (or wrapped-around interval)

$$\left[0, \frac{Q}{N} \right) \cup \left(1 - \frac{Q}{N}, 1 \right]. \tag{13.68}$$

Exercise 13.3.11. *Show that if N is large then the Major arcs $\mathcal{M}_{a,q}$ are disjoint for $q \le Q$ and $a \le q$, a and q relatively prime.*

We define the Major arcs to be the union of the arcs $\mathcal{M}_{a,q}$:

$$\mathcal{M} = \bigcup_{q=1}^{Q} \bigcup_{\substack{a=1 \\ (a,q)=1}}^{q} \mathcal{M}_{a,q}, \tag{13.69}$$

where (a, q) is the greatest common divisor of a and q.

Remark 13.3.12. As the Major arcs depend on N and B, we should really write $\mathcal{M}_{a,q}(N, B)$ and $\mathcal{M}(N, B)$; however, for notational convenience N and B are often suppressed.

Exercise 13.3.13. *Show* $|\mathcal{M}| \leq \frac{2Q^3}{N}$. *As* $Q = \log^B N$, *this implies* $|\mathcal{M}| \to 0$ *as* $N \to \infty$. *Thus in the limit most of* $[0,1]$ *is contained in the Minor arcs; the choice of terminology reflects where* $F_N(x)$ *is large, and not which subset of* $[0,1]$ *is larger.*

Note that the above choice for the Major arcs has two advantages. First, recall that we required the denominator q to be small relative to N: $q \leq Q = \log^B N$. Once a denominator is small for some N we can apply the Siegel-Walfisz Theorem and we can evaluate $F_N(\frac{a}{q})$ well (see §13.3.3). Second, each Major arc $\mathcal{M}_{a,q}$ has length $\frac{2Q}{N} = \frac{2\log^B N}{N}$; as these intervals are small, we expect $F_N(x) \approx F_N(\frac{a}{q})$. It should be possible to estimate the integral over $\mathcal{M}_{a,q}$. Thus, for a fixed $\frac{a}{q}$, the size of the arc about it tends to zero as N tends to infinity, but $F_N(x)$ becomes better and better understood in a smaller windows about $\frac{a}{q}$.

Exercise 13.3.14. *For large* N, *find a good asymptotic formula for* $|\mathcal{M}|$.

Exercise 13.3.15. *For a fixed* B, *how large must* N *be for the Major arcs to be disjoint?*

13.3.4.2 Minor Arcs

The Minor arcs, m, are whatever is *not* in the Major arcs. Thus,

$$\text{m} = [0,1] - \mathcal{M}. \tag{13.70}$$

Clearly, as $N \to \infty$ almost all of $[0,1]$ is in the Minor arcs. The hope is that by staying away from rationals with small denominator, we will be able to obtain significant cancellation in $F_N(x)$.

13.3.5 The Major Arcs and the Singular Series

We are trying to write m as a sum of s primes. Let us consider the case $m = N$ and $s = 3$. We have shown the (weighted) answer is given by

$$\int_0^1 F_N(x)^3 e(-Nx) dx = \sum_{\substack{p_1,p_2,p_3 \leq N \\ p_1+p_2+p_3=N}} \log p_1 \log p_2 \log p_3; \tag{13.71}$$

the weights can easily be removed by partial summation. We merely sketch what happens now; we handle a Major arc calculation in full detail in Chapter 14.

First one shows that for $x \in \mathcal{M}_{a,q}$, $F_N(x)$ is very close to $F_N\left(\frac{a}{q}\right)$. While one could calculate the Taylor series expansion (see §A.2.3), in practice it is technically easier to find a function which is non-constant, agrees with $F_N(x)$ at $x = \frac{a}{q}$ and is easily integrated. As the Major arcs are disjoint for large N,

$$\int_{\mathcal{M}} F_N(x)^3 e(-Nx) dx = \sum_{q=1}^{Q} \sum_{\substack{a=1 \\ (a,q)=1}}^{q} \int_{\mathcal{M}_{a,q}} F_N(x)^3 e(-Nx) dx. \tag{13.72}$$

For heuristic purposes, we approximate $F_N(x)^3 e(-Nx)$ by $F_N\left(\frac{a}{q}\right)^3 e\left(-N\frac{a}{q}\right)$. After reading Chapter 14 the reader is encouraged to do these calculations correctly. Therefore

$$\int_{\mathcal{M}} F_N(x)^3 e(-Nx) dx \approx \int_{\mathcal{M}} F_N\left(\frac{a}{q}\right)^3 e\left(-N\frac{a}{q}\right) dx$$

$$= F_N\left(\frac{a}{q}\right)^3 e\left(-N\frac{a}{q}\right) \cdot \frac{2Q}{N}. \qquad (13.73)$$

In (13.66) we used the Siegel-Walfisz Theorem to evaluate $F_N\left(\frac{a}{q}\right)$. Again, for heuristic purposes we suppress the lower order error terms, and find that the contribution from the Major arcs is

$$\sum_{\substack{p_1,p_2,p_3 \le N \\ p_1+p_2+p_3=N}} \log p_1 \log p_2 \log p_3$$

$$= \frac{2Q^3}{N} \sum_{q=1}^{Q} \sum_{\substack{a=1 \\ (a,q)=1}}^{q} \left(\frac{N}{\phi(q)} \sum_{\substack{r=1 \\ (r,q)=1}}^{q} e\left(\frac{ar}{q}\right)\right)^3 e\left(\frac{-Na}{q}\right)$$

$$= \left[2Q^3 \sum_{q=1}^{Q} \frac{1}{\phi(q)^3} \sum_{\substack{a=1 \\ (a,q)=1}}^{q} \left(\sum_{\substack{r=1 \\ (r,q)=1}}^{q} e\left(\frac{ar}{q}\right)\right)^3 e\left(\frac{-Na}{q}\right) \right] N^2.$$

$$(13.74)$$

To complete the proof, we need to show that what multiplies N^2 is positive and not too small. If N^2 were multiplied by $\frac{1}{N^3}$, for example, the main term from the Major arcs would be of size $\frac{1}{N}$, which could easily be canceled by the contribution from the Minor arcs. An elementary analysis often bounds the factor away from 0 and infinity.

Note that, up to factors of $\log N$ (which are important!), the contribution from the Major arcs is of size N^2. A more careful analysis, where we do not just replace $F_N(x)^3 e(-Nx)$ with $F_N(\frac{a}{q})^3 e\left(-N\frac{a}{q}\right)$ on $\mathcal{M}_{a,q}$, would show that the Major arcs contribute

$$\mathfrak{S}(N)\frac{N^2}{2} + o(N^2), \qquad (13.75)$$

with

$$\mathfrak{S}(N) = \sum_{q=1}^{\infty} \frac{1}{\phi(q)^3} \sum_{\substack{a=1 \\ (a,q)=1}}^{q} \left(\sum_{\substack{r=1 \\ (r,q)=1}}^{q} e\left(\frac{ar}{q}\right)\right)^3 e\left(-N\frac{a}{q}\right). \qquad (13.76)$$

$\mathfrak{S}(N)$ is called the **Singular Series**; in all Circle Method investigations, the contribution from the Major arcs is given by such a series. The singular series for Germain primes will be discussed in detail in Chapter 14; for complete details on the

singular series for sums of three primes, the interested reader should see [EE, Na].
If we set

$$c_p(N) = \begin{cases} p-1 & \text{if } p|N \\ 0 & \text{otherwise,} \end{cases}$$ (13.77)

then one can show

$$\mathfrak{S}(N) = \prod_p \left(1 - \frac{c_p(N)}{\phi(p)^3}\right).$$ (13.78)

The product expansion is a much more useful expression for the factor multiplying
N^2 than the series expansion. If N is odd, there exist constants c_1 and c_2 such that

$$0 < c_1 < \mathfrak{S}(N) < c_2 < \infty.$$ (13.79)

This allows us to conclude the Major arcs' contribution is of order N^2.

We do not go into great detail concerning the arithmetic properties of $\mathfrak{S}(N)$, and
content ourselves with an important observation. If $\mathfrak{S}(N) > c_1$ for all N, then the
main term will be greater than the error term for N sufficiently large. Can $\mathfrak{S}(N)$
ever vanish?

Consider N even. Then $c_2(N) = 1$, $\phi(2) = 1$, and the factor in $\mathfrak{S}(N)$ corresponding to $p = 2$ vanishes! Thus, for even N, the main term from the Circle
Method is zero. In hindsight this is not surprising. Assume an even $N > 6$ can be
written as the sum of three primes. Exactly one of the primes must be even (if all
or exactly one were odd, then N would be odd; if all were even, N would be 6).
Therefore, if the Circle Method tells us that we can write an even N as the sum of
three primes, we could immediately conclude that $N - 2$ is the sum of two primes.

The Singular Series "knows" about the difficulty of Goldbach's conjecture. For
many Circle Method problems, one is able to write the main term from the Major
arcs (up to computable constants and factors of $\log N$) as $\mathfrak{S}(N)N^a$, with $\mathfrak{S}(N)$
a product over primes. The factors at each prime often encode information about
obstructions to solving the original problem. For more on obstructions, see §4.4.

Exercise 13.3.16. *For N odd, show there exist positive constants c_1, c_2 (independent of N) such that $0 < c_1 < \mathfrak{S}(N) < c_2 < \infty$.*

Exercise 13.3.17. *In the spirit of Exercise 13.1.13, we sketch a heuristic for the
expected average value of the number of ways of writing n as a sum of k primes.
Consider $n \in [N, 2N]$ for N large. Count the number of k-tuples of primes with
$p_1 + \cdots + p_k \in [N, 2N]$. As there are approximately N even numbers in the
interval, deduce the average number of representations for such n. What if we
instead considered short intervals, such as $n \in [N, N + N^{1-\delta}]$ for some $\delta > 0$?*

Exercise 13.3.18. *Prove (13.76) implies the product representation in (13.78).
Hint: Many of the sums of arithmetic functions in (13.76) arise in the Germain
prime investigations; see §14.6 for results about sums of these functions.*

Remark 13.3.19 (Goldbach's Conjecture). If instead we investigate writing even
numbers as the sum of two primes, we would integrate $F_N(x)^2$ and obtain a new
singular series, say $\widetilde{\mathfrak{S}}(N)$. The Major arcs would then contribute $\widetilde{\mathfrak{S}}(N)N$.

13.3.6 Contribution from the Minor Arcs

We bound the contribution from the Minor arcs to $R_{N,s}(N)$, the weighted number of ways to write N as a sum of s primes:

$$\left| \int_{\mathrm{m}} F_N(x)^3 e(-Nx) dx \right| \leq \int_{\mathrm{m}} |F_N(x)|^3 dx$$

$$\leq \left(\max_{x \in \mathrm{m}} |F_N(x)| \right) \int_{\mathrm{m}} |F_N(x)|^2 dx$$

$$\leq \left(\max_{x \in \mathrm{m}} |F_N(x)| \right) \int_0^1 F_N(x) F_N(-x) dx$$

$$\leq \left(\max_{x \in \mathrm{m}} |F_N(x)| \right) N \log N. \tag{13.80}$$

As the Minor arcs are most of the unit interval, replacing $\int_{\mathrm{m}}$ with $\int_0^1$ does not introduce much of an over-estimation. *In order for the Circle Method to succeed, we need a non-trivial, good bound for*

$$\max_{x \in \mathrm{m}} |F_N(x)|. \tag{13.81}$$

This is where most of the difficulty arises, showing that there is some cancellation in $F_N(x)$ if we stay away from rationals with small denominator. We need an estimate such as

$$\max_{x \in \mathrm{m}} |F_N(x)| \leq \frac{N}{\log^{1+\epsilon} N}, \tag{13.82}$$

or even

$$\max_{x \in \mathrm{m}} |F_N(x)| \ll o\left(\frac{N}{\log N} \right). \tag{13.83}$$

Relative to the average size of $|F_N(x)|^2$, which is $N \log N$, this is significantly larger. Unfortunately, as we have inserted absolute values, it is not enough to bound $|F_N(x)|$ on average; we need to obtain a good bound uniformly in x. We know such a bound cannot be true for all $x \in [0,1]$ because $F_N(x)$ is large on the Major arcs! The hope is that if x is not near a rational with small denominator, we will obtain at least moderate cancellation. While this is reasonable to expect, it is not easy to prove; the interested reader should see [EE, Na]. Following Vinogradov [Vin1, Vin2] one shows

$$\max_{x \in \mathrm{m}} |F_N(x)| \ll \frac{N}{\log^D N}, \tag{13.84}$$

which allows one to deduce any large odd number is the sum of three primes. While (13.61) gives us significantly better cancellation on average, telling us that $|F_N(x)|^2$ is usually of size N, bounds such as (13.84) are the best we can do if we require the bound to hold for *all* $x \in \mathrm{m}$.

Exercise 13.3.20. *Using the definition of the Minor arcs, bound*

$$\left| \int_0^1 |F_N(x)|^2 dx - \int_{\mathrm{m}} |F_N(x)|^2 dx \right|. \tag{13.85}$$

Show, therefore, that there is little harm in extending the integral of $|F_N(x)|^2$ to all of $[0,1]$. In general, there is very little loss of information in integrating $|F_N(x)|^{2^k}$.

13.3.7 Why Goldbach's Conjecture Is Hard

We give some arguments which indicate the difficulty of applying the Circle Method to Goldbach's conjecture. To investigate $R_{N,s}(N)$, the number of weighted ways of writing N as the sum of s primes, we considered the generating function

$$F_N(x) = \sum_{p \leq N} \log p \cdot e\,(px)\,, \tag{13.86}$$

which led to

$$R_{N,s}(N) = \int_0^1 F_N(x)^s e(-Nx)dx. \tag{13.87}$$

Remember that the average size of $|F_N(x)|^2$ is $N \log N$.

We have seen that, up to logarithms, the contribution from the Major arcs is of size N^2 for $s = 3$. Similar arguments show that the Major arcs contribute on the order of N^{s-1} for sums of s primes. We now investigate why the Circle Method works for $s = 3$ but fails for $s = 2$.

When $s = 3$, we can bound the Minor arcs' contribution by

$$\left| \int_m F_N(x)^3 e(-Nx)dx \right| \leq \max_{x \in m} |F_N(x)| \int_0^1 |F_N(x)|^2 dx$$

$$\leq \max_{x \in m} |F_N(x)| \cdot N \log N. \tag{13.88}$$

As the Major arcs contribute $\mathfrak{S}(N)N^2$, one needs only a small savings on the Minor arcs; Vinogradov's bound

$$\max_{x \in m} |F_N(x)| \ll \frac{N}{\log^D N} \tag{13.89}$$

suffices. What goes wrong when $s = 2$? The Major arcs' contribution is now expected to be of size N. How should we estimate the contribution from the Minor arcs? We have $F_N(x)^2 e(-Nx)$. The simplest estimate to try is to just insert absolute values, which gives

$$\left| \int_m F_N(x)^2 e(-Nx)dx \right| \leq \int_0^1 |F_N(x)|^2 dx = N. \tag{13.90}$$

Note, unfortunately, that this is the same size as the expected contribution from the Major arcs!

We could try pulling a $\max_{x \in m} |F_N(x)|$ outside the integral, and hope to get a good saving (pulling out $|F_N(x)|^2$ clearly cannot work as the maximum of this expression is at least $N \log N$). The problem is this leaves us with having to bound $\int_m |F_N(x)|dx$. Note $F_N(x)$ on average is approximately of size $\sqrt{N \log N}$; this is not quite right as we have only shown $|F_N(x)|^2$ on average is $N \log N$; however, let us ignore this complication and see what bound we obtain. Replacing $|F_N(x)|$ in the integral with its average value leads us to

$$\left| \int_m F_N(x)^2 e(-Nx)dx \right| \leq \max_{x \in m} |F_N(x)| \cdot \sqrt{N \log N}. \tag{13.91}$$

As the Major arcs' contribution is of size N, we would need

$$\max_{x \in \mathfrak{m}} |F_N(x)| \ll o\left(\sqrt{\frac{N}{\log N}}\right). \tag{13.92}$$

There is no chance of such cancellation; this is better than square root cancellation, and contradicts the average value of $|F_N(x)|^2$ from (13.61).

Another approach is to use the Cauchy-Schwarz inequality (see Lemma A.6.9):

$$\int_0^1 |f(x)g(x)| dx \le \left(\int_0^1 |f(x)|^2 dx\right)^{\frac{1}{2}} \cdot \left(\int_0^1 |g(x)|^2 dx\right)^{\frac{1}{2}}. \tag{13.93}$$

Thus

$$\left|\int_{\mathfrak{m}} F_N(x)^2 e(-mx) dx\right| \le \max_{x \in \mathfrak{m}} |F_N(x)| \int_0^1 |F_N(x)| dx$$

$$\le \max_{x \in \mathfrak{m}} |F_N(x)| \left(\int_0^1 |F_N(x)|^2 dx\right)^{\frac{1}{2}} \cdot \left(\int_0^1 1^2 dx\right)^{\frac{1}{2}}$$

$$\le \max_{x \in \mathfrak{m}} |F_N(x)| \cdot (N \log N)^{\frac{1}{2}} \cdot 1. \tag{13.94}$$

Unfortunately, this is the same bound as (13.91), which was too large.

Remark 13.3.21. Even though it failed, it was a good idea to use the Cauchy-Schwartz inequality. The reason is we are integrating over a finite interval; thus $\int_0^1 1^2 dx$ is harmless; if the size of the interval depended on N (or was all of $\mathbb{R}$), applying Cauchy-Schwartz might be a mistake.

While the above sketch shows the Circle Method is not, at present, powerful enough to handle the Minor arcs' contribution, all is not lost. The quantity we *need* to bound is

$$\left|\int_{\mathfrak{m}} F_N(x)^2 e(-mx) dx\right|. \tag{13.95}$$

However, we have instead been studying

$$\int_{\mathfrak{m}} |F_N(x)|^2 dx \tag{13.96}$$

and

$$\max_{x \in \mathfrak{m}} |F_N(x)| \int_0^1 |F_N(x)| dx. \tag{13.97}$$

We are ignoring the probable oscillation and cancellation in the integral $\int_{\mathfrak{m}} F_N(x)^2 e(-mx) dx$. It is this expected cancellation that would lead to the Minor arcs contributing significantly less than the Major arcs.

However, showing there is cancellation in the above integral is very difficult. It is a lot easier to work with absolute values. Further, just because we cannot prove that the Minor arc contribution is small, does not mean the Circle Method is not useful or may not some day be able to prove Goldbach's conjecture. Numerical simulations confirm, for many problems, that the Minor arcs do not contribute for many N. For example, for $N \le 10^9$ the observed values are in excellent agreement with the Major arc predictions (see [Ci, Sch, Weir]).

Chapter Fourteen

Circle Method: Heuristics for Germain Primes

We apply the Circle Method to investigate Germain primes. As current techniques are unable to adequately bound the Minor arc contributions, we concentrate on the Major arcs, where we perform the calculations in great detail The methods of this chapter immediately generalize to other standard problems, such as investigating twin primes or prime tuples.

We have chosen to describe the Circle Method for Germain primes as this problem highlights many of the complications that arise in applications. Unlike the previous investigations of writing N as a sum of s primes, our generating function $F_N(x)$ is the product of two different generating functions. To approximate $F_N(x)$ on the Major arc $\mathcal{M}_{a,q}$, we could try to Taylor expand $F_N(x)$; however, the derivative is not easy to analyze or integrate. Instead we construct a new function which is easy to integrate on $[-\frac{1}{2}, \frac{1}{2}]$, has most of its mass concentrated near $\frac{a}{q}$, and is a good approximation to $F_N(x)$ on $\mathcal{M}_{a,q}$. To show the last claim requires multiple applications of partial summation. For numerical investigations of the Minor arcs, as well as spacing properties of Germain primes, see [Weir].

In §14.1 and §14.2 we define Germain primes, the generating function $F_N(x)$, and the Major and Minor arcs. In §14.3 we estimate $F_N(x)$ and find an easily integrable function $u(x)$ which should be close to $F_N(x)$ on the Major arcs. We prove $u(x)$ is a good approximation to $F_N(x)$ in §14.4; this is a technical section and can easily be skimmed or skipped on a first reading. We then determine the contribution from the Major arcs by performing the integration in §14.5 and then analyzing the singular series in §14.6. Finally in §14.7 we remove the $\log p$ weights and then conclude with some exercises and open problems.

14.1 GERMAIN PRIMES

Consider an odd prime $p > 3$. Clearly $p - 1$ cannot be prime, as it is even; however, $\frac{p-1}{2}$ could be prime, and sometimes is as $p = 5, 7$ and 11 show.

Definition 14.1.1 (Germain Prime). *A prime p is a Germain prime if both p and $\frac{p-1}{2}$ are prime. We often call $(p, \frac{p-1}{2})$ a Germain prime pair. An alternate definition is to have p and $2p + 1$ both prime.*

Germain primes have many wonderful properties. Around 1825 Sophie Germain proved that if p is a Germain prime then the first case of Fermat's Last Theorem, which states the only integer solutions of $x^p + y^p = z^p$ have $p|xyz$, is true for exponent p; see Lecture III of [vdP6] for more details. For more on Fermat's Last

Theorem, see §4.1.3. As another application, recent advances in cryptography are known to run faster if there are many Germain primes (see [AgKaSa]).

Germain primes are just one example of the following type of problem: given relatively prime positive integers a and b, for $p \leq N$ how often are p and $ap + b$ prime? Or, more generally, how often are $p, a_1p + b_1, \ldots, a_kp + b_k$ prime? One well known example is the famous Twin Prime Conjecture, which states that there are infinitely many primes p such that $p + 2$ is also prime. It is not known if this is true. Unlike the sum of the reciprocals of the primes, which diverges, Brun has shown that the sum of the reciprocal of the twin primes converges (to approximately 1.90216058). See [Na] for a proof. Interestingly, it was by numerically investigating the sum of the reciprocals of the twin primes that the bug in the Pentium processor was discovered (which is not surprising, as the bug involves errors in large numbers of division, which is exactly what we need to do to evaluate this sum). See [Ni1, Ni2] for details.

Therefore, if there are infinitely many twin primes, there are in some sense fewer twin primes than primes. Explicitly, Brun proved that there exists an N_0 such that for all $N > N_0$ the number of twin primes less than N is at most $\frac{100N}{\log^2 N}$. This should be compared to the number of primes less than N, which is of size $\frac{N}{\log N}$. Using the Circle Method, Hardy and Littlewood [HL3, HL4] were led to conjectures on the number of such primes, and their Major arc calculations agree beautifully with numerical investigations.

We have chosen to go through the calculation of the number of Germain primes less than N rather than twin primes as these other problems are well documented in the literature ([Da2, EE, Est2, Na]). Note the Germain problem is slightly different from the original formulation of the Circle Method. Here we are investigating how often $p_1 - 2p_2 = 1$, with $p_2 < p_1 < N$. Let

$$A_1 = \{p : p \text{ prime}\}$$
$$A_2 = \{-2p : p \text{ prime}\}. \tag{14.1}$$

To construct generating functions that converge, we consider the truncated sets

$$A_{1N} = \{p : p \text{ prime}, p \leq N\}$$
$$A_{2N} = \{-2p : p \text{ prime}, p \leq N\}. \tag{14.2}$$

We are interested in

$$\{(a_1, a_2) : a_1 + a_2 = 1, a_i \in A_{iN}\}. \tag{14.3}$$

In the original applications of the Circle Method, we were just interested in whether or not a number m was in $A_N + \cdots + A_N$. To show m could be written as the sum of elements $a_i \in A_N$, we counted the number of ways to write it as such a sum, and showed this is positive.

For Germain primes and related problems, we are no longer interested in determining all numbers that can be written as the sum $a_1 + a_2$. We only want to find pairs with $a_1 + a_2 = 1$. The common feature with our previous investigations is showing how many ways certain numbers can be written as the sum of elements in A_{iN}. Such knowledge gives estimates for the number of Germain primes at most N.

For any $N \geq 5$ we know $1 \in A_{1N} + A_{2N}$ as 5 is a Germain prime. Note the number of ways of writing 1 as $a_1 + a_2$ with $a_i \in A_{iN}$ is the number of Germain primes at most N; similar to before, we need to compute for this problem $r(1; A_{1N}, A_{2N})$, with the obvious notation.

Exercise 14.1.2. *Looking at tables of primes less than* 100, *do you think there will be more Germain primes or twin primes in the limit? What if you study primes up to* 10^4? *Up to* 10^8? *What percentage of primes less than* 100 $(10^4, 10^8)$ *are Germain primes? Twin primes? How many primes less than* N *(for* N *large) do you expect to be Germain primes? Twin primes?*

Exercise 14.1.3. *By the Prime Number Theorem, for primes near* x *the average spacing between primes is* $\log x$. *One can interpret this as the probability a number near* x *is prime is* $\frac{1}{\log x}$. *We flip a biased coin with probability* $\frac{1}{\log x}$ *of being a prime,* $1 - \frac{1}{\log x}$ *of being composite; this is called the* **Cramér model**. *Using such a model predict how many Germain primes and twin primes are less than* N.

Remark 14.1.4 (Remark on the Previous Exercise). The Cramér model of the previous exercise cannot be correct — knowledge that p is prime gives some information about the potential primality of nearby numbers. One needs to correct the model to account for congruence information. See §8.2.3 and [Rub1].

Exercise 14.1.5. *Use Brun's result that the number of twin primes at most* N *is at most* $\frac{100N}{\log^2 N}$ *to prove that the sum of the reciprocals of the twin primes converges. See §2.3.3 for additional problems of this nature.*

14.2 PRELIMINARIES

We use the Circle Method to calculate the contribution from the Major arcs for the Germain problem, namely, how many primes $p \leq N$ there are such that $\frac{p-1}{2}$ is also prime. As pointed out earlier, for this problem the Minor arc calculations cannot be determined with sufficient accuracy to be shown to be smaller than the Major arc contributions. We will, however, do the Major arc calculations in complete detail. As usual let

$$e(x) = e^{2\pi i x}$$

$$\lambda(n) = \begin{cases} \log p & \text{if } n = p \text{ is prime} \\ 0 & \text{otherwise.} \end{cases} \tag{14.4}$$

We will use the integral version of partial summation (§2.2.2) and the Siegel-Walfisz Theorem (Theorem 13.2.13). We have introduced the arithmetic function $\lambda(n)$ for notational convenience. In applying partial summation, we will have sums over integers, but our generating function is defined as a sum over primes; $\lambda(n)$ is a convenient notation which allows us to write the sum over primes as a sum over integers.

14.2.1 Germain Integral

Define

$$F_{1N}(x) = \sum_{p_1 \leq N} \log p_1 \cdot e(p_1 x)$$

$$F_{2N}(x) = \sum_{p_2 \leq N} \log p_2 \cdot e(-2p_2 x)$$

$$F_N(x) = \sum_{p_1 \leq N} \sum_{p_2 \leq N} \log p_1 \log p_2 \cdot e\left((p_1 - 2p_2)x\right) = F_{1N}(x)F_{2N}(x).$$

$$(14.5)$$

$F_N(x)$ is the generating function for the Germain primes. As $F_N(x)$ is periodic with period 1, we can integrate either over $[0, 1]$ or $[-\frac{1}{2}, \frac{1}{2}]$. We choose the latter because the main contribution to the integral is from x near 0, although both choices obviously yield the same result. Letting $R(1; A_{1N}, A_{2N})$ denote the weighted number of Germain primes, we have

$$R(1; A_{1N}, A_{2N}) = \int_{-\frac{1}{2}}^{\frac{1}{2}} F_N(x)e(-x)dx$$

$$= \sum_{p_1 \leq N} \sum_{p_2 \leq N} \log p_1 \log p_2 \int_{-\frac{1}{2}}^{\frac{1}{2}} e\left((p_1 - 2p_2 - 1)x\right) dx.$$

$$(14.6)$$

By Exercise 13.2.7

$$\int_{-\frac{1}{2}}^{\frac{1}{2}} e\left((p_1 - 2p_2 - 1)x\right) dx = \begin{cases} 1 & \text{if } p_1 - 2p_2 - 1 = 0 \\ 0 & \text{if } p_1 - 2p_2 - 1 \neq 0. \end{cases} \qquad (14.7)$$

In (14.6) we have a contribution of $\log p_1 \log p_2$ if p_1 and $p_2 = \frac{p_1 - 1}{2}$ are both prime and 0 otherwise. Thus

$$R(1; A_{1N}, A_{2N}) = \int_{-\frac{1}{2}}^{\frac{1}{2}} F_N(x)e(-x)dx = \sum_{\substack{p_1 \leq N \\ p_1, p_2 = \frac{p_1 - 1}{2} \text{ prime}}} \log p_1 \log p_2.$$

$$(14.8)$$

The above is a weighted counting of Germain primes. We have introduced these weights to facilitate applying the Siegel-Walfisz formula; it is easy to pass from bounds for $r(1; A_{1N}, A_{2N})$ to bounds for the number of Germain primes (see §14.7).

Remark 14.2.1. Using the λ-function from (14.4), we can rewrite the generating function as a sum over pairs of integers instead of pairs of primes:

$$F_N(x) = \sum_{m_1=1}^{N} \sum_{m_2=1}^{N} \lambda(m_1)\lambda(m_2) \cdot e((m_1 - 2m_2)x). \qquad (14.9)$$

Of course, the two functions are the same; sometimes it is more convenient to use one notation over the other. When we apply partial summation, it is convenient if our terms are defined for all integers, and not just at primes.

Exercise[(h)] 14.2.2. *Determine (or at least bound) the average values of $F_{1N}(x)$, $F_{2N}(x)$ and $F_N(x)$. Hint: For $F_N(x)$, use the Cauchy-Schwartz inequality.*

14.2.2 The Major and Minor Arcs

Let B, D be positive integers with $D > 2B$. Set $Q = \log^D N$. Define the Major arc $\mathcal{M}_{a,q}$ for each pair (a, q) with a and q relatively prime and $1 \le q \le \log^B N$ by

$$\mathcal{M}_{a,q} = \left\{ x \in \left(-\frac{1}{2}, \frac{1}{2} \right) : \left| x - \frac{a}{q} \right| < \frac{Q}{N} \right\} \tag{14.10}$$

if $\frac{a}{q} \ne \frac{1}{2}$ and

$$\mathcal{M}_{1,2} = \left[-\frac{1}{2}, -\frac{1}{2} + \frac{Q}{N} \right) \cup \left(\frac{1}{2} - \frac{Q}{N}, \frac{1}{2} \right]. \tag{14.11}$$

Remember, as our generating function is periodic with period 1, we can work on either $[0, 1]$ or $[-\frac{1}{2}, \frac{1}{2}]$. As the Major arcs depend on N and D, we should write $\mathcal{M}_{a,q}(N, D)$ and $\mathcal{M}(N, D)$; however, for notational convenience N and D are often suppressed. Note we are giving ourselves a little extra flexibility by having $q \le \log^B N$ and each $\mathcal{M}_{a,q}$ of size $\frac{\log^D N}{N}$. We see in §14.5 why we need to have $D > 2B$.

By definition, the Minor arcs m are whatever is not in the Major arcs. Thus the Major arcs are the subset of $[-\frac{1}{2}, \frac{1}{2}]$ near rationals with small denominators, and the Minor arcs are what is left. Here near and small are relative to N. Then

$$R(1; A_{1N}, A_{2N}) = \int_{-\frac{1}{2}}^{\frac{1}{2}} F_N(x) e(-x) dx$$

$$= \int_{\mathcal{M}} F_N(x) e(-x) dx + \int_{\mathrm{m}} F_N(x) e(-x) dx. \tag{14.12}$$

We will calculate the contribution to $R(1; A_{1N}, A_{2N})$ from the Major arcs, and then in §14.7 we remove the $\log p_i$ weights.

We chose the above definition for the Major arcs because our main tool for evaluating $F_N(x)$ is the Siegel-Walfisz formula (Theorem 13.2.13), which states that given any $B, C > 0$, if $q \le \log^B N$ and $(r, q) = 1$ then

$$\sum_{\substack{p \le N \\ p \equiv r(q)}} \log p = \frac{N}{\phi(q)} + O\left(\frac{N}{\log^C N} \right). \tag{14.13}$$

For C very large, the error term leads to small, manageable errors on the Major arcs. For a more detailed explanation of this choice for the Major arcs, see §13.3.3 and §13.3.4.

We show the Major arcs contribute, up to lower order terms, $T_2 N$, where T_2 is a constant independent of N. By choosing B, C and D sufficiently large we can ensure that the errors from the Major arc calculations are less than the main term from the Major arcs. We cannot yet control what happens on the Minor arcs. Similar to Chapter 13 (see §13.3.2), up to powers of $\log N$ we have $F_N(x)$ on average is of size N, but is of size N^2 on the Major arcs. As there is a lot of oscillation in the generating function $F_N(x)$, for generic $x \in [-\frac{1}{2}, \frac{1}{2}]$ we expect a lot of cancellation in the size of $F_N(x)$. Unfortunately, we are unable to prove that this oscillation yields the Minor arcs contributing less than the Major arcs.

We highlight the upcoming calculations. On the Major arcs $\mathcal{M}_{a,q}$, we find a function u of size N^2 such that the error from u to F_N on $\mathcal{M}_{a,q}$ is much smaller than N^2, say N^2 divided by a large power of $\log N$. When we integrate u over the Major arcs, we find the main term is of size N (because up to powers of $\log N$ the Major arcs are of size $\frac{1}{N}$), and we succeed if we can show the errors in the approximations are much smaller than N, say N divided by a large power of $\log N$. Numerical simulations for x up to 10^9 and higher support the conjecture that the Minor arcs do not contribute for the Germain problem. Explicitly, the observed number of Germain prime pairs in this range agrees with the prediction from the Major arcs (see [Weir]). We content ourselves with calculating the contribution from the Major arcs,

$$\int_{\mathcal{M}} F_N(x)e(-x)dx. \tag{14.14}$$

14.3 THE FUNCTIONS $F_N(x)$ AND $u(x)$

After determining F_N on $\mathcal{M}_{a,q}$, we describe an easily integrable function which is close to F_N on $\mathcal{M}_{a,q}$. We calculate $F_N\left(\frac{a}{q}\right)$ for $q \le \log^B N$. Define the **Ramanujan sum** $c_q(a)$ by

$$c_q(a) = \sum_{\substack{r=1 \\ (r,q)=1}}^{q} e\left(r\frac{a}{q}\right). \tag{14.15}$$

As usual, we evaluate $F_N\left(\frac{a}{q}\right)$ with the Siegel-Walfisz Theorem (Theorem 13.2.13). We restrict below to $(r_i, q) = 1$ because if $(r_i, q) > 1$, there is at most one prime $p_i \equiv r_i \bmod q$, and one prime will give a negligible contribution as $N \to \infty$. See also §13.3.5. We have

$$
\begin{aligned}
F_N\left(\frac{a}{q}\right) &= \sum_{p_1 \le N} \log p_1 \cdot e\left(p_1 \frac{a}{q}\right) \sum_{p_2 \le N} \log p_2 \cdot e\left(-2p_2 \frac{a}{q}\right) \\
&= \sum_{r_1=1}^{q} \sum_{\substack{p_1 \le N \\ p_1 \equiv r_1(q)}} \log p_1 \cdot e\left(p_1 \frac{a}{q}\right) \sum_{r_2=1}^{q} \sum_{\substack{p_2 \le N \\ p_2 \equiv r_1(q)}} \log p_2 \cdot e\left(-2p_2 \frac{a}{q}\right) \\
&= \sum_{r_1=1}^{q} e\left(r_1 \frac{a}{q}\right) \sum_{r_2=1}^{q} e\left(r_2 \frac{-2a}{q}\right) \sum_{\substack{p_1 \le N \\ p_1 \equiv r_1(q)}} \log p_1 \sum_{\substack{p_2 \le N \\ p_2 \equiv r_2(q)}} \log p_2 \\
&= \sum_{\substack{r_1=1 \\ (r_1,q)=1}}^{q} e\left(r_1 \frac{a}{q}\right) \sum_{\substack{r_2=1 \\ (r_2,q)=1}}^{q} e\left(r_2 \frac{-2a}{q}\right) \left[\frac{N}{\phi(q)} + O\left(\frac{N}{\log^C N}\right)\right]^2 \\
&= \frac{c_q(a)c_q(-2a)}{\phi(q)^2} N^2 + O\left(\frac{N^2}{\log^{C-B} q}\right). \tag{14.16}
\end{aligned}
$$

Exercise 14.3.1. *Show that the contribution from* (r_i, q) *may safely be absorbed by the error term.*

Let

$$u(x) = \sum_{m_1=1}^{N} \sum_{m_2=1}^{N} e\left((m_1 - 2m_2)x\right). \tag{14.17}$$

As $u(0) = N^2$, it is natural to compare $F_N(x)$ on the Major arcs $\mathcal{M}_{a,q}$ to

$$\frac{c_q(a)c_q(-2a)}{\phi(q)^2} u\left(x - \frac{a}{q}\right), \tag{14.18}$$

as these two functions agree at $x = \frac{a}{q}$. The function $u(x)$ is a lot easier to analyze than $F_N(x)$. We show that for $x \in \mathcal{M}_{a,q}$ there is negligible error in replacing $F_N(x)$ with $\frac{c_q(a)c_q(-2a)}{\phi(q)^2} u(x - \frac{a}{q})$. We then integrate over $\mathcal{M}_{a,q}$, and then sum over all Major arcs. We describe in great detail in Remark 14.5.10 why it is natural to consider $u(x)$.

14.4 APPROXIMATING $F_N(x)$ ON THE MAJOR ARCS

In this technical section we apply partial summation multiple times to show u is a good approximation to F_N on the Major arcs $\mathcal{M}_{a,q}$. Define

$$C_q(a) = \frac{c_q(a)c_q(-2a)}{\phi(q)^2}. \tag{14.19}$$

We show

Theorem 14.4.1. *For* $\alpha \in \mathcal{M}_{a,q}$,

$$F_N(\alpha) = C_q(a)u\left(\alpha - \frac{a}{q}\right) + O\left(\frac{N^2}{\log^{C-2D} N}\right). \tag{14.20}$$

For $\alpha \in \mathcal{M}_{a,q}$ we write α as $\beta + \frac{a}{q}$, $\beta \in \left[-\frac{Q}{N}, \frac{Q}{N}\right]$. Remember $Q = \log^D N$ and $q \leq \log^B N$. Note $F_N(x)$ is approximately $C_q(a)N^2$ for x near $\frac{a}{q}$, and from our definitions of F_N, u and $C_q(a)$, (14.20) is immediate for $\alpha = \frac{a}{q}$. The reader interested in the main ideas of the Circle Method may skip to §14.5 (though see Exercise 14.4.3 for an alternate proof), where we integrate $u(x)$ over the Major arcs. The rest of this section is devoted to rigorously showing that $|F_N(x) - C_q(a)u(x - \frac{a}{q})|$ is small.

The calculation below is a straightforward application of partial summation. The difficulty is that we must apply partial summation twice. Each application yields two terms, a boundary term and an integral term. We will have four pieces to analyze. The problem is to estimate the difference

$$S_{a,q}(\alpha) = F_N(\alpha) - C_q(a)u\left(\alpha - \frac{a}{q}\right) = F_N\left(\beta + \frac{a}{q}\right) - C_q(a)u(\beta). \tag{14.21}$$

Recall that $q \leq \log^B N$ and $F_N(\frac{a}{q}) = C_q(a)u(0)$ is of size $\frac{N^2}{\phi(q)^2}$. To prove Theorem 14.4.1 we must show that $|S_{a,q}(\alpha)| \leq \frac{N^2}{\log^{C-2D} N}$. As mentioned in Remark 14.2.1, it is easier to apply partial summation if we use the λ-formulation of the generating function F_N because now both F_N and u will be sums over $m_1, m_2 \leq N$. Thus

$$S_{a,q}(\alpha)$$
$$= \sum_{m_1, m_2 \leq N} \lambda(m_1)\lambda(m_2)e\left((m_1 - 2m_2)\beta\right) - C_q(a) \sum_{m_1, m_2 \leq N} e\left((m_1 - 2m_2)\beta\right)$$
$$= \sum_{m_1, m_2 \leq N} \left[\lambda(m_1)\lambda(m_2)e\left(\frac{(m_1 - 2m_2)a}{q}\right) - C_q(a)\right] e\left((m_1 - 2m_2)\beta\right)$$
$$= \sum_{m_1 \leq N} \sum_{m_2 \leq N} a_{m_2}(m_1, N)e(-2m_2\beta)e(m_1\beta)$$
$$= \sum_{m_1 \leq N} \left[\sum_{m_2 \leq N} a_{m_2}(m_1, N)b_{m_2}(m_1, N)\right] e(m_1\beta)$$
$$= \sum_{m_1 \leq N} S_{a,q}(\alpha; m_1)e(m_1\beta), \tag{14.22}$$

where

$$a_{m_2}(m_1, N) = \lambda(m_1)\lambda(m_2)e\left(\frac{(m_1 - 2m_2)a}{q}\right) - C_q(a)$$
$$b_{m_2}(m_1, N) = e(-2m_2\beta)$$
$$S_{a,q}(\alpha; m_1) = \sum_{m_2 \leq N} a_{m_2}(m_1, N)b_{m_2}(m_1, N). \tag{14.23}$$

We have written $S_{a,q}(\alpha)$ as above to illuminate the application of partial summation. We hold m_1 fixed and then use partial summation on the m_2-sum. This generates two terms, a boundary and an integral term. We then apply partial summation to the m_1-sum. The difficulty is not in evaluating the sums, but rather in the necessary careful book-keeping required.

Recall the integral version of partial summation (Theorem 2.2.6) states

$$\sum_{m=1}^{N} a_m b(m) = A(N)b(N) - \int_1^N A(u)b'(u)du, \tag{14.24}$$

where b is a differentiable function and $A(u) = \sum_{m \leq u} a_m$. We apply this to $a_{m_2}(m_1, N)$ and $b_{m_2}(m_1, N)$. As $b_{m_2} = b(m_2) = e(-2\beta m_2) = e^{-4\pi i \beta m_2}$, $b'(m_2) = -4\pi i \beta e(-2\beta m_2)$.

Applying the integral version of partial summation to the m_2-sum gives

$S_{a,q}(\alpha; m_1)$

$$
= \sum_{m_2 \le N} \left[\lambda(m_1)\lambda(m_2) e\left((m_1 - 2m_2)\frac{a}{q} \right) - C_q(a) \right] e(-2m_2\beta)
$$

$$
= \sum_{m_2 \le N} a_{m_2}(m_1, N) b_{m_2}(m_1, N)
$$

$$
= \left[\sum_{m_2 \le N} a_{m_2}(m_1, N) \right] e(-2N\beta)
$$

$$
+ 4\pi i\beta \int_{u=1}^{N} \left[\sum_{m_2 \le u} a_{m_2}(m_1, N) \right] e(-u\beta) du. \tag{14.25}
$$

The first term is called the boundary term, the second the integral term. We substitute these into (14.22) and find

$$
S_{a,q}(\alpha) = \sum_{m_1 \le N} \left[\left[\sum_{m_2 \le N} a_{m_2}(m_1, N) \right] e(-2N\beta) \right] e(m_1\beta)
$$

$$
+ \sum_{m_1 \le N} \left[4\pi i\beta \int_{u=1}^{N} \left[\sum_{m_2 \le u} a_{m_2}(m_1, N) \right] e(-u\beta) du \right] e(m_1\beta)
$$

$$
= S_{a,q}(\alpha; \text{boundary}) + S_{a,q}(\alpha; \text{integral}). \tag{14.26}
$$

The proof of Theorem 14.4.1 is completed by showing $S_{a,q}(\alpha; \text{boundary})$ and $S_{a,q}(\alpha; \text{integral})$ are small. This is done in Lemmas 14.4.4 and 14.4.9 by straightforward partial summation.

Remark 14.4.2. The factor of $4\pi i\beta$ in (14.26) is from differentiating $b(m_2)$. Remember $\alpha = \frac{a}{q} + \beta$ is in the Major arc $\mathcal{M}_{a,q} = \left[\frac{a}{q} - \frac{Q}{N}, \frac{a}{q} + \frac{Q}{N} \right]$. Thus, $|\beta| \le \frac{Q}{N} = \frac{\log^D N}{N}$. Even though the integral in (14.26) is over a range of length N, it is multiplied by β, which is small. If β were not present, this term would yield a contribution greater than the expected main term.

Exercise 14.4.3. *We prove Theorem 14.4.1 by applying partial summation twice to study $|F_N(x) - u(x)|$. We chose this proof as it illustrates techniques useful to handle complicated sums. Alternatively we could proceed by finding a good approximation $u_1(x)$ (resp., $u_2(x)$) to the generating function $F_{1N}(x)$ (resp., $F_{2N}(x)$) from (14.5) on $\mathcal{M}_{a,q}$ such that $u_1(x)u_2(x) = u(x)$ and $|F_{iN}(x) - u_i(x)|$ is small for $x \in \mathcal{M}_{a,q}$. The proof follows from the common technique of adding zero in a clever way:*

$$
|F_{1N}(x)F_{2N}(x) - u_1(x)u_2(x)|
$$

$$
= |F_{1N}(x)F_{2N}(x) - u_1(x)F_{2N}(x) + u_1(x)F_{2N}(x) - u_1(x)u_2(x)|
$$

$$
\le |F_{1N}(x) - u_1(x)| \max_{x \in \mathcal{M}_{a,q}} |F_{2N}(x)| + |F_{2N}(x) - u_2(x)| \max_{x \in \mathcal{M}_{a,q}} |u_1(x)|.
$$

$$
\tag{14.27}
$$

The fact that our generating function $F_N(x)$ factors as the product of two generating functions allows us to avoid multiple partial summations. Find such functions $u_i(x)$ and use these and the above arguments to complete this alternate proof of Theorem 14.4.1.

14.4.1 Boundary Term

We first deal with the boundary term from the first partial summation on m_2, the $S_{a,q}(\alpha; \text{boundary})$ term from (14.26).

Lemma 14.4.4. *We have*

$$S_{a,q}(\alpha; \text{boundary}) \;=\; O\left(\frac{N^2}{\log^{C-D} N}\right). \tag{14.28}$$

Proof. Recall that

$$S_{a,q}(\alpha; \text{boundary}) \;=\; \sum_{m_1 \leq N}\left[\left[\sum_{m_2 \leq N} a_{m_2}(m_1, N)\right] e(-2N\beta)\right] e(m_1\beta)$$

$$=\; e(-2N\beta) \sum_{m_1 \leq N}\left[\sum_{m_2 \leq N} a_{m_2}(m_1, N)\right] e(m_1\beta). \tag{14.29}$$

As $|(e(-2N\beta)| = 1$, we can ignore it in the bounds below. We again apply the integral version of partial summation with

$$a_{m_1} \;=\; \sum_{m_2 \leq N} a_{m_2}(m_1, N)$$

$$=\; \sum_{m_2 \leq N}\left[\lambda(m_1)\lambda(m_2)e\left((m_1 - 2m_2)\frac{a}{q}\right) - C_q(a)\right]$$

$$b_{m_1} \;=\; e(m_1\beta). \tag{14.30}$$

We find

$$e(2N\beta)S_{a,q}(\alpha; \text{boundary}) \;=\; \sum_{m_1 \leq N}\left[\sum_{m_2 \leq N} a_{m_2}(m_1, N)\right] e(N\beta)$$

$$-\; 2\pi i\beta \int_{t=0}^{N} \sum_{m_1 \leq t}\left[\sum_{m_2 \leq N} a_{m_2}(m_1, N)\right] e(t\beta)dt. \tag{14.31}$$

The proof of Lemma 14.4.4 is completed by bounding the two terms in (14.31), which we do in Lemmas 14.4.5 and 14.4.6. $\qquad\square$

Lemma 14.4.5.

$$\sum_{m_1 \leq N}\left[\sum_{m_2 \leq N} a_{m_2}(m_1, N)\right] e(N\beta) \;=\; O\left(\frac{N^2}{\log^C N}\right). \tag{14.32}$$

Proof. As $|e(N\beta)| = 1$, this factor is harmless, and the m_1, m_2-sums are bounded by the Siegel-Walfisz Theorem.

$$\sum_{m_1 \leq N} \sum_{m_2 \leq N} a_{m_2}(m_1, N)$$

$$= \sum_{m_1 \leq N} \sum_{m_2 \leq N} \left[\lambda(m_1)\lambda(m_2) e\left((m_1 - 2m_2)\frac{a}{q} \right) - C_q(a) \right]$$

$$= \left[\sum_{m_1 \leq N} \lambda(m_1) e\left(m_1 \frac{a}{q} \right) \right] \left[\sum_{m_2 \leq N} \lambda(m_2) e\left(-m_2 \frac{a}{q} \right) \right] - C_q(a)N^2$$

$$= \left[\frac{c_q(a)N}{\phi(q)} + O\left(\frac{N}{\log^C N} \right) \right] \cdot \left[\frac{c_q(-2a)N}{\phi(q)} + O\left(\frac{N}{\log^C N} \right) \right] - C_q(a)N^2$$

$$= O\left(\frac{N^2}{\log^C N} \right) \tag{14.33}$$

as $C_q(a) = \frac{c_q(a)c_q(-2a)}{\phi(q)^2}$ and $|c_q(b)| \leq \phi(q)$. $\qquad\square$

Lemma 14.4.6.

$$2\pi i \beta \int_{t=0}^{N} \sum_{m_1 \leq t} \left[\sum_{m_2 \leq N} a_{m_2}(m_1, N) \right] e(t\beta) dt = O\left(\frac{N^2}{\log^{C-D} N} \right). \tag{14.34}$$

Proof. Note $|\beta| \leq \frac{Q}{N} = \frac{\log^D N}{N}$, and $C_q(a) = \frac{c_q(a)c_q(-2a)}{\phi(q)^2}$. For $t \leq \sqrt{N}$ we trivially bound the m_2-sum by $2N$. Thus these t contribute at most

$$|\beta| \int_{t=0}^{\sqrt{N}} \sum_{m_1 \leq t} 2N dt = 2|\beta|N \int_{t=0}^{\sqrt{N}} t \, dt = |\beta|N^2 \leq N \log^D N. \tag{14.35}$$

An identical application of Siegel-Walfisz as in the proof of Lemma 14.4.5 yields for $t \geq \sqrt{N}$,

$$\sum_{m_1 \leq t} \sum_{m_2 \leq N} a_{m_2}(m_1, N)$$

$$= \left[\frac{c_q(a)t}{\phi(q)} + O\left(\frac{t}{\log^C N} \right) \right] \cdot \left[\frac{c_q(-2a)N}{\phi(q)} + O\left(\frac{N}{\log^C N} \right) \right] - C_q(a)tN$$

$$= O\left(\frac{tN}{\log^C N} \right). \tag{14.36}$$

Therefore

$$|\beta| \int_{t=\sqrt{N}}^{N} \left| \sum_{m_1 \leq t} \sum_{m_2 \leq N} a_{m_2}(m_1, N) \right| dt = O\left(\frac{N^3 \beta}{\log^C N} \right)$$

$$= O\left(\frac{N^2}{\log^{C-D} N} \right) \tag{14.37}$$

$\qquad\square$

Remark 14.4.7. Note, of course, that the contribution is only negligible while $|\beta| \leq \frac{Q}{N}$. We see a natural reason to take the Major arcs small in length.

Remark 14.4.8. The above argument illustrates a very common technique. Namely, if $t \in [0, N]$ and N is large, the interval $[0, \sqrt{N}]$ has negligible relative length. It is often useful to break the problem into two such regions, as different bounds are often available when t is large and small. For our problem, the Siegel-Walfisz formula requires that $q \leq \log^B t$; this condition fails if t is small compared to q. For small t, the bounds may not be as good; however, the length of such an interval is so small that weak bounds suffice. See also the example in §2.3.4.

14.4.2 Integral Term

We now analyze the integral term from the first partial summation on m_2, the $S_{a,q}(\alpha; \text{integral})$ term from (14.26).

Lemma 14.4.9.

$$S_{a,q}(\alpha; \text{integral}) = O\left(\frac{N^2}{\log^{C-2D} N}\right). \tag{14.38}$$

Proof. Recall

$$S_{a,q}(\alpha; \text{integral})$$

$$= 4\pi i \beta \sum_{m_1 \leq N} \left[\int_{u=1}^{N} \left[\sum_{m_2 \leq u} a_{m_2}(m_1, N)\right] e(-u\beta) du\right] e(m_1 \beta) \tag{14.39}$$

where

$$a_{m_2}(m_1, N) = \lambda(m_1)\lambda(m_2)e\left((m_1 - 2m_2)\frac{a}{q}\right) - C_q(a). \tag{14.40}$$

We apply the integral version of partial summation, with

$$a_{m_1} = \int_{u=1}^{N} \left[\sum_{m_2 \leq u} a_{m_2}(m_1, N)\right] e(-u\beta) du$$

$$b_{m_1} = e(m_1 \beta). \tag{14.41}$$

We find

$$S_{a,q}(\alpha; \text{integral})$$

$$= 4\pi i \beta \left[\sum_{m_1 \leq N} \int_{u=1}^{N} \sum_{m_2 \leq u} a_{m_2}(m_1, N)e(-u\beta) du\right] e(N\beta)$$

$$+ 8\pi \beta^2 \int_{t=1}^{N} \left[\sum_{m_1 \leq t} \int_{u=1}^{N} \sum_{m_2 \leq u} a_{m_2}(m_1, N)e(-u\beta) du\right] e(m_1 t) dt. \tag{14.42}$$

The factor of $8\pi\beta^2 = -(4\pi i\beta) \cdot (2\pi i\beta)$ and comes from the derivative of $e(m_1\beta)$. Arguing in a similar manner as in §14.4.1, in Lemmas 14.4.10 and 14.4.11 we show the two terms in (14.42) are small, which will complete the proof. □

Lemma 14.4.10.

$$4\pi i\beta \left[\sum_{m_1 \leq N} \int_{u=1}^{N} \sum_{m_2 \leq u} a_{m_2}(m_1, N)e(-u\beta)du \right] e(N\beta) = O\left(\frac{N^2}{\log^{C-D} N} \right).$$

$$(14.43)$$

Proof. Arguing along the lines of Lemma 14.4.6, one shows the contribution from $u \leq \sqrt{N}$ is bounded by $N \log^{D} N$. For $u \geq \sqrt{N}$ we apply the Siegel-Walfisz formula as in Lemma 14.4.6, giving a contribution bounded by

$$4|\beta| \int_{u=\sqrt{N}}^{N} \left(\left[\frac{c_q(a)u}{\phi(q)} + O\left(\frac{u}{\log^{C} N} \right) \right] \right.$$

$$\left. \cdot \left[\frac{c_q(-2a)N}{\phi(q)} + O\left(\frac{N}{\log^{C} N} \right) \right] - C_q(a)uN \right) du$$

$$\ll |\beta| \int_{u=\sqrt{N}}^{N} \frac{uN}{\log^{C} N} du$$

$$\ll \frac{N^3 |\beta|}{\log^{C} N}.$$

$$(14.44)$$

As $|\beta| \leq \frac{Q}{N} = \frac{\log^{D} N}{N}$, the above is $O\left(\frac{N^2}{\log^{C-D} N} \right)$. $\square$

Lemma 14.4.11. *We have*

$$8\pi\beta^2 \int_{t=1}^{N} \left[\sum_{m_1 \leq t} \int_{u=1}^{N} \sum_{m_2 \leq u} a_{m_2}(m_1, N)e(-u\beta)du \right] e(m_1 t)dt$$

$$= O\left(\frac{N^2}{\log^{C-2D} N} \right).$$

$$(14.45)$$

Proof. The proof proceeds along the same lines as the previous lemmas. Arguing as in Lemma 14.4.6, one shows that the contribution when either $t \leq \sqrt{N}$ or $u \leq \sqrt{N}$ is $O\left(\frac{N}{\log^{C-2D} N} \right)$. We then apply the Siegel-Walfisz Theorem as before, and find the contribution when both $t, u \geq \sqrt{N}$ is

$$\ll 8\beta^2 \int_{t=\sqrt{N}}^{N} \int_{u=\sqrt{N}}^{N} \frac{ut}{\log^{C} N} dudt \ll \frac{N^4 \beta^2}{\log^{C} N}.$$

$$(14.46)$$

As $|\beta| \leq \frac{\log^{D} N}{N}$, the above is $O\left(\frac{N^2}{\log^{C-2D} N} \right)$. $\square$

This completes the proof of Theorem 14.4.1.

14.5 INTEGRALS OVER THE MAJOR ARCS

We first compute the integral of $u(x)e(-x)$ over the Major arcs and then use Theorem 14.4.1 to deduce the corresponding integral of $F_N(x)e(-x)$.

14.5.1 Integrals of $u(x)$

By Theorem 14.4.1 we know for $x \in \mathcal{M}_{a,q}$ that

$$\left| F_N(x) - C_q(a)u\left(x - \frac{a}{q}\right) \right| \ll O\left(\frac{N^2}{\log^{C-2D} N}\right). \tag{14.47}$$

We now evaluate the integral of $u(x - \frac{a}{q})e(-x)$ over $\mathcal{M}_{a,q}$; by Theorem 14.4.1 we then obtain the integral of $F_N(x)e(-x)$ over $\mathcal{M}_{a,q}$. Remember (see (14.17)) that

$$u(x) = \sum_{m_1,m_2 \leq N} e\left((m_1 - 2m_2)x\right). \tag{14.48}$$

Theorem 14.5.1. *We have*

$$\int_{\mathcal{M}_{a,q}} u\left(\alpha - \frac{a}{q}\right) \cdot e(-\alpha)d\alpha = e\left(-\frac{a}{q}\right)\frac{N}{2} + O\left(\frac{N}{\log^D N}\right). \tag{14.49}$$

Theorem 14.5.1 will follow from a string of lemmas on various integrals of u. We first determine the integral of u over all of $[-\frac{1}{2}, \frac{1}{2}]$, and then show that the integral of $u(x)$ is small if $|x| > \frac{Q}{N}$.

Lemma 14.5.2. *We have*

$$\int_{-\frac{1}{2}}^{\frac{1}{2}} u(x)e(-x)dx = \frac{N}{2} + O(1). \tag{14.50}$$

Proof. We start with the observation that

$$\int_{-\frac{1}{2}}^{\frac{1}{2}} u(x)e(-x)dx = \int_{-\frac{1}{2}}^{\frac{1}{2}} \sum_{m_1 \leq N}\sum_{m_2 \leq N} e\left((m_1 - 2m_2)x\right) \cdot e(-x)dx$$

$$= \sum_{m_1 \leq N}\sum_{m_2 \leq N} \int_{-\frac{1}{2}}^{\frac{1}{2}} e\left((m_1 - 2m_2 - 1)x\right)dx. \tag{14.51}$$

By Exercise 13.2.7 the integral is 1 if $m_1 - 2m_2 - 1 = 0$ and 0 otherwise. For $m_1, m_2 \in \{1, \dots, N\}$, there are $\left[\frac{N}{2}\right] = \frac{N}{2} + O(1)$ solutions to $m_1 - 2m_2 - 1 = 0$, which completes the proof. $\square$

Define

$$I_1 = \left[-\frac{1}{2} + \frac{Q}{N}, -\frac{Q}{N}\right], \quad I_2 = \left[\frac{Q}{N}, \frac{1}{2} - \frac{Q}{N}\right]. \tag{14.52}$$

The following bound is crucial in our investigations.

Lemma 14.5.3. *For $x \in I_1$ or I_2, $\frac{1}{1-e(ax)} \ll \frac{1}{x}$ for a equal to 1 or -2.*

Exercise 14.5.4. *Prove Lemma 14.5.3 and with a as above, find a constant C_i such that for $x \in I_i$ we have $\left|\frac{1}{1-e(ax)}\right| \leq \frac{C_i}{|x|}$.*

Lemma 14.5.5. *We have*

$$\int_{x \in I_1 \cup I_2} u(x)e(-x)dx = O\left(\frac{N}{\log^D N}\right). \tag{14.53}$$

Proof. We have

$$
\int_{I_i} u(x)e(-x)dx = \int_{I_i} \sum_{m_1,m_2 \le N} e\left((m_1 - 2m_2 - 1)x\right) dx
$$

$$
= \int_{I_i} \sum_{m_1 \le N} e(m_1 x) \sum_{m_2 \le N} e(-2m_2 x) \cdot e(-x)dx
$$

$$
= \int_{I_i} \left[\frac{e(x) - e((N+1)x)}{1 - e(x)} \right] \left[\frac{e(-2x) - e(-2(N+1)x)}{1 - e(-2x)} \right] e(-x)dx
$$

$$
\tag{14.54}
$$

because these are geometric series. By Lemma 14.5.3 each geometric series is
$O\left(\frac{1}{x}\right)$ for $x \in I_i$. We thus have

$$
\int_{I_i} u(x)e(-x)dx \ll \int_{I_i} \frac{1}{x}\frac{1}{x}dx \ll \frac{N}{Q} = \frac{N}{\log^D N},
\tag{14.55}
$$

which completes the proof of Lemma 14.5.5.   □

Exercise 14.5.6. *Show*

$$
\int_{I_i} \frac{1}{x^2}dx \ll \frac{N}{Q}.
\tag{14.56}
$$

Remark 14.5.7. It is because the error term in Lemma 14.5.5 is $O\left(\frac{N}{\log^D N}\right)$ that
we must take $D > 2B$.

Lemma 14.5.8. *We have*

$$
\int_{\frac{1}{2}-\frac{Q}{N}}^{\frac{1}{2}+\frac{Q}{N}} u(x)e(-x)dx = O\left(\log^D N\right).
\tag{14.57}
$$

Proof. The argument is similar to the proof of Lemma 14.5.5. The difference is we
use the geometric series formula only for the m_1-sum, which is $\frac{e(x)-e((N+1)x)}{1-e(x)} \ll$
$\frac{1}{x}$. As x is near $\frac{1}{2}$, the m_1-sum is $O(1)$. There are N terms in the m_2-sum. As
each term is at most 1, we may bound the m_2-sum by N. Thus the integrand
is $O(N)$. We integrate over a region of length $\frac{2Q}{N}$ and see that the integral is
$O(Q) = O(\log^D N)$ for N large.   □

Note in the above proof we could not use the geometric series for both m-sums,
as near $x = \frac{1}{2}$ the second sum is quite large. Fortunately, we still have significant cancellation in the first sum, and we are integrating over a small region. The
situation is different in the following lemma. There, *both* m-sums are large. Not
surprisingly, this is where most of the mass of u is concentrated.

Lemma 14.5.9.

$$
\int_{-\frac{Q}{N}}^{\frac{Q}{N}} u(x)e(-x)dx = \frac{N}{2} + O\left(\frac{N}{\log^D N}\right).
\tag{14.58}
$$

CIRCLE METHOD: HEURISTICS FOR GERMAIN PRIMES

Proof. This is immediate from Lemma 14.5.2 (which shows the integral of u over $[-\frac{1}{2}, \frac{1}{2}]$ is $\frac{N}{2} + O(1)$) and Lemmas 14.5.5 and 14.5.8 (which show the integral of u over $|x| > \frac{Q}{N}$ is small). $\qquad\square$

It is now trivial to prove Theorem 14.5.1.

Proof of Theorem 14.5.1. We have

$$
\int_{\mathcal{M}_{a,q}} u\left(\alpha - \frac{a}{q}\right) \cdot e(-\alpha)d\alpha = \int_{\frac{a}{q}-\frac{Q}{N}}^{\frac{a}{q}+\frac{Q}{N}} u\left(\alpha - \frac{a}{q}\right) \cdot e(-\alpha)d\alpha
$$
$$
= \int_{-\frac{Q}{N}}^{\frac{Q}{N}} u(\beta) \cdot e\left(-\frac{a}{q} - \beta\right) d\beta
$$
$$
= e\left(-\frac{a}{q}\right) \int_{-\frac{Q}{N}}^{\frac{Q}{N}} u(\beta)e(-\beta)d\beta
$$
$$
= e\left(-\frac{a}{q}\right)\frac{N}{2} + O\left(\frac{N}{\log^D N}\right). \quad (14.59)
$$

$\qquad\square$

Note there are two factors in Theorem 14.5.1. The first, $e\left(-\frac{a}{q}\right)$, is an arithmetic factor which depends on which Major arc $\mathcal{M}_{a,q}$ we are in. The second factor is universal, and is the size of the contribution.

Remark 14.5.10. We remark once more on the utility of finding a function $u(x)$ to approximate $F_N(x)$, as opposed to a Taylor series expansion. We found a function that is easy to integrate and by straightforward applications of partial summation is close to our generating function. Further, most of the mass of $u(x)$ is concentrated in a neighborhood of size $\frac{2Q}{N}$ about 0. Hence integrating u (or its translates) over a Major arc is approximately the same as integrating u over the entire interval $[-\frac{1}{2}, \frac{1}{2}]$. While there are a few points where we need to be careful in analyzing the behavior of u, the slight complications are worth the effort because of how easy it is to work with $u(x)$. For this problem, it was $x = 0$ giving the main contribution, and $x = \pm\frac{1}{2}$ was a potential trouble point which turned out to give a small contribution. The reason we need to check $x = \pm\frac{1}{2}$ is due to the definition of Germain primes, namely the **2** in $F_{2N}(x) = \sum_{p_2 \leq N} e(-2p_2 x)$. Because of this 2, when x is near $\frac{1}{2}$, $F_{2N}(x)$ is near N.

14.5.2 Integrals of $F_N(x)$

An immediate consequence of Theorem 14.5.1 is

Theorem 14.5.11. *We have*
$$
\int_{\mathcal{M}_{a,q}} F_N(x)e(-x)dx = C_q(a)e\left(-\frac{a}{q}\right)\frac{N}{2} + O\left(\frac{N}{\log^D N}\right) + O\left(\frac{N}{\log^{C-3D} N}\right).
$$
$$
(14.60)
$$

Exercise 14.5.12. *Prove Theorem 14.5.11.*

From Theorem 14.5.11 we immediately obtain the integral of $F_N(x)e(-x)$ over the Major arcs $\mathcal{M}$:

Theorem 14.5.13. *We have*

$$\int_{\mathcal{M}} F_N(x)e(-x)dx = \mathfrak{S}_N \frac{N}{2} + O\left(\frac{N}{\log^{D-2B} N} + \frac{N}{\log^{C-3D-2B} N}\right),$$

$$(14.61)$$

where

$$\mathfrak{S}_N = \sum_{q=1}^{\log^B N} \sum_{\substack{a=1 \\ (a,q)=1}}^{q} C_q(a)e\left(-\frac{a}{q}\right)$$

$$(14.62)$$

is the truncated singular series for the Germain primes.

Proof. We have shown that

$$F_N(x)e(-x)dx$$

$$= \sum_{q=1}^{\log^B N} \sum_{\substack{a=1 \\ (a,q)=1}}^{q} \left[C_q(a)e\left(-\frac{a}{q}\right)\frac{N}{2} + O\left(\frac{N}{\log^{D-2B} N} + \frac{N}{\log^{C-3D-2B} N}\right)\right].$$

$$(14.63)$$

By definition the sum over the Major arcs of the main term is the truncated singular series $\mathfrak{S}_N$. We are left with bounding the sum of the error terms over the Major arcs. As

$$\mathcal{M} = \bigcup_{q=1}^{\log^B N} \bigcup_{\substack{a=1 \\ (a,q)=1}}^{q} \mathcal{M}_{a,q},$$

$$(14.64)$$

the number of Major arcs $\mathcal{M}_{a,q}$ is bounded by $\log^{2B} N$. In summing over the Major arcs, the error terms in Theorem 14.5.11 are multiplied by at most $\log^{2B} N$, and the claim now follows. $\square$

We will show the main term in Theorem 14.5.13 is of size N; thus we need to take $D > 2B$ and $C > 3D + 2B$. We study $\mathfrak{S}_N$ in §14.6, and remove the $\log p_i$ weights in §14.7.

14.6 MAJOR ARCS AND THE SINGULAR SERIES

If we can show that there exists a constant $c_0 > 0$ (independent of N) such that

$$\mathfrak{S}_N > c_0,$$

$$(14.65)$$

then for $D > 2B$ and $C > 3D + 2B$ by Theorem 14.5.13 the contribution from the Major arcs is positive and of size $\mathfrak{S}_N \frac{N}{2}$ for N sufficiently large. Recall

$$c_q(a) = \sum_{\substack{r=1 \\ (r,q)=1}}^{q} e\left(r\frac{a}{q}\right)$$

$$C_q(a) = \frac{c_q(a)c_q(-2a)}{\phi(q)^2}. \qquad (14.66)$$

Substituting

$$\rho_q = \sum_{\substack{a=1 \\ (a,q)=1}}^{q} C_q(a)e\left(-\frac{a}{q}\right), \qquad (14.67)$$

into the series expansion of $\mathfrak{S}_N$ in (14.62), we find that

$$\mathfrak{S}_N = \sum_{q=1}^{\log^B N} \rho_q. \qquad (14.68)$$

The singular series for the Germain primes is

$$\mathfrak{S} = \sum_{q=1}^{\infty} \rho_q. \qquad (14.69)$$

We show $\mathfrak{S}$ is given by a multiplicative product and is positive in Theorem 14.6.18, and in Theorem 14.6.20 we show $|\mathfrak{S} - \mathfrak{S}_N| = O\left(\frac{1}{\log^{(1-2\epsilon)B} N}\right)$ for any $\epsilon > 0$. This will complete our evaluation of the contribution from the Major arcs.

Many of the arithmetical functions we investigate below were studied in Chapter 2. Recall a function f is multiplicative if $f(mn) = f(m)f(n)$ for m, n relatively prime, and completely multiplicative if $f(mn) = f(m)f(n)$; see Definition 2.1.2. The reader should consult Chapter 2 as necessary.

14.6.1 Properties of Arithmetic Functions

We follow the presentation of [Na] (Chapter 8 and Appendix A), where many of the same functions arise from studying a related Circle Method problem. Below we determine simple formulas for the arithmetic functions we have encountered, which then allows us to prove our claims about $\mathfrak{S}_N$ and $\mathfrak{S}$ (see §14.6.2).

Lemma 14.6.1. *If $(q, q') = 1$ then we can write the congruence classes relatively prime to qq' as $rq' + r'q$, with $1 \le r \le q$, $1 \le r' \le q'$ and $(r, q) = (r', q') = 1$.*

Exercise 14.6.2. *Prove Lemma 14.6.1.*

Lemma 14.6.3. $c_q(a)$ *is multiplicative.*

Proof. Using Lemma 14.6.1 we have

$$
\begin{aligned}
c_q(a)c_{q'}(a) &= \sum_{\substack{r=1 \\ (r,q)=1}}^{q} e\left(r\frac{a}{q}\right) \sum_{\substack{r'=1 \\ (r',q')=1}}^{q'} e\left(r'\frac{a}{q'}\right) \\
&= \sum_{\substack{r=1 \\ (r,q)=1}}^{q} \sum_{\substack{r'=1 \\ (r',q')=1}}^{q'} e\left(\frac{(rq'+r'q)a}{qq'}\right) \\
&= \sum_{\substack{\tilde{r}=1 \\ (\tilde{r},qq')=1}}^{qq'} e\left(\tilde{r}\frac{a}{q}\right) = c_{qq'}(a),
\end{aligned}
\tag{14.70}
$$

which completes the proof. $\qquad\square$

We will soon determine $c_q(a)$ for $(a,q) = 1$. We first state some needed results. Let

$$
h_d(a) := \sum_{r=1}^{d} e\left(r\frac{a}{d}\right).
\tag{14.71}
$$

Then

Lemma 14.6.4. *We have*

$$
h_d(a) = \begin{cases} d & \text{if } d|a \\ 0 & \text{otherwise.} \end{cases}
\tag{14.72}
$$

Exercise 14.6.5. *Prove the above lemma.*

Recall the Möbius function (see §2.1):

$$
\mu(d) = \begin{cases} (-1)^r & \text{if } d \text{ is the product of } r \text{ distinct primes} \\ 0 & \text{otherwise.} \end{cases}
\tag{14.73}
$$

The following sum is a nice indicator of whether or not r and q are relatively prime. By Lemma 2.1.11,

$$
\sum_{d|(r,q)} \mu(d) = \begin{cases} 1 & \text{if } (r,q) = 1 \\ 0 & \text{otherwise.} \end{cases}
\tag{14.74}
$$

Lemma 14.6.6. *If* $(a,q) = 1$ *then* $c_q(a) = \mu(q)$.

Proof. We have

$$
c_q(a) = \sum_{\substack{r=1 \\ (r,q)=1}}^{q} e\left(r\frac{a}{q}\right) = \sum_{r=1}^{q} e\left(r\frac{a}{q}\right) \sum_{d|(r,q)} \mu(d),
\tag{14.75}
$$

where we used (14.73) to expand the sum from $(r,q) = 1$ to all $r \bmod q$. Further

$$
c_q(a) = \sum_{d|q} \mu(d) \sum_{\substack{r=1 \\ d|r}}^{q} e\left(r\frac{a}{q}\right).
\tag{14.76}
$$

This is because $d|(r,q)$ implies $d|r$ and $d|q$, which allows us to rewrite the conditions above. We change variables and replace r with ℓ; as r ranges from 1 to q through values divisible by ℓ, ℓ ranges from 1 to $\frac{q}{d}$. We use Lemma 14.6.4 to evaluate this sum. Therefore

$$c_q(a) = \sum_{d|q} \mu(d) \sum_{\ell=1}^{q/d} e\left(\ell \frac{a}{q/d}\right)$$

$$= \sum_{d|q} \mu(d) h_{q/d}(a)$$

$$= \sum_{d|q} \mu\left(\frac{q}{d}\right) h_d(a)$$

$$= \sum_{\substack{d|q \\ d|a}} \mu\left(\frac{q}{d}\right) d$$

$$= \sum_{d|(a,q)} \mu\left(\frac{q}{d}\right) d. \qquad (14.77)$$

If $(a, q) = 1$ then the only term above is $d = 1$, which yields $c_q(a) = \mu(q)$. $\qquad \square$

Corollary 14.6.7. *If $q = p^k$, $k \geq 2$ and $(a, q) = 1$, then $c_q(a) = 0$.*

We have shown $c_{qq'}(a) = c_q(a)c_{q'}(a)$ if $(q, q') = 1$. Recall that the Euler ϕ-function, $\phi(q)$, is the number of numbers less than q which are relatively prime to q and is a multiplicative function (see §2.1 for more details). We now have

Lemma 14.6.8. *$C_q(a)$ is multiplicative in q.*

Proof. Assume $(q, q') = 1$. We have

$$C_{qq'}(a) = \frac{c_{qq'}(a)c_{qq'}(-2a)}{\phi(qq')^2}$$

$$= \frac{c_q(a)c_{q'}(a)c_q(-2a)c_{q'}(-2a)}{\phi(q)^2\phi(q')^2}$$

$$= \frac{c_q(a)c_q(-2a)}{\phi(q)^2} \cdot \frac{c_{q'}(a)c_{q'}(-2a)}{\phi(q')^2}$$

$$= C_q(a)C_{q'}(a). \qquad (14.78)$$

$\qquad \square$

We now prove ρ_q is multiplicative. We first prove a needed lemma.

Lemma 14.6.9. *If $(q_1, q_2) = 1$, $C_{q_1}(a_1 q_2) = C_{q_1}(a_1)$.*

Proof. As $C_{q_1}(a_1 q_2) = \frac{c_{q_1}(a_1 q_2)c_{q_1}(-2a_1 q_2)}{\phi(q_1)^2}$, we see it suffices to show $c_{q_1}(a_1 q_2) = c_{q_1}(a_1)$ and $c_{q_1}(-2a_1 q_2) = c_{q_1}(-2a_1)$. As the proofs are similar, we only prove the first statement. From the definition of $c_q(a)$, (14.66), we have

$$
\begin{aligned}
c_{q_1}(a_1 q_2) &= \sum_{\substack{r_1=1 \\ (r_1,q_1)=1}}^{q_1} e\left(r_1 \frac{a_1 q_2}{q_1}\right) \\
&= \sum_{\substack{r_1=1 \\ (r_1,q_1)=1}}^{q_1} e\left(r_1 q_2 \frac{a_1}{q_1}\right) \\
&= \sum_{\substack{r=1 \\ (r,q_1)=1}}^{q_1} e\left(r \frac{a_1}{q_1}\right) = c_{q_1}(a),
\end{aligned}
\tag{14.79}
$$

because $(q_1, q_2) = 1$ implies that as r_1 goes through all residue classes that are relatively prime to q_1, so too does $r = r_1 q_2$. $\square$

Lemma 14.6.10. ρ_q *is multiplicative.*

Proof. Recall

$$
\rho_q = \sum_{\substack{a=1 \\ (a,q)=1}}^{q} C_q(a) e\left(-\frac{a}{q}\right).
\tag{14.80}
$$

Assume $(q_1, q_2) = 1$. By Lemma 14.6.1 we can write the congruence classes relatively prime to $q_1 q_2$ as $a_1 q_2 + a_2 q_1$, with $1 \le a_1 \le q_1$, $1 \le a_2 \le q_2$ and $(a_1, q_1) = (a_2, q_2) = 1$. Then

$$
\begin{aligned}
\rho_{q_1 q_2} &= \sum_{\substack{a=1 \\ (a,q_1 q_2)=1}}^{q_1 q_2} C_{q_1 q_2}(a) e\left(-\frac{a}{q_1 q_2}\right) \\
&= \sum_{\substack{a=1 \\ (a,q_1 q_2)=1}}^{q_1 q_2} C_{q_1}(a) C_{q_2}(a) e\left(-\frac{a}{q_1 q_2}\right) \\
&= \sum_{\substack{a_1=1 \\ (a_1,q_1)=1}}^{q_1} \sum_{\substack{a_2=1 \\ (a_2,q_2)=1}}^{q_2} C_{q_1}(a_1 q_2 + a_2 q_1) C_{q_2}(a_1 q_2 + a_2 q_1) e\left(-\frac{a_1 q_2 + a_2 q_1}{q_1 q_2}\right).
\end{aligned}
\tag{14.81}
$$

A straightforward calculation shows $C_{q_1}(a_1 q_2 + a_2 q_1) = C_{q_1}(a_1 q_2)$ and $C_{q_2}(a_1 q_2 + a_2 q_1) = C_{q_2}(a_2 q_1)$, which implies

$$
\begin{aligned}
\rho_{q_1 q_2} &= \sum_{\substack{a_1=1 \\ (a_1,q_1)=1}}^{q_1} \sum_{\substack{a_2=1 \\ (a_2,q_2)=1}}^{q_2} C_{q_1}(a_1 q_2) C_{q_2}(a_2 q_1) e\left(-\frac{a_1 q_2 + a_2 q_1}{q_1 q_2}\right) \\
&= \left[\sum_{\substack{a_1=1 \\ (a_1,q_1)=1}}^{q_1} C_{q_1}(a_1 q_2) e\left(-\frac{a_1}{q_1}\right)\right]\left[\sum_{\substack{a_2=1 \\ (a_2,q_2)=1}}^{q_2} C_{q_2}(a_2 q_1) e\left(-\frac{a_2}{q_2}\right)\right] \\
&= \left[\sum_{\substack{a_1=1 \\ (a_1,q_1)=1}}^{q_1} C_{q_1}(a_1) e\left(-\frac{a_1}{q_1}\right)\right]\left[\sum_{\substack{a_2=1 \\ (a_2,q_2)=1}}^{q_2} C_{q_2}(a_2) e\left(-\frac{a_2}{q_2}\right)\right] \\
&= \rho_{q_1} \cdot \rho_{q_2},
\end{aligned}
\tag{14.82}
$$

where we used Lemma 14.6.9 to replace $C_{q_1}(a_1 q_2)$ with $C_{q_1}(a_1)$, and similarly for $C_{q_2}(a_2 q_1)$. Thus, ρ_q is multiplicative. □

Exercise 14.6.11. *Prove* $C_{q_1}(a_1 q_2 + a_2 q_1) = C_{q_1}(a_1 q_2)$ *and* $C_{q_2}(a_1 q_2 + a_2 q_1) = C_{q_2}(a_2 q_1)$.

We now determine ρ_q.

Lemma 14.6.12. *If* $k \geq 2$ *and* p *is prime then* $\rho_{p^k} = 0$.

Proof. This follows immediately from $C_{p^k}(a) = 0$ (see Corollary 14.6.7 and the definition of $C_q(a)$). □

Lemma 14.6.13. *If* $p > 2$ *is prime then* $\rho_p = -\frac{1}{(p-1)^2}$.

Proof. Note for p prime, $(a, p) = 1$ means a runs from 1 to $p - 1$. Therefore

$$
\begin{aligned}
\rho_p &= \sum_{\substack{a=1 \\ (a,p)=1}}^{p} C_p(a) e\left(-\frac{a}{p}\right) \\
&= \sum_{a=1}^{p-1} \frac{c_p(a) c_p(-2a)}{\phi(p)^2} e\left(-\frac{a}{p}\right).
\end{aligned}
\tag{14.83}
$$

For $p > 2$, $(a, p) = 1$ implies $(-2a, p) = 1$ as well. By Lemma (14.6.6), $c_p(a) = c_p(-2a) = \mu(p)$. As $\mu(p)^2 = 1$ and $\phi(p) = p - 1$ we have

$$
\begin{aligned}
\rho_p &= \sum_{a=1}^{p-1} \frac{1}{(p-1)^2} e\left(-\frac{a}{p}\right) \\
&= \frac{1}{(p-1)^2}\left[-e\left(-\frac{0}{p}\right) + \sum_{a=0}^{p-1} e\left(-\frac{a}{p}\right)\right] \\
&= -\frac{1}{(p-1)^2}.
\end{aligned}
\tag{14.84}
$$

□

Lemma 14.6.14. $\rho_2 = 1$.

Proof.

$$\rho_2 = \sum_{\substack{a=1 \\ (a,2)=1}}^{2} C_2(a)e\left(-\frac{a}{2}\right)$$

$$= C_2(1)e\left(-\frac{1}{2}\right)$$

$$= \frac{c_2(1)c_2(-2)}{\phi(2)^2} \cdot e^{-\pi i}$$

$$= \frac{e^{\pi i}e^{-2\pi i}}{1^2} \cdot e^{-\pi i} = 1, \tag{14.85}$$

where we have used Exercise 14.6.15, which gives $c_2(1) = e^{\pi i}$ and $c_2(-2) = e^{-2\pi i}$. $\qquad\square$

Exercise 14.6.15. *Prove $c_2(1) = e^{\pi i}$ and $c_2(-2) = e^{-2\pi i}$.*

14.6.2 Determination of $\mathfrak{S}_N$ and $\mathfrak{S}$

We use the results from §14.6.1 to study $\mathfrak{S}_N$ and $\mathfrak{S}$, which from (14.68) and (14.69) are

$$\mathfrak{S}_N = \sum_{q=1}^{\log^B N} \rho_q, \quad \mathfrak{S} = \sum_{q=1}^{\infty} \rho_q. \tag{14.86}$$

We show that $|\mathfrak{S} - \mathfrak{S}(N)|$ is small by first determining $\mathfrak{S}$ (Theorem 14.6.18) and then estimating the difference (Theorem 14.6.20).

Exercise 14.6.16. *Let h_q be any multiplicative sequence (with whatever growth conditions are necessary to ensure the convergence of all sums below). Prove*

$$\sum_{q=1}^{\infty} h_q = \prod_{p \text{ prime}} \left(1 + \sum_{k=1}^{\infty} h_{p^k}\right). \tag{14.87}$$

Determine what growth conditions ensure convergence.

Definition 14.6.17 (Twin Prime Constant).

$$T_2 = \prod_{p>2}\left[1 - \frac{1}{(p-1)^2}\right] \approx .6601618158 \tag{14.88}$$

is the twin prime constant. Using the Circle Method, Hardy and Littlewood [HL3, HL4] were led to the conjecture that the number of twin primes at most x is given by

$$\pi_2(x) = 2T_2\frac{x}{\log^2 x} + o\left(\frac{x}{\log^2 x}\right). \tag{14.89}$$

The techniques of this chapter suffice to determine the contribution from the Major arcs to this problem as well; however, again the needed bounds on the Minor arcs are unknown.

Theorem 14.6.18. $\mathfrak{S}$ *has a product representation and satisfies*

$$\mathfrak{S} = 2T_2. \tag{14.90}$$

Proof. By Exercise 14.6.16 we have

$$\mathfrak{S} = \sum_{q=1}^{\infty} \rho_q$$

$$= \prod_{p \text{ prime}} \left(1 + \sum_{k=1}^{\infty} \rho_{p^k}\right)$$

$$= \prod_{p \text{ prime}} (1 + \rho_p), \tag{14.91}$$

because $\rho_{p^k} = 0$ for $k \geq 2$ and p prime by Lemma 14.6.12. The product is easily shown to converge (see Exercise 14.6.19). By Lemmas 14.6.13 and 14.6.14, $\rho_2 = 1$ and $\rho_p = -\frac{1}{(p-1)}$ for $p > 2$ prime. Therefore

$$\mathfrak{S} = \prod_{p} (1 + \rho_p)$$

$$= (1 + \rho_2) \prod_{p>2} (1 + \rho_p)$$

$$= 2 \prod_{p>2} \left[1 - \frac{1}{(p-1)^2}\right]$$

$$= 2T_2. \tag{14.92}$$

$\square$

We need to estimate $|\mathfrak{S} - \mathfrak{S}_N|$. As ρ_q is multiplicative and zero if $q = p^k$ ($k \geq 2$), we need only look at sums of ρ_p. As $\rho_p = -\frac{1}{(p-1)^2}$, it follows that the difference between $\mathfrak{S}$ and $\mathfrak{S}_N$ tends to zero as $N \to \infty$.

Exercise[h] **14.6.19.** *Show the product in* (14.91) *converges.*

Theorem 14.6.20. *For any $\epsilon > 0$ and B, N such that $\log^B N > 2$,*

$$|\mathfrak{S} - \mathfrak{S}_N| \ll O\left(\frac{1}{\log^{(1-2\epsilon)B} N}\right). \tag{14.93}$$

Proof.

$$\mathfrak{S} - \mathfrak{S}_N = \sum_{q=\log^B N}^{\infty} \rho_q. \tag{14.94}$$

By Lemma 14.6.12, $\rho_{p^k} = 0$ for $k \geq 2$ and p prime. By Lemma 14.6.13, $\rho_p = -\frac{1}{(p-1)^2}$ if $p > 2$ is prime. By Lemma 14.6.10, ρ_q is multiplicative. Therefore for any $\epsilon > 0$

$$|\rho_q| \leq \frac{1}{\phi(q)^2} \ll \frac{1}{q^{2-2\epsilon}}; \tag{14.95}$$

if q is not square-free this is immediate, and for q square-free we note $\phi(p) = p - 1$ and $\phi(q) \gg q^{1-\epsilon}$. Hence

$$|\mathfrak{S} - \mathfrak{S}_N| \ll \sum_{q=\log^B N}^{\infty} \frac{1}{q^{2-2\epsilon}} = O\left(\frac{1}{\log^{(1-2\epsilon)B} N}\right). \qquad (14.96)$$

$\square$

Exercise 14.6.21. *For q square-free, prove that for any $\epsilon > 0$, $\phi(q) \gg q^{1-\epsilon}$. Does a similar bound hold for all q?*

Combining the results above, we have finally determined the contribution from the Major arcs:

Theorem 14.6.22. *Let $D > 2B$, $C > 3D + 2B$, $\epsilon > 0$ and $\log^B N > 2$. Then*

$$\int_{\mathcal{M}} F_N(x)e(-x)dx$$

$$= \mathfrak{S}\frac{N}{2} + O\left(\frac{N}{\log^{(1-2\epsilon)BN}} + \frac{N}{\log^{D-2B} N} + \frac{N}{\log^{C-3D-2B} N}\right), \qquad (14.97)$$

where $\mathfrak{S}$ is twice the twin prime constant T_2.

In the binary and ternary Goldbach problems, to see if N could be written as the sum of two or three primes we evaluated the Singular Series at N (see §13.3.5). Thus, even after taking limits, we still evaluated the Singular Series at multiple points as we were trying to see *which* integers can be written as a sum of two or three primes; the answer told us in how many ways this was possible. Here we really have $\mathfrak{S}(1)$; knowing how large this is tells us information about what percentage of primes at most N are Germain primes (see §14.7). As things stand, it does not make sense to evaluate this Singular Series at additional points. However, if we were interested in a more general problem, such as $p, \frac{p-b}{2}$ are both prime, b odd, this would lead to $p_1 - 2p_2 = b$. We would replace $e(-x)$ in (14.6) with $e(-bx)$. Working in such generality would lead to a Singular Series depending on b. More generally, we could consider prime pairs of the form $p, \frac{ap+b}{c}$. If we take $a = c$ and $b = 2ck$ we have the special case of prime pairs, and the Singular Series will depend on the factorization of $2k$ (see [HL3, HL4]).

Exercise 14.6.23. *Redo the calculations of this chapter for one of the problems described above or in §14.1.*

14.7 NUMBER OF GERMAIN PRIMES AND WEIGHTED SUMS

We now remove the $\log p_i$ weights in our counting function. By Theorem 14.6.22, we know the contribution from the Major arcs. If we assume the minor arcs contribute $o(N)$ then we would have

$$\sum_{\substack{p \leq N \\ p, \frac{p-1}{2} \text{ prime}}} \log p \cdot \log \frac{p-1}{2} = \mathfrak{S}\frac{N}{2} + o(N) = T_2 N + o(N). \qquad (14.98)$$

We can pass from this weighted sum to a count of the number of Germain prime pairs $\left(\frac{p-1}{2}, p\right)$ with $p \leq N$. Again we follow [Na], Chapter 8; for more on weighted sums, see §2.3.4. Define

$$\pi_G(N) = \sum_{\substack{p \leq N \\ p, \frac{p-1}{2} \text{ prime}}} 1$$

$$G(N) = \sum_{\substack{p \leq N \\ p, \frac{p-1}{2} \text{ prime}}} \log p \cdot \log \frac{p-1}{2}. \qquad (14.99)$$

Theorem 14.7.1.

$$\frac{G(N)}{\log^2 N} \leq \pi_G(N) \leq \frac{G(N)}{\log^2 N} + O\left(N \frac{\log \log N}{\log N}\right). \qquad (14.100)$$

Proof. In (14.99), $\log p \log \frac{p-1}{2} < \log^2 N$. Thus $G(N) \leq \log^2 N \cdot \pi_G(N)$, proving the first inequality in (14.100).

The other inequality is more involved, and illustrates a common technique in analytic number theory. As there clearly are less Germain primes than primes, for any $\delta > 0$

$$\pi_G(N^{1-\delta}) = \sum_{\substack{p \leq N^{1-\delta} \\ p, \frac{p-1}{2} \text{ prime}}} 1 \leq \pi(N^{1-\delta}) = \frac{N^{1-\delta}}{\log N^{1-\delta}} \ll \frac{N^{1-\delta}}{\log N}. \qquad (14.101)$$

In the above inequality, the implied constant in $\pi_G(N^{1-\delta}) \ll \frac{N^{1-\delta}}{\log N}$ would seem to depend on δ; namely, we should write $\ll_\delta$ to indicate this dependence, or explicitly write the constant $\frac{1}{1-\delta}$. However, we do not want the implied constant to depend on δ as later we will take δ to be a function of N. As δ will be small, we note $\frac{1}{1-\delta} \leq 2$ and hence we can take the implied constant to be independent of δ.

We now obtain a good upper bound for $\pi_G(N)$. If $p \in [N^{1-\delta}, N]$ then

$$\log \frac{p-1}{2} = \log p + \log\left(1 - \frac{1}{2p}\right)$$

$$\geq (1-\delta)\log N + O\left(\frac{1}{p}\right)$$

$$\geq (1-\delta)\log N + O\left(\frac{1}{N}\right). \qquad (14.102)$$

In the arguments below, the error from (14.102) is negligible and is smaller than the other errors we encounter. We therefore suppress this error for convenience. Thus, up to lower order terms,

$$G(N) \geq \sum_{\substack{p \geq N^{1-\delta} \\ p, \frac{p-1}{2} \text{ prime}}} \log p \cdot \log \frac{p-1}{2}$$

$$= (1-\delta)^2 \log^2 N \sum_{\substack{p \geq N^{1-\delta} \\ p, \frac{p-1}{2} \text{ prime}}} 1$$

$$= (1-\delta)^2 \log^2 N \left(\pi_G(N) - \pi_G(N^{1-\delta}) \right)$$

$$\geq (1-\delta)^2 \log^2 N \cdot \pi_G(N) + O\left((1-\delta)^2 \log^2 N \cdot \frac{N^{1-\delta}}{\log N} \right).$$

$$(14.103)$$

Therefore

$$\log^2 N \cdot \pi_G(N) \leq (1-\delta)^{-2} \cdot G(N) + O\left(\log^2 N \cdot \frac{N^{1-\delta}}{\log N} \right)$$

$$0 \leq \log^2 N \cdot \pi_G(N) - G(N) \leq \left[(1-\delta)^{-2} - 1 \right] G(N) + O\left(\log N \cdot N^{1-\delta} \right).$$

$$(14.104)$$

If $0 < \delta < \frac{1}{2}$, then $(1-\delta)^{-2} - 1 \ll \delta$. We thus have

$$0 \leq \log^2 N \cdot \pi_G(N) - G(N) \ll N\left[\delta + O\left(\frac{\log N}{N^\delta} \right) \right]. \qquad (14.105)$$

Choose $\delta = \frac{2 \log \log N}{\log N}$. Then we get

$$0 \leq \log^2 N \cdot \pi_G(N) - G(N) \leq O\left(N \frac{\log \log N}{\log N} \right), \qquad (14.106)$$

which completes the proof. $\square$

Combining our results of this section, we have proved

Theorem 14.7.2. *Assuming there is no contribution to the main term from the Minor arcs, up to lower order terms we have*

$$\pi_G(N) = \frac{T_2 N}{\log^2 N}, \qquad (14.107)$$

where T_2 is the twin prime constant (see Definition 14.6.17).

Remark 14.7.3 (Important). It is a common technique in analytic number theory to choose an auxiliary parameter such as δ. Note how crucial it was in the proof for δ to depend (albeit very weakly) on N. Whenever one makes such approximations, it is good to get a feel for how much information is lost in the estimation. For $\delta = \frac{2 \log \log N}{\log N}$ we have

$$N^{1-\delta} = N \cdot N^{-2 \log \log N / \log N} = N \cdot e^{-2 \log \log N} = \frac{N}{\log^2 N}. \qquad (14.108)$$

Hence there is little cost in ignoring the Germain primes less than $N^{1-\delta}$. Our final answer is of size $\frac{N}{\log^2 N}$. As $N^{1-\delta} = \frac{N}{\log^2 N}$ and there are $O\left(\frac{N}{\log^3 N} \right)$ primes less than $N^{1-\delta}$, there are at most $O\left(\frac{N}{\log^3 N} \right)$ Germain primes at most $N^{1-\delta}$.

Exercise 14.7.4. *Let $\Lambda(n)$ be the von Magnoldt function (see §2.1). Prove*

$$\sum_{n\leq x} \Lambda(n) = \sum_{p\leq x} \log p + O(x^{\frac{1}{2}} \log x). \tag{14.109}$$

As $\sum_{p\leq x}$ is of size x, there is negligible loss in ignoring prime powers.

Exercise 14.7.5. *The number of weighted ways to write an odd N as the sum of three primes is*

$$G_3(N) = \sum_{p_1+p_2+p_3=N} \log p_1 \log p_2 \log p_3 = \mathfrak{S}_3(N)N^2 + o(N^2), \tag{14.110}$$

where there exist positive constants c_1, c_2 such that $0 < c_1 < \mathfrak{S}_3(N) < c_2 < \infty$; see Exercises 14.8.4 and 14.8.5. Assuming the above, approximately how many unweighted ways are there to write an odd N as the sum of three primes?

14.8 EXERCISES

The following problems are known questions (either on the Circle Method, or needed results to prove some of these claims).

Exercise 14.8.1. *Prove $\forall \epsilon > 0$, $q^{1-\epsilon} \ll \phi(q) \ll q$.*

Exercise 14.8.2. *Let*

$$c_q(N) = \sum_{\substack{a=1 \\ (a,q)=1}}^{q} e^{2\pi i Na/q}.$$

Prove $c_q(N)$ is multiplicative. Further, show

$$c_q(N) = \begin{cases} p-1 & \text{if } p|N \\ -1 & \text{otherwise.} \end{cases} \tag{14.111}$$

Exercise 14.8.3. *Prove $\mu(q)c_q(N)/\phi(q)^3$ is multiplicative.*

Exercise 14.8.4. *Using the above exercises and the methods of this chapter, calculate the contribution from the Major arcs to writing any integer N as the sum of three primes. Deduce for writing numbers as the sum of three primes that*

$$\mathfrak{S}_3(N) = \sum_{q=1}^{\infty} \frac{\mu(q)c_q(N)}{\phi(q)^3}$$

$$= \prod_{p} \left(1 + \sum_{j=1}^{\infty} \frac{\mu(p^j)c_{p^j}(N)}{\phi(p^j)^3}\right)$$

$$= \prod_{p} \left(1 + \frac{1}{(p-1)^3}\right) \prod_{p|N} \left(1 - \frac{1}{p^2 - 3p + 3}\right). \tag{14.112}$$

*Note $\mathfrak{S}_3(N) = 0$ if N is even; thus the Circle Method "knows" that Goldbach is hard. We call $\mathfrak{S}(N)$ the **Singular Series**.*

Exercise 14.8.5. *Let* $\mathfrak{S}_{3,Q}(N)$ *be the first* Q *terms of* $\mathfrak{S}_3(N)$. *Bound* $\mathfrak{S}_3(N) -$ $\mathfrak{S}_{3,Q}(N)$. *Show for* N *odd there exist constants* c_1, c_2 *such that* $0 < c_1 <$ $\mathfrak{S}_3(N) < c_2 < \infty$.

Exercise 14.8.6. *Assume every large integer is the sum of three primes. Prove every large even integer is the sum of two primes. Conversely, show if every large even integer is the sum of two primes, every large integer is the sum of three primes.*

Exercise 14.8.7 (Non-Trivial). *Calculate the Singular Series* $\mathfrak{S}_2(N)$ *and* $\mathfrak{S}_{2,Q}(N)$ *for the Goldbach problem (even numbers as the sum of two primes), and* $\mathfrak{S}_{W,k,s}(N)$ *and* $\mathfrak{S}_{W,k,s,Q}(N)$ *for Waring's problem (writing numbers as the sum of* s *perfect* k-*powers*). **Warning:** $\mathfrak{S}_2(N) - \mathfrak{S}_{2,Q}(N)$ **cannot be shown to be small for all even** N **in the Goldbach problem.** *Do* $\mathfrak{S}_{2,Q}(N)$ *and* $\mathfrak{S}_2(N)$ *vanish for* N *odd? See [Law2].*

14.9 RESEARCH PROJECTS

One can use the Circle Method to predict the number of primes (or prime tuples) with given properties, and then investigate these claims numerically; see, for example, [Law2, Sch, Weir] (for additional Circle Method investigations, see [Ci]). After counting the number of such primes (or prime tuples), the next natural question is to investigate the spacings between adjacent elements (see Chapter 12, in particular Research Project 12.9.5).

Research Project 14.9.1. For many questions in number theory, the Cramér model (see §8.2.3 and Exercise 14.1.2) leads to good heuristics and predictions; recently, however, [MS] have shown that this model is inconsistent with certain simple numerical investigations of primes. On the other hand the Random Matrix Theory model of the zeros of the Riemann zeta function and the Circle Method give a prediction which agrees beautifully with experiments. There are many additional interesting sequences of primes to investigate and see which model is correct. Candidates include primes in arithmetic progression, twin primes, generalized twin primes (fix an integer k, look for primes such that p and p+2k are prime), prime tuples (fix integers k_1 through k_r such that $p, p+2k_1, \ldots, p+2k_r$ are all prime), Germain primes, and so on. A natural project is to investigate the statistics from [MS] for these other sequences of primes, using the Circle Method and the Cramér model to predict two answers, and then see which agrees with numerics. See [Bre1, Bre2] for some numerical investigations of primes.

Research Project 14.9.2. In many successful applications of the Circle Method, good bounds are proved for the generating function on the Minor arcs. From these bounds it is then shown that the Minor arcs' contribution is significantly smaller than that from the Major arcs. However, to *prove* that the Major arcs are the main term does not require one to obtain good cancellation at every point in the Minor arcs; all that is required is that the *integral* is small.

For problems such as Goldbach's conjecture or Germain primes, the needed estimates on the Minor arcs are conjectured to hold; by counting the number of solutions, we see that the integral over the Minor arcs is small (at least up to about 10^9). A good investigation is to numerically calculate the generating function at various points on the Minor arcs for several of these problems, and see how often large values are obtained. See the student report [Law2] as well as [BS, Moz1]. Warning: Calculations of this nature are very difficult. The Major arcs are defined as intervals of size $\frac{2\log^D N}{N}$ about rationals with denominators at most $\log^B N$. For example, if $D = 10$ than $\log^D N > N$ until N is about 3.4×10^{15}, and there will not be any Minor arcs! For $N \approx 10^{15}$, there are too many primes to compute the generating function in a reasonable amount of time. Without resorting to supercomputers, one must assume that we may take B small for such numerical investigations.

See also Project 4.5.2.

PART 5
Random Matrix Theory and L-Functions

Chapter Fifteen

From Nuclear Physics to L-Functions

In attempting to describe the energy levels of heavy nuclei ([Wig1, Wig3, Po, BFFMPW]), researchers were confronted with daunting calculations for a many bodied system with extremely complicated interaction forces. Unable to explicitly calculate the energy levels, physicists developed Random Matrix Theory to predict general properties of the systems. Surprisingly, similar behavior is seen in studying the zeros of L-functions!

In this chapter we give a brief introduction to classical Random Matrix Theory, Random Graphs and L-Functions. Our goal is to show how diverse systems exhibit similar universal behaviors, and introduce the techniques used in the proofs. In some sense, this is a continuation of the Poissonian behavior investigations of Chapter 12. The survey below is meant to only show the broad brush strokes of this rich landscape; detailed proofs will follow in later chapters. We assume familiarity with the basic concepts of L-functions (Chapter 3), probability theory (Chapter 8) and linear algebra (a quick review of the needed background is provided in Appendix B).

While we assume the reader has some familiarity with the basic concepts in physics for the historical introduction in §15.1, no knowledge of physics is required for the detailed expositions. After describing the physics problems, we describe several statistics of eigenvalues of sets of matrices. It turns out that the spacing properties of these eigenvalues is a good model for the spacings between energy levels of heavy nuclei and zeros of L-functions; exactly why this is so is still an open question. For those interested in learning more (as well as a review of recent developments), we conclude this chapter with a brief summary of the literature.

15.1 HISTORICAL INTRODUCTION

A central question in mathematical physics is the following: given some system with observables $t_1 \leq t_2 \leq t_3 \leq \ldots$, describe how the t_i are spaced. For example, we could take the t_i to be the energy levels of a heavy nuclei, or the prime numbers, or zeros of L-functions, or eigenvalues of real symmetric or complex Hermitian matrices (or as in Chapter 12 the fractional parts $\{n^k \alpha\}$ arranged in increasing order). If we completely understood the system, we would know exactly where all the t_i are; in practice we try and go from knowledge of how the t_i are spaced to knowledge of the underlying system.

15.1.1 Nuclear Physics

In classical mechanics it is possible to write down closed form solutions to the two body problem: given two points with masses m_1 and m_2 and initial velocities $\vec{v}_1$ and $\vec{v}_2$ and located at $\vec{r}_1$ and $\vec{r}_2$, describe how the system evolves in time given that gravity is the only force in play. The three body problem, however, defies closed form solutions (though there are known solutions for special arrangements of special masses, three bodies in general position is still open; see [Wh] for more details). From physical grounds we know of course a solution must exist; however, for our solar system we cannot analyze the solution well enough to determine whether or not billions of years from now Pluto will escape from the sun's influence! In some sense this is similar to the problems with the formula for counting primes in Exercise 2.3.18.

Imagine how much harder the problems are in understanding the behavior of heavy nuclei. Uranium, for instance, has over 200 protons and neutrons in its nucleus, each subject to and contributing to complex forces. If the nucleus were completely understood, one would know the energy levels of the nucleus. Physicists were able to gain some insights into the nuclear structure by shooting high-energy neutrons into the nucleus, and analyzing the results; however, a complete understanding of the nucleus was, and still is, lacking. Later, when we study zeros of L-functions from number theory, we will find analogues of high-energy neutrons!

One powerful formulation of physics is through infinite dimensional linear algebra. The fundamental equation for a system becomes

$$H\psi_n = E_n\psi_n, \tag{15.1}$$

where H is an operator (called the **Hamiltonian**) whose entries depend on the physical system and the ψ_n are the energy eigenfunctions with eigenvalues E_n. Unfortunately for nuclear physics, H is too complicated to write down and solve; however, a powerful analogy with Statistical Mechanics leads to great insights.

15.1.2 Statistical Mechanics

For simplicity consider N particles in a box where the particles can only move left or right and each particle's speed is v; see Figure 15.1.

If we want to calculate the pressure on the left wall, we need to know how many particles strike the wall in an infinitesimal time. Thus we need to know how many particles are close to the left wall and moving towards it. Without going into all of the physics (see for example [Re]), we can get a rough idea of what is happening. The complexity, the enormous number of configurations of positions of the molecules, actually helps us. For each configuration we can calculate the pressure due to that configuration. We then *average* over all configurations, and hope that a generic configuration is, in some sense, close to the system average.

Wigner's great insight for nuclear physics was that similar tools could yield useful predictions for heavy nuclei. He modeled the nuclear systems as follows: instead of the infinite dimensional operator H whose entries are given by the physical laws, he considered collections of $N \times N$ matrices where the entries were independently chosen from some probability distribution p. The eigenvalues of these

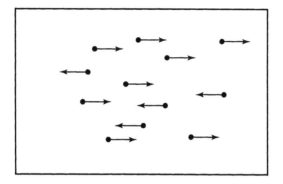

Figure 15.1 Molecules in a box

matrices correspond to the energy levels of the physical system. Depending on physical symmetries, we consider different collections of matrices (real symmetric, complex Hermitian). For any given finite matrix we can calculate statistics of the eigenvalues. We then average over all such matrices, and look at the limits as $N \to \infty$. The main result is that *the behavior of the eigenvalues of an arbitrary matrix is often well approximated by the behavior obtained by averaging over all matrices, and this is a good model for the energy levels of heavy nuclei.* This is reminiscent of the Central Limit Theorem (§8.4). For example, if we average over all sequences of tossing a fair coin $2N$ times, we obtain N heads, and *most* sequences of $2N$ tosses will have approximately N heads.

Exercise 15.1.1. *Consider $2N$ identical, indistinguishable particles, which are in the left (resp., right) half of the box with probability $\frac{1}{2}$. What is the expected number of particles in each half? What is the probability that one half has more than $(2N)^{\frac{3}{4}}$ particles than the other half? As $(2N)^{\frac{3}{4}} \ll N$, most systems will have similar behavior although of course some will not. The point is that a* typical *system will be close to the system average.*

Exercise 15.1.2. *Consider $4N$ identical, indistinguishable particles, which are in the left (resp., right) half of the box with probability $\frac{1}{2}$; each particle is moving left (resp., right) with probability $\frac{1}{2}$. Thus there are four possibilities for each particle, and each of the 4^{4N} configurations of the $4N$ particles is equally likely. What is the expected number of particles in each possibility (left-left, left-right, right-left, right-right)? What is the probability that one possibility has more than $(4N)^{\frac{3}{4}}$ particles than the others? As $(4N)^{\frac{3}{4}} \ll N$, most systems will have similar behavior.*

15.1.3 Random Matrix Ensembles

The first collection of matrices we study are $N \times N$ real symmetric matrices, with the entries independently chosen from a fixed probability distribution p on $\mathbb{R}$. Given such a matrix A,

$$A = \begin{pmatrix} a_{11} & a_{12} & a_{13} & \cdots & a_{1N} \\ a_{21} & a_{22} & a_{23} & \cdots & a_{2N} \\ \vdots & \vdots & \vdots & \ddots & \vdots \\ a_{N1} & a_{N2} & a_{N3} & \cdots & a_{NN} \end{pmatrix} = A^T \tag{15.2}$$

(so $a_{ij} = a_{ji}$), the probability density of observing A is

$$\text{Prob}(A)dA = \prod_{1 \leq i \leq j \leq N} p(a_{ij})da_{ij}. \tag{15.3}$$

We may interpret this as giving the probability of observing a real symmetric matrix where the probability of the ij^{th} entry lying in $[a_{ij}, a_{ij} + da_{ij}]$ is $p(a_{ij})da_{ij}$. More explicitly,

$$\text{Prob}(A : a_{ij} \in [\alpha_{ij}, \beta_{ij}]) = \prod_{1 \leq i \leq j \leq N} \int_{\alpha_{ij}}^{\beta_{ij}} p(a_{ij})da_{ij}. \tag{15.4}$$

Example 15.1.3. *For a 2×2 real symmetric matrix we would have*

$$A = \begin{pmatrix} a_{11} & a_{12} \\ a_{12} & a_{22} \end{pmatrix}, \quad \text{Prob}(A) = p(a_{11})p(a_{12})p(a_{22})da_{11}da_{12}da_{22}. \tag{15.5}$$

An $N \times N$ real symmetric matrix is determined by specifying $\frac{N(N+1)}{2}$ entries: there are N entries on the main diagonal, and $N^2 - N$ off-diagonal entries (for these entries, only half are needed, as the other half are determined by symmetry). We say such a matrix has $\frac{N(N+1)}{2}$ **degrees of freedom**. Because p is a probability density, it integrates to 1. Thus

$$\int \text{Prob}(A)dA = \prod_{1 \leq i \leq j \leq N} \int_{a_{ij}=-\infty}^{\infty} p(a_{ij})da_{ij} = 1; \tag{15.6}$$

this corresponds to the fact that we must choose some matrix.

For convergence reasons we often assume that the moments of p are finite. We mostly study $p(x)$ satisfying

$$p(x) \geq 0$$
$$\int_{-\infty}^{\infty} p(x)dx = 1$$
$$\int_{-\infty}^{\infty} |x|^k p(x)dx < \infty. \tag{15.7}$$

The last condition ensures that the probability distribution is not too spread out (i.e., there is not too much probability near infinity). Many times we normalize p so that the mean (first moment) is 0 and the variance (second moment if the mean is zero) is 1.

Exercise 15.1.4. *For the kth moment $\int_{\mathbb{R}} x^k p(x)dx$ to exist, we require $\int_{\mathbb{R}} |x|^k p(x)dx < \infty$; if this does not hold, the value of the integral could depend on how we approach infinity. Find a probability function $p(x)$ and an integer k such that*

$$\lim_{A \to \infty} \int_{-A}^{A} x^k p(x)dx = 0 \quad but \quad \lim_{A \to \infty} \int_{-A}^{2A} x^k p(x)dx = \infty. \tag{15.8}$$

Exercise 15.1.5. *Let p be a probability density such that all of its moments exist. If p is an even (resp., odd) function, show all the odd (resp., even) moments vanish.*

Exercise 15.1.6. *Let p be a continuous probability density on* $\mathbb{R}$. *Show there exist constants* a, b *such that* $q(x) = a \cdot p(ax + b)$ *has mean 0 and variance 1. Thus in some sense the third and the fourth moments are the first "free" moments as the above transformation is equivalent to translating and rescaling the initial scale.*

Exercise 15.1.7. *It is not necessary to choose each entry from the same probability distribution. Let the* ij^{th} *entry be chosen from a probability distribution* p_{ij}. *What is the probability density of observing A? Show this also integrates to 1.*

Definition 15.1.8 (Ensembles). *A collection of matrices, along with a probability density describing how likely it is to observe a given matrix, is called an **ensemble** of matrices (or a **random matrix ensemble**).*

Example 15.1.9. *Consider the ensemble of* 2×2 *real symmetric matrices A where for a matrix* $A = \left(\begin{smallmatrix} x & y \\ y & z \end{smallmatrix} \right)$,

$$p(A) = \begin{cases} \frac{3}{4\pi} & \text{if } x^2 + y^2 + z^2 \leq 1 \\ 0 & \text{otherwise.} \end{cases} \tag{15.9}$$

Note the entries are not independent. We can parametrize these matrices by using spherical coordinates. For a sphere of radius r we have

$$\begin{aligned} x = x(r, \theta, \phi) &= r\cos(\theta)\sin(\phi) \\ y = y(r, \theta, \phi) &= r\sin(\theta)\sin(\phi) \\ z = z(r, \theta, \phi) &= r\cos(\phi), \end{aligned} \tag{15.10}$$

where $\theta \in [0, 2\pi]$ *is the azimuthal angle,* $\phi \in [0, \pi]$ *is the polar angle and the volume of the sphere is* $\frac{4}{3}\pi r^3$.

In this introduction we confine ourselves to real symmetric matrices, although many other ensembles of matrices are important. Complex Hermitian matrices (the generalization of real symmetric matrices) also play a fundamental role in the theory. Both of these types of matrices have a very important property: *their eigenvalues are real*; this is what allows us to ask questions such as how are the spacings between eigenvalues distributed.

In constructing our real symmetric matrices, we have not said much about the probability density p. In Chapter 17 we show for that some physical problems, additional assumptions about the physical systems force p to be a Gaussian. For many of the statistics we investigate, it is either known or conjectured that the answers should be independent of the specific choice of p; however, in this method of constructing random matrix ensembles, there is often no unique choice of p. Thus, for this method, there is no unique answer to what it means to choose a matrix *at random*.

Remark 15.1.10 (Advanced). We would be remiss if we did not mention another notion of randomness, which leads to a more natural method of choosing a matrix at random. Let $U(N)$ be the space of $N \times N$ unitary matrices, and consider its compact subgroups (for example, the $N \times N$ orthogonal matrices). There is a natural (canonical) measure, called the **Haar measure**, attached to each of these compact groups, and we can use this measure to choose matrices *at random*. Further, the eigenvalues of unitary matrices have modulus 1. They can be written as $e^{i\theta_j}$, with the θ_j real. We again obtain a sequence of real numbers, and can again ask many questions about spacings and distributions. This is the notion of random matrix ensemble which has proven the most useful for number theory.

Exercise 15.1.11. *Prove the eigenvalues of real symmetric and complex Hermitian matrices are real.*

Exercise 15.1.12. *How many degrees of freedom does a complex Hermitian matrix have?*

15.2 EIGENVALUE PRELIMINARIES

15.2.1 Eigenvalue Trace Formula

Our main tool to investigate the eigenvalues of matrices will be the Eigenvalue Trace Formula. Recall the trace of a matrix is the sum of its diagonal entries:

$$\text{Trace}(A) = a_{11} + \cdots + a_{NN}. \tag{15.11}$$

We will also need the trace of powers of matrices. For example, a 2×2 matrices

$$A = \begin{pmatrix} a_{11} & a_{12} \\ a_{21} & a_{22} \end{pmatrix}, \tag{15.12}$$

we have

$$\text{Trace}(A^2) = a_{11}a_{11} + a_{12}a_{21} + a_{12}a_{21} + a_{22}a_{22} = \sum_{i=1}^{2}\sum_{j=1}^{2} a_{ij}a_{ji}. \tag{15.13}$$

In general we have

Theorem 15.2.1. *Let A be an $N \times N$ matrix. Then*

$$\text{Trace}(A^k) = \sum_{i_1=1}^{N} \cdots \sum_{i_k=1}^{N} a_{i_1 i_2} a_{i_2 i_3} \cdots a_{i_{k-1} i_k} a_{i_k i_1}. \tag{15.14}$$

For small values of k, instead of using $i_1, i_2, i_3, \ldots$ we often use $i, j, k, \ldots$. For example, $\text{Trace}(A^3) = \sum_i \sum_j \sum_k a_{ij} a_{jk} a_{ki}$.

Exercise 15.2.2. *Show (15.13) is consistent with Theorem 15.2.1.*

Exercise 15.2.3. *Prove Theorem 15.2.1.*

Theorem 15.2.4 (Eigenvalue Trace Formula). *For any non-negative integer k, if A is an $N \times N$ matrix with eigenvalues $\lambda_i(A)$, then*

$$\text{Trace}(A^k) = \sum_{i=1}^{N} \lambda_i(A)^k. \tag{15.15}$$

The importance of this formula is that it relates the *eigenvalues* of a matrix (which is what we *want* to study) to the *entries* of A (which is what we *choose* at random). The importance of this formula cannot be understated – it is what makes the whole subject possible.

Sketch of the proof. The case $k = 1$ follows from looking at the characteristic polynomial $\det(A - \lambda I) = 0$. For higher k, note any matrix A can be conjugated to an upper triangular matrix: $U^{-1}AU = T$ where T is upper triangular and U is unitary. The eigenvalues of A equal those of T and are given by the diagonal entries of T. Further the eigenvalues of A^k equal those of T^k. If $\lambda_i(A)$ and $\lambda_i(A^k)$ are the eigenvalues of A and A^k, note $\lambda_i(A^k) = \lambda_i(A)^k$. The claim now follows by applying the $k = 1$ result to the matrix A^k:

$$\text{Trace}(A^k) = \sum_{i=1}^{N} \lambda_i(A^k) = \sum_{i=1}^{N} \lambda_i(A)^k. \qquad (15.16)$$

$\square$

Exercise 15.2.5. *Prove all the claims used in the proof of the Eigenvalue Trace Formula. If A is real symmetric, one can use the diagonalizability of A. To show any matrix can be triangularized, start with every matrix has at least one eigenvalue-eigenvector pair. Letting $\overrightarrow{v_1}$ be the eigenvector, using Gram-Schmidt one can find an orthonormal basis. Let these be the columns of U_1, which will be a unitary matrix. Continue by induction.*

15.2.2 Normalizations

Before we can begin to study fine properties of the eigenvalues, we first need to figure out what is the correct scale to use in our investigations. For example, the celebrated Prime Number Theorem (see Theorem 13.2.10 for an exact statement of the error term) states that $\pi(x)$, the number of primes less than x, satisfies

$$\pi(x) = \frac{x}{\log x} + \text{lower order terms}. \qquad (15.17)$$

Remark 15.2.6. If we do not specify exactly how much smaller the error terms are, we do not need the full strength of the Prime Number Theorem; Chebyshev's arguments (Theorem 2.3.9) are sufficient to get the order of magnitude of the scale.

The average spacing between primes less than x is about $\frac{x}{x/\log x} = \log x$, which tends to infinity as $x \to \infty$. Asking for primes that differ by 2 is a very hard question: as $x \to \infty$, this becomes insignificant on the "natural" scale. Instead, a more natural question is to inquire how often two primes are twice the average spacing apart. This is similar to our investigations in Chapter 12 where we needed to find the correct scale.

If we fix a probability density p, how do we expect the sizes of the eigenvalues $\lambda_i(A)$ to depend on N as we vary A? A good estimate falls out immediately from the Eigenvalue Trace Formula; this formula will be exploited numerous times in the arguments below, and is essential for all investigations in the subject.

We give a heuristic for the eigenvalues of our $N \times N$ ensembles of matrices being roughly of size $\sqrt{N}$. Fix a matrix A whose entries a_{ij} are randomly and independently chosen from a fixed probability distribution p with mean 0 and variance 1. By Theorem 15.2.1, for $A = A^T$ we have that

$$\text{Trace}(A^2) = \sum_{i=1}^{N}\sum_{j=1}^{N} a_{ij}a_{ji} = \sum_{i=1}^{N}\sum_{j=1}^{N} a_{ij}^2. \tag{15.18}$$

From our assumptions on p, we expect each a_{ij}^2 to be of size 1. By the Central Limit Theorem (Theorem 8.4.1) or Chebyshev's Theorem (Exercise 8.1.55), we expect with high probability

$$\sum_{i=1}^{N}\sum_{j=1}^{N} a_{ij}^2 \sim N^2 \cdot 1, \tag{15.19}$$

with an error of size $\sqrt{N^2} = N$ (as each a_{ij}^2 is approximately of size 1 and there are N^2 of them, with high probability their sum should be approximately of size N^2). Thus

$$\sum_{i=1}^{N} \lambda_i(A)^2 \sim N^2, \tag{15.20}$$

which yields

$$N \cdot \text{Ave}(\lambda_i(A)^2) \sim N^2. \tag{15.21}$$

For heuristic purposes we shall pass the square root through to get

$$|\text{Ave}(\lambda_i(A))| \sim \sqrt{N}. \tag{15.22}$$

In general the square root of an average need not be the same as the average of the square root; however, our purpose here is merely to give a heuristic as to the correct scale. Later in our investigations we shall see that $\sqrt{N}$ is the correct normalization.

Thus it is natural to guess that the correct scale to study the eigenvalues of an $N \times N$ real symmetric matrix is $c\sqrt{N}$, where c is some constant independent of N. This yields normalized eigenvalues $\widetilde{\lambda}_1(A) = \frac{\lambda_i(A)}{c\sqrt{N}}$; choosing $c = 2$ leads to clean formulas. One could of course normalize the eigenvalues by $f(N)$, with f an undetermined function, and see which choices of f give good results; eventually one would find $f(N) = c\sqrt{N}$.

Exercise 15.2.7. *Consider real $N \times N$ matrices with entries independently chosen from a probability distribution with mean 0 and variance 1. How large do you expect the average eigenvalue to be?*

Exercise 15.2.8. *Use Chebyshev's Theorem (Exercise 8.1.55) to bound the probability that $|\sum_i \sum_j a_{ij}^2 - N^2| \geq N \log N$. Conclude that with high probability that the sum of the squares of the eigenvalues is of size N^2 for large N.*

15.2.3 Eigenvalue Distribution

We quickly review the theory of point masses and induced probability distributions (see §11.2.2 and §12.5). Let δ_{x_0} represent a unit point mass at x_0. We define its action on functions by

$$\delta_{x_0}(f) := \int_{-\infty}^{\infty} f(x)\delta(x-x_0)dx = f(x_0). \qquad (15.23)$$

δ_{x_0}, called the **Dirac delta functional** at x_0, is similar to our approximations to the identity. There is finite mass (its integral is 1), the density is 0 outside x_0 and infinite at x_0. As its argument is a function and not a complex number, δ_{x_0} is a **functional** and not a function. To each A, we attach a probability measure (the **eigenvalue probability distribution**):

$$\mu_{A,N}(x)dx = \frac{1}{N}\sum_{i=1}^{N}\delta\left(x - \frac{\lambda_i(A)}{2\sqrt{N}}\right)dx. \qquad (15.24)$$

At each normalized eigenvalue $\frac{\lambda_i(A)}{2\sqrt{N}}$ we have placed a mass of weight $\frac{1}{N}$; there are N masses, thus we have a probability distribution. If $p(x)$ is a probability distribution then $\int_a^b p(x)dx$ is the probability of observing a value in $[a,b]$. For us, $\int_a^b \mu_{A,N}(x)dx$ is the percentage of normalized eigenvalues in $[a,b]$:

$$\int_a^b \mu_{A,N}(x)dx = \frac{\#\{i : \frac{\lambda_i(A)}{2\sqrt{N}} \in [a,b]\}}{N}. \qquad (15.25)$$

We can calculate the moments of $\mu_{A,N}(x)$.

Definition 15.2.9. *Let* $\mathbb{E}[x^k]_A$ *denote the* k^{th} *moment of* $\mu_{A,N}(x)$. *We often denote this* $M_{N,k}(A)$.

The following corollary of the Eigenvalue Trace formula is the starting point of many of our investigations; we see in Remark 15.3.15 why it is so useful.

Lemma 15.2.10. $M_{N,k}(A) = \frac{\text{Trace}(A^k)}{2^k N^{\frac{k}{2}+1}}$.

Proof. As $\text{Trace}(A^k) = \sum_i \lambda_i(A)^k$ we have

$$M_{N,k}(A) = \mathbb{E}[x^k]_A = \int x^k \mu_{A,N}(x)dx$$

$$= \frac{1}{N}\sum_{i=1}^{N}\int_{\mathbb{R}} x^k \delta\left(x - \frac{\lambda_i(A)}{2\sqrt{N}}\right)dx$$

$$= \frac{1}{N}\sum_{i=1}^{N}\frac{\lambda_i(A)^k}{(2\sqrt{N})^k}$$

$$= \frac{\text{Trace}(A^k)}{2^k N^{\frac{k}{2}+1}}. \qquad (15.26)$$

$\square$

Exercise 15.2.11. *Let A be an $N \times N$ real symmetric matrix with $|a_{ij}| \leq B$. In terms of B, N and k bound $|\text{Trace}(A^k)|$ and $M_{N,k}(A)$. How large can $\max_i |\lambda_i(A)|$ be?*

15.3 SEMI-CIRCLE LAW

15.3.1 Statement

A natural question to ask concerning the eigenvalues of a matrix is: *What percentage of the normalized eigenvalues lie in an interval* $[a, b]$? Let $\mu_{A,N}(x)$ be the eigenvalue probability distribution. For a given A, the answer is

$$\int_a^b \mu_{A,N}(x)dx. \tag{15.27}$$

How does the above behave as we vary A? We have the following startling result, which is almost independent of the underlying probability density p we used to choose the entries of A:

Theorem 15.3.1 (Semi-Circle Law). *Consider the ensemble of $N \times N$ real symmetric matrices with entries independently chosen from a fixed probability density $p(x)$ with mean 0, variance 1, and finite higher moments. As $N \to \infty$, for almost all A, $\mu_{A,N}(x)$ converges to the semi-circle density $\frac{2}{\pi}\sqrt{1 - x^2}$.*

Thus the percentage of normalized eigenvalues of A in $[a, b] \subset [-1, 1]$ for a typical A as $N \to \infty$ is

$$\int_a^b \frac{2}{\pi}\sqrt{1 - x^2}dx. \tag{15.28}$$

Later in §15.3.4 we discuss what happens if the higher moments are infinite.

15.3.2 Moment Problem

We briefly describe a needed result from Probability Theory: the solution to the Moment Problem. See page 110 of [Du] for details; see [ShTa] for a connection between the moment problem and continued fractions!

Let k be a non-negative integer; below we always assume $m_0 = 1$. We are interested in when numbers m_k determine a unique probability distribution P whose k^{th} moment is m_k. If the m_k do not grow too rapidly, there is at most one continuous probability density with these moments (see [Bi, CaBe, Fe]). A sufficient condition is Carleman's Condition that $\sum_{j=1}^{\infty} m_{2j}^{-1/2j} = \infty$. Another is that $\sum_{j=1}^{\infty} \frac{m_j t^j}{j!}$ has a positive radius of convergence. This implies the moment generating function (see Exercise 15.3.2) exists in an interval and the distribution is uniquely determined.

Exercise 15.3.2 (Non-uniqueness of moments). *For $x \in [0, \infty)$, consider*

$$f_1(x) = \frac{1}{\sqrt{2\pi}x}e^{-(\log x)^2/2}$$
$$f_2(x) = f_1(x)\left[1 + \sin(2\pi \log x)\right]. \tag{15.29}$$

*Show that for $r \in \mathbb{N}$, the rth moment of f_1 and f_2 is $e^{r^2/2}$. The reason for the non-uniqueness of moments is that the **moment generating function***

$$M_f(t) = \int_{-\infty}^{\infty} e^{tx} f(x)dx \tag{15.30}$$

does not converge in a neighborhood of the origin. See [CaBe], Chapter 2. See also Exercise A.2.7.

For us the numbers m_k arise from averaging the moments $M_{N,k}(A)$ of the $\mu_{A,N}(x)$'s and taking the limit as $N \to \infty$. Let

$$M_{N,k} = \int_A M_{N,k}(A)\text{Prob}(A)dA, \quad m_k = \lim_{N \to \infty} M_{N,k}. \quad (15.31)$$

For each N the moments $M_{N,k}$ yield a probability distribution P_N, and $\lim_{N \to \infty} M_{N,k} = m_k$. If the m_k grow sufficiently slowly, there is a unique continuous probability density P with k^{th} moment m_k. It is reasonable to posit that as for each k, $\lim_{N \to \infty} M_{N,k} = m_k$, then "most" $\mu_{A,N}(x)$ converge (in some sense) to the probability density $P(x)$.

Remark 15.3.3 (Warning). For each N, consider N numbers $\{a_{n,N}\}_{n=1}^N$ defined by $a_{n,N} = 1$ if n is even and -1 if N is odd. For N even, note the average of the $a_{n,N}$'s is 0, but each $|a_{n,N}| = 1$; thus, no element is close to the system average. Therefore, it is not always the case that a typical element is close to the system average. What is needed in this case is to consider the variance of the moments (see Exercise 15.3.5).

Remark 15.3.4. While it is not true that every sequence of numbers m_k that grow sufficiently slowly determines a continuous probability density (see Exercise 15.3.8), as our m_k arise from limits of moments of probability distributions, we do obtain a unique limiting probability density. This is similar to determining when a Taylor series converges to a unique function. See also Exercise A.2.7.

Exercise 15.3.5. Let $\{b_{n,N}\}_{n=1}^N$ be a sequence with mean $\mu(N) = \frac{1}{N}\sum_{n=1}^N b_{n,N}$ and variance $\sigma^2(N) = \frac{1}{N}\sum_{n=1}^N |b_{n,N} - \mu(N)|^2$. Assume that as $N \to \infty$, $\mu(N) \to \mu$ and $\sigma^2(N) \to 0$. Prove for any $\epsilon > 0$ as $N \to \infty$ for a fixed N at most ϵ percent of $b_{n,N}$ are not within ϵ of μ. Therefore, if the mean of a sequence converges and we have control over the variance, then we have control over the limiting behavior of most elements.

In this text we content ourselves with calculating the average moments $m_k = \lim_{N \to \infty} \int_A M_{N,k}(A)dA$. In many cases we derive simple expressions for the probability density P with moments m_k; however, we do not discuss the probability arguments needed to show that as $N \to \infty$, a "typical" matrix A has $\mu_{A,n}(x)$ close to P. The interested reader should see [CB, HM] for an application to moment arguments in random matrix theory.

Some care is needed in formulating what it means for two probability distributions to be close. For us, $\mu_{A,N}(x)$ is the sum of N Dirac delta functionals of mass $\frac{1}{N}$. Note $|P(x) - \mu_{A,N}(x)|$ can be large for individual x. For example, if $P(x)$ is the semi-circle distribution, then $|P(x) - \mu_{A,N}(x)|$ will be of size 1 for almost all $x \in [-1, 1]$. We need to define what it means for two probability distributions to be close.

One natural measure is the Kolmogoroff-Smirnov discrepancy. For a probability distribution $f(x)$ its **Cumulative Distribution Function** $C_f(x)$ is defined to be the probability of $[-\infty, x]$:

$$C_f(x) = \int_{-\infty}^{x} f(x)dx. \tag{15.32}$$

If our distribution is continuous, note this is the same as the probability of $[-\infty, x)$; however, for distributions arising from Dirac delta functionals like our $\mu_{A,N}(x)$, there will be finite, non-zero jumps in the cumulative distribution function at the normalized eigenvalues. For example, for $\mu_{A,N}(x)$ we have

$$C_{\mu_{A,N}}(x) = \frac{1}{N} \sum_{\frac{\lambda_i(A)}{2\sqrt{N}} < x} 1, \tag{15.33}$$

which jumps by at least $\frac{1}{N}$ at each normalized eigenvalue. For two probability distributions f and g we define the **Kolmogoroff-Smirnov discrepency of f and g** to be $\sup_x |C_f(x) - C_g(x)|$. Note as $N \to \infty$ each normalized eigenvalue contributes a smaller percentage of the total probability. Using the Kolmogoroff-Smirnov discrepancy for when two probability distributions are close, one can show that as $N \to \infty$ "most" $\mu_{A,N}(x)$ are close to P.

Remark 15.3.6. It is not true that all matrices A yield $\mu_{A,N}(x)$ that are close to P; for example, consider multiples of the identity matrix. All the normalized eigenvalues are the same, and these real symmetric matrices will clearly not have $\mu_{A,N}(x)$ close to $P(x)$. Of course, as $N \to \infty$ the probability of A being close to a multiple of the identity matrix is zero.

Exercise 15.3.7. *Fix a probability distribution p, and consider $N \times N$ real symmetric matrices with entries independently chosen from p. What is the probability that a matrix in this ensemble has all entries within ϵ of a multiple of the $N \times N$ identity matrix? What happens as $N \to \infty$ for fixed ϵ? How does the answer depend on p?*

Exercise 15.3.8. *Let m_k be the kth moment of a probability density P. Show $m_2 m_0 - m_1^2 \geq 0$. Note this can be interpreted as $\left| \begin{smallmatrix} m_0 & m_1 \\ m_1 & m_2 \end{smallmatrix} \right| \geq 0$. Thus, if $m_2 m_0 - m_1^2 < 0$, the m_k cannot be the moments of a probability distribution. Find a similar relation involving m_0, m_1, m_2, m_3 and m_4 and a determinant. See [Chr] and the references therein for more details, as well as [ShTa, Si] (where the determinant condition is connected to continued fraction expansions!).*

Exercise 15.3.9. *If $p(x) = 0$ for $|x| > R$, show the kth moment satisfies $m_k \leq R^k$. Hence $\lim_{j \to \infty} m_{2j}^{1/2j} < \infty$. Therefore, if a probability distribution has $\lim_{j \to \infty} m_{2j}^{1/2j} = \infty$, then for any R there is a positive probability of observing $|x| > R$. Alternatively, we say such p has unbounded support. Not surprisingly, the Gaussian moments (see Exercise 15.3.10) grow sufficiently rapidly so that the Gaussian has unbounded support.*

Exercise 15.3.10 (Moments of the Gaussian). *Calculate the moments of the Gaussian $g(x) = \frac{1}{\sqrt{2\pi}} e^{-x^2/2}$. Prove the odd moments vanish and the even moments are $m_{2k} = (2k - 1)!!$, where $n!! = n(n - 2)(n - 4) \cdots$. This is also the number of* *ways to match $2k$ objects in pairs. Show the moments grow sufficiently slowly to determine a unique continuous probability density.*

Exercise 15.3.11. *Consider two probability distributions f and g on $[0,1]$ where $f(x) = 1$ for all x and $g(x) = 1$ for $x \notin \mathbb{Q}$ and 0 otherwise. Note both f and g assign the same probability to any $[a,b]$ with $b \neq a$. Show $\sup_{x \in [0,1]} |f(x) - g(x)| = 1$ but the Kolmogoroff-Smirnov discrepancy is zero. Thus looking at the pointwise difference could incorrectly cause us to conclude that f and g are very different.*

Exercise 15.3.12. *Do there exist two probability distributions that have a large Kolmogoroff-Smirnov discrepancy but are close pointwise?*

15.3.3 Idea of the Proof of the Semi-Circle Law

We give a glimpse of the proof of the Semi-Circle Law below; a more detailed sketch will be provided in Chapter 16. We use Moment Method from §15.3.2.

For each $\mu_{A,N}(x)$, we calculate its kth-moment, $M_{N,k}(A) = \mathbb{E}[x^k]_A$. Let $M_{N,k}$ be the average of $M_{N,k}(A)$ over all A. We must show as $N \to \infty$, $M_{N,k}$ converges to the kth moment of the semi-circle. We content ourselves with just the second moment below, and save the rest for §16.1. By Lemma 15.2.10,

$$M_{N,2} = \int_A M_{N,k}(A)\mathrm{Prob}(A)dA$$

$$= \frac{1}{2^2 N^{\frac{2}{2}+1}} \int_A \mathrm{Trace}(A^2)\mathrm{Prob}(A)dA. \qquad (15.34)$$

We use Theorem 15.2.1 to expand $\mathrm{Trace}(A^2)$ and find

$$M_{N,2} = \frac{1}{2^2 N^2} \int_A \sum_{i=1}^{N} \sum_{j=1}^{N} a_{ij}^2 \, \mathrm{Prob}(A)dA. \qquad (15.35)$$

We now expand $\mathrm{Prob}(A)dA$ by (15.3):

$M_{N,2}$

$$= \frac{1}{2^2 N^2} \int_{a_{11}=-\infty}^{\infty} \cdots \int_{a_{NN}=-\infty}^{\infty} \sum_{i=1}^{N}\sum_{j=1}^{N} a_{ij}^2 \cdot p(a_{11})da_{11} \cdots p(a_{NN})da_{NN}$$

$$= \frac{1}{2^2 N^2} \sum_{i=1}^{N}\sum_{j=1}^{N} \int_{a_{11}=-\infty}^{\infty} \cdots \int_{a_{NN}=-\infty}^{\infty} a_{ij}^2 \cdot p(a_{11})da_{11} \cdots p(a_{NN})da_{NN};$$

$$(15.36)$$

we may interchange the summations and the integrations as there are finitely many sums. For each of the N^2 pairs (i,j), we have terms like

$$\int_{a_{ij}=-\infty}^{\infty} a_{ij}^2 p(a_{ij})da_{ij} \cdot \prod_{\substack{(k,l) \neq (ij) \\ k \leq l}} \int_{a_{kl}=-\infty}^{\infty} p(a_{kl})da_{kl}. \qquad (15.37)$$

The above equals 1. The first factor is 1 because it is the variance of a_{ij}, which was assumed to be 1. The second factor is a product of integrals where each integral is 1 (because p is a probability density). As there are N^2 summands in (15.36), we find $M_{N,2} = \frac{1}{4}$ (so $\lim_{N\to\infty} M_{N,2} = \frac{1}{4}$), which is the second moment of the semi-circle.

Exercise 15.3.13. *Show the second moment of the semi-circle is $\frac{1}{4}$.*

Exercise 15.3.14. *Calculate the third and fourth moments, and compare them to those of the semi-circle.*

Remark 15.3.15 (Important). Two features of the above proof are worth highlighting, as they appear again and again below. First, note that we want to answer a question about the *eigenvalues* of A; however, our notion of randomness gives us information on the *entries* of A. The key to converting information on the entries to knowledge about the eigenvalues is having some type of Trace Formula, like Theorem 15.2.4.

The second point is the Trace Formula would be useless, merely converting us from one hard problem to another, if we did not have a good Averaging Formula, some way to average over all random A. In this problem, the averaging is easy because of how we defined randomness.

Remark 15.3.16. While the higher moments of p are not needed for calculating $M_{N,2} = \mathbb{E}[x^2]$, their finiteness comes into play when we study higher moments.

15.3.4 Examples of the Semi-Circle Law

First we look at the density of eigenvalues when p is the standard Gaussian, $p(x) = \frac{1}{\sqrt{2\pi}} e^{-x^2/2}$. In Figure 15.2 we calculate the density of eigenvalues for 500 such matrices (400×400), and note a great agreement with the semi-circle.

What about a density where the higher moments are infinite? Consider the Cauchy distribution,

$$p(x) = \frac{1}{\pi(1 + x^2)}. \tag{15.38}$$

The behavior is clearly not semi-circular (see Figure 15.3). The eigenvalues are unbounded; for graphing purposes, we have put all eigenvalues greater than 300 in the last bin, and less than -300 in the first bin.

Exercise 15.3.17. *Prove the Cauchy distribution is a probability distribution by showing it integrates to 1. While the distribution is symmetric, one cannot say the mean is 0, as the integral $\int |x| p(x)dx = \infty$. Regardless, show the second moment is infinite.*

15.3.5 Summary

Note the universal behavior: though the proof is not given here, the Semi-Circle Law holds for all mean zero, finite moment distributions. The independence of the behavior on the exact nature of the underlying probability density p is a common feature of Random Matrix Theory statements, as is the fact that as $N \to \infty$ most

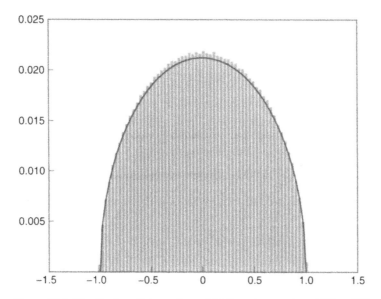

Figure 15.2 Distribution of eigenvalues: 500 Gaussian matrices (400×400)

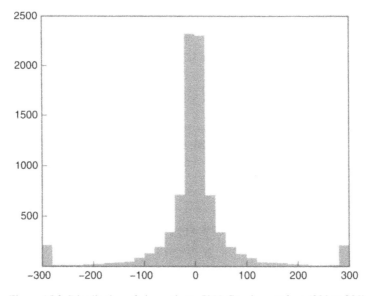

Figure 15.3 Distribution of eigenvalues: 5000 Cauchy matrices (300×300)

A yield $\mu_{A,N}(x)$ that are close (in the sense of the Kolmogoroff-Smirnov discrepancy) to P (where P is determined by the limit of the average of the moments $M_{N,k}(A)$). For more on the Semi-Circle Law, see [Bai, BK].

15.4 ADJACENT NEIGHBOR SPACINGS

15.4.1 GOE Distribution

The Semi-Circle Law (when the conditions are met) tells us about the density of eigenvalues. We now ask a more refined question:

Question 15.4.1. *How are the spacings between adjacent eigenvalues distributed?*

For example, let us write the eigenvalues of A in increasing order; as A is real symmetric, the eigenvalues will be real:

$$\lambda_1(A) \leq \lambda_2(A) \leq \cdots \leq \lambda_N(A). \tag{15.39}$$

The spacings between adjacent eigenvalues are the $N-1$ numbers

$$\lambda_2(A) - \lambda_1(A),\ \lambda_3(A) - \lambda_2(A),\ \ldots,\ \lambda_N(A) - \lambda_{N-1}(A). \tag{15.40}$$

As before (see Chapter 12), it is more natural to study the spacings between adjacent normalized eigenvalues; thus, we have

$$\frac{\lambda_2(A)}{2\sqrt{N}} - \frac{\lambda_1(A)}{2\sqrt{N}},\ \ldots,\ \frac{\lambda_N(A)}{2\sqrt{N}} - \frac{\lambda_{N-1}(A)}{2\sqrt{N}}. \tag{15.41}$$

Similar to the probability distribution $\mu_{A,N}(x)$, we can form another probability distribution $\nu_{A,N}(s)$ to measure spacings between adjacent normalized eigenvalues.

Definition 15.4.2.

$$\nu_{A,N}(s)ds = \frac{1}{N-1} \sum_{i=2}^{N} \delta\left(s - \frac{\lambda_i(A) - \lambda_{i-1}(A)}{2\sqrt{N}}\right) ds. \tag{15.42}$$

Based on experimental evidence and some heuristical arguments, it was conjectured that as $N \to \infty$, the limiting behavior of $\nu_{A,N}(s)$ is independent of the probability density p used in randomly choosing the $N \times N$ matrices A.

Conjecture 15.4.3 (GOE Conjecture:). *As $N \to \infty$, $\nu_{A,N}(s)$ approaches a universal distribution that is independent of p.*

Remark 15.4.4. GOE stands for Gaussian Orthogonal Ensemble; the conjecture is known if p is (basically) a Gaussian. We explain the nomenclature in Chapter 17.

Remark 15.4.5 (Advanced). The universal distribution is $\frac{\pi^2}{4}\frac{d^2\Psi}{dt^2}$, where $\Psi(t)$ is (up to constants) the Fredholm determinant of the operator $f \to \int_{-t}^{t} K * f$ with kernel $K = \frac{1}{2\pi}\left(\frac{\sin(\xi-\eta)}{\xi-\eta} + \frac{\sin(\xi+\eta)}{\xi+\eta}\right)$. This distribution is well approximated by $p_W(s) = \frac{\pi}{2}s\exp\left(-\frac{\pi s^2}{4}\right)$.

Exercise 15.4.6. *Prove* $p_W(s) = \frac{\pi}{2} s \exp\left(-\frac{\pi s^2}{4}\right)$ *is a probability distribution with mean 1. What is its variance?*

We study the case of $N = 2$ and p a Gaussian in detail in Chapter 17.

Exercise[hr] **15.4.7** (Wigner's surmise). *In 1957 Wigner conjectured that as $N \to \infty$ the spacing between adjacent normalized eigenvalues is given by*

$$p_W(s) = \frac{\pi}{2} s \exp\left(-\frac{\pi s^2}{4}\right). \tag{15.43}$$

He was led to this formula from the following assumptions:

- *Given an eigenvalue at x, the probability that another one lies s units to its right is proportional to s.*

- *Given an eigenvalue at x and $I_1, I_2, I_3, \ldots$ any disjoint intervals to the right of x, then the events of observing an eigenvalue in I_j are independent for all j.*

- *The mean spacing between consecutive eigenvalues is 1.*

Show these assumptions imply (15.43).

15.4.2 Numerical Evidence

We provide some numerical support for the GOE Conjecture. In all the experiments below, we consider a large number of $N \times N$ matrices, where for each matrix we look at a small (small relative to N) number of eigenvalues in the **bulk of the eigenvalue spectrum** (eigenvalues near 0), not near the **edge** (for the semi-circle, eigenvalues near ± 1). We do not look at all the eigenvalues, as the average spacing changes over such a large range, nor do we consider the interesting case of the largest or smallest eigenvalues. We study a region where the average spacing is approximately constant, and as we are in the middle of the eigenvalue spectrum, there are no edge effects. These edge effects lead to fascinating questions (for random graphs, the distribution of eigenvalues near the edge is related to constructing good networks to rapidly transmit information; see for example [DSV, Sar]).

First we consider 5000 300×300 matrices with entries independently chosen from the uniform distribution on $[-1, 1]$ (see Figure 15.4). Notice that even with N as low as 300, we are seeing a good fit between conjecture and experiment.

What if we take p to be the Cauchy distribution? In this case, the second moment of p is infinite, and the alluded to argument for semi-circle behavior is not applicable. Simulations showed the density of eigenvalues did not follow the Semi-Circle Law, which does not contradict the theory as the conditions of the theorem were not met. What about the spacings between adjacent normalized eigenvalues of real symmetric matrices, with the entries drawn from the Cauchy distribution?

We study 5000 100×100 and then 5000 300×300 Cauchy matrices (see Figures 15.5 and 15.6. We note good agreement with the conjecture, and as N increases the fit improves.

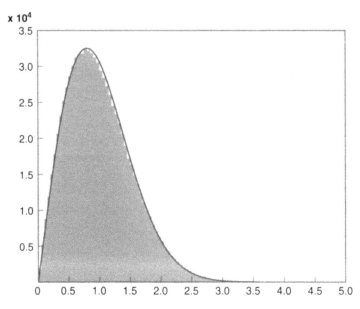

Figure 15.4 The local spacings of the central three-fifths of the eigenvalues of 5000 matrices
(300×300) whose entries are drawn from the Uniform distribution on $[-1, 1]$

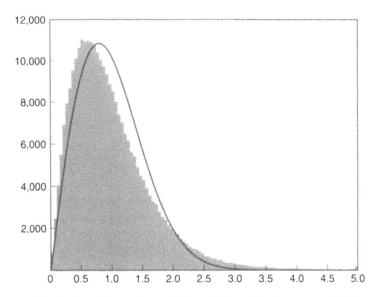

Figure 15.5 The local spacings of the central three-fifths of the eigenvalues of 5000 matrices
(100×100) whose entries are drawn from the Cauchy distribution

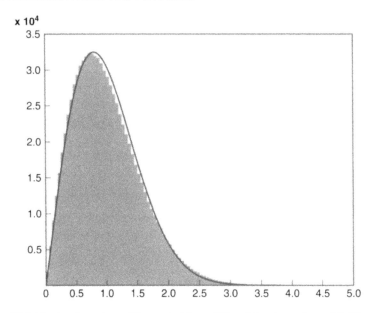

Figure 15.6 The local spacings of the central three-fifths of the eigenvalues of 5000 matrices
(300×300) whose entries are drawn from the Cauchy distribution

We give one last example. Instead of using continuous probability distribution,
we investigate a discrete case. Consider the Poisson Distribution:

$$p(n) = \frac{\lambda^n}{n!} \, e^{-\lambda}. \tag{15.44}$$

We investigate 5000 300×300 such matrices, first with $\lambda = 5$, and then with
$\lambda = 20$, noting again excellent agreement with the GOE Conjecture (see Figures
15.7 and 15.8):

15.5 THIN SUB-FAMILIES

Before moving on to connections with number theory, we mention some very important subsets of real symmetric matrices. The subsets will be large enough so
that there are averaging formulas at our disposal, but thin enough so that sometimes
we see new behavior. Similar phenomena will resurface when we study zeros of
Dirichlet *L*-functions.

As motivation, consider as our initial set all even integers. Let $N_2(x)$ denote the
number of even integers at most x. We see $N_2(x) \sim \frac{x}{2}$, and the spacing between
adjacent integers is 2. If we look at *normalized* even integers, we would have
$y_i = \frac{2i}{2}$, and now the spacing between adjacent normalized even integers is 1.

Now consider the subset of even squares. If $N_\square(x)$ is the number of even squares
at most x, then $N_\square(x) \sim \frac{\sqrt{x}}{2}$. For even squares of size x, say $x = (2m)^2$, the next

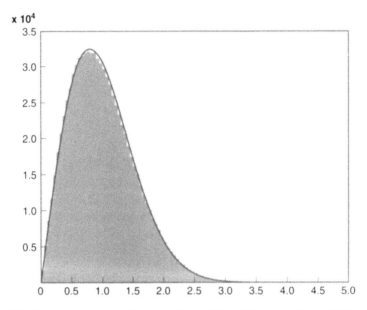

Figure 15.7 The local spacings of the central three-fifths of the eigenvalues of 5000 matrices
(300×300) whose entries are drawn from the Poisson distribution ($\lambda = 5$)

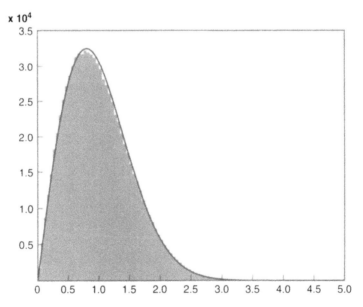

Figure 15.8 The local spacings of the central three fifths of the eigenvalues of 5000 matrices
(300×300) whose entries are drawn from the Poisson distribution ($\lambda = 20$)

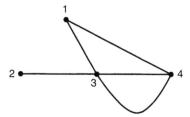

Figure 15.9 A typical graph

even square is at $(2m + 2)^2 = x + 8m + 4$. Note the spacing between adjacent even squares is about $8m \sim 4\sqrt{x}$ for m large.

Exercise 15.5.1. *By appropriately normalizing the even squares, show we obtain a new sequence with a similar distribution of spacings between adjacent elements as in the case of normalized even integers. Explicitly, look at the spacings between N consecutive even squares with each square of size x and $N \ll x$.*

Remark 15.5.2. A far more interesting example concerns prime numbers. For the first set, consider all prime numbers. For the subset, fix an integer m and consider all prime numbers p such that $p + 2m$ is also prime; if $m = 1$ we say p and $p + 2$ are a twin prime pair. It is unknown if there are infinitely many elements in the second set for any m, though there are conjectural formulas (using the techniques of Chapter 13). It is fascinating to compare these two sets; for example, what is the spacing distribution between adjacent (normalized) primes look like, and is that the same for normalized twin prime pairs? See Research Project 12.9.5.

15.5.1 Random Graphs: Theory

A **graph** G is a collection of points (the **vertices** V) and lines connecting pairs of points (the **edges** E). While it is possible to have an edge from a vertex to itself (called a **self-loop**), we study the subset of graphs where this does not occur. We will allow multiple edges to connect the same two vertices (if there are no multiple edges, the graph is **simple**). The **degree of a vertex** is the number of edges leaving (or arriving at) that vertex. A graph is d-**regular** if every vertex has exactly d edges leaving (or arriving).

For example, consider the graph in Figure 15.9: The degrees of vertices are 2, 1, 4 and 3, and vertices 3 and 4 are connected with two edges.

To each graph with N vertices we can associate an $N \times N$ real symmetric matrix, called the **adjacency matrix**, as follows: First, label the vertices of the graph from 1 to N (see Exercise 15.5.3). Let a_{ij} be the number of edges from vertex i to vertex j. For the graph above, we have

$$A = \begin{pmatrix} 0 & 0 & 1 & 1 \\ 0 & 0 & 1 & 0 \\ 1 & 1 & 0 & 2 \\ 1 & 0 & 2 & 0 \end{pmatrix}. \tag{15.45}$$

For each N, consider the space of all d-regular graphs. To each graph G we associate its adjacency matrix $A(G)$. We can build the eigenvalue probability distributions (see §15.2.3) as before. We can investigate the density of the eigenvalues and spacings between adjacent eigenvalues. We are no longer choosing the matrix elements at random; once we have chosen a graph, the entries are determined. Thus we have a more combinatorial type of averaging to perform: we average over all graphs, not over matrix elements. Even though these matrices are all real symmetric and hence a subset of the earlier ensembles, the probability density for these matrices are very different, and lead to different behavior (see also Remark 16.2.13 and §7.10).

One application of knowledge of eigenvalues of graphs is to network theory. For example, let the vertices of a graph represent various computers. We can transmit information between any two vertices that are connected by an edge. We desire a well connected graph so that we can transmit information rapidly through the system. One solution, of course, is to connect all the vertices and obtain the **complete graph**. In general, there is a cost for each edge; if there are N vertices in a simple graph, there are $\frac{N(N-1)}{2}$ possible edges; thus the complete graph quickly becomes very expensive. For N vertices, d-regular graphs have only $\frac{dN}{2}$ edges; now the cost is linear in the number of vertices. The distribution of eigenvalues (actually, the second largest eigenvalue) of such graphs provide information on how well connected it is. For more information, as well as specific constructions of such well connected graphs, see [DSV, Sar].

Exercise 15.5.3. *For a graph with N vertices, show there are $N!$ ways to label the vertices. Each labeling gives rise to an adjacency matrix. While a graph could potentially have $N!$ different adjacency matrices, show all adjacency matrices have the same eigenvalues, and therefore the same eigenvalue probability distribution.*

Remark 15.5.4. Fundamental quantities should not depend on presentation. Exercise 15.5.3 shows that the eigenvalues of a graph do not depend on how we label the graph. This is similar to the eigenvalues of an operator $T : \mathbb{C}^n \to \mathbb{C}^n$ do not depend on the basis used to represent T. Of course, the eigenvectors *will* depend on the basis.

Exercise 15.5.5. *If a graph has N labeled vertices and E labeled edges, how many ways are there to place the E edges so that each edge connects two distinct vertices? What if the edges are not labeled?*

Exercise 15.5.6 (Bipartite graphs). *A graph is bipartite if the vertices V can be split into two distinct sets, A_1 and A_2, such that no vertices in an A_i are connected by an edge. We can construct a d-regular bipartite graph with $\#A_1 = \#A_2 = N$. Let A_1 be vertices $1, \ldots, N$ and A_2 be vertices $N + 1, \ldots, 2N$. Let $\sigma_1, \ldots, \sigma_d$ be permutations of $\{1, \ldots, N\}$. For each σ_j and $i \in \{1, \ldots, N\}$, connect vertex $i \in A_1$ to vertex $N + \sigma_j(i) \in A_2$. Prove this graph is bipartite and d-regular. If $d = 3$, what is the probability (as $N \to \infty$) that two vertices have two or more edges connecting them? What is the probability if $d = 4$?*

Remark 15.5.7. Exercise 15.5.6 provides a method for sampling the space of bipartite d-regular graphs, but does this construction sample the space uniformly (i.e., is every d-regular bipartite graph equally likely to be chosen by this method)? Further, is the behavior of eigenvalues of d-regular bipartite graphs the same as the behavior of eigenvalues of d-regular graphs? See [Bol], pages 50–57 for methods to sample spaces of graphs uniformly.

Exercise 15.5.8. *The **coloring number** of a graph is the minimum number of colors needed so that no two vertices connected by an edge are colored the same. What is the coloring number for the complete graph on N? For a bipartite graph with N vertices in each set?*

Consider now the following graphs. For any integer N let G_N be the graph with vertices the integers $2, 3, \ldots, N$, and two vertices are joined if and only if they have a common divisor greater than 1. Prove the coloring number of G_{10000} is at least 13. Give good upper and lower bounds as functions of N for the coloring number of G_N.

15.5.2 Random Graphs: Results

The first result, due to McKay [McK], is that while the density of states is *not* the semi-circle there is a universal density for each d.

Theorem 15.5.9 (McKay's Law). *Consider the ensemble of all d-regular graphs with N vertices. As $N \to \infty$, for almost all such graphs G, $\mu_{A(G),N}(x)$ converges to Kesten's measure*

$$f(x) = \begin{cases} \frac{d}{2\pi(d^2-x^2)}\sqrt{4(d-1)-x^2}, & |x| \le 2\sqrt{d-1} \\ 0 & \text{otherwise.} \end{cases} \tag{15.46}$$

Exercise 15.5.10. *Show that as $d \to \infty$, by changing the scale of x, Kesten's measure converges to the semi-circle distribution.*

Below (Figures 15.10 and 15.11) we see excellent agreement between theory and experiment for $d = 3$ and 6; the data is taken from [QS2].

The idea of the proof is that locally almost all of the graphs almost always look like trees (connected graphs with no loops), and for trees it is easy to calculate the eigenvalues. One then does a careful book-keeping. Thus, this sub-family is thin enough so that a new, universal answer arises. Even though all of these adjacency matrices are real symmetric, it is a very thin subset. It is *because* it is such a thin subset that we are able to see new behavior.

Exercise 15.5.11. *Show a general real symmetric matrix has $\frac{N(N+1)}{2}$ independent entries, while a d-regular graph's adjacency matrix has $\frac{dN}{2}$ non-zero entries.*

What about spacings between normalized eigenvalues? Figure 15.12 shows that, surprisingly, the result *does* appear to be the same as that from all real symmetric matrices. See [JMRR] for more details.

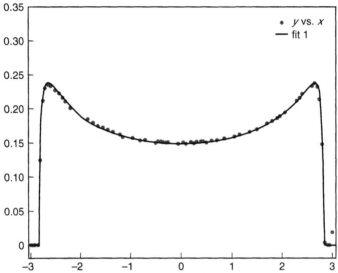

Figure 15.10 Comparison between theory (solid line) and experiment (dots) for 1000 eigenvalues of 3-regular graphs (120 bins in the histogram)

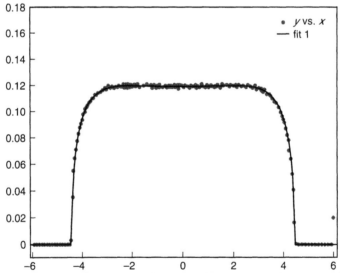

Figure 15.11 Comparison between theory (solid line) and experiment (dots) for 1000 eigenvalues of 6-regular graphs (240 bins in the histogram)

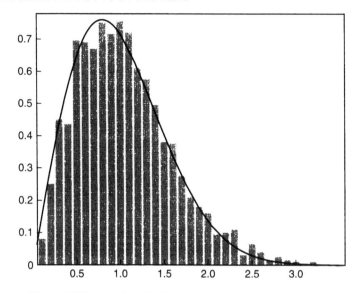

Figure 15.12 3-regular, 2000 vertices (graph courtesy of [JMRR])

15.6 NUMBER THEORY

We assume the reader is familiar with the material and notation from Chapter 3. For us an *L*-**function** is given by a **Dirichlet series** (which converges if $\Re s$ is sufficiently large), has an **Euler product**, and the coefficients have arithmetic meaning:

$$L(s,f) \;=\; \sum_{n=1}^{\infty} \frac{a_n(f)}{n^s} \;=\; \prod_{p} L_p(p^{-s},f)^{-1}, \quad \Re s > s_0. \qquad (15.47)$$

The **Generalized Riemann Hypothesis** asserts that all non-trivial zeros have $\Re s = \frac{1}{2}$; i.e., they are on the **critical line** $\Re s = \frac{1}{2}$ and can be written as $\frac{1}{2} + i\gamma$, $\gamma \in \mathbb{R}$.

The simplest example is $\zeta(s)$, where $a_n(\zeta) = 1$ for all n; in Chapter 3 we saw how information about the distribution of zeros of $\zeta(s)$ yielded insights into the behavior of primes. The next example we considered were Dirichlet *L*-functions, the *L*-functions from Dirichlet characters χ of some conductor m. Here $a_n(\chi) = \chi(n)$, and these functions were useful in studying primes in arithmetic progressions.

For a fixed m, there are $\phi(m)$ Dirichlet *L*-functions modulo m. This provides our first example of a **family** of *L*-functions. We will not rigorously define a family, but content ourselves with saying a family of *L*-functions is a collection of "similar" *L*-functions.

The following examples will be considered families: (1) all Dirichlet *L*-functions with conductor m; (2) all Dirichlet *L*-functions with conductor $m \in [N, 2N]$; (3) all Dirichlet *L*-functions arising from quadratic characters with prime conductor $p \in [N, 2N]$. In each of the cases, each *L*-function has the same conductor, similar functional equations, and so on. It is not unreasonable to think they might share other properties.

Another example comes from elliptic curves. We commented in §4.2.2 that given a cubic equation $y^2 = x^3 + A_f x + B_f$, if $a_p(f) = p - N_p$ (where N_p is the number of solutions to $y^2 \equiv x^3 + A_f x + B_f \bmod p$), we can construct an L-function using the $a_p(f)$'s. We construct a family as follows. Let $A(T), B(T)$ be polynomials with integer coefficients in T. For each $t \in \mathbb{Z}$, we get an elliptic curve E_t (given by $y^2 = x^3 + A(t)x + B(t)$), and can construct an L-function $L(s, E_t)$. We can consider the family where $t \in [N, 2N]$.

Remark 15.6.1. Why are we considering "restricted" families, for example Dirichlet L-functions with a fixed conductor m, or $m \in [N, 2N]$, or elliptic curves with $t \in [N, 2N]$? The reason is similar to our random matrix ensembles: we do not consider infinite dimensional matrices: we study $N \times N$ matrices, and take the limit as $N \to \infty$. Similarly in number theory, it is easier to study finite sets, and then investigate the limiting behavior.

Assuming the zeros all lie on the line $\Re s = \frac{1}{2}$, similar to the case of real symmetric or complex Hermitian matrices, we can study spacings between zeros. We now describe some results about the distribution of zeros of L-functions. Two classical ensembles of random matrices play a central role: the Gaussian Orthogonal Ensemble **GOE** (resp., Gaussian Unitary Ensemble **GUE**), the space of real symmetric (complex Hermitian) matrices where the entries are chosen independently from Gaussians; see Chapter 17. It was observed that the spacings of energy levels of heavy nuclei are in excellent agreement with those of eigenvalues of real symmetric matrices; thus, the GOE became a common model for the energy levels. In §15.6.1 we see there is excellent agreement between the spacings of normalized zeros of L-functions and those of eigenvalues of complex Hermitian matrices; this led to the belief that the GUE is a good model for these zeros.

15.6.1 n-Level Correlations

In an amazing set of computations starting at the $10^{20\text{th}}$ zero, Odlyzko [Od1, Od2] observed phenomenal agreement between the spacings between adjacent normalized zeros of $\zeta(s)$ and spacings between adjacent normalized eigenvalues of complex Hermitian matrices. Specifically, consider the set of $N \times N$ random Hermitian matrices with entries chosen from the Gaussian distribution (the GUE). As $N \to \infty$ the limiting distribution of spacings between adjacent eigenvalues is indistinguishable from what Odlyzko observed in zeros of $\zeta(s)$!

His work was inspired by Montgomery [Mon2], who showed that for suitable test functions the pair correlation of the normalized zeros of $\zeta(s)$ agree with that of normalized eigenvalues of complex Hermitian matrices. Let $\{\alpha_j\}$ be an increasing sequence of real numbers, $B \subset \mathbb{R}^{n-1}$ a compact box. Define the n-**level correlation** by

$$\lim_{N \to \infty} \frac{\#\left\{\left(\alpha_{j_1} - \alpha_{i_2}, \ldots, \alpha_{j_{n-1}} - \alpha_{j_n}\right) \in B, j_i \leq N; j_i \neq j_k\right\}}{N}. \quad (15.48)$$

For example, the 2-level (or pair) correlation provides information on how often two normalized zeros (not necessarily adjacent zeros) have a difference in a given interval. One can show that if all the n-level correlations could be computed, then we would know the spacings between adjacent zeros.

We can regard the box B as a product of $n-1$ characteristic functions of intervals (or binary indicator variables). Let

$$I_{a_i,b_i}(x) = \begin{cases} 1 & \text{if } x \in [a_i, b_i], \\ 0 & \text{otherwise.} \end{cases} \tag{15.49}$$

We can represent the condition $x \in B$ by $I_B(x) = \prod_{i=1}^n I_{a_i,b_i}(x_i)$. Instead of using a box B and its function I_B, it is more convenient to use an infinitely differentiable test function (see [RS] for details). In addition to the pair correlation and the numerics on adjacent spacings, Hejhal [Hej] showed for suitable test functions the 3-level (or triple) correlation for $\zeta(s)$ agrees with that of complex Hermitian matrices, and Rudnick-Sarnak [RS] proved (again for suitable test functions) that the n-level correlations of *any* "nice" L-function agree with those of complex Hermitian matrices.

The above work leads to the **GUE conjecture**: in the limit (as one looks at zeros with larger and larger imaginary part, or $N \times N$ matrices with larger and larger N), the spacing between zeros of L-functions is the same as that between eigenvalues of complex Hermitian matrices. In other words, the GUE is a good model of zeros of L-functions.

Even if true, however, the above cannot be the complete story.

Exercise 15.6.2. *Assume that the imaginary parts of the zeros of $\zeta(s)$ are unbounded. Show that if one removes any finite set of zeros, the n-level correlations are unchanged. Thus this statistic is insensitive to finitely many zeros.*

The above exercise shows that the n-level correlations are not sufficient to capture all of number theory. For many L-functions, there is reason to believe that there is different behavior near the central point $s = \frac{1}{2}$ (the center of the critical strip) than higher up. For example, the **Birch and Swinnerton-Dyer conjecture** (see §4.2.2) says that if $E(\mathbb{Q})$ (the group of rational solutions for an elliptic curve E; see §4.2.1) has rank r, then there are r zeros at the central point, and we might expect different behavior if there are more zeros.

Katz and Sarnak [KS1, KS2] proved that the n-level correlations of complex Hermitian matrices are also equal to the n-level correlations of the **classical compact groups**: unitary matrices (and its subgroups of symplectic and orthogonal matrices) with respect to Haar measure. Haar measure is the analogue of fixing a probability distribution p and choosing the entries of our matrices randomly from p; it should be thought of as specifying how we "randomly" chose a matrix from these groups. As a unitary matrix U satisfies $U^*U = I$ (where U^* is the complex conjugate transpose of U), we see each entry of U is at most 1 in absolute value, which shows unitary matrices are a compact group. A similar argument shows the set of orthogonal matrices Q such that $Q^T Q = I$ is compact.

What this means is that *many* different ensembles of matrices have the same n-level correlations – there is not one unique ensemble with these values. This led to a new statistic which is different for different ensembles, and allows us to "determine" which matrix ensemble the zeros follow.

Remark 15.6.3 (Advanced). Consider the following classical compact groups: $U(N)$, $USp(2N)$, SO, SO(even) and SO(odd) with their Haar measure. Fix a group and choose a generic matrix element. Calculating the n-level correlations of its eigenvalues, integrating over the group, and taking the limit as $N \to \infty$, Katz and Sarnak prove the resulting answer is universal, independent of the particular group chosen. In particular, we cannot use the n-level correlations to distinguish the other classical compact groups from each other.

15.6.2 1-Level Density

In the n-level correlations, given an L-function we studied differences between zeros. It can be shown that any "nice" L-function has infinitely many zeros on the line $\Re s = \frac{1}{2}$; thus, if we want to study "high" zeros (zeros very far above the central point $s = \frac{1}{2}$), each L-function has enough zeros to average over.

The situation is completely different if we study "low" zeros, zeros near the central point. Now each L-function only has a few zeros nearby, and there is nothing to average: wherever the zeros are, that's where they are! This led to the introduction of families of L-functions. For example, consider Dirichlet L-functions with characters of conductor m. There are $\phi(m)$ such L-functions. For each L-function we can calculate properties of zeros near the central point and then we can *average* over the $\phi(m)$ L-functions, taking the limit as $m \to \infty$.

Explicitly, let $h(x)$ be a continuous function of rapid decay. For an L-function $L(s, f)$ with non-trivial zeros $\frac{1}{2} + i\gamma_f$ (assuming GRH, each $\gamma_f \in \mathbb{R}$), consider

$$D_f(h) = \sum_j h\left(\gamma_f \frac{\log c_f}{2\pi}\right). \tag{15.50}$$

Here c_f is the **analytic conductor**; basically, it rescales the zeros near the central point. As h is of rapid decay, almost all of the contribution to (15.50) will come from zeros very close to the central point. We then average over all f in a family $\mathcal{F}$. We call this statistic the 1-**level density**:

$$D_{\mathcal{F}}(h) = \frac{1}{|\mathcal{F}|} \sum_{f \in \mathcal{F}} D_f(h). \tag{15.51}$$

Katz and Sarnak conjecture that the distribution of zeros near the **central point** $s = \frac{1}{2}$ in a family of L-functions should agree (in the limit) with the distribution of eigenvalues near 1 of a classical compact group (unitary, symplectic, orthogonal); *which* group depends on underlying symmetries of the family. The important point to note is that the GUE is not the entire story: other ensembles of matrices naturally arise. These conjectures, for suitable test functions, have been verified for a variety of families: we sketch the proof for Dirichlet L-functions in Chapter 18 and give an application as well.

Remark 15.6.4. Why does the central point $s = \frac{1}{2}$ correspond to the eigenvalue 1? As the classical compact groups are subsets of the unitary matrices, their eigenvalues can be written $e^{i\theta}$, $\theta \in (-\pi, \pi]$. Here $\theta = 0$ (corresponding to an eigenvalue of 1) is the center of the "critical line." Note certain such matrices have a forced eigenvalue at 1 (for example, any $N \times N$ orthogonal matrix with N odd); this is expected to be similar to L-functions with a forced zeros at the central point. The situation with multiple forced zeros at the central point is very interesting; while in some cases the corresponding random matrix models are known, other cases are still very much open. See [Mil6, Sn] for more details.

Exercise[h] 15.6.5. *U is a unitary matrix if $U^*U = I$, where U^* is the complex conjugate transpose of U. Prove the eigenvalues of unitary matrices can be written as $e^{i\theta_j}$ for $\theta_j \in \mathbb{R}$. An orthogonal matrix is a real unitary matrix; thus $Q^T Q = I$ where Q^T is the transpose of Q. Must the eigenvalues of an orthogonal matrix be real?*

Remark 15.6.6 (Advanced). In practice, one takes h in (15.50) to be a Schwartz function whose Fourier transform has finite support (see §11.4.1). Similar to the n-level correlations, one can generalize the above and study n-level densities. The determination of which classical compact group can sometimes be calculated by studying the monodromy groups of function field analogues.

We sketch an interpretation of the 1-level density. Again, the philosophy is that to each family of L-functions $\mathcal{F}$ there is an ensemble of random matrices $G(\mathcal{F})$ (where $G(\mathcal{F})$ is one of the classical compact groups), and to each $G(\mathcal{F})$ is attached a density function $W_{G(\mathcal{F})}$. Explicitly, consider the family of all non-trivial Dirichlet L-functions with prime conductor m, denoted by $\mathcal{F}_m$. We study this family in detail in Chapter 18. Then for suitable test functions h, we prove

$$\lim_{m \to \infty} D_{\mathcal{F}_m}(h) = \lim_{m \to \infty} \frac{1}{|\mathcal{F}_m|} \sum_{\chi \in \mathcal{F}_m} \sum_{\gamma_\chi} h\left(\gamma_\chi \frac{\log c_\chi}{2\pi}\right)$$

$$= \int_{-\infty}^{\infty} h(x) W_{G(\mathcal{F})}(x) dx. \tag{15.52}$$

We see that summing a test function of rapid decay over the scaled zeros is equivalent to integrating that test function against a family-dependent density function. We can see a similar phenomenon if we study sums of test functions at primes. For simplicity of presentation, we assume the Riemann Hypothesis to obtain better error estimates, though it is not needed (see Exercise 15.6.8).

Theorem 15.6.7. *Let F and its derivative F' be continuously differentiable functions of rapid decay; it suffices to assume $\int |F(x)|dx$ and $\int |F'(x)|dx$ are finite. Then*

$$\sum_p \frac{\log p}{p \log N} F\left(\frac{\log p}{\log N}\right) = \int_0^\infty F(x)dx + O\left(\frac{1}{\log N}\right). \tag{15.53}$$

Sketch of the proof. By the Riemann Hypothesis and partial summation (Theorem 2.2.6), we have

$$\sum_{p \le x} \log p = x + O(x^{\frac{1}{2}} \log^2(x)). \tag{15.54}$$

See [Da2] for how this bound follows from RH. We apply the integral version of partial summation (Theorem 2.2.6) to

$$\sum_{p\leq x} \log p \cdot \frac{1}{p}. \tag{15.55}$$

In the notation of Theorem 2.2.6, $a_n = \log p$ if p is prime and 0 otherwise, and $h(x) = \frac{1}{x}$. We find

$$\sum_{p\leq x} \frac{\log p}{p} = O(1) - \int_2^x (u + O(u^{\frac{1}{2}} \log^2 u)) \frac{-1}{u^2} du = \log x + O(1). \tag{15.56}$$

We again use the integral version of partial summation, but now on $\frac{\log p}{p} \cdot F\left(\frac{\log p}{\log N}\right)$ where $a_n = \frac{\log p}{p}$ for p prime and $h(x) = F\left(\frac{\log x}{\log N}\right)$. Let $u_0 = \frac{\log 2}{\log N}$. Then

$$\sum_{p\geq 2} \frac{\log p}{p} F\left(\frac{\log p}{\log N}\right) = -\int_2^\infty (\log x + O(1)) \frac{d}{dx} F\left(\frac{\log x}{\log N}\right) dx$$

$$= \int_2^\infty \left[\frac{1}{x} F\left(\frac{\log x}{\log N}\right) + O\left(\frac{1}{x \log N} \left|F'\left(\frac{\log x}{\log N}\right)\right|\right)\right] dx$$

$$= \log N \int_{u_0}^\infty \left[F(u) + O\left(\frac{1}{\log N} |F'(u)|\right)\right] du$$

$$= \log N \int_0^\infty \left[F(u) + O\left(\frac{|F'(u)|}{\log N}\right)\right] du + O(u_0 \log N \max_{t\in[0,u_0]} F(t))$$

$$= \log N \int_0^\infty F(u)du + O\left(\int_0^\infty |F'(u)|du\right) + O\left(u_0 \log N \max_{t\in[0,u_0]} F(t)\right)$$

$$= \log N \int_0^\infty F(u)du + O(1), \tag{15.57}$$

as $u_0 = \frac{\log 2}{\log N}$ and our assumption that F' is of rapid decay ensures that the F' integral is $O(1)$. Dividing by $\log N$ yields the theorem. Using the Prime Number Theorem instead of RH yields the same result, but with a worse error term.   □

Exercise 15.6.8. *Redo the above arguments using the bounds from §13.2.6, which eliminate the need to assume the Riemann Hypothesis.*

The above shows that summing a nice test function at the primes is related to integrating that function against a density; here the density is just 1. The 1-level density is a generalization of this to summing weighted zeros of L-functions, and the density we integrate against depends on properties of the family of L-functions. See §4.3.3 for more on distribution of points.

Exercise 15.6.9. *How rapidly must F decay as $x \to \infty$ to justify the arguments above? Clearly if F has compact support (i.e., if $F(x)$ is zero if $|x| > R$ for some R), F decays sufficiently rapidly, and this is often the case of interest.*

Exercise 15.6.10. *Why is the natural scale for Theorem 15.6.7 $\log N$ (i.e., why is it natural to evaluate the test function at $\frac{\log p}{\log N}$ and not p)?*

Exercise 15.6.11. *Instead of studying all primes, fix m and b with $(b, m) = 1$, and consider the set of primes $p \equiv b \bmod m$ (recall such p are called **primes in an arithmetic progression**); see §3.3. Modify the statement and proof of Theorem 15.6.7 to calculate the density for primes in arithmetic progression. If instead we consider twin primes, and we assume the number of twin primes at most x satisfies $\pi_2(x) = T_2 \frac{x}{\log^2 x} + O(x^{\frac{1}{2}+\epsilon})$ for some constant T_2, what is the appropriate normalization and density? See Definition 14.6.17 for the conjectured value of T_2.*

15.7 SIMILARITIES BETWEEN RANDOM MATRIX THEORY AND L-FUNCTIONS

The following (conjectural) correspondence has led to many fruitful predictions: in some sense, the zeros of L-functions behave like the eigenvalues of matrices which in turn behave like the energy levels of heavy nuclei. To study the energy levels of heavy nuclei, physicists bombard them with neutrons and study what happens; however, physical constraints prevent them from using neutrons of arbitrary energy. Similarly, we want to study zeros of L-functions. We "bombard" the zeros with a test function, but not an arbitrary one (*advanced:* the technical condition is the support of the Fourier transform of the test function must be small; the test function's support corresponds to the neutron's energy). To evaluate the sums of the test function at the zeros, similar to physicists restricting the neutrons they can use, number theorists can evaluate the sums for only a small class of test functions.

Similar to our proofs of the Semi-Circle Law, we again have three key ingredients. The first is we average over a collection of objects. Before it was the probability measures built from the normalized eigenvalues, now it is the $D_f(h)$ for each L-function f in the family for a fixed test function h. Second, we need some type of Trace Formula, which tells us what the correct scale is to study our problem and allows us to pass from knowledge of what we can sum to knowledge about what we want to understand. For matrices, we passed from sums over eigenvalues (which we wanted to understand) to sums over the matrix elements (which we were given and could execute). For number theory, using what are known as Explicit Formulas (see §18.1), we pass from sums over zeros in (15.50) to sums over the coefficients $a_n(f)$ in the L-functions. Finally, the Trace Formula is useless if we do not have some type of Averaging Formula. For matrices, because of how we generated matrices at random, we were able to average over the matrix elements; for number theory, one needs powerful theorem concerning averages of $a_n(f)$ as f ranges over a family. We have already seen a special case where there is an averaging relation: the orthogonality relations for Dirichlet characters (see Lemma 3.3.15). In §18.3 we summarize the similarities between Random Matrix Theory and Number Theory calculations. We give an application of the 1-level density to number theory in Theorem 18.2.7, namely bounding the number of characters χ such that $L(s, \chi)$ is non-zero at the central point. See [IS1, IS2] for more on non-vanishing of L-functions at the central point and applications of such results.

15.8 SUGGESTIONS FOR FURTHER READING

In addition to the references in this and subsequent chapters, we provide a few
starting points to the vast literature; the interested reader should consult the bibliographies of the references for additional reading.

A terrific introduction to classical random matrix theory is [Meh2], whose exposition has motivated our approach and many others; see also [For]. We recommend
reading at least some of the original papers of Wigner [Wig1, Wig2, Wig3, Wig4,
Wig5] and Dyson [Dy1, Dy2]. For a more modern treatment via Haar measure,
see [KS2]. Many of the properties of the classical compact groups can be found in
[Weyl]. See [Ha] for an entertaining account of the first meeting of Random Matrix Theory and Number Theory, and [Roc] for an accessible tour of connections
between $\zeta(s)$ and much of mathematics.

In Chapter 16 we sketch a proof of the Semi-Circle Law. See [CB] for a rigorous
treatment (including convergence issues and weaker conditions on the distribution
p). For more information, we refer the reader to [Bai, BK]. In Chapter 17 we
investigate the spacings of eigenvalues of 2×2 matrices. See [Gau, Meh1, Meh2]
for the spacings of $N \times N$ matrices as $N \to \infty$.

In Chapter 18 we study the 1-level density for all Dirichlet characters with conductor m, and state that as $m \to \infty$ the answer agrees with the similar statistic for unitary matrices (see [HuRu, Mil2]). If we look just at quadratic Dirichlet characters (Legendre symbols), then instead of seeing unitary symmetry one
finds agreement with eigenvalues of symplectic matrices (see [Rub2]). This is
similar to the behavior of eigenvalues of adjacency matrices of d-regular graphs,
which are a very special subset of real symmetry matrices but have different behavior. For more on connections between random graphs and number theory, see
[DSV] and Chapter 3 of [Sar]; see [Bol, McK, MW, Wor] and the student reports
[Cha, Gold, Nov, Ric, QS2] for more on random graphs.

The 1-level density (see also [ILS, Mil1]) and n-level correlations [Hej, Mon2,
RS] are but two of many statistics where random matrices behave similarly as L-
functions. We refer the reader to the survey articles [Con1, Dia, FSV, KS2, KeSn],
Chapter 25 of [IK] and to the research works [CFKRS, DM, FSV, KS1, Mil6, Od1,
Od2, Sn, TrWi] for more information.

Chapter Sixteen

Random Matrix Theory: Eigenvalue Densities

In this chapter we study the eigenvalue densities for many collections of random matrices. We concentrate on the density of normalized eigenvalues, though we mention a few questions regarding the spacings between normalized eigenvalues (which we investigate further in Chapter 17). We use the notation of Chapter 15.

16.1 SEMI-CIRCLE LAW

Consider an ensemble of $N \times N$ real symmetric matrices, where for simplicity we choose the entries independently from some fixed probability distribution p. One very important question we can ask is: given an interval $[a, b]$, how many eigenvalues do we expect to lie in this interval? We must be careful, however, in phrasing such questions. We have seen in §15.2.2 that the average size of the eigenvalues grows like $\sqrt{N}$. Hence it is natural to look at the density of normalized eigenvalues.

For example, the Prime Number Theorem states that the number of primes $p \leq x$ is $\frac{x}{\log x}$ plus lower order terms; see Theorem 13.2.10 for an exact statement. Thus the average spacing between primes $p \leq x$ is $\frac{x}{x/\log x} = \log x$. Consider two intervals $[10^5, 10^5 + 1000]$ and $[10^{200}, 10^{200} + 1000]$. The average spacing between primes in the first is about 11.5; the average spacing between primes in the second is about 460.5. We expect to find about 87 primes in the first interval, and about 2 in the second. In order to obtain a universal answer, we instead look at the density of normalized primes.

The appropriate question to ask is not what is the density of eigenvalues or primes in an interval $[a, b]$, but rather in an interval $[a \cdot (\text{Ave Spacing}), b \cdot (\text{Ave Spacing})]$.

Exercise 16.1.1. *As $x \to \infty$ how many numbers at most x are square-free (m is **square-free** if $n^2 | m$ implies $n = \pm 1$)? What is the average spacing between square-free numbers?*

16.1.1 Moments of the Semi-Circle Density

Consider

$$P(x) = \begin{cases} \frac{2}{\pi}\sqrt{1 - x^2} & \text{if } |x| \leq 1 \\ 0 & \text{otherwise.} \end{cases} \tag{16.1}$$

Exercise 16.1.2. *Show that $P(x)$ is a probability density (show that it is non-negative and integrates to 1). Graph $P(x)$.*

We call $P(x)$ the semi-circle density. We calculate the moments of the semi-circle. We prove that for $k \le 3$, the kth moment of the semi-circle $C(k)$ equals the expected k^{th} moment of $\mu_{A,N}(x)$ as $N \to \infty$, and sketch the proof for higher moments; see §15.2.3 for the definition of $\mu_{A,N}(x)$. We have

$$C(k) = \int_{-\infty}^{\infty} x^k P(x)dx = \frac{2}{\pi}\int_{-1}^{1} x^k \sqrt{1-x^2}dx. \tag{16.2}$$

We note that, by symmetry, $C(k) = 0$ for k odd, and $C(0) = 1$ as $P(x)$ is a probability density. For $k = 2m$ even, we change variables. Letting $x = \sin\theta$,

$$C(2m) = \frac{2}{\pi}\int_{-\frac{\pi}{2}}^{\frac{\pi}{2}} \sin^{2m}(\theta) \cdot \cos^2(\theta)d\theta. \tag{16.3}$$

Using $\sin^2(\theta) = 1 - \cos^2(\theta)$ gives

$$C(2m) = \frac{2}{\pi}\int_{-\frac{\pi}{2}}^{\frac{\pi}{2}} \sin^{2m}(\theta)d\theta - \frac{2}{\pi}\int_{-\frac{\pi}{2}}^{\frac{\pi}{2}} \sin^{2m+2}(\theta)d\theta. \tag{16.4}$$

The above integrals can be evaluated exactly. We repeatedly use

$$\cos^2(\phi) = \frac{1}{2} + \frac{1}{2}\cos(2\phi)$$

$$\sin^2(\phi) = \frac{1}{2} - \frac{1}{2}\cos(2\phi). \tag{16.5}$$

Repeated applications of the above allow us to write $\sin^{2m}(\theta)$ as a linear combination of $1, \cos(2\theta), \dots, \cos(2m\theta)$. Let

$$n!! = \begin{cases} n \cdot (n-2)\cdots 2 & \text{if } n \text{ is even} \\ n \cdot (n-2)\cdots 1 & \text{if } n \text{ is odd.} \end{cases} \tag{16.6}$$

We find (either prove directly or by induction) that

$$\frac{2}{\pi}\int_{-\frac{\pi}{2}}^{\frac{\pi}{2}} \sin^{2m}(\theta)d\theta = 2\frac{(2m-1)!!}{(2m)!!}. \tag{16.7}$$

Exercise 16.1.3. *Calculate $C(2)$ and $C(4)$ and show that in general*

$$C(2m) = 2\frac{(2m-1)!!}{(2m+2)!!}. \tag{16.8}$$

To each $N \times N$ real symmetric matrix A we associate a probability distribution $\mu_{A,N}(x)$ (see §15.2.3). We now sketch the proof that as $N \to \infty$ most of the $\mu_{A,N}(x)$ are close to $P(x)$, the semi-circle density.

16.1.2 Moment Preliminaries

Definition 16.1.4. *$M_{N,k}(A)$ is the kth moment of the probability measure attached to $\mu_{A,N}(x)$:*

$$M_{N,k}(A) = \int x^k \mu_{A,N}(x)dx = \frac{1}{N}\sum_{i=1}^{N}\left(\frac{\lambda_i(A)}{2\sqrt{N}}\right)^k. \tag{16.9}$$

As always, the starting point for our investigations is Theorem 15.2.4, which says $\sum \lambda_i(A)^k = \text{Trace}(A^k)$. By Lemma 15.2.10,

$$M_{N,k}(A) = \frac{1}{2^k N^{1+\frac{k}{2}}} \text{Trace}(A^k). \tag{16.10}$$

We show that as $N \to \infty$ the expected value of the moments $M_{N,k}(A)$ of the $\mu_{A,N}(x)$ converge to the moments of the semi-circle. This does not prove Wigner's Semi-Circle Law; we need some results from Probability Theory to complete the proof (see §15.3.2 for an explanation of the needed technical arguments, and [CB] for a rigorous derivation of the Semi-Circle Law).

See §15.3.3 for a review of notation. Let $M_{N,k} = \mathbb{E}[M_{N,k}(A)]$ be the average over all A (appropriately weighted by the probability density) of $M_{N,k}(A)$. Explicitly, the probability density of observing a matrix A with entries a_{ij} is $P(A)dA = \prod_{1 \le i \le j \le N} p(a_{ij})da_{ij}$, and averaging over all matrices gives the expected value of $M_{N,k}(A)$ is

$$M_{N,k} = \int_{a_{11}=-\infty}^{\infty} \cdots \int_{a_{NN}=-\infty}^{\infty} M_{N,k}(A) \prod_{1 \le i \le j \le N} p(a_{ij})da_{ij}. \tag{16.11}$$

From Theorem 15.2.1

$$\text{Trace}(A^k) = \sum_{1 \le i_1,\dots,i_k \le N} a_{i_1 i_2} a_{i_2 i_3} \cdots a_{i_k i_1}. \tag{16.12}$$

This and (16.10) yield

$$M_{N,k} = \frac{1}{2^k N^{1+\frac{k}{2}}} \sum_{1 \le i_1,\dots,i_k \le N} \mathbb{E}[a_{i_1 i_2} a_{i_2 i_3} \cdots a_{i_k i_1}], \tag{16.13}$$

where

$$\mathbb{E}[a_{i_1 i_2} a_{i_2 i_3} \cdots a_{i_k i_1}]$$
$$= \int_{a_{11}=-\infty}^{\infty} \cdots \int_{a_{NN}=-\infty}^{\infty} a_{i_1 i_2} a_{i_2 i_3} \cdots a_{i_k i_1} \prod_{1 \le i \le j \le N} p(a_{ij})da_{ij}. \tag{16.14}$$

There are N^k terms in $\text{Trace}(A^k)$, each term is a product of k factors a_{ij}. We use the notation $\mathbb{E}[a_{i_1 i_2} a_{i_2 i_3} \cdots a_{i_k i_1}]$ as we integrate each term $a_{i_1 i_2} a_{i_2 i_3} \cdots a_{i_k i_1}$ against $P(A)dA$, and this gives the expected value of the term.

We can write (16.14) in a more useful manner. While the above formula is correct, by grouping terms together we can rewrite it in such a way that it will be easier to evaluate. For small k, by brute force one can determine these integrals; however, as k increases, the computation becomes significantly harder and good combinatorics are needed, and the reformulation emphasizes the important parts of the calculation. Further, writing out the integrals each time leads to long formulas; by using expectation we have more concise formulas (though of course they convey the same information).

In the product $a_{i_1 i_2} a_{i_2 i_3} \cdots a_{i_k i_1}$, group a_{ij} together that have the same subscripts; as the matrices are symmetric, $a_{ij} = a_{ji}$ and we consider the pairs (i, j) and (j, i) equal. Say we can write

$$a_{i_1 i_2} a_{i_2 i_3} \cdots a_{i_k i_1} = a_{x_1 y_1}^{r_1} \cdots a_{x_\ell y_\ell}^{r_\ell}, \tag{16.15}$$

where all pairs (x_j, y_j) are distinct (remember, we consider the pairs (x, y) and (y, x) equal). For example,

$$a_{13}a_{34}a_{45}a_{53}a_{31}a_{14}a_{43}a_{31} = a_{13}^3 a_{34}^2 a_{14}a_{45}a_{35}.$$

As $a_{ij} = a_{ji}$, we have chosen to write the lower subscript first, especially as $P(A)dA = \prod_{1 \le i \le j \le N} p(a_{ij})da_{ij}$ has $i \le j$. We then obtain

$$\mathbb{E}[a_{i_1 i_2} a_{i_2 i_3} \cdots a_{i_k i_1}]$$

$$= \int_{a_{11}=-\infty}^{\infty} \cdots \int_{a_{NN}=-\infty}^{\infty} a_{x_1 y_1}^{r_1} \cdots a_{x_\ell y_\ell}^{r_\ell} \prod_{1 \le i \le j \le N} p(a_{ij})da_{ij}. \qquad (16.16)$$

As all entries are *independently* drawn from the *same* distribution, this integral greatly simplifies. Let p_k be the kth moment of p:

$$p_k = \int_{a=-\infty}^{\infty} a^k p(a)da. \qquad (16.17)$$

Then (16.16) becomes

Lemma 16.1.5. *Let $a_{i_1 i_2} a_{i_2 i_3} \cdots a_{i_k i_1} = a_{x_1 y_1}^{r_1} \cdots a_{x_\ell y_\ell}^{r_\ell}$, where all pairs (x_j, y_j) are distinct, remembering that we consider (i, j) the same as (j, i). Then*

$$\mathbb{E}[a_{i_1 i_2} a_{i_2 i_3} \cdots a_{i_k i_1}] = p_{r_1} \cdots p_{r_\ell}. \qquad (16.18)$$

Note we could also write

$$\mathbb{E}[a_{i_1 i_2} a_{i_2 i_3} \cdots a_{i_k i_1}] = \mathbb{E}[a_{x_1 y_1}^{r_1}] \cdots \mathbb{E}[a_{x_\ell y_\ell}^{r_\ell}] = p_{r_1} \cdots p_{r_\ell}. \qquad (16.19)$$

As we assume p has mean 0, variance 1 and finite higher moments, if any $r_j = 1$ then the above product vanishes. If each $r_j = 2$ the above product is 1.

Instead of proving Lemma 16.1.5, we handle the contribution from one of the N^k terms; the general proof proceeds similarly. Let us calculate the contribution from the term in (16.16), assuming $N > 5$. Let

$$\mathcal{S} = \{(1,3),(3,4),(1,4),(4,5),(3,5)\}$$

$$\mathcal{T} = \{(i,j) : 1 \le i \le j \le N, (i,j) \notin \mathcal{S}\}. \qquad (16.20)$$

For each $(i,j) \in \mathcal{S}$, let $r(i,j)$ be the exponent of $a_{i,j}$ in (16.16):

$$r(1,3) = 3, \quad r(3,4) = 2, \quad r(1,4) = r(4,5) = r(3,5) = 1. \qquad (16.21)$$

We have $\frac{N(N+1)}{2}$ integrals over a_{ij}, with $1 \le i \le j \le N$. Thus the integral in (16.16) from the term in (16.16) becomes

$$\prod_{(i,j)\in\mathcal{S}} \int_{a_{ij}=-\infty}^{\infty} a_{ij}^{r(i,j)} p(a_{ij})da_{ij} \prod_{(i,j)\in\mathcal{T}} \int_{a_{ij}=-\infty}^{\infty} p(a_{ij})da_{ij}. \qquad (16.22)$$

Each integral over an $(i,j) \in \mathcal{T}$ gives 1, and the integrals over $(i,j) \in \mathcal{S}$ give $p_{r(i,j)}$. Explicitly,

$$\prod_{(i,j)\in\mathcal{S}} \int_{a_{ij}=-\infty}^{\infty} a_{ij}^{r(i,j)} p(a_{ij})da_{ij}$$

$$= \int_{a_{13}=-\infty}^{\infty} a_{13}^3 p(a_{13})da_{13} \int_{a_{34}=-\infty}^{\infty} a_{34}^2 p(a_{34})da_{34} \int_{a_{14}=-\infty}^{\infty} a_{14}p(a_{14})da_{14}$$

$$\cdot \int_{a_{45}=-\infty}^{\infty} a_{45}p(a_{45})da_{45} \int_{a_{35}=-\infty}^{\infty} a_{35}p(a_{35})da_{35}$$

$$= p_3 p_2 p_1 p_1 p_1. \qquad (16.23)$$

Therefore, the contribution from the term in (16.16) is $p_3 p_2 p_1^3 \cdot 1^{\frac{N(N+1)}{2} - 5}$; the exponent of 1 is $|T| = \frac{N(N+1)}{2} - 5$. This is zero as p has mean zero, implying $p_1 = 0$.

Exercise 16.1.6. *Prove (16.12), (16.13) and Lemma 16.1.5.*

16.1.3 The First Few Moments

We use the expansions from §16.1.2 to calculate the first few moments. See §15.3 for a review of the formulation of Wigner's Semi-Circle Law. We must show that $\lim_{N \to \infty} M_{N,k} = C(k)$, where $C(k)$ is the kth moment of the semi-circle distribution.

Lemma 16.1.7. *The expected value of $M_{N,0}(A) = 1$, thus $\lim_{N \to \infty} M_{N,0} = C(0)$.*

Proof. We have

$$M_{N,0} = \mathbb{E}\left[M_{N,0}(A)\right] = \frac{1}{N}\mathbb{E}\left[\text{Trace}(I)\right] = \frac{1}{N}\mathbb{E}[N] = \frac{N}{N}\mathbb{E}[1] = (16.24)$$

$\square$

Lemma 16.1.8. *The expected value of $M_{N,1}(A) = 0$, thus $\lim_{N \to \infty} M_{N,1} = C(1)$.*

Proof. We have

$$M_{N,1} = \mathbb{E}\left[M_{N,1}(A)\right] = \frac{1}{2N^{3/2}}\mathbb{E}\left[\text{Trace}(A)\right]$$

$$= \frac{1}{2N^{3/2}}\mathbb{E}\left[\sum_{i=1}^{N} a_{ii}\right]$$

$$= \frac{1}{2N^{3/2}}\sum_{i=1}^{N}\mathbb{E}[a_{ii}]. \tag{16.25}$$

As each a_{ij} is drawn from a probability distribution with mean zero, each $\mathbb{E}[a_{ii}] = 0$.

$\square$

Lemma 16.1.9. *The expected value of $M_{N,2}(A) = \frac{1}{4}$, thus $\lim_{N \to \infty} M_{N,2} = C(2)$.*

Proof. By Theorem 15.2.1

$$\text{Trace}(A^2) = \sum_{i=1}^{N}\sum_{j=1}^{N} a_{ij}a_{ji}. \tag{16.26}$$

As A is symmetric, $a_{ij} = a_{ji}$. Thus, the trace is $\sum_i \sum_j a_{ij}^2$. Now

$$
M_{N,2} = \mathbb{E}[M_{N,2}(A)] = \frac{1}{4N^2}\mathbb{E}\left[\text{Trace}(A^2)\right]
$$

$$
= \frac{1}{4N^2}\mathbb{E}\left[\sum_{i=1}^{N}\sum_{j=1}^{N}a_{ij}^2\right]
$$

$$
= \frac{1}{4N^2}\sum_{i=1}^{N}\sum_{j=1}^{N}\mathbb{E}[a_{ij}^2]. \tag{16.27}
$$

Each $\mathbb{E}[a_{ij}^2] = 1$ because we have assumed p has mean 0 and variance 1 (which implies the second moment of p is 1). There are N^2 pairs (i,j). Thus, we have $\frac{1}{4N^2} \cdot (N^2 \cdot 1) = \frac{1}{4}$. □

Lemma 16.1.10. $\lim_{N \to \infty} M_{N,3} = C(3) = 0$.

Proof. By Theorem 15.2.1

$$
\text{Trace}(A^3) = \sum_{i=1}^{N}\sum_{j=1}^{N}\sum_{k=1}^{N}a_{ij}a_{jk}a_{ki}. \tag{16.28}
$$

Therefore

$$
M_{N,3} = \mathbb{E}[M_{N,3}(A)] = \frac{1}{8N^{2.5}}\mathbb{E}\left[\text{Trace}(A^3)\right]
$$

$$
= \frac{1}{8N^{2.5}}\mathbb{E}\left[\sum_{i=1}^{N}\sum_{j=1}^{N}\sum_{k=1}^{N}a_{ij}a_{jk}a_{ki}\right]
$$

$$
= \frac{1}{8N^{2.5}}\sum_{i=1}^{N}\sum_{j=1}^{N}\sum_{k=1}^{N}\mathbb{E}[a_{ij}a_{jk}a_{ki}]. \tag{16.29}
$$

There are three cases. If the subscripts i, j and k are all distinct, then a_{ij}, a_{jk}, and a_{ki} are three independent variables (in other words, these are three distinct pairs). As p has mean zero, by Lemma 16.1.5

$$
\mathbb{E}[a_{ij}a_{jk}a_{ki}] = \mathbb{E}[a_{ij}] \cdot \mathbb{E}[a_{jk}] \cdot \mathbb{E}[a_{ki}] = 0. \tag{16.30}
$$

If two of the subscripts are the same (say $i = j$) and the third is distinct, by Lemma 16.1.5

$$
\mathbb{E}[a_{ii}a_{ik}a_{ki}] = \mathbb{E}[a_{ii}] \cdot \mathbb{E}[a_{ik}^2] = 0 \cdot 1 = 0 \tag{16.31}
$$

because p has mean zero and variance 1. If all three subscripts are the same, we have

$$
\mathbb{E}[a_{ii}^3]. \tag{16.32}
$$

This is the third moment of p. It is the same for all pairs (i,i), equal to p_3 by Lemma 16.1.5. This is where we use the assumption that the higher moments of p are finite. There are N triples where $i = j = k$. Thus,

$$
M_{N,3} = \mathbb{E}[M_{N,3}(A)] = \frac{1}{8N^{2.5}} \cdot Np_3 = \frac{p_3}{8N^{1.5}}. \tag{16.33}
$$

Letting $N \to \infty$ we see that the expected value of the third moment is zero in the limit. □

Remark 16.1.11. Many of the above calculations are unnecessary. We are dividing by $N^{2.5}$. There are N^3 triples $a_{ij}a_{jk}a_{ki}$. If i, j and k are distinct, we showed by Lemma 16.1.5 the contribution is zero. If the indices are *not* distinct, there are at most $3N^2$ such triples, and as all moments of p are finite, by Lemma 16.1.5 each such triple contributes a bounded amount (independent of N). As we divide by $N^{2.5}$, the total contribution is at most some universal constant times $\frac{1}{\sqrt{N}}$, which tends to zero as $N \to \infty$. This illustrates a general principle: often order of magnitude calculations are sufficient to show certain terms do not contribute in the limit.

16.1.4 The Higher Moments

Lemma 16.1.12. *For odd k, the expected value of $M_{N,k}(A)$ as $N \to \infty$ is zero.*

Exercise[hr] **16.1.13.** *Prove Lemma 16.1.12.*

We are left with calculating the limit of the averages of $M_{N,k}(A)$ for $k = 2m$ even.

Lemma 16.1.14. *Notation as before, the only tuples which contribute as $N \to \infty$ to the main term of the average of $M_{N,2m}(A)$ are those where each $r_j = 2$.*

Exercise[hr] **16.1.15.** *Prove Lemma 16.1.14.*

We are reduced to calculating the contributions to the average of $M_{N,2m}(A)$ from tuples with each $r_j = 2$. By Lemma 16.1.5, a tuple

$$a_{i_1 i_2} \cdots a_{i_{2m} i_1} = a_{x_1 y_1}^2 \cdots a_{x_m y_m}^2 \qquad (16.34)$$

contributes 1^m (because we have a product of m second moments of p, and the second moment of p is 1). The above arguments and (16.13) yield, up to lower order terms,

$$M_{N,2m} = \mathbb{E}[M_{N,2m}(A)] = \frac{1}{2^m N^{1+m}} \sum_{1 \le 1_i, \dots, i_{2m} \le N}^{*} 1, \qquad (16.35)$$

where $\sum^{*}$ means we restrict to tuples $(i_1, \dots, i_{2m})$ such that the corresponding r_j's are all 2. The determination of the limits of the even moments is completed by showing

$$\frac{1}{2^m N^{1+m}} \sum_{1 \le 1_i, \dots, i_{2m} \le N}^{*} 1 = C(2m) = 2 \frac{(2m-1)!!}{(2m+2)!!}. \qquad (16.36)$$

The solution of this counting problem involves the Catalan number (see [CG]) $c_k = \frac{1}{k+1}\binom{2k}{k}$. See [Leh] for details on these calculations.

Exercise 16.1.16. *Compute the fourth and sixth moments directly, and compare them to $C(4)$ and $C(6)$.*

Exercise 16.1.17. *For each m show there exists a constant $c_m > 0$ such that*

$$\sum_{1 \le 1_i, \dots, i_{2m} \le N}^{*} 1 \ge c_m N^{1+m}. \qquad (16.37)$$

This implies that the even moments do not vanish.

Exercise 16.1.18. *For each m show there exists a constant d_m such that*

$$\sideset{}{^*}\sum_{1\leq 1_i,\ldots,i_{2m}\leq N} 1 \leq d_m N^{1+M}. \tag{16.38}$$

This implies that the even moments are finite.

Exercise[h] **16.1.19.** *Prove* (16.36).

16.2 NON-SEMI-CIRCLE BEHAVIOR

In our investigations of random d-regular graphs, we showed the density of normalized eigenvalues do not converge to the semi-circle (Theorem 15.5.9). We give several more examples of ensembles of matrices where the density of eigenvalues is provably *not* given by the Semi-Circle Law. The d-regular graphs are combinatorial objects, and we are not constructing our matrices by choosing entries at random from a fixed probability distribution p. Now we give several examples where we do choose the entries randomly, but with additional structure (otherwise we would of course just have the ensemble of all real symmetric matrices). A generic real symmetric matrix has $\frac{N(N+1)}{2}$ independent entries. We now consider subsets with far fewer independent entries, often of size N. The hope is that these thin ensembles might exhibit new, interesting behavior.

16.2.1 Band Matrices

Definition 16.2.1 (Band Matrix (of width r)). *We say a real symmetric matrix is a band matrix (of width r) if $a_{ij} = 0$ whenever $|i - j| > r$.*

A band matrix of width 0 is a diagonal matrix and of width 1 has non-zero entries only along the main diagonal and the diagonals immediately above and below. In general the number of independent entries is of size $(2r + 1)N$.

Exercise 16.2.2. *Calculate exactly how many entries can be non-zero if the band width is r.*

While band matrices are a subset of real symmetric matrices, they are a very thin subset for $r \ll N$. Do they obey the Semi-Circle Law? Is the spacing between adjacent eigenvalues the GOE?

If the band width $r = N - 1$, then the matrix is "full"; in other words, every entry can be non-zero and the density of normalized eigenvalues converges to the semi-circle. What about the opposite extreme, when $r = 0$? Consider $N \times N$ real symmetric band matrices of width 0, each entry which can be non-zero is chosen randomly and independently from some fixed probability distribution p. For $r = 0$, we do not need to assume anything about the moments of p.

Theorem 16.2.3. *The normalized eigenvalue density is* not *the semi-circle; it is just p.*

Proof. There is no need to normalize the eigenvalues. As we have a diagonal matrix, the entries *are* the eigenvalues! Asking how many eigenvalues are in $[a, b]$ is equivalent to calculating the probability that an $a_{ii} \in [a, b]$, which is given by $\int_a^b p(x)dx$. $\qquad\square$

Exercise 16.2.4. *Let A be an $N \times N$ band matrix of width 1 with real entries, but not necessarily symmetric. Which entries can be non-zero in $A^T A$?*

16.2.2 Toeplitz Matrices

We consider another ensemble of random matrices with far fewer independent entries than the ensemble of all real symmetric matrices.

Definition 16.2.5. *A Toeplitz matrix A is of the form*

$$A = \begin{pmatrix} b_0 & b_1 & b_2 & b_3 & \cdots \\ b_{-1} & b_0 & b_1 & b_2 & \cdots \\ b_{-2} & b_{-1} & b_0 & b_1 & \cdots \\ b_{-3} & b_{-2} & b_{-1} & b_0 & \cdots \\ \vdots & \vdots & \vdots & \vdots & \ddots \end{pmatrix}. \tag{16.39}$$

That is, A is constant along its diagonals. Note $a_{ij} = b_{j-i}$.

We consider real symmetric Toeplitz matrices whose entries are chosen according to some distribution p with mean 0 and variance 1. Thus $b_{i-j} = b_{j-i}$. It is convenient to normalize the eigenvalues of these Toeplitz matrices by $\frac{1}{\sqrt{N}}$ rather than $\frac{1}{2\sqrt{N}}$. Thus

$$\mu_{A,N}(x) = \frac{1}{N} \sum_{i=1}^{N} \delta \left(x - \frac{\lambda_i(A)}{\sqrt{N}} \right). \tag{16.40}$$

Remark 16.2.6. As the main diagonal is constant, the effect of the main diagonal being b_0 is simply to shift all eigenvalues by b_0. For simplicity, we take $b_0 = 0$. Note there are $N - 1$ independent entries $b_1, \dots, b_{N-1}$.

Exercise 16.2.7. *If $B = A + mI$, prove the eigenvalues of B are m plus the eigenvalues of A.*

The eigenvalue distribution is again not the semi-circle. As long as the probability distribution p has mean 0, variance 1 and finite higher moments, the answer is universal (i.e., independent of all other properties of p). It is *almost* the standard Gaussian. Its moments are bounded by the moments of the standard Gaussian. Its fourth moment is $2\frac{2}{3}$, while the standard Gaussian's is 3.

Exercise 16.2.8. *Show $M_{N,1} = 0$ and $M_{N,2} = 1 - \frac{1}{N}$. Thus as $N \to \infty$ the expected value of the first two moments are 0 and 1, respectively. Recall the second moment of the semi-circle is $\frac{1}{4}$.*

Just because $\lim_{N\to\infty} M_{N,2} \neq \frac{1}{4}$ does not imply that the eigenvalue probability distribution does not converge to a semi-circle; it only implies it does not converge to the *standard* semi-circle — we need to examine the fourth moment. See Exercise 15.1.6.

It turns out that it is not the semi-circle that this distribution is trying to approach, but rather the Gaussian. The odd moments of the Gaussian vanish, and the even moments are $G(2m) = (2m-1)!!$. The limits of the average of the moments want to be $G(2m)$; however, to calculate these moments involves solving a system of Diophantine equations (see Chapter 4). Obstructions to these equations arise due to the fact that the indices must be in $\{1, \ldots, N\}$, and this prevents the limits from equalling the Gaussian's moments.

The fourth moment calculation highlights the Diophantine obstructions, which bound the moments away from the Gaussian. As $a_{ij} = b_{j-i} = b_{i-j}$, the trace expansion becomes

$$M_{N,4}(A) = \frac{1}{N^3} \sum_{1 \leq i_1 i_2 i_3, i_4 \leq N} \mathbb{E}(b_{i_1-i_2} b_{i_2-i_3} b_{i_3-i_4} b_{i_4-i_1}). \tag{16.41}$$

Let $x_j = |i_j - i_{j+1}|$. If any b_{x_j} occurs to the first power, its expected value is zero (since the mean of p is zero, and the b's are drawn from p), and these tuples do not contribute. Thus either the x_j's are matched in pairs (with different values), or all four are equal (in which case they are still matched in pairs). There are 3 possible matchings; however, by symmetry (simply relabel), we see the contribution from $x_1 = x_2, x_3 = x_4$ is the same as the contribution from $x_1 = x_4, x_2 = x_3$.

If $x_1 = x_2, x_3 = x_4$, we have

$$i_1 - i_2 = \pm(i_2 - i_3) \quad \text{and} \quad i_3 - i_4 = \pm(i_4 - i_1). \tag{16.42}$$

Exercise 16.2.9. *Show the number of tuples (i_1, i_2, i_3, i_4) satisfying the pair of equations in (16.42) is $O(N^2)$ if a + sign holds in either equation. As we divide by N^3, in the limit these terms do not contribute and the main contribution arises when both equations have the minus sign.*

If both signs are negative in (16.42), then $i_1 = i_3$ and i_2 and i_4 are arbitrary. We see there are N^3 such tuples. Almost all of these will have $x_1 \neq x_3$, and contribute 1; the rest will contribute a smaller term. Explicitly, let p_4 denote the fourth moment of p. Given i_1 and i_2, $N-1$ choices of i_4 yield $x_1 \neq x_3$ (contributing $\mathbb{E}[b_{x_1}^2 b_{x_3}^2] = 1$), and one choice yields the two equal (contributing $\mathbb{E}[b_{x_1}^4] = p_4$). Therefore this case contributes

$$\frac{1}{N^3}\left(N^2(N-1)\cdot 1 + N^2(1)\cdot p_4\right) = 1 - \frac{1}{N} + \frac{p_4}{N} = 1 + O\left(\frac{1}{N}\right). \tag{16.43}$$

The case of $x_1 = x_4$ and $x_2 = x_3$ is handled identically, and contributes $1 + O\left(\frac{1}{N}\right)$.

The other possibility is for $x_1 = x_3$ and $x_2 = x_4$. Non-adjacent pairing is what leads to Diophantine obstructions, which decreases the contribution to the moment. We call this a non-adjacent pairing as the neighbors of x_1 are x_2 and x_4, but x_1 is paired with x_3. Now we have

$$i_1 - i_2 = \pm(i_3 - i_4) \quad \text{and} \quad i_2 - i_3 = \pm(i_4 - i_1). \tag{16.44}$$

Exercise 16.2.10. *Show the number of tuples* (i_1, i_2, i_3, i_4) *satisfying the pair of equations in* (16.44) *is* $O(N^2)$ *if a + sign holds in either equation. As we divide by* N^3, *in the limit these terms do not contribute and the main contribution arises when both equations have the minus sign.*

If both signs are negative in (16.44), then we have

$$i_1 = i_2 + i_4 - i_3, \quad i_1, i_2, i_3, i_4 \in \{1, \ldots, N\}. \tag{16.45}$$

The fact that each $i_j \in \{1, \ldots, N\}$ is what leads to the Diophantine obstructions. In the first case (when $x_1 = x_2$ and $x_3 = x_4$), we saw we had three independent variables and $N^3 + O(N^2)$ choices that were mutually consistent. Now it is possible for choices of i_2, i_3 and i_4 to lead to impossible values for i_1. For example, if $i_2, i_4 \geq \frac{2N}{3}$ and $i_3 < \frac{N}{3}$, this forces $i_1 > N$, which is not allowed, which implies there are at most $(1 - \frac{1}{27})N^3$ valid choices. This is enough to show the Gaussian's moment is strictly greater; we have lost a positive percentage of solutions. The following lemma shows this case contributes $\frac{2}{3}$ to the fourth moment.

Lemma 16.2.11. *Let* $I_N = \{1, \ldots, N\}$. *Then* $\#\{x, y, z \in I_N : 1 \leq x + y - z \leq N\} = \frac{2}{3}N^3 + \frac{1}{3}N$.

Proof. Say $x + y = S \in \{2, \ldots, 2N\}$. For $2 \leq S \leq N$, there are $S - 1$ choices of z, and for $S \geq N + 1$, there are $2N - S + 1$. Similarly, the number of $x, y \in I_N$ with $x + y = S$ is $S - 1$ if $S \leq N + 1$ and $2N - S + 1$ otherwise. The number of triples is therefore

$$\sum_{S=2}^{N}(S-1)^2 + \sum_{S=N+1}^{2N}(2N - S + 1)^2 = \frac{2}{3}N^3 + \frac{1}{3}N. \tag{16.46}$$

$\square$

Collecting all the pieces, we have shown

Theorem 16.2.12 (Fourth Moment). $M_{N,4} = 2\frac{2}{3} + O\left(\frac{1}{N}\right)$.

In [BDJ, HM] the Toeplitz ensemble is investigated and shown to be non-Semi-Circular and non-Gaussian. See [HM] for upper and lower bounds for the moments of the new distribution that the densities $\mu_{A,N}(x)$ converge to.

Remark 16.2.13. Similar to our previous arguments, one can show that the odd moments vanish and the main contribution to the even moments occur when the b_x's are matched in pairs. For $2m$ objects there are $(2m - 1)!!$ ways to match in pairs. Each matching wants to contribute 1 (and if they all did, then we would have the standard Gaussian's moments); however, not all matchings contribute 1. For some matchings, a positive percentage of tuples are inaccessible. Explicitly, for each matching we divide by N^{m+1}. It turns out that of the $2m$ indices $i_1, \ldots, i_{2m}$, once $m + 1$ are specified the others are determined. If we could choose $m + 1$ indices freely, we would have N^{m+1} tuples for each matching, and a contribution of 1. It is here that the loss of a positive percent is felt. Interestingly, if we add additional symmetries, all the moments are Gaussian. Explicitly, assume the first row is a palindrome; for $N = 2M$ this means the first row is

$$(0 \; b_1 \; b_2 \; b_3 \; \ldots \; b_{M-2} \; b_{M-1} \; b_{M-1} \; b_{M-2} \; \ldots \; b_3 \; b_2 \; b_1 \; 0). \tag{16.47}$$

Instead of $N - 1$ free variables, there are now just $\frac{N-2}{2}$. Similar to the density of states of d-regular graphs (§15.5.1), we have a sub-ensemble with different behavior. See [MMS] for a proof that the moments are Gaussian.

16.2.3 Truncated Cauchy Distribution

In §15.3.4 we saw that numerical simulations of eigenvalues of matrices with entries independently chosen from the Cauchy distribution appeared not to satisfy the Semi-Circle Law. For $N \times N$ matrices, instead of choosing the entries from the Cauchy distribution, choose the entries from a *truncated* Cauchy distribution, where the truncation depends on N. Explicitly, let

$$
p_N(x) = \begin{cases} A_N \frac{1}{\pi(1+x^2)} & \text{if } |x| \leq f(N) \\ 0 & \text{otherwise,} \end{cases} \tag{16.48}
$$

where A_N is chosen to make $\int_{\mathbb{R}} p_N(x)dx = 1$. By appropriately choosing the cut-off $f(N)$ and normalizing the eigenvalues, one obtains a new distribution. See [Za] for complete statements and proofs, as well as generalizations to other distributions.

16.3 SPARSE MATRICES

A common theme of some of the above problems (band matrices, random graphs) is that we are considering **sparse matrices**: real symmetric matrices where most entries are zero. Such matrices open up fascinating possibilities to see new behavior. In general, the following heuristic principle is a good guide: if you consider a very small subset of objects, you can see very special behavior. However, in mathematical proofs, we need to average over many similar objects. Thus, if we have too few objects, we cannot perform the averaging; if we have too many objects, non-standard behavior (which occurs rarely) could be washed away.

For example, as most matrices are not band symmetric of small width, even though they have different eigenvalue statistics, this difference will not be noticed when we look at all symmetric matrices. The goal, therefore, is to find an ensemble that is large enough so that we can do the averaging, yet small enough so that new interesting behavior is still present.

The **generalized coin toss matrices** provide another candidate. For $q_N \in [0, \frac{1}{2}]$, let $p_N(1) = \frac{q_N}{2}$, $p_N(-1) = \frac{q_N}{2}$, and $p_N(0) = 1 - q_N$. We use the probability function p_N to construct real symmetric matrices A by choosing the independent entries from p_N. We expect to have about $q_N \cdot \frac{N(N+1)}{2}$ non-zero entries in a typical A. If q_N is small relative to N, these matrices are sparse, and there is the possibility for new behavior. Note, of course, that if q_N is independent of N then the standard proof of the Semi-Circle Law is applicable. See [Liu] for more details.

16.4 RESEARCH PROJECTS

For more on connections between random graphs and number theory, see [DSV] and Chapter 3 of [Sar].

Research Project 16.4.1 (Band Matrices). Investigate how the eigenvalue density depends on the band width. When do we observe the transition from p to the semi-circle? In other words, how large must r be in order to see semi-circle behavior. Does this critical r depend on p? It has been observed for many systems that transitions occur around $r = \sqrt{N}$.

Research Project 16.4.2 (Band, Sparse, d-Regular). Compare the eigenvalue distributions and spacing distributions (see Chapter 17) of band matrices of width r, generalized coin toss matrices, and d-regular random graphs. If we choose r, q and d so that

$$\frac{(r+1)(2N-r)}{2} \sim \frac{qN(N+1)}{2} \sim \frac{dN}{2}, \tag{16.49}$$

are the distributions similar? All three ensembles have approximately the same number of non-zero entries, but they differ greatly in *where* the non-zero entries may lie.

Research Project 16.4.3. In the above project we considered sparse matrices with entries in $\{-1, 0, 1\}$. As the probability distribution depends on N, the arguments used to prove Wigner's Semi-Circle Law are not applicable. The adjacency matrix of a simple d-regular graph with no self-loops has $\frac{dN}{2}$ of the a_{ij} (with $1 \le i < j \le N$) equal to 1 (and the rest are zero). Let now

$$p_{N,d}(x) = \begin{cases} \frac{d}{N-1} & \text{if } x = 1 \\ 1 - \frac{d}{N-1} & \text{if } x = 0. \end{cases} \tag{16.50}$$

If we choose the entries a_{ij} (with $1 \le i < j \le N$) from $p_{N,d}$ and consider the graph of such a matrix, the expected number of edges from each vertex is d. Thus it is natural to see whether or not such an ensemble approximates d-regular graphs. How are the eigenvalues distributed? See also Remark 15.5.7.

Research Project 16.4.4 (Self-Adjoint Matrices). Fix a probability distribution p and choose *all* the entries of A randomly and independently from p. Consider the matrix $A^T A$. This matrix is real symmetric, but has N^2 degrees of freedom. What is the density of its eigenvalues, or at least what are the first few moments? Are the eigenvalues real? Are they non-negative? What is the density of the *square root* of the eigenvalues? Matrices of this form, called Wishart matrices, are well studied by other techniques. See for example [SM, Wis].

Research Project 16.4.5 (Weighted Random Graphs). Consider the space of d-regular graphs. To each graph we attach an adjacency matrix, and we can study the distribution of the eigenvalues. Consider the following generalization: fix a probability distribution p. Let A be the adjacency matrix of a d-regular graph G. Construct a matrix B as follows: if $a_{ij} = 1$, choose b_{ij} randomly from p; if $a_{ij} = 0$, set

$b_{ij} = 0$. How does the distribution of eigenvalues of B depend on p? The density of eigenvalues of d-regular graphs is not the semi-circle; however, is there a choice of p that leads to semi-circular behavior? These are called weighted graphs; one can regard these weights (especially if p is positive) as encoding different information about the system (for example, how far apart different vertices are, or how long or how expensive it is to transmit information between vertices). See [Gold, QS2] for more details.

Research Project 16.4.6 (Complex Hermitian). Investigate the eigenvalue densities for some of the ensembles for complex Hermitian rather than real symmetric matrices. For example, consider complex Hermitian Toeplitz matrices.

Research Project 16.4.7 (Spherical Ensemble: Non-Independent Entries). In the spirit of Example 15.1.9, consider the ensemble of $N \times N$ real symmetric matrices where

$$\sum_{i=1}^{N}\sum_{j=i}^{N} a_{ij}^2 = \frac{N(N+1)}{2}. \tag{16.51}$$

Note the entries are not chosen independently from a fixed probability distribution, but rather we choose a point on a sphere of radius $\sqrt{N(N+1)/2}$; we do this so each a_{ij} is of size 1. What is the density of eigenvalues? Warning: the integrals will probably be too difficult to directly evaluate (except possibly for low N), though one can numerically investigate the eigenvalues. If we let $x_1 = a_{11}, \ldots,$ $x_{N(N+1)/2} = a_{NN}$, then we have

$$x_1^2 + \cdots + x_{N(N+1)/2}^2 = R^2, \tag{16.52}$$

where $R = \sqrt{N(N+1)/2}$ is the radius of the $\frac{N(N+1)}{2}$-dimensional sphere. The following coordinate transformations are useful to generate points on an n-sphere of radius r:

$$
\begin{aligned}
x_1 &= x_1(r, \phi_1, \ldots, \phi_{n-1}) &&= r\cos(\phi_1) \\
x_2 &= x_2(r, \phi_1, \ldots, \phi_{n-1}) &&= r\sin(\phi_1)\cos(\phi_2) \\
x_3 &= x_3(r, \phi_1, \ldots, \phi_{n-1}) &&= r\sin(\phi_1)\sin(\phi_2)\cos(\phi_3) \\
&\ \ \vdots \\
x_{n-1} &= x_{n-1}(r, \phi_1, \ldots, \phi_{n-1}) &&= r\sin(\phi_1)\cdots\sin(\phi_{n-2})\cos(\phi_{n-1}) \\
x_n &= x_n(r, \phi_1, \ldots, \phi_{n-1}) &&= r\sin(\phi_1)\cdots\sin(\phi_{n-2})\sin(\phi_{n-1}),
\end{aligned}
$$

where $\phi_1, \ldots, \phi_{n-2} \in [0, \pi]$, $\phi_{n-1} \in [0, 2\pi]$ and the volume is $\frac{\pi^{n/2} r^n}{\Gamma(\frac{n}{2}+1)}$. One can also consider other ensembles where the entries are not chosen independently; the point is to find ensembles that are easy to work with (either in determining the eigenvalues or in generating the matrices).

Chapter Seventeen

Random Matrix Theory: Spacings between Adjacent Eigenvalues

We sketch the proof of how the spacings between adjacent normalized eigenvalues of real symmetric matrices are distributed. We concentrate on the case $N = 2$ and refer the reader to [For, Gau, Meh1, Meh2] for general N. We assume the reader is familiar with basic linear algebra at the level of Appendix B.

17.1 INTRODUCTION TO THE 2×2 GOE MODEL

Consider the space of 2×2 real symmetric matrices:

$$\left\{ \begin{pmatrix} a_{11} & a_{12} \\ a_{12} & a_{22} \end{pmatrix} : a_{11}, a_{12}, a_{22} \in \mathbb{R} \right\}. \tag{17.1}$$

Let p be a probability density on the space of 2×2 real symmetric matrices. Then the probability of observing a matrix $\begin{pmatrix} x & y \\ y & z \end{pmatrix}$ with $x \in [a_{11}, a_{11} + \Delta a_{11}]$, $y \in [a_{12}, a_{12} + \Delta a_{12}]$ and $z \in [a_{22}, a_{22} + \Delta a_{22}]$ is approximately $p(a_{11}, a_{12}, a_{22})$ $\Delta a_{11} \Delta a_{12} \Delta a_{22}$. Alternatively, we could choose three different probability functions P_{11}, P_{12} and P_{13}, and independently choose the three entries of our matrix. In this case, the probability of observing such a matrix $P_{11}(a_{11}) P_{12}(a_{12}) P_{22}(a_{22})$ $\Delta a_{11} \Delta a_{12} \Delta a_{22}$.

While we shall only investigate the spacings between eigenvalues of 2×2 matrices, in practice one studies the spacings of $N \times N$ matrices as $N \to \infty$. Many results are conjectured to hold as $N \to \infty$ for all nice probability functions p; however, we can only prove results for a very small number of models. Below we describe the most common model, the Gaussian Orthogonal Ensemble or GOE. In addition to being mathematically tractable, it is also the natural model to work with physically.

17.1.1 Definition of the GOE Ensemble

Consider the ensemble of 2×2 symmetric matrices

$$\left\{ A : A = \begin{pmatrix} a_{11} & a_{12} \\ a_{12} & a_{22} \end{pmatrix} \right\} \tag{17.2}$$

where the probability of the matrix A, $P(A)$, satisfies

1. $P(A) = P_{11}(a_{11}) P_{12}(a_{12}) P_{22}(a_{22})$ (the entries of the matrix are independent);

2. for any orthogonal transformation (i.e., a good change of basis) Q, $P(QAQ^T)$ $= P(A)$.

We call the above model the GOE, the Gaussian Orthogonal Ensemble. The name is derived from the fact that the probability of a matrix in the ensemble is invariant under orthogonal rotations of basis, and further this then implies that the P_{ij} are Gaussian probability distributions.

The second condition is very natural. Remember that a matrix is a representation of a transformation; however, in order to write down a matrix, we *must* choose a basis. Thus, the same transformation "looks different" in different bases. As the physical world does not care how we orient the x- and y-axes, the probability of a transformation A should not depend on the basis we are using.

For example, let $\vec{e_1} = \binom{1}{0}$ and $\vec{e_2} = \binom{0}{1}$ be the standard basis. Any orthonormal basis $\vec{u_1} = \binom{u_{11}}{u_{21}}$, $\vec{u_2} = \binom{u_{12}}{u_{22}}$ can be obtained through an orthogonal transformation Q. Let $Q = \left(\begin{smallmatrix} u_{11} & u_{12} \\ u_{21} & u_{22} \end{smallmatrix} \right)$. Note $Q\vec{e_i} = \vec{u_i}$ and $Q^T \vec{u_i} = \vec{e_i}$. Then if a transformation T is written as $\left(\begin{smallmatrix} x & y \\ y & z \end{smallmatrix} \right)$ in the $\vec{e_i}$ basis, it would be written as $Q \left(\begin{smallmatrix} x & y \\ y & z \end{smallmatrix} \right) Q^T$ in the $\vec{u_i}$ basis. Why? Let $\vec{v}$ be a vector written in terms of the $\vec{u_i}$ basis. Then $Q^T \vec{v}$ converts $\vec{v}$ to a vector written in the $\vec{e_i}$ basis. We can now apply the transformation T, which is $\left(\begin{smallmatrix} x & y \\ y & z \end{smallmatrix} \right)$ in this basis. The result is a vector written in terms of the $\vec{e_i}$ basis; by applying Q we convert the vector to the $\vec{u_i}$ basis. We see that the transformation T is $Q \left(\begin{smallmatrix} x & y \\ y & z \end{smallmatrix} \right) Q^T$ in the $\vec{u_i}$ basis, as claimed.

Therefore, we see that the second condition merely states that for our model, the probability of a matrix should be **invariant** (i.e., the same) under any change of basis.

Exercise 17.1.1. *Consider the probability density on 2×2 matrices from Example 15.1.9. Does this p satisfy $p(QAQ^T) = p(A)$?*

17.1.2 Probabilities P_{11}, P_{12}, P_{22}

The two defining conditions of the GOE ensemble determine which probability distributions P_{11}, P_{12} and P_{22} are possible; we obtain differential equations for the P_{ij}, which force the P_{ij}'s to be Gaussians.

Let us consider an infinitesimal rotation by an angle ϵ (see §B.3), which by Exercise B.3.7 is

$$Q = \begin{pmatrix} \cos\epsilon & -\sin\epsilon \\ \sin\epsilon & \cos\epsilon \end{pmatrix} = \begin{pmatrix} 1 + O(\epsilon^2) & -\epsilon + O(\epsilon^3) \\ \epsilon + O(\epsilon^3) & 1 + O(\epsilon^2) \end{pmatrix}. \tag{17.3}$$

For a matrix

$$A = \begin{pmatrix} a_{11} & a_{12} \\ a_{12} & a_{22} \end{pmatrix}, \tag{17.4}$$

we must have $P(QAQ^T) = P(A)$. Direct calculation gives

$$QAQ^T = \begin{pmatrix} a_{11} - 2\epsilon a_{12} + O(\epsilon^2) & a_{12} - \epsilon(a_{22} - a_{11}) + O(\epsilon^2) \\ a_{12} - \epsilon(a_{22} - a_{11}) + O(\epsilon^2) & a_{22} + 2\epsilon a_{12} + O(\epsilon^2) \end{pmatrix}. \tag{17.5}$$

Taylor expanding the probabilities (see §A.2.3) yields

$$P_{11}(a_{11} - 2\epsilon a_{12} + O(\epsilon^2)) = P_{11}(a_{11}) - 2\epsilon a_{12}\frac{dP_{11}}{da_{11}} + O(\epsilon^2)$$

$$P_{12}(a_{12} - \epsilon(a_{22} - a_{11}) + O(\epsilon^2)) = P_{12}(a_{12}) - \epsilon(a_{22} - a_{11})\frac{dP_{12}}{da_{12}} + O(\epsilon^2)$$

$$P_{22}(a_{22} + 2\epsilon a_{12} + O(\epsilon^2)) = P_{22}(a_{22}) + 2\epsilon a_{12}\frac{dP_{22}}{da_{22}} + O(\epsilon^2). \quad (17.6)$$

As the probability of a matrix is the product of the probabilities of each entry,

$$P(QAQ^T)$$
$$= P_{11}(a_{11})P_{12}(a_{12})P_{22}(a_{22})$$
$$- \left[2a_{12}P_{12}(a_{12})P_{22}(a_{22})\frac{dP_{11}}{da_{11}} + (a_{22} - a_{11})P_{11}(a_{11})P_{22}(a_{22})\frac{dP_{12}}{da_{12}} \right.$$
$$\left. - 2a_{12}P_{11}(a_{11})P_{12}(a_{12})\frac{dP_{22}}{da_{22}} \right]\epsilon + O(\epsilon^2). \quad (17.7)$$

Noting that $P(A) = P_{11}(a_{11})P_{12}(a_{12})P_{22}(a_{22})$, we find that

$$\frac{P(A) - P(QAQ^T)}{P_{11}(a_{11})P_{12}(a_{12})P_{22}(a_{22})}$$
$$= \left[\frac{2a_{12}}{P_{11}(a_{11})}\frac{dP_{11}}{da_{11}} + \frac{a_{22} - a_{11}}{P_{12}(a_{12})}\frac{dP_{12}}{da_{12}} - \frac{2a_{12}}{P_{22}(a_{22})}\frac{dP_{22}}{da_{22}} \right]\epsilon$$
$$+ O\left(\frac{\epsilon^2}{P_{11}(a_{11})P_{12}(a_{12})P_{22}(a_{22})} \right). \quad (17.8)$$

As $P(A) = P(QAQ^T)$ for all orthogonal Q, the coefficient of ϵ must vanish in (17.8); if not, we obtain a contradiction as $\epsilon \to 0$. This is an extremely useful method for solving such problems. Therefore

$$\frac{2a_{12}}{P_{11}(a_{11})}\frac{dP_{11}}{da_{11}} + \frac{a_{22} - a_{11}}{P_{12}(a_{12})}\frac{dP_{12}}{da_{12}} - \frac{2a_{12}}{P_{22}(a_{22})}\frac{dP_{22}}{da_{22}} = 0. \quad (17.9)$$

We rewrite (17.9) as

$$\frac{1}{a_{12}P_{12}(a_{12})}\frac{dP_{12}}{da_{12}} = -\frac{2}{a_{22} - a_{11}}\left(\frac{1}{P_{11}(a_{11})}\frac{dP_{11}}{da_{11}} - \frac{1}{P_{22}(a_{22})}\frac{dP_{22}}{da_{22}} \right).$$
$$(17.10)$$

It is easy to solve (17.10). The left hand side is a function of only a_{12}, while the right hand side is a function of only a_{11} and a_{22}. The only way equality can hold for all choices of a_{11}, a_{12}, a_{22} is for each side to be constant, which for convenience we write as $-C$. The left hand side now gives

$$\frac{dP_{12}}{da_{12}} = -Ca_{12}P(a_{12}). \quad (17.11)$$

The solution to this, as can be checked by differentiation, is a Gaussian:

$$P_{12}(a_{12}) = \sqrt{\frac{C}{2\pi}}\, e^{-\frac{Ca_{12}^2}{2}}. \quad (17.12)$$

As (17.11) is a first order differential equation, it should have one arbitrary parameter. Of course, if $f(a_{12})$ solves (17.11), so does any multiple $\alpha f(a_{12})$; however, P_{12} is to be a probability distribution. Hence its integral must be one, which fixes the normalization constant (see Exercise 8.2.12). Further, as it is a probability distribution, $C > 0$.

We now analyze the right hand side of (17.10), which must also equal $-C$. After some straightforward algebra we find

$$\frac{2}{P_{11}(a_{11})}\frac{dP_{11}}{da_{11}} + Ca_{11} = \frac{2}{P_{22}(a_{22})}\frac{dP_{22}}{da_{22}} + Ca_{22}. \qquad (17.13)$$

Again, note the left hand side depends on only one variable (this time a_{11}) and the right hand side depends only on a_{22}. Thus each side must equal a constant, say CD. Hence

$$2\frac{dP_{11}}{da_{11}} = -C(a_{11} - D)P_{11}(a_{11}). \qquad (17.14)$$

Arguing as before, we find the solution is

$$P_{11}(a_{11}) = \sqrt{\frac{C}{4\pi}} e^{-\frac{C(a_{11}-D)^2}{4}}, \qquad (17.15)$$

and similarly

$$P_{22}(a_{22}) = \sqrt{\frac{C}{4\pi}} e^{-\frac{C(a_{22}-D)^2}{4}}. \qquad (17.16)$$

By changing where we label zero (think of a_{11} as representing some quantity such as potential, where we have the freedom to choose what will be zero potential), we may set $D = 0$ above. Alternatively, if we want our probability densities to have mean 0 then we must have $D = 0$.

We have shown the GOE conditions force the following:

$$P_{11}(a_{11}) = \sqrt{\frac{C}{4\pi}} e^{-\frac{Ca_{11}^2}{4}}$$

$$P_{12}(a_{12}) = \sqrt{\frac{C}{2\pi}} e^{-\frac{Ca_{12}^2}{2}}$$

$$P_{22}(a_{22}) = \sqrt{\frac{C}{4\pi}} e^{-\frac{Ca_{22}^2}{4}}. \qquad (17.17)$$

Further,

$$\begin{aligned} P(A) &= P_{11}(a_{11})P_{12}(a_{12})P_{22}(a_{22}) \\ &= \frac{C\sqrt{C}}{4\pi\sqrt{2\pi}} e^{-\frac{C}{4}(a_{11}^2+2a_{12}^2+a_{22}^2)} \\ &= \frac{C\sqrt{C}}{4\pi\sqrt{2\pi}} e^{-\frac{C}{4}\operatorname{Trace}(A^2)}. \end{aligned} \qquad (17.18)$$

We have shown

Theorem 17.1.2. *For the ensemble of 2×2 real symmetric matrices satisfying the GOE assumptions, the entries are chosen independently from Gaussians, and the probability of a matrix is proportional to $e^{-\frac{C}{4}\operatorname{Trace}(A^2)}$.*

Remark 17.1.3. Our assumptions force the entries to be drawn from Gaussians. The variance for the diagonal entries is twice that of the non-diagonal entry; this should be understood by noting that each diagonal entry occurs once, but by symmetry each non-diagonal entry occurs twice.

Exercise 17.1.4. *Prove the various formulas from this subsection. In particular, show that $p(x) = ae^{-\frac{Cx^2}{2}}$ satisfies $p'(x) = -Cxp(x)$, and determine α such that $\int_{-\infty}^{\infty} p(x)dx = 1$.*

Exercise 17.1.5. *Prove that if $P(QAQ^T) = P(A)$, then the coefficient of ϵ must vanish in (17.8).*

Exercise 17.1.6. *Let $f : \mathbb{R} \to \mathbb{R}$ be a polynomial of degree k. For 2×2 matrices $\left(\begin{smallmatrix} x & y \\ y & z \end{smallmatrix}\right)$, let $p(A) = c_f e^{-f(x^2+2y^2+z^2)}$, where c_f is chosen so that p is a probability distribution:*

$$\int p(A)dA = \int_{-\infty}^{\infty}\int_{-\infty}^{\infty}\int_{-\infty}^{\infty} c_f e^{-f(x^2+2y^2+z^2)}dxdydz = 1. \qquad (17.19)$$

For what f does $p(QAQ^T) = p(A)$? Does this contradict Theorem 17.1.2?

17.2 DISTRIBUTION OF EIGENVALUES OF 2×2 GOE MODEL

We are interested in the distribution of spacings between eigenvalues of 2×2 matrices chosen from the GOE ensemble. How likely is it for such matrices to have two eigenvalues close? Far apart? We order the eigenvalues so that $\lambda_1 > \lambda_2$. As we are just studying 2×2 matrices (N is fixed), we do not need to worry about renormalizing the eigenvalues.

17.2.1 Eigenvalues of the 2×2 GOE

For notational convenience we denote our matrices by $A = \left(\begin{smallmatrix} x & y \\ y & z \end{smallmatrix}\right)$. By the Spectral Theorem of Linear Algebra (Theorem B.5.1), for any real symmetric matrix A there is an orthogonal matrix Q such that $QAQ^T = \Lambda$ is diagonal:

$$Q^T \begin{pmatrix} x & y \\ y & z \end{pmatrix} Q = \begin{pmatrix} \lambda_1 & 0 \\ 0 & \lambda_2 \end{pmatrix} = \Lambda. \qquad (17.20)$$

To find the eigenvalues we solve the **characteristic equation** $\det(A - \lambda I) = 0$:
$$\lambda^2 - \text{Trace}(A)\lambda + \det(A) = 0, \qquad (17.21)$$

where

$$\text{Trace}(A) = x + z, \quad \det(A) = xz - y^2. \qquad (17.22)$$

The solutions are

$$\lambda_1 = \frac{x+z}{2} + \sqrt{\left(\frac{x-z}{2}\right)^2 + y^2}$$

$$\lambda_2 = \frac{x+z}{2} - \sqrt{\left(\frac{x-z}{2}\right)^2 + y^2}. \qquad (17.23)$$

Remark 17.2.1. If the two eigenvalues are equal, we say the matrix is **degenerate**. Initially we are in a three-dimensional space (as x, y and z are arbitrary). Degeneracy requires that x, y and z satisfy

$$\left(\frac{x-z}{2}\right)^2 + y^2 = 0, \tag{17.24}$$

or equivalently

$$x - z = 0, \quad y = 0. \tag{17.25}$$

We lose two degrees of freedom because there are two equations which must be satisfied. Thus the set of solutions is $\{(x, y, z) = (x, 0, x)\}$, and the matrices are $A = xI$. If we choose the entries independently from a fixed probability distribution p, the above implies there is zero probability of having a degenerate matrix. Another way to interpret this is the probability of $y, z - x \in [-\epsilon, \epsilon]$ and x arbitrary is proportional to ϵ^2.

The eigenvectors of $A = \left(\begin{smallmatrix} x & y \\ y & z \end{smallmatrix}\right)$ are

$$\vec{v_1} = \begin{pmatrix} \cos\theta \\ \sin\theta \end{pmatrix}, \quad \vec{v_2} = \begin{pmatrix} -\sin\theta \\ \cos\theta \end{pmatrix} \tag{17.26}$$

for some $\theta \in [0, 2\pi]$. Why can we write the eigenvectors as above? We can always normalize the eigenvector attached to a given eigenvalue to have length 1. If the eigenvalues are distinct, then the eigenvectors of a real symmetric matrix are perpendicular (see §B.5.1). This forces the above form for the two eigenvectors, at least when $\lambda_1 \neq \lambda_2$. If $\lambda_1 = \lambda_2$, then A is a multiple of the identity matrix. All vectors are then eigenvectors, and the above form still holds.

A 2×2 **rotation** is defined by an angle θ:

$$Q = Q(\theta) = \begin{pmatrix} \cos\theta & -\sin\theta \\ \sin\theta & \cos\theta \end{pmatrix}. \tag{17.27}$$

The structure of the eigenvectors of the real symmetric matrix A is actually quite rich.

Exercise[h] 17.2.2. *Find θ in terms of x, y, z.*

Exercise 17.2.3. *Another way to write a general real symmetric matrix A is by*

$$A = \alpha \begin{pmatrix} \cos\beta & \sin\beta \\ \sin\beta & -\cos\beta \end{pmatrix} + \gamma \begin{pmatrix} 1 & 0 \\ 0 & 1 \end{pmatrix}. \tag{17.28}$$

Find λ_1, λ_2 in terms of α, β, γ. Show that the eigenvector angle is given by $\theta = \frac{\beta}{2}$.

Remark 17.2.4 (Important). We would be remiss if we did not mention that this method of proof cannot be generalized to all $N \times N$ matrices, even at the cost of more involved computations. Because our matrices are 2×2, the characteristic equation for the eigenvalues λ is quadratic. For polynomials of degree at most four, there are formulas which allow us to write down the roots explicitly in terms of the coefficients; however, there are no such formulas for degree 5 and higher (see, for example, [Art]), and other methods are needed.

17.2.2 Joint Distribution Function of the Eigenvalues

To understand the distribution of $\lambda_1 - \lambda_2$, we change variables. Initially we are given a matrix A with entries x, y and z and by Theorem 17.1.2 a probability density $p(x, y, z) = \frac{C\sqrt{C}}{4\pi\sqrt{2\pi}} e^{-C\text{Trace}(A^2)}$. We can diagonalize A and write it as $Q^T \Lambda Q$ for a diagonal Λ (with parameters λ_1 and λ_2) and an orthogonal Q (with parameter θ). We determine the probability density $\widetilde{p}(\lambda_1, \lambda_2, \theta)$ in this new coordinate system.

As we are interested *only* in properties of the eigenvalues (one can consider θ to be a consequence of how we orient our coordinate axes), we integrate out the θ dependence, leaving ourselves with a joint probability distribution for λ_1 and λ_2. The change of variables is 1-to-1 apart from a set of measure zero (i.e., the lower dimensional subspace where $\lambda_1 = \lambda_2$) corresponding to degenerate eigenvalues.

$$
\begin{aligned}
A &= Q^T \Lambda Q \\
&= \begin{pmatrix} \cos\theta & \sin\theta \\ -\sin\theta & \cos\theta \end{pmatrix} \begin{pmatrix} \lambda_1 & 0 \\ 0 & \lambda_2 \end{pmatrix} \begin{pmatrix} \cos\theta & -\sin\theta \\ \sin\theta & \cos\theta \end{pmatrix} \\
&= \begin{pmatrix} \cos^2(\theta)\lambda_1 + \sin^2(\theta)\lambda_2 & -\cos(\theta)\sin(\theta) \cdot (\lambda_1 - \lambda_2) \\ -\cos(\theta)\sin(\theta) \cdot (\lambda_1 - \lambda_2) & \sin^2(\theta)\lambda_1 + \cos^2(\theta)\lambda_2 \end{pmatrix} \\
&= \begin{pmatrix} x & y \\ y & z \end{pmatrix}.
\end{aligned}
\tag{17.29}
$$

The above shows that the change of variable transformation is linear in the eigenvalues λ_1 and λ_2, though this was clear without explicitly performing the matrix multiplication. We are interested in the **marginal** or **joint distribution of the eigenvalues**,

$$
\widetilde{p}(\lambda_1, \lambda_2) = \int \widetilde{p}(\lambda_1, \lambda_2, \theta) d\theta.
\tag{17.30}
$$

In other words, we do not care what θ is. We want to know how to transform the probability density from (x, y, z) space to $(\lambda_1, \lambda_2, \theta)$ space. This is because it is significantly easier in the second space to see the spacings between eigenvalues. By the Change of Variables Theorem from Multivariable Calculus (Theorem A.2.11), if J is the transformation from (x, y, z) to $(\lambda_1, \lambda_2, \theta)$, then

$$
\iiint p(x, y, z) dx\, dy\, dz = \iiint \widetilde{p}(\lambda_1, \lambda_2, \theta) d\lambda_1\, d\lambda_2\, d\theta
\tag{17.31}
$$

gives

$$
p(x, y, z) = |\det(J)| \widetilde{p}(\lambda_1, \lambda_2, \theta).
\tag{17.32}
$$

As $\text{Trace}(A^2) = \lambda_1^2 + \lambda_2^2$ is independent of the angle θ, the probability density of observing a matrix with eigenvalues λ_1 and λ_2 is

$$
\widetilde{p}(\lambda_1, \lambda_2, \theta) = \frac{C\sqrt{C}}{4\pi\sqrt{2\pi}} e^{-C(\lambda_1^2 + \lambda_2^2)}.
\tag{17.33}
$$

For simplicity, let us rescale so that $C = 1$. Then our volume elements are

$$
p(x, y, z) dx\, dy\, dz = |\det(J)| \frac{1}{4\pi\sqrt{2\pi}} e^{-(\lambda_1^2 + \lambda_2^2)} d\lambda_1\, d\lambda_2\, d\theta,
\tag{17.34}
$$

where J is the Jacobian matrix

$$
J = \begin{pmatrix} \frac{\partial x}{\partial \lambda_1} & \frac{\partial y}{\partial \lambda_1} & \frac{\partial z}{\partial \lambda_1} \\ \frac{\partial x}{\partial \lambda_2} & \frac{\partial y}{\partial \lambda_2} & \frac{\partial z}{\partial \lambda_2} \\ \frac{\partial x}{\partial \theta} & \frac{\partial y}{\partial \theta} & \frac{\partial z}{\partial \theta} \end{pmatrix}. \tag{17.35}
$$

We could explicitly determine each J_{ij} by differentiating (17.29); however, we note the following: after differentiation, λ_1 and λ_2 occur only in the last row, and they always occur as a multiple of $\lambda_1 - \lambda_2$:

$$
J = \begin{pmatrix} * & * & * \\ * & * & * \\ -\frac{\sin(2\theta)}{2} \cdot (\lambda_1 - \lambda_2) & \cos(2\theta) \cdot (\lambda_1 - \lambda_2) & \frac{\sin(2\theta)}{2} \cdot (\lambda_1 - \lambda_2) \end{pmatrix}. \tag{17.36}
$$

As $\lambda_1 - \lambda_2$ appears in each entry of the bottom row, we can pull it out of the matrix and we are left with the determinant of a 3×3 matrix which *does not depend on either* λ_1 *or* λ_2. Let us call the absolute value of the determinant of this θ-matrix $g(\theta)$. Then

$$
\det(J) = g(\theta) \cdot (\lambda_1 - \lambda_2). \tag{17.37}
$$

Warning: As the Jacobian is the absolute value of the determinant, we need $|\lambda_1 - \lambda_2|$ above. As we have chosen $\lambda_1 > \lambda_2$, we may drop the absolute value.

The only dependence on the λ's is given by the second factor. Substituting into (17.34) yields

$$
p(x, y, z)dxdydz = \frac{g(\theta)}{4\pi\sqrt{2\pi}}(\lambda_1 - \lambda_2)e^{-(\lambda_1^2 + \lambda_2^2)}d\lambda_1 d\lambda_2 d\theta. \tag{17.38}
$$

As we are concerned with the joint probability distribution of λ_1 and λ_2, we integrate out θ:

$$
\begin{aligned}
\widetilde{p}(\lambda_1, \lambda_2) &= \frac{1}{4\pi\sqrt{2\pi}} \int (\lambda_1 - \lambda_2)e^{-(\lambda_1^2 + \lambda_2^2)}g(\theta)d\theta \\
&= C'(\lambda_1 - \lambda_2)e^{-(\lambda_1^2 + \lambda_2^2)}, \tag{17.39}
\end{aligned}
$$

where C' is a computable constant based on the θ-integral. Note C' must be such that $\widetilde{p}(\lambda_1, \lambda_2)$ is a probability distribution. This is the famous joint density of the eigenvalues of the 2×2 GOE. In §17.2.3 we use this to determine the spacing properties of $\lambda_1 - \lambda_2$.

Exercise 17.2.5. *Determine* C' *in (17.39). It is remarkable that we do not need to know* $g(\theta)$ *to determine* C'.

Exercise[h] **17.2.6.** *For* 2×2 *matrices* $A = \begin{pmatrix} x & y \\ y & z \end{pmatrix}$, *let*

$$
p(A) = Ce^{-(x^4 + 4x^2y^2 + 2y^4 + 4xy^2z + 4y^2z^2 + z^4)}. \tag{17.40}
$$

Calculate $\widetilde{p}(\lambda_1, \lambda_2)$ *for this ensemble.*

17.2.3 Spacing Distribution

We now have all the tools required to study the spacings between the eigenvalues. Let us calculate the probability that $\lambda_1 - \lambda_2 \in [\lambda, \lambda + \Delta\lambda]$. In the end we send $\Delta\lambda \to 0$, so we only keep terms to first order in $\Delta\lambda$. We need to integrate over all pairs (λ_1, λ_2) with $\lambda_1 > \lambda_2$ and $\lambda_1 - \lambda_2 \in [\lambda, \lambda + \Delta\lambda]$. We use $\lambda_1 = \lambda_2 + \lambda + O(\Delta\lambda)$ a few times in the algebra below, and find

$$
\begin{aligned}
&\mathrm{Prob}\Big(\lambda_1 - \lambda_2 \in [\lambda, \lambda + \Delta\lambda]\Big) \\
&= \int_{\lambda_2=-\infty}^{\infty} \int_{\lambda_1=\lambda_2+\lambda}^{\lambda_2+\lambda+\Delta\lambda} \widehat{p}(\lambda_1, \lambda_2) d\lambda_1 d\lambda_2 \\
&= \int_{\lambda_2=-\infty}^{\infty} \int_{\lambda_1=\lambda_2+\lambda}^{\lambda_2+\lambda+\Delta\lambda} C'(\lambda_1 - \lambda_2) e^{-(\lambda_1^2+\lambda_2^2)} d\lambda_1 d\lambda_2 \\
&= C' \int_{\lambda_2=-\infty}^{\infty} \int_{\lambda_1=\lambda_2+\lambda}^{\lambda_2+\lambda+\Delta\lambda} (\lambda + O(\Delta\lambda)) e^{-((\lambda_2+\lambda+O(\Delta\lambda))^2+\lambda_2^2)} d\lambda_1 d\lambda_2.
\end{aligned}
$$

$$(17.41)$$

As the λ_1-integral is over a region of size $\Delta\lambda$, the λ_1-integral gives a contribution of $\lambda e^{-(\lambda^2+2\lambda_2^2+2\lambda\lambda_2)}\Delta\lambda + O((\Delta\lambda)^2)$. Therefore

$$
\begin{aligned}
&\mathrm{Prob}\Big(\lambda_1 - \lambda_2 \in [\lambda, \lambda + \Delta\lambda]\Big) \\
&= C' \int_{\lambda_2=-\infty}^{\infty} \lambda e^{-(\lambda^2+2\lambda_2^2+2\lambda\lambda_2)} \Delta\lambda d\lambda_2 + O((\Delta\lambda)^2) \\
&= C' \lambda e^{-\lambda^2} \Delta\lambda \int_{\lambda_2=-\infty}^{\infty} e^{-2(\lambda_2^2+\lambda\lambda_2)} d\lambda_2 + O((\Delta\lambda)^2) \\
&= C' \lambda e^{-\lambda^2} \Delta\lambda \int_{\lambda_2=-\infty}^{\infty} e^{-2(\lambda_2^2+\lambda\lambda_2+\frac{\lambda^2}{4})} e^{\frac{\lambda^2}{2}} d\lambda_2 + O((\Delta\lambda)^2) \\
&= C' \lambda e^{-\frac{\lambda^2}{2}} \Delta\lambda \int_{\lambda_2=-\infty}^{\infty} e^{-2(\lambda_2+\frac{\lambda}{2})^2} d\lambda_2 + O((\Delta\lambda)^2) \\
&= C' \lambda e^{-\frac{\lambda^2}{2}} \Delta\lambda \int_{\lambda_2=-\infty}^{\infty} e^{-2\lambda_2^2} d\lambda_2 + O((\Delta\lambda)^2) \\
&= C'' \lambda e^{-\frac{\lambda^2}{2}} \Delta\lambda + O((\Delta\lambda)^2),
\end{aligned}
$$

$$(17.42)$$

where C'' is the new normalization constant obtained by multiplying C' by the λ_2-integration (which is a constant independent of λ). While we can evaluate the λ_2-integration (see Example 8.2.11) and C' directly, there is no need. As $p_{\mathrm{GOE},2}(\lambda)$ is a probability distribution, the product C'' must be such that it integrates to 1. This simple observation allows us to avoid dealing with many complicated integrals. This is amazing, and a powerful technique in probability theory to avoid having to evaluate numerous constants throughout a proof. We have shown

Lemma 17.2.7.

$$
\mathrm{Prob}\Big(\lambda_1 - \lambda_2 \in [\lambda, \lambda + \Delta\lambda]\Big) = C'' \lambda e^{-\frac{\lambda^2}{2}} \Delta\lambda + O\left((\Delta\lambda)^2\right). \quad (17.43)
$$

Taking the limit as $\Delta\lambda \to 0$, we obtain the probability density that the difference in eigenvalues is λ is $p_{\text{GOE},2}(\lambda) = C''\lambda e^{-\frac{\lambda^2}{2}}$.

Exercise[h] 17.2.8. *Evaluate C''. Hint: As $p_{\text{GOE},2}(\lambda)$ is a probability distribution, $\int_0^\infty p_{\text{GOE},2}(\lambda)d\lambda = 1$. What are the mean and standard deviation of $p_{\text{GOE},2}$? Approximately what is the probability that the difference is less than half the mean?*

The fact that the probability of observing a difference of λ is proportional to $\lambda e^{-\frac{\lambda^2}{2}}$ is called the **Wigner Surmise** for the eigenvalue spacing density (see Exercise 15.4.7). Remarkably, in the $N \times N$ case, even for large N this density is very close to the true spacing distribution of adjacent eigenvalues.

Remark 17.2.9 (Repulsion of Eigenvalues). For λ near 0, $p_{\text{GOE},2}(\lambda) \approx C''\lambda$. Thus, unlike the Poisson distribution (where small spacings are the most likely; see Chapter 12), here we find small spacings are unlikely: the eigenvalues "repel" each other.

Exercise 17.2.10. *The probability density $p_{\text{GOE},2}(\lambda)$ can be derived faster through use of the Dirac delta function (see §11.2.2 and §15.2.3):*

$$p_{\text{GOE},2}(\lambda) = \int_{\lambda_2=-\infty}^\infty \int_{\lambda_1=\lambda_2}^\infty \widetilde{p}(\lambda_1,\lambda_2)\delta(\lambda-(\lambda_1-\lambda_2))d\lambda_1 d\lambda_2. \quad (17.44)$$

Along these lines, derive $p_{\text{GOE},2}(\lambda) = C''\lambda e^{-\frac{\lambda^2}{2}}$.

Exercise 17.2.11. *Derive similar results (from the last few sections) for 2×2 complex Hermitian matrices. These matrices have four free parameters,*

$$A = \begin{pmatrix} x_{11} & x_{12}+iy_{12} \\ x_{12}+iy_{12} & x_{22} \end{pmatrix}. \quad (17.45)$$

Instead of $p(Q^T AQ) = p(A)$, basis independence yields $p(U^* AU) = p(A)$, where U is an arbitrary unitary transformation and U^* is the complex conjugate transpose of U. Such an assumption again leads to the entries being chosen from Gaussian distributions. In particular, show the distribution of $\lambda = \lambda_1 - \lambda_2$ is $C\lambda^2 e^{-B\lambda^2}$ for some constants B and C. This is called the Gaussian Unitary Ensemble or GUE.

Exercise 17.2.12. *Try to generalize the arguments to determine the spacing between adjacent eigenvalues of 3×3 matrices. For each matrix we can order the eigenvalues $\lambda_1 \geq \lambda_2 \geq \lambda_3$. Note each matrix has two differences: $\lambda_1 - \lambda_2$ and $\lambda_2 - \lambda_3$.*

Exercise 17.2.13. *Can you calculate the probability of observing a spacing of λ using the probability distribution from Example 15.1.9? The difficulty is not in writing down the integrals, but in evaluating them.*

17.3 GENERALIZATION TO $N \times N$ GOE

17.3.1 Dimension of Space

We discuss the generalizations to the $N \times N$ GOE ensemble. In the 2×2 case, a real symmetric matrix had 3 degrees of freedom, written either as (a_{11}, a_{12}, a_{22})

or $(\lambda_1, \lambda_2, \theta)$. For an $N \times N$ real symmetric matrix there are $\frac{N(N+1)}{2}$ independent parameters. This follows from the fact that any real symmetric matrix is determined once we specify the N diagonal elements and the $\frac{N^2-N}{2}$ upper diagonal elements. Thus there are $\frac{N(N+1)}{2}$ elements as claimed. This gives the standard basis $(a_{11}, \ldots, a_{1N}, a_{22}, \ldots, a_{NN})$. By the Spectral Theorem for real symmetric matrices, we may write $A = Q^T \Lambda Q$, where Λ is the diagonal matrix with entries $\lambda_1 \geq \cdots \geq \lambda_N$.

Lemma 17.3.1. *Another coordinate system for $N \times N$ real symmetric matrices is given by $(\lambda_1, \ldots, \lambda_N, \theta_1, \ldots, \theta_n)$, where $\theta_1, \ldots, \theta_n$ specify the orthogonal matrix Q. Here $n = \frac{N(N-1)}{2}$.*

Sketch of the proof. Consider an $N \times N$ orthogonal matrix. The columns are mutually perpendicular, and each column has length 1. Thus we may choose $N - 1$ entries of the first column, and the last entry is forced (the length must be 1).

There are two conditions on the second column: it must have length 1, and it must be perpendicular to the first column. There will be $N - 2$ free entries. Continuing along these lines, we see we can freely choose $N - i$ entries in column i. Therefore, the number of free entries in an orthogonal matrix is

$$\sum_{i=1}^{N}(N - i) = N^2 - \frac{N(N+1)}{2} = \frac{N(N-1)}{2}. \tag{17.46}$$

See §A.1.1 for an evaluation of the sum in (17.46). The astute reader will notice that we do have the freedom to multiply all entries of any column by ± 1, but this will not change the dimension of the space. $\square$

We let λ stand for the λ_i variables and θ for the θ_j variables.

Exercise 17.3.2. *How many free parameters are there in an $N \times N$ complex Hermitian matrix? How many parameters are needed to specify an $N \times N$ unitary matrix?*

17.3.2 Probability of Individual Matrices

We saw that for the 2×2 GOE ensemble, the assumptions that the probability of a matrix is the product of the probabilities of the entries and the probability is base-independent (or $p(QAQ^T) = p(A)$) led to

1. the probabilities of the individual entries are Gaussian with mean zero and standard deviation $\sqrt{\frac{1}{C}}$ for the off-diagonal and $\sqrt{\frac{2}{C}}$ for the diagonal entries;

2. the probability of A is proportional to $e^{-\frac{C}{4}\operatorname{Trace}(A^2)}$.

These formulas also hold in the $N \times N$ case. Let $i < j$ and consider the entries $a_{ii}, a_{ij}, a_{ji} = a_{ij}, a_{jj}$. Similar to the 2×2 case, we can consider a rotation by ϵ. Thus, Q is the identity matrix with the following changes: $q_{ii} = \cos \epsilon$, $q_{ij} = \sin \epsilon$, $q_{ji} = -\sin \epsilon$, and $q_{jj} = \cos \epsilon$.

Using $p(QAQ^T) = p(A)$, we will be led to the same results as before, namely that

$$P_{ii}(a_{ii}) = \sqrt{\frac{C}{4\pi}}\, e^{-\frac{Ca_{ii}^2}{4}}$$

$$P_{ij}(a_{ij}) = \sqrt{\frac{C}{2\pi}}\, e^{-\frac{Ca_{ij}^2}{2}}$$

$$P_{jj}(a_{jj}) = \sqrt{\frac{C}{4\pi}}\, e^{-\frac{Ca_{jj}^2}{4}}. \tag{17.47}$$

There are several cleaner ways to write the above calculation. We may write $Q = I + Q'$, where Q' is the zero matrix except that $q'_{ii} = \cos\epsilon$, $q'_{ij} = -\sin\epsilon$, $q'_{ji} = \sin\epsilon$, and $q'_{jj} = \cos\epsilon$. Taylor expanding we have

$$Q = I + Q'' + O(\epsilon^2), \quad q''_{ii} = q''_{jj} = 1, \quad q''_{ij} = -q''_{ji} = -\epsilon. \tag{17.48}$$

Then, up to corrections of size $O(\epsilon^2)$,

$$\begin{aligned} QAQ^T &= (I + Q'')A(I + Q'')^T \\ &= A + IAQ''^T + Q''AI + Q''AQ''^T. \end{aligned} \tag{17.49}$$

The above highlights how the entries of A change under Q; see [Meh2] for complete details.

Finally, one could argue without loss of generality that we may relabel the axes so that $i = 1$ and $j = 2$, in which case the calculations reduce to the 2×2 case (and then argue by symmetry for the other entries). We again find

Theorem 17.3.3. *The* GOE *conditions force the probabilities of the entries to be Gaussian:*

$$P_{ii}(a_{ii}) = \sqrt{\frac{C}{4\pi}}\, e^{-\frac{Ca_{ii}^2}{4}}$$

$$P_{ij}(a_{ij}) = \sqrt{\frac{C}{2\pi}}\, e^{-\frac{Ca_{ij}^2}{2}}. \tag{17.50}$$

As the probability of a matrix A is the product of the probabilities of its diagonal and upper diagonal entries, we have

$$p(A) = 2^{-\frac{N}{2}} \left(\frac{C}{2\pi}\right)^{\frac{N(N+1)}{4}} e^{-\frac{C}{4}\operatorname{Trace}(A^2)}. \tag{17.51}$$

17.3.3 Joint Distribution of the Eigenvalues

As in the 2×2 case, we want to integrate out the θ variables to obtain the joint distribution of the eigenvalues. Let $n = \frac{N(N-1)}{2}$ and choose Q so that $QAQ^T = \Lambda$, where Λ is diagonal with entries $\lambda_1 \geq \cdots \geq \lambda_N$ and Q is an orthogonal entries specified by parameters $\theta_1, \ldots, \theta_n$:

$$A = \begin{pmatrix} Q(\theta_1,\ldots,\theta_n) \end{pmatrix}^T \begin{pmatrix} \lambda_1 & & \\ & \ddots & \\ & & \lambda_N \end{pmatrix} \begin{pmatrix} Q(\theta_1,\ldots,\theta_n) \end{pmatrix}. \tag{17.52}$$

We are changing variables:

$$(a_{11}, \ldots, a_{1N}, a_{22}, \ldots, a_{NN}) \longleftrightarrow (\lambda_1, \ldots, \lambda_N, \theta_1, \ldots, \theta_n). \qquad (17.53)$$

Let $a(j)$ be the jth entry on the left. Thus, $a(1) = a_{11}$, $a(2) = a_{12}$, $\ldots$, $a(N) = a_{1N}$, $a(N+1) = a_{2,2}, \ldots, a(\frac{N(N+1)}{2}) = a_{NN}$. Let J be the Jacobian, with entries

$$J_{ij} = \begin{cases} \frac{\partial a(j)}{\partial \lambda_i} & \text{if } j \le N \\ \frac{\partial a(j)}{\partial \theta_{i-N}} & \text{if } N < j \le \frac{N(N+1)}{2}. \end{cases} \qquad (17.54)$$

Multiplying the right hand side of (17.52), we see that $a_{\alpha\beta}$ is a *linear* function of the λ_i's and a complicated function of the θ_j's. Thus, for $i \le N$, J_{ij} (which is obtained by differentiating with respect to λ_i) will be independent of the eigenvalues λ_i. For $N < i \le \frac{N(N+1)}{2}$, J_{ij} is obtained by differentiating with respect to θ_{i-N}, and will be a linear function of the eigenvalues.

Thus, $|\det J|$ is a polynomial in the eigenvalues of degree at most $\frac{N(N-1)}{2}$, as there are only $\frac{N(N-1)}{2}$ rows of J (the rows where we differentiate with respect to an angle) which contain eigenvalues, and all these rows are linear in the eigenvalues. We now show

Theorem 17.3.4.

$$|\det J| = g(\theta_1, \ldots, \theta_n) \prod_{1 \le i < j \le N} (\lambda_i - \lambda_j). \qquad (17.55)$$

Proof. If two eigenvalues are equal, the change of variables from $(a_{\alpha\beta})$ to (λ_i, θ_j) is undetermined. For example, in the 2×2 case, if $\lambda_1 = \lambda_2 = \lambda$, we would have

$$A = Q^T \lambda I Q = \lambda Q^T Q = \lambda I, \qquad (17.56)$$

and any rotation $Q = Q(\theta)$ works. As there is no unique Q, J is not invertible here. A similar calculation holds in the $N \times N$ case. By the Change of Variables Theorem (Theorem A.2.11), the determinant of the change of variables transformation therefore vanishes whenever two eigenvalues are equal. Hence $|\det J|$ is divisible by $\lambda_i - \lambda_j$ for any i and j. Consider the product

$$\prod_{1 \le i < j \le N} (\lambda_i - \lambda_j) = \prod_{i=1}^{N-1} \prod_{j=i+1}^{N} (\lambda_i - \lambda_j). \qquad (17.57)$$

We have just seen the above polynomial must divide $|\det J|$. It is a polynomial of degree $\frac{N(N-1)}{2}$ in the λ_i's, as (see §A.1.1)

$$\sum_{i=1}^{N-1} \sum_{j=i+1}^{N} 1 = \sum_{i=1}^{N-1} i = \frac{(N-1)N}{2}. \qquad (17.58)$$

Thus, $|\det(J)|$ and $\prod_{i<j}(\lambda_i - \lambda_j)$ both vanish whenever two eigenvalues are equal, and they have the same degree. Therefore they must be multiples of each other, where the multiple is independent of the λ's. Hence

$$|\det J| = g(\theta_1, \ldots, \theta_n) \prod_{1 \le i < j \le N} (\lambda_i - \lambda_j) \qquad (17.59)$$

as claimed.   □

Note that $\left|\prod_{i<j}(\lambda_i - \lambda_j)\right| = \prod_{i<j}(\lambda_i - \lambda_j)$, as we have ordered the eigenvalues so that $\lambda_1 \geq \cdots \geq \lambda_N$. A similar argument as in the 2×2 case now yields

Theorem 17.3.5 (Joint Distribution of $\lambda_1 \geq \cdots \geq \lambda_N$).

$$p(\lambda_1, \ldots, \lambda_N) = C' \prod_{1 \leq i < j \leq N} (\lambda_i - \lambda_j) e^{-(\lambda_1^2 + \cdots + \lambda_N^2)} \qquad (17.60)$$

Again, note the probability of two eigenvalues being equal tends to zero. The vanishing of this probability density as any two eigenvalues converge to a common value is called **level repulsion**. We see in the $N \times N$ GOE ensemble that eigenvalues repel each other.

It is significantly harder to analyze the spacings between adjacent eigenvalues in the $N \times N$ case than in the 2×2 case; however, the starting point for all such investigations is the joint probability distribution which we have just determined. Remarkably, the answer from the 2×2 GOE case, namely that the probability of observing two adjacent eigenvalues differing by λ is proportional to $\lambda e^{-\frac{\lambda^2}{2}}$, is very close to the answer in the $N \times N$ GOE case. For complete details, see [Gau, Meh1, Meh2].

Exercise 17.3.6. *How many degrees of freedom are lost if two eigenvalues are equal?*

Exercise 17.3.7. *Repeat these calculations for $N \times N$ complex Hermitian matrices.*

Exercise 17.3.8. *Can you use the formulas for the roots of a cubic or quartic to write down algebraically tractable formulas for the spacings between the eigenvalues of the 3×3 or 4×4 GOE Ensemble?*

Exercise[h] **17.3.9.** *Let $\lambda_1, \ldots, \lambda_n$ be complex numbers. Then define an $n \times n$ matrix A by $A = (\lambda_i^{j-1})_{ij}$; matrices of this form are called Vandermonde matrices. Prove that*

$$\det A = \prod_{i>j}(\lambda_i - \lambda_j). \qquad (17.61)$$

It is often very difficult to determine the normalization constants that arise; see Chapter 17 of [Meh2].

17.4 CONJECTURES AND RESEARCH PROJECTS

The limiting spacing distributions for the $N \times N$ GOE ensemble can be determined in closed form. We saw that two very natural physical assumptions led to the GOE model, namely (1) the probability of a matrix was the product of the probabilities of its entries (chosen independently) and (2) the probability is basis independent.

We can consider more general ensembles of matrices. Fix some probability distributions p_{ij} and for each N consider all real symmetric $N \times N$ matrices where the probability of A is

$$P(A) = \prod_{1 \leq i \leq j \leq N} p_{ij}(a_{ij}). \tag{17.62}$$

It is conjectured that for nice choices of p_{ij}, as $N \to \infty$ we should see the same behavior as in the GOE ensemble. Additionally, consider the ensemble of d-regular graphs; note we are not choosing the entries from a fixed probability distribution. While the density of states is *not* given by the semi-circle, numerical experiments suggest the spacings between normalized eigenvalues is the same as the GOE.

Research Project 17.4.1. Investigate the spacings between normalized eigenvalues for a variety of choices of p_{ij}, including the Cauchy distribution: $p_{ij}(x) = \frac{1}{\pi} \frac{1}{1+x^2}$.

Not all ensembles of real symmetric matrices have spacings given by the GOE. Note $N \times N$ real symmetric matrices have $\frac{N(N+1)}{2}$ free parameters; for ensembles with significantly fewer degrees of freedom, very different behavior is possible.

Research Project 17.4.2 (Band Matrices). Investigate the spacings between adjacent normalized eigenvalues of band matrices of width r. For $r = 0$, using the methods of Chapter 12 (or Exercise 12.7.4 or Appendix A of [MN]), prove that the distribution of eigenvalues is Poissonian; for $r = N - 1$, we have a full real symmetric matrix, and we expect to see the GOE distribution. Investigate the transition from one type of spacings to the other. How does it depend on p? It has been observed for many systems that transitions occur around $\sqrt{N}$. See [Liu, Mil2].

Research Project 17.4.3 (Toeplitz Matrices). How are the spacings between adjacent eigenvalues of Toeplitz matrices distributed? The number of independent entries is so small (size N rather than size N^2), that one predicts the spacing to be Poissonian, and not GOE; computer simulations support this conjecture. We chose 1000 Toeplitz matrices (1000×1000), with entries independently (according to the Toeplitz condition) chosen from the standard normal. We looked at the spacings between the middle 11 normalized eigenvalues for each matrix, giving 10 spacings per matrix. Figure 17.1 shows that the plot of the spacings between normalized eigenvalues looks Poissonian. We conjecture that in the limit as $N \to \infty$ the local spacings between adjacent normalized eigenvalues will be Poissonian. It is interesting to note that random d-regular graphs have a comparable number non-zero entries, but in §15.5.2 we saw the spacings between eigenvalues seem to follow the GOE distribution. In their adjacency matrices, there is significantly more independence in the a_{ij}; for the Toeplitz ensemble, we have a strict structure, namely a_{ij} depends only on $|i - j|$. What if we look at Palindromic Toeplitz matrices (see Remark 16.2.13).

Research Project 17.4.4 (Hankel-Markov). There are other ensembles similar to Toeplitz matrices, the Hankel and Markov, where one has order N independent entries. In [BDJ] the eigenvalue density distribution is determined (it is not the semi-circle), but the spacings between adjacent eigenvalues is still open.

Research Project 17.4.5 (Weighted Graph, Self-Adjoint). Consider the other ensembles from §16.4. Do the spacings between their eigenvalues agree with the

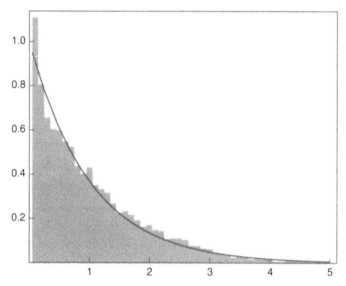

Figure 17.1 Spacings between normalized eigenvalues from 1000 Real Symmetric Toeplitz matrices (1000 × 1000)

GOE? How do the spacings from Project 16.4.3 compare to those of d-regular graphs?

Research Project 17.4.6 (Spherical Ensemble: Dependent Entries). Study the spacings between eigenvalues from the ensemble in Project 16.4.7 (as well as other ensembles with dependent entries).

Chapter Eighteen

The Explicit Formula and Density Conjectures

We sketch the 1-level density (see §15.6.2) calculation for the family of Dirichlet L-functions of prime conductor m. This statistic provides information about the zeros of these L-functions near the **central point** $s = \frac{1}{2}$. According to the conjectures of Katz and Sarnak [KS1, KS2], the behavior of zeros near $s = \frac{1}{2}$ as $m \to \infty$ is the same as the behavior of eigenvalues near 1 of a classical compact subgroup of $U(N)$ (the $N \times N$ unitary matrices) as $N \to \infty$. In this case the subgroup is actually the full group $U(N)$. We do not calculate the random matrix quantities (we refer the reader to [KS1, HuRu] for details of the random matrix calculations); we concentrate on the calculations on the number theory side. As an application of our 1-level density calculations, in Theorem 18.2.7 we bound the number of characters that have a zero at the central point (assuming GRH).

The 1-level density calculations highlight many of the features of investigations of zeros of L-functions. To understand the zeros of an L-function one uses an explicit formula (see §3.2.5 and §18.1) to relate sums of a test function h over zeros to sums of the product of the L-function's coefficients with the Fourier transform $\widehat{h}$ at primes. This is the number theory analogue of the Eigenvalue Trace Formula (Theorem 15.2.4), and the starting point of our investigations. Just as the Eigenvalue Trace Formula allows us to pass from knowledge about the matrix coefficients (which we randomly choose) to knowledge about the eigenvalues (what we want to study), the Explicit Formula allows us to pass from sums over the coefficients of the L-functions (which we can evaluate) to properties of the zeros (which we want to know). Further, like the Eigenvalue Trace Lemma, the Explicit Formula tells us what the appropriate scale is for studying the zeros.

In Random Matrix Theory we were able to evaluate the sums over the matrix coefficients because of how we randomly chose our matrices. For number theory investigations we need some type of Averaging Formula to execute the prime sums. Unfortunately the results for number theory in general are not as strong as those from Random Matrix Theory; however, for Dirichlet L-functions, reasonably good summation formulas are available because of the orthogonality relations of Dirichlet characters (see Lemma 3.3.15).

We assume the reader is familiar with L-functions at the level of Chapter 3; while we need some complex analysis for the proofs, for heuristic purposes the complex analysis exposition in Chapter 3 suffices. We strongly encourage the reader to consult [KS2] for more details. Again our goal is to show how similar techniques are used in Random Matrix Theory and Number Theory. While they are different statistics, the structure of the calculation of the density of states of normalized eigenvalues in Random Matrix Theory is very similar to looking at zeros near the central point of L-functions. After performing the number theory calculations, we highlight these similarities in great detail in §18.3.

18.1 EXPLICIT FORMULA

An identity of the form stated in Theorem 3.2.31, which relates a sum over the prime numbers of coefficients of an L-function to sums over the zeros of the L-function, is called an **explicit formula**. We derive such a formula for $\zeta(s)$ and sketch the modifications needed for Dirichlet L-functions. The rest of this section assumes familiarity with contour integration (see §3.2.2) and the Fourier transform (see §11.4.1); all that will be used later are the statements of the explicit formula, Theorem 18.1.15.

18.1.1 Explicit Formula for $\zeta(s)$

Recall from Theorem 3.1.19 that the completed zeta function $\xi(s)$ satisfies

$$\xi(s) \;=\; \frac{1}{2}s(s-1)\Gamma\left(\frac{s}{2}\right)\pi^{-\frac{s}{2}}\zeta(s) \;=\; \xi(1-s). \qquad (18.1)$$

The following derivation is a modification of Proposition 2.1 of [RS]. See also [Da2], especially Chapter 17. We start with some preliminary computations. The logarithmic derivative of $\xi(s)$ is

$$\frac{\xi'(s)}{\xi(s)} \;=\; -\frac{1}{2}\log\pi + \frac{1}{s} + \frac{1}{s-1} + \frac{1}{2}\frac{\Gamma'(\frac{s}{2})}{\Gamma(\frac{s}{2})} + \frac{\zeta'(s)}{\zeta(s)} \;=\; -\frac{\xi'(1-s)}{\xi(1-s)}. \qquad (18.2)$$

By contour integration, the above will yield information about the zeros and poles of $\xi(s)$ (and these are related to the zeros and poles of $\zeta(s)$). By Exercise 3.1.18 the Γ-function has simple poles with residue $\frac{(-1)^k}{k!}$ at $s = -k$ (k a non-negative integer) and $\Gamma(s)$ is never zero. Note $\zeta(s)$ has only one pole; it is at $s = 1$ and has residue 1 (see Chapter 17 of [Da2] for a proof that $\frac{\zeta'(0)}{\zeta(0)} = \log 2\pi$; one can also deduce that $\zeta(s)$ has neither a pole nor a zero at $s = 0$ by the integral representation of $\zeta(s)$ in (3.21) and the fact that $\Gamma(s)$ has a simple pole with residue 1 at $s = 0$).

By Theorem 3.1.19, $\xi(s)$ has an analytic continuation to all $s \in \mathbb{C}$; one consequence of this is that $\xi(s)$ has no poles. Therefore whenever one factor in the definition of $\xi(s)$ has a pole there must be at least one other factor that has a zero. In taking the logarithmic derivative in (18.2), there are four terms on the right hand side that can contribute a zero or pole: $\frac{1}{s}$, $\frac{1}{s-1}$, $\frac{\Gamma'(\frac{s}{2})}{\Gamma(\frac{s}{2})}$ and $\frac{\zeta'(s)}{\zeta(s)}$; these come from the factors s, $s - 1$, $\Gamma\left(\frac{s}{2}\right)$ and $\zeta(s)$. As $\xi(s)$ has no poles, the pole from $\zeta(s)$ at $s = 1$ is canceled by the zero of $s - 1$. Further the pole from $\Gamma\left(\frac{s}{2}\right)$ at $s = 0$ is canceled by the zero of s. Finally, as the Γ-function has poles at the negative integers, the poles from $\Gamma\left(\frac{s}{2}\right)$ at $s = -k$ ($k < 0$ an integer) are canceled by zeros of $\zeta(s)$ at the negative even integers. The implications of all of this is the following:

Lemma 18.1.1. *When we contour integrate $\frac{\xi'(s)}{\xi(s)}$, the only zeros and poles of $\xi(s)$ are the zeros of $\zeta(s)$ in the critical strip, namely those zeros $\rho = \sigma + i\gamma$ with $\sigma \in [0, 1]$.*

We express $\frac{\zeta'(s)}{\zeta(s)}$ in a form convenient for deriving the explicit formula. By the Euler product expansion, for $\Re s > 1$

$$\zeta(s) = \prod_{p}(1 - p^{-s})^{-1}. \tag{18.3}$$

A common method to deal with products is to take the logarithm; see Theorem 10.2.4 for another example. The reason is this converts a product to a sum, and we have many methods to understand sums. Hence

$$\frac{d \log \zeta(s)}{ds} = \frac{\zeta'(s)}{\zeta(s)} = -\sum_{p} \frac{d}{ds} \log(1 - p^{-s}) = -\sum_{p} \sum_{k=1}^{\infty} \frac{\log p}{(p^k)^s}. \tag{18.4}$$

We see the essential role the Euler product plays in these investigations; by contour integration the left hand side will be related to sums over zeros of $\zeta(s)$ and the right hand side will be related to sums of the coefficients of $\zeta(s)$. Without an Euler product we would not have such a connection; it is not by accident that all L-functions we study have an Euler product!

Let g be an even Schwartz function of compact support (see §11.4.1), and let

$$\phi(r) = \int_{-\infty}^{\infty} g(u)e^{iru} \, du. \tag{18.5}$$

Define

$$H(s) := \phi\left(\frac{s - \frac{1}{2}}{i}\right). \tag{18.6}$$

Exercise 18.1.2 (Important). *Show $H(s)$ is well defined for all s. If $s = x + iy$ then we may interpret $H(s)$ as*

$$H(x + iy) = \int_{-\infty}^{\infty} \left[g(u)e^{(x - \frac{1}{2})u} \right] e^{iuy} \, du. \tag{18.7}$$

For fixed x this is the Fourier transform of $g(u)e^{(x - \frac{1}{2})u}$ evaluated at $-\frac{y}{2\pi}$; see §11.4 for a review of the Fourier transform. Because of the compact support of $g(u)$ we can multiply g by a (possibly exponentially increasing) factor and still have a nice function. Show $g_x(u)$ is a Schwartz function with compact support. If we let $\widehat{g}_x$ denote the Fourier transform of this function, show $\widehat{g}_x\left(\frac{y}{2\pi}\right) \to 0$ as $|y| \to \infty$ (see also Exercise 11.4.5).

Exercise 18.1.3. *Prove $\phi(r) = \widehat{g}\left(\frac{r}{2\pi}\right)$. Show that if g is even, so is ϕ.*

Since $H(s)$ and $\xi(s)$ are entire functions (see Theorem 3.1.19 for a proof that $\xi(s)$ is entire), and by Exercise 18.1.2 $H(s)$ is rapidly decreasing in the y-direction, we can consider

$$\mathcal{I} = \frac{1}{2\pi i} \int_{\Re s = 1\frac{1}{2}} \frac{\xi'(s)}{\xi(s)} H(s) \, ds. \tag{18.8}$$

We shift contours from $\Re s = 1\frac{1}{2}$ to $-\frac{1}{2}$. This picks up poles coming from the zeros and poles of $\xi(s)$. By Lemma 18.1.1 we know that $\xi(s)$ has no poles, and its zeros are just the zeros of $\zeta(s)$ inside the critical strip (i.e., $\rho = \sigma + i\gamma$ with $\sigma \in [0,1]$ and $\zeta(\rho) = 0$). The residue of $\frac{\xi'(s)}{\xi(s)}H(s)$ at a zero ρ is then equal to $H(\rho)$. This follows from similar arguments as Exercise 3.2.13; if a zero ρ of $\zeta(s)$ has multiplicity greater than 1, this is incorporated by the fact that there will be several zeros that are equal. For example, let us say $\rho_n = \rho_{n+1} = \sigma + i\gamma$. Then the contribution from the zeros at $\sigma + i\gamma$ would be

$$H(\rho_n) + H(\rho_{n+1}) = 2H(\sigma + i\gamma) = H(\sigma + i\gamma)\mathrm{ord}_\zeta(\sigma + i\gamma). \qquad (18.9)$$

Therefore

$$\mathcal{I} = \sum_\rho H(\rho) + \frac{1}{2\pi i}\int_{\Re s = -\frac{1}{2}} \frac{\xi'(s)}{\xi(s)}H(s)ds, \qquad (18.10)$$

where the sum over ρ includes the multiplicity of each zero (conjectured to always be one). By (18.2), $\frac{\xi'(s)}{\xi(s)} = -\frac{\xi'(1-s)}{\xi(1-s)}$. Substituting this into (18.10) and changing s to $1 - s$, we obtain

$$\mathcal{I} = \sum_\rho H(\rho) - \frac{1}{2\pi i}\int_{\Re s = 1\frac{1}{2}} \frac{\xi'(s)}{\xi(s)}H(1-s)ds. \qquad (18.11)$$

Bringing $\sum H(\rho)$ over to the left hand side and recalling the definition of $\mathcal{I}$ yields

$$\sum_\rho H(\rho)$$

$$= \frac{1}{2\pi i}\int_{\Re s = 1\frac{1}{2}} \frac{\xi'(s)}{\xi(s)}\left[H(s) + H(1-s)\right]ds$$

$$= \frac{1}{2\pi i}\int_{\Re s = 1\frac{1}{2}} \frac{\zeta'(s)}{\zeta(s)}\left[H(s) + H(1-s)\right]ds$$

$$+ \frac{1}{2\pi i}\int_{\Re s = 1\frac{1}{2}} \left(\frac{1}{s-1} + \frac{\Gamma'(\frac{1}{2}s+1)}{\Gamma(\frac{1}{2}s+1)} - \frac{1}{2}\log\pi\right)\left[H(s) + H(1-s)\right]ds.$$

$$(18.12)$$

We can shift the contour in the second integral to $\Re s = \frac{1}{2}$. We do so, and because ϕ is even we note that for $s = \frac{1}{2} + iy$, $H(1-s) = H(s) = \phi(y)$. If $s = \frac{1}{2} + iy$ then $ds = idy$ and we have

$$\sum_\rho H(\rho) = \frac{1}{2\pi i}\int_{\Re s = 1\frac{1}{2}} \frac{\zeta'(s)}{\zeta(s)}\left[H(s) + H(1-s)\right]ds + \phi\left(\frac{i}{2}\right) + \phi\left(\frac{-i}{2}\right)$$

$$+ \frac{1}{\pi}\int_{-\infty}^{\infty} \left(\frac{1}{iy - \frac{1}{2}} + \frac{\Gamma'(\frac{iy}{2} + \frac{5}{4})}{\Gamma(\frac{iy}{2} + \frac{5}{4}))} - \frac{1}{2}\log\pi\right)\phi(y)\,dy. \qquad (18.13)$$

We now concentrate on the integrals involving $\frac{\zeta'}{\zeta}$. We have

$$\frac{1}{2\pi i}\int_{\Re s = 1\frac{1}{2}} \frac{\zeta'(s)}{\zeta(s)}H(s)ds = -\frac{1}{2\pi}\sum_p\sum_{k\geq 1} \frac{\log p}{p^{\frac{3}{2}k}}\int_{\mathbb{R}} \frac{H(\frac{3}{2} + iy)}{p^{iyk}}\,dy$$

$$= -\frac{1}{2\pi}\sum_p\sum_{k\geq 1} \frac{\log p}{p^{\frac{3}{2}k}}\int_{\mathbb{R}} \frac{\phi(y+i)}{p^{iyk}}\,dy. \qquad (18.14)$$

Exercise 18.1.4. *Justify changing the order of the integration and summation.*

We now consider the innermost integral. Consider the contour determined by the following four points: $(\pm R, 0), (\pm R, -i)$. Then Cauchy's theorem gives

$$\int_{(-R,0)}^{(R,0)} \frac{\phi(z+i)}{p^{izk}} \, dz + \int_{(R,0)}^{(R,-i)} \frac{\phi(z+i)}{p^{izk}} \, dz$$

$$+ \int_{(R,-i)}^{(-R,-i)} \frac{\phi(z+i)}{p^{izk}} \, dz + \int_{(-R,-i)}^{(-R,0)} \frac{\phi(z+i)}{p^{izk}} \, dz \; = \; 0.$$

(18.15)

Exercise 18.1.5. *Show that as $R \to \infty$*

$$\int_{(R,0)}^{(R,-i)} \frac{\phi(z+i)}{p^{izk}} \, dz \longrightarrow 0, \qquad \int_{(-R,-i)}^{(-R,0)} \frac{\phi(z+i)}{p^{izk}} \, dz \longrightarrow 0. \qquad (18.16)$$

Hint: Use the rapid decay of ϕ.

Therefore, by (18.15) and Exercise 18.1.5 we have

$$\lim_{R \to \infty} \int_{(-R,0)}^{(R,0)} \frac{\phi(z+i)}{p^{izk}} \, dz \; = \; \lim_{R \to \infty} \int_{(-R,-i)}^{(R,-i)} \frac{\phi(z+i)}{p^{izk}} \, dz. \qquad (18.17)$$

Consequently

$$\frac{1}{2\pi} \int_{-\infty}^{\infty} \frac{\phi(y+i)}{p^{iyk}} \, dy \; = \; \frac{p^k}{2\pi} \int_{-\infty}^{\infty} \frac{\phi(y)}{p^{iyk}} \, dy$$

$$= \; \frac{p^k}{2\pi} \int_{-\infty}^{\infty} \phi(y) e^{iyk \log p} \, dy \qquad (18.18)$$

$$= \; p^k g(k \log p).$$

This implies

$$\frac{1}{2\pi i} \int_{\Re s = 1\frac{1}{2}} \frac{\zeta'(s)}{\zeta(s)} H(s) ds \; = \; \sum_{p} \sum_{k \geq 1} \frac{\log p}{p^{\frac{1}{2}k}} g(k \log p). \qquad (18.19)$$

Similarly

$$\frac{1}{2\pi i} \int_{\Re s = 1\frac{1}{2}} \frac{\zeta'(s)}{\zeta(s)} H(1-s) ds \; = \; \sum_{p} \sum_{k \geq 1} \frac{\log p}{p^{\frac{1}{2}k}} g(-k \log p). \qquad (18.20)$$

For a zero ρ of $\zeta(s)$ in the critical strip ($\Re(\rho) \in [0,1]$), we write $\rho = \frac{1}{2} + i\gamma_\rho$ where γ_ρ is a complex number. The truth of the Riemann Hypothesis would, of course, imply that $\gamma_\rho \in \mathbb{R}$. We have shown

Theorem 18.1.6 (Explicit Formula). *Let $\sum_\rho$ denote the sum over the zeros of $\zeta(s)$ in the critical strip. With g an even Schwartz function of compact support and ϕ as in (18.5),*

$$\sum_\rho \phi(\gamma_\rho) \; = \; 2\phi\left(\frac{i}{2}\right) - \sum_p \sum_{k=1}^{\infty} \frac{2 \log p}{p^{\frac{k}{2}}} g\left(k \log p\right)$$

$$+ \frac{1}{\pi} \int_{-\infty}^{\infty} \left(\frac{1}{iy - \frac{1}{2}} + \frac{\Gamma'(\frac{iy}{2} + \frac{5}{4})}{\Gamma(\frac{iy}{2} + \frac{5}{4})} - \frac{1}{2} \log \pi \right) \phi(y) \, dy.$$

(18.21)

18.1.2 Explicit Formula for Dirichlet L-Functions

We *sketch* the proof of the explicit formula for Dirichlet L-function. For us the significance of the explicit formula lies in applying it to a family of Dirichlet L-functions. As the proof is similar to that of Theorem 18.1.6, we merely highlight the modifications. For simplicity we assume that $\chi(-1) = 1$; our final result holds even when $\chi(-1) = -1$, though the algebra is slightly different. From (3.149) the logarithmic derivative of $\Lambda(s, \chi)$ is

$$\frac{\Lambda'(s,\chi)}{\Lambda(s,\chi)} = \frac{1}{2}\log\left(\frac{m}{\pi}\right) + \frac{1}{2}\frac{\Gamma'(\frac{s}{2})}{\Gamma(\frac{s}{2})} + \frac{L'(s,\chi)}{L(s,\chi)} = -\frac{\Lambda'(1-s,\overline{\chi})}{\Lambda(1-s,\overline{\chi})}. \tag{18.22}$$

By the Euler product expansion, for $\Re s > 1$

$$L(s,\chi) = \prod_p (1 - \chi(p)p^{-s})^{-1}. \tag{18.23}$$

Hence

$$\frac{L'(s,\chi)}{L(s,\chi)} = -\sum_p \frac{d}{ds}\log(1 - \chi(p)p^{-s}) = -\sum_p \sum_{k=1}^{\infty} \frac{\chi(p^k)(\log p)}{(p^k)^s}. \tag{18.24}$$

We again see the utility of the Euler product; by taking the logarithmic derivative we will be able to relate sums over zeros of $L(s,\chi)$ to sums over primes of its coefficients.

Let g, ϕ, H be as in Theorem 18.1.6, and define

$$\mathcal{I} = \frac{1}{2\pi i}\int_{\Re s=1\frac{1}{2}} \frac{\Lambda'(s,\chi)}{\Lambda(s,\chi)}H(s)ds. \tag{18.25}$$

We shift contours from $\Re s = 1\frac{1}{2}$ to $-\frac{1}{2}$. This picks up poles coming from the zeros of $\Lambda(s,\chi)$. By a similar argument as in Lemma 18.1.1, it is clear that the only zeros of $\Lambda(s,\chi)$ are the zeros of $L(s,\chi)$ in the critical strip. The residue of $\frac{\Lambda'(s,\chi)}{\Lambda(s,\chi)}H(s)$ at a zero ρ is then equal to $H(\rho)$. Therefore

$$\mathcal{I} = \sum_\rho H(\rho) + \frac{1}{2\pi i}\int_{\Re s=-\frac{1}{2}} \frac{\Lambda'(s,\chi)}{\Lambda(s,\chi)}H(s)ds. \tag{18.26}$$

By (18.22), $\frac{\Lambda'(s,\chi)}{\Lambda(s,\chi)} = -\frac{\Lambda'(1-s,\overline{\chi})}{\Lambda(1-s,\overline{\chi})}$. Substituting this above and changing s to $1-s$ we obtain

$$\mathcal{I} = \sum_\rho H(\rho) - \frac{1}{2\pi i}\int_{\Re s=1\frac{1}{2}} \frac{\Lambda'(s,\overline{\chi})}{\Lambda(s,\overline{\chi})}H(1-s)ds. \tag{18.27}$$

We bring $\sum_\rho H(\rho)$ over to the left hand side, recall the definition of $\mathcal{I}$, and then shift the contour in the second integral to $\Re s = \frac{1}{2}$. As ϕ is even, note that for $s = \frac{1}{2} + iy$, $H(1-s) = H(s) = \phi(y)$. If $s = \frac{1}{2} + iy$, $ds = idy$, and we have

$$\sum_\rho H(\rho) = \frac{1}{2\pi i}\int_{\Re s=1\frac{1}{2}} \left[\frac{L'(s,\chi)}{L(s,\chi)}H(s) + \frac{L'(s,\overline{\chi})}{L(s,\overline{\chi})}H(1-s)\right]ds$$

$$+\frac{1}{\pi}\int_{-\infty}^{\infty}\left(\frac{1}{2}\log\left(\frac{m}{\pi}\right) + \frac{1}{2}\frac{\Gamma'(\frac{iy}{2}+\frac{1}{4})}{\Gamma(\frac{iy}{2}+\frac{1}{4})}\right)\phi(y)dy. \tag{18.28}$$

We now concentrate on the integrals involving $\frac{L'}{L}$. By an argument similar to before, we have

$$\frac{1}{2\pi i}\int_{\Re s = 1\frac{1}{2}} \frac{L'(s,\chi)}{L(s,\chi)} H(s)ds = \sum_p \sum_{k\geq 1} \frac{\chi(p^k)\log p}{p^{\frac{1}{2}k}} g(k\log p). \quad (18.29)$$

Similarly

$$\frac{1}{2\pi i}\int_{\Re s = 1\frac{1}{2}} \frac{L'(s,\overline{\chi})}{L(s,\overline{\chi})} H(1-s)ds = \sum_p \sum_{k\geq 1} \frac{\overline{\chi}(p^k)\log p}{p^{\frac{1}{2}k}} g(k\log p). \quad (18.30)$$

For a zero ρ of $L(s,\chi)$ we write $\rho = \frac{1}{2} + i\gamma_\rho$ and γ_ρ is a complex number. The truth of the Generalized Riemann Hypothesis would, of course, imply that $\gamma_\rho \in \mathbb{R}$. We have shown

Theorem 18.1.7 (Explicit Formula; First Form). *Let $L(s,\chi)$ be a Dirichlet L-function from a non-trivial character χ with conductor m and zeros $\rho = \frac{1}{2} + i\gamma$; if the Generalized Riemann Hypothesis is true then $\gamma \in \mathbb{R}$. Let $\sum_\rho$ denote the sum over the zeros of $L(s,\chi)$ in the critical strip. With g an even Schwartz function of compact support and ϕ as in (18.5),*

$$\sum_\rho \phi(\gamma_\rho) = -\sum_p \sum_{k=1}^\infty \frac{\log p}{p^{\frac{k}{2}}} g\left(k\log p\right)\left(\chi(p^k) + \overline{\chi}(p^k)\right)$$

$$+ \frac{1}{\pi}\int_{-\infty}^\infty \left(\frac{1}{2}\log\left(\frac{m}{\pi}\right) + \frac{1}{2}\frac{\Gamma'(\frac{iy}{2}+\frac{1}{4})}{\Gamma(\frac{iy}{2}+\frac{1}{4})}\right) \phi(y)dy. \quad (18.31)$$

As $\phi(r) = \widehat{g}\left(\frac{r}{2\pi}\right)$, if we let $g = \widehat{A}$ (this is possible because the Fourier transform is an isometry on the Schwartz space and g is even, so we can find a Schwartz function A satisfying the above), we may rewrite (18.31) as

$$\sum_\rho A\left(\frac{\gamma_\rho}{2\pi}\right) = -\sum_p \sum_{k=1}^\infty \frac{\log p}{p^{\frac{k}{2}}} \widehat{A}\left(k\log p\right)\left(\chi(p^k) + \overline{\chi}(p^k)\right)$$

$$+ \frac{1}{\pi}\int_{-\infty}^\infty \left(\frac{1}{2}\log\left(\frac{m}{\pi}\right) + \frac{1}{2}\frac{\Gamma'(\frac{iy}{2}+\frac{1}{4})}{\Gamma(\frac{iy}{2}+\frac{1}{4})}\right) A(y)dy \quad (18.32)$$

The left hand side of (18.32) is close to the 1-level density (see §15.6.2, in particular (15.50)). The difference is that we have not scaled the zeros of $L(s,\chi)$ in (18.32). As we have seen throughout the book, it is imperative that we investigate quantities on the correct scale.

Exercise 18.1.8. *Show (18.31) of Theorem 18.1.7 is true even when $\chi(-1) = -1$.*

Exercise 18.1.9. *Prove the analogue of Lemma 18.1.1 for Dirichlet L-functions. An analogue of the integral representation of $\zeta(s)$ in (3.21) might be useful.*

Exercise 18.1.10. *For h an even Schwartz function, let $A(x) = h(x\log R)$. Show*

$$\widehat{A}(y) = \frac{1}{\log R}\widehat{h}\left(\frac{y}{\log R}\right). \quad (18.33)$$

Using Exercise 18.1.10 with $R = \frac{m}{\pi}$ in (18.32) yields

Theorem 18.1.11 (Explicit Formula; Second Form). *Let h be an even Schwartz function and $L(s, \chi)$ a Dirichlet L-function from a non-trivial character χ with conductor m and zeros $\rho = \frac{1}{2} + i\gamma_\chi$; if the Generalized Riemann Hypothesis is true then $\gamma \in \mathbb{R}$. Then*

$$\sum_\rho h\left(\gamma_\rho \frac{\log\frac{m}{\pi}}{2\pi}\right)$$

$$= -\sum_p \sum_{k=1}^\infty \frac{\log p}{p^{\frac{k}{2}} \log\frac{m}{\pi}} \hat{h}\left(\frac{k\log p}{\log\frac{m}{\pi}}\right) (\chi(p^k) + \overline{\chi}(p^k))$$

$$+ \frac{1}{\pi} \int_{-\infty}^\infty \left(\frac{1}{2}\log\left(\frac{m}{\pi}\right) + \frac{1}{2}\frac{\Gamma'(\frac{iy}{2} + \frac{1}{4})}{\Gamma(\frac{iy}{2} + \frac{1}{4})}\right) h\left(\frac{y\log\frac{m}{\pi}}{2\pi}\right) dy. \qquad (18.34)$$

Note the left side of (18.34) is the 1-level density for the L-function $L(s, \chi)$. We are interested in the $m \to \infty$ limits. In this limit, many of the terms in (18.34) will not contribute.

Exercise 18.1.12. *As $|\chi(p^k)| \le 1$, show*

$$\sum_p \sum_{k\ge 3} \frac{\log p}{p^{\frac{k}{2}} \log\frac{m}{\pi}} h\left(\frac{k\log p}{\log\frac{m}{\pi}}\right) (\chi(p^k) + \overline{\chi}(p^k)) = O\left(\frac{1}{\log m}\right). \qquad (18.35)$$

Exercise 18.1.13. *Prove*

$$\frac{1}{\pi} \int_{-\infty}^\infty \frac{1}{2}\log\left(\frac{m}{\pi}\right) h\left(\frac{y\log\frac{m}{\pi}}{2\pi}\right) dy = \int_{-\infty}^\infty h(y)dy. \qquad (18.36)$$

Exercise 18.1.14. *Prove*

$$\frac{1}{\pi} \int_{-\infty}^\infty \frac{1}{2}\frac{\Gamma'(\frac{iy}{2} + \frac{1}{4})}{\Gamma(\frac{iy}{2} + \frac{1}{4})} h\left(\frac{y\log\frac{m}{\pi}}{2\pi}\right) dy = O\left(\frac{1}{\log m}\right). \qquad (18.37)$$

The following identity (see, for example, [Al]) may be useful:

$$\frac{\Gamma'(z)}{\Gamma(z)} = C + \log z - \frac{1}{2z} - \int_0^\infty \frac{2u}{u^2 + z^2}\frac{du}{e^{2\pi u} - 1}, \quad \Re z > 0. \qquad (18.38)$$

Combining the previous three exercises with (18.34) yields

Theorem 18.1.15 (Explicit Formula, Third Form). *Let h be an even Schwartz function and $L(s, \chi)$ a Dirichlet L-function from a non-trivial character χ with conductor m and zeros $\rho = \frac{1}{2} + i\gamma_\chi$; if the Generalized Riemann Hypothesis is true then $\gamma \in \mathbb{R}$. Then*

$$\sum_\rho h\left(\gamma_\rho \frac{\log(m/\pi)}{2\pi}\right) = -\sum_p \frac{\log p}{\log(m/\pi)} \hat{h}\left(\frac{\log p}{\log(m/\pi)}\right) [\chi(p) + \overline{\chi}(p)]p^{-\frac{1}{2}}$$

$$- \sum_p \frac{\log p}{\log(m/\pi)} \hat{h}\left(2\frac{\log p}{\log(m/\pi)}\right) [\chi^2(p) + \overline{\chi}^2(p)]p^{-1}$$

$$+ \int_{-\infty}^\infty h(y)dy + O\left(\frac{1}{\log m}\right). \qquad (18.39)$$

Remark 18.1.16. The explicit formula is a tool to study the zeros of L-functions; depending on what one wants to study, some formulations are more useful than others. Theorem 18.1.15 is no longer exact; however, as we only care about the limit as $m \to \infty$, for our purposes it is worthwhile to estimate certain terms. In some families of L-functions, these terms provide lower-order corrections to the "universal" random matrix theory answers. See also §18.3.

Remark 18.1.17. We changed notation to have a test function h evaluated at the scaled zeros and its Fourier transform $\widehat{h}$ evaluated at primes. It does not matter which which we call the test function and which the Fourier transform. As we are interested here in the zeros and not the primes, we chose notation to have the test function on the zero side.

In any problem it is essential that we determine the appropriate scale. The Explicit Formula suggests that the appropriate scaling of the zeros near the central point $s = \frac{1}{2}$ is to multiply the zeros by $\frac{\log(m/\pi)}{2\pi}$. See §18.3 for more on determining the appropriate scale.

18.2 DIRICHLET CHARACTERS FROM A PRIME CONDUCTOR

We calculate the 1-level density for the family of non-trivial Dirichlet characters of prime conductor m as $m \to \infty$. In order to talk about the spacings between adjacent zeros we need the all zeros to lie on the same line. This is because there is no way to sequentially order general points in the complex plane; see Exercise 18.2.1. We assume the Generalized Riemann Hypothesis for such interpretation purposes; however, at the cost of losing our interpretation of the 1-level density as a statistic about the spacings of zeros, we may drop this assumption.

This is the simplest of all such families, but it highlights the key features. The reader is strongly encouraged to review §15.6.2, especially Theorem 15.6.7. Assuming GRH, our result implies bounds for the number of characters such that $L(\frac{1}{2}, \chi) \neq 0$.

Exercise[h] 18.2.1. *Show there is no way to order the complex plane. Specifically, show there is no way to generalize $<$ to a transitive relation on $\mathbb{C}$: if $x < y$ and $y < z$ then $x < z$.*

18.2.1 Density Conjecture

The 1-level density is obtained by summing (18.39) over all $\chi \bmod m$, $\chi \neq \chi_0$. Assuming GRH, the zeros can be written as $\frac{1}{2} + i\gamma_\chi$ with $\gamma_\chi \in \mathbb{R}$. As m is prime there are $\phi(m) - 1 = m - 2$ such characters. Denoting this family by $\mathcal{F}_m$ we have

$$D_{\mathcal{F}_m}(h) = \frac{1}{m-2} \sum_{\substack{\chi \bmod m \\ \chi \neq \chi_0}} \sum_{\gamma_\chi} h\left(\gamma_\chi \frac{\log(m/\pi)}{2\pi}\right)$$

$$= -\frac{1}{m-2} \sum_{\substack{\chi \bmod m \\ \chi \neq \chi_0}} \sum_p \frac{\log p}{\log(m/\pi)} \hat{h}\left(\frac{\log p}{\log(m/\pi)}\right) [\chi(p) + \overline{\chi}(p)] p^{-\frac{1}{2}}$$

$$-\frac{1}{m-2} \sum_{\substack{\chi \bmod m \\ \chi \neq \chi_0}} \sum_p \frac{\log p}{\log(m/\pi)} \hat{h}\left(2\frac{\log p}{\log(m/\pi)}\right) [\chi^2(p) + \overline{\chi}^2(p)] p^{-1}$$

$$+ \int_{-\infty}^{\infty} h(y) dy + O\left(\frac{1}{\log m}\right). \tag{18.40}$$

The Density Conjecture states that the family average should converge to the Unitary Density. In [KS2, HuRu] it is shown that the 1-level density of $N \times N$ unitary matrices as $N \to \infty$ converges to

$$\int_{-\infty}^{\infty} h(y) dy. \tag{18.41}$$

We do not say *why* we expect the behavior of the low zeros of the Dirichlet L-functions to have unitary symmetry; we merely show they do. Thus we must show the two prime sums in (18.40) tend to zero as $m \to \infty$; we call these the **First Sum** S_1 and **Second Sum** S_2.

18.2.2 Review of Dirichlet Characters

We quickly review Dirichlet characters; see §3.3 for details. As m is prime, $(\mathbb{Z}/m\mathbb{Z})^*$ is cyclic of order $m-1$ with generator g (so any element is of the form g^a for some a). Let $\zeta_{m-1} = e^{2\pi i/(m-1)}$. The principal character χ_0 is given by

$$\chi_0(k) = \begin{cases} 1 & \text{if } (k,m) = 1, \\ 0 & \text{if } (k,m) > 1. \end{cases} \tag{18.42}$$

Each of the $m - 2$ primitive (i.e., non-trivial) characters are determined (because they are multiplicative) once their action on a generator g is specified. As each $\chi : (\mathbb{Z}/m\mathbb{Z})^* \to \mathbb{C}^*$, for each χ there exists an l such that $\chi(g) = \zeta_{m-1}^l$. Hence for each $l \in \{1, \dots, m-2\}$ we have

$$\chi_l(k) = \begin{cases} \zeta_{m-1}^{la} & \text{if } k \equiv g^a \bmod m, \\ 0 & \text{if } (k,m) > 0. \end{cases} \tag{18.43}$$

Hence $\{\chi_l\}_{1 \leq l \leq m-2}$ are all the non-trivial characters mod m, and as each χ_l is primitive, we may use the explicit formula (Theorem 18.39) to obtain (18.40). The analysis is completed by showing the First and Second Sums tend to zero as $m \to \infty$.

18.2.3 The First Sum

We must analyze (for m prime)

$$S_1 = \frac{1}{m-2} \sum_{\chi \neq \chi_0} \sum_p \frac{\log p}{\log(m/\pi)} \widehat{h}\left(\frac{\log p}{\log(m/\pi)}\right) [\chi(p) + \overline{\chi}(p)] p^{-\frac{1}{2}}. \quad (18.44)$$

As remarked, the explicit formula is useless if we do not have a good Averaging Formula. In random matrix theory, the method of choosing the matrices led to formulas to deal with integrals of products of the matrix coefficients. Here the analogue is Lemma 3.3.15, which gives

$$\sum_{\chi \bmod m} \chi(k) = \begin{cases} m-1 & \text{if } k \equiv 1 \bmod m \\ 0 & \text{otherwise,} \end{cases} \quad (18.45)$$

which allows us to sum the coefficients $\chi(p)$. Therefore, for any prime $p \neq m$,

$$\sum_{\chi \neq \chi_0} \chi(p) = \begin{cases} m-2 & \text{if } p \equiv 1 \bmod m \\ -1 & \text{otherwise,} \end{cases} \quad (18.46)$$

and the same result holds for summing $\overline{\chi}(p)$. Let

$$\delta_m(p, 1) = \begin{cases} 1 & \text{if } p \equiv 1 \bmod m \\ 0 & \text{otherwise.} \end{cases} \quad (18.47)$$

The contribution to S_1 from $p = m$ is zero; thus if we substitute -1 for $\sum_{\chi \neq \chi_0} \chi(m)$, our error is $O(\frac{1}{\log m})$ and hence negligible.

We now calculate S_1, suppressing the errors of $O(\frac{1}{\log m})$. Note $\widehat{h}$ is an even Schwartz function with support in $(-\sigma, \sigma)$. This implies that if $|y| > \sigma$, $\widehat{h}(y) = 0$. Hence while the sums are initially over all primes, $\widehat{h}$ being of compact support restricts the prime sums to $p \leq \left(\frac{m}{\pi}\right)^\sigma$.

$$\begin{aligned}
S_1 &= \frac{1}{m-2} \sum_{\chi \neq \chi_0} \sum_p \frac{\log p}{\log(m/\pi)} \widehat{h}\left(\frac{\log p}{\log(m/\pi)}\right) [\chi(p) + \overline{\chi}(p)] p^{-\frac{1}{2}} \\
&= \frac{1}{m-2} \sum_p \frac{\log p}{\log(m/\pi)} \widehat{h}\left(\frac{\log p}{\log(m/\pi)}\right) \sum_{\chi \neq \chi_0} [\chi(p) + \overline{\chi}(p)] p^{-\frac{1}{2}}.
\end{aligned}$$
$$(18.48)$$

Remark 18.2.2. We have interchanged the summation over the characters (or L-functions) in our family and the summation over the primes. Explicitly, we have passed the sum over the family of Dirichlet L-functions *past* the summation over primes and then *past* the test functions *to* the coefficients in the L-series. The point of interchanging the order of summation is that we can exploit averaging formulas for the sums of the coefficients over the family; this is a common trick in number theory.

If the Dirichlet characters had different conductors, this would not be possible. While in many families the conductors are constant (see for example [ILS]), this is not always the case, and can cause technical complications (see [Mil1]).

Continuing the proof, we find

$$
\begin{aligned}
S_1 &= \frac{2}{m-2} \sum_p \frac{\log p}{\log(m/\pi)} \widehat{h}\left(\frac{\log p}{\log(m/\pi)}\right) p^{-\frac{1}{2}}(-1 + (m-1)\delta_m(p,1)) \\
&= \frac{-2}{m-2} \sum_p^{m^\sigma} \frac{\log p}{\log(m/\pi)} \widehat{h}\left(\frac{\log p}{\log(m/\pi)}\right) p^{-\frac{1}{2}} \\
&\quad + 2\frac{m-1}{m-2} \sum_{p\equiv 1(m)}^{m^\sigma} \frac{\log p}{\log(m/\pi)} \widehat{h}\left(\frac{\log p}{\log(m/\pi)}\right) p^{-\frac{1}{2}} \\
&\ll \frac{1}{m} \sum_p^{m^\sigma} p^{-\frac{1}{2}} + \sum_{p\equiv 1(m)}^{m^\sigma} p^{-\frac{1}{2}} \\
&\ll \frac{1}{m} \sum_k^{m^\sigma} k^{-\frac{1}{2}} + \sum_{\substack{k\equiv 1(m) \\ k \geq m+1}}^{m^\sigma} k^{-\frac{1}{2}} \\
&\ll \frac{1}{m} \sum_k^{m^\sigma} k^{-\frac{1}{2}} + \frac{1}{m} \sum_k^{m^\sigma} k^{-\frac{1}{2}} \\
&\ll \frac{1}{m} m^{\sigma/2}.
\end{aligned}
\tag{18.49}
$$

Remark 18.2.3. One must be careful with the estimates of the second sum. Each residue class of $k \bmod m$ has approximately the same sum, with the difference between two classes bounded by the first term of whichever class has the smallest element. Since we are dropping the first term $(k = 1)$ because 1 is not prime, the class of $k \equiv 1(m)$ has the smallest sum of the m classes. Hence if we add all the classes and divide by m we increase the sum, so the above arguments are valid.

We have shown $S_1 = \frac{1}{m}m^{\sigma/2} + O\left(\frac{1}{\log m}\right)$, implying that there is no contribution from the First Sum if $\sigma < 2$. If the limiting behavior agrees with Random Matrix Theory, this sum should be negligible for *all* σ and not just $\sigma < 2$. Such results are common in the field: generally, our averaging formulas lead to sums that we can only evaluate or bound in certain regions. See [ILS] for another family of L-functions where more refined analysis of averaging formulas lead to improved results.

The above arguments are very crude: we inserted absolute values, and did not take into account that there was a weight of -1 attached to all primes, and a weight of $+\phi(m)$ to primes congruent to 1 mod m. Exploiting this should lead to S_1 being small for larger σ; unfortunately, even assuming GRH one does not have the required cancellation. See Remark 3.3.1 and [GV, Mon1, Va] and the references there for more on how the primes are distributed among the different $\phi(m)$ residue classes, and [Mil2] for how such bounds and conjectures can be used to extend the support.

18.2.4 The Second Sum

We must analyze (for m prime)

$$S_2 = \frac{1}{m-2} \sum_{\chi \neq \chi_0} \sum_p \frac{\log p}{\log(m/\pi)} \hat{h}\left(2\frac{\log p}{\log(m/\pi)}\right) [\chi^2(p) + \overline{\chi}^2(p)] p^{-1}. \quad (18.50)$$

If $p \equiv \pm 1(m)$ then $\sum_{\chi \neq \chi_0} [\chi^2(p) + \overline{\chi}^2(p)] = 2(m-2)$. If $p \not\equiv \pm 1 \bmod m$ then fix a generator g and write $p \equiv g^a(m)$. As $p \not\equiv \pm 1$, $a \not\equiv 0, \frac{m-1}{2} \bmod (m-1)$, as $(\mathbb{Z}/m\mathbb{Z})^*$ is cyclic of order $m-1$. Hence $e^{4\pi i a/(m-1)} \neq 1$. Recall $\zeta_{m-1} = e^{2\pi i/(m-1)}$. Note $\zeta_{m-1}^{2a} = e^{4\pi i a/(m-1)} \neq 1$. By the geometric series formula

$$\sum_{\chi \neq \chi_0} [\chi^2(p) + \overline{\chi}^2(p)] = \sum_{l=1}^{m-2} [\chi_l^2(p) + \overline{\chi}_l^2(p)]$$

$$= \sum_{l=1}^{m-2} [\chi_l^2(g^a) + \overline{\chi}_l^2(g^a)]$$

$$= \sum_{l=1}^{m-2} [(\chi_l(g))^{2a} + (\overline{\chi}_l(g))^{2a}]$$

$$= \sum_{l=1}^{m-2} [(\zeta_{m-1}^l)^{2a} + (\zeta_{m-1}^l)^{-2a}]$$

$$= \sum_{l=1}^{m-2} [(\zeta_{m-1}^{2a})^l + (\zeta_{m-1}^{-2a})^l]$$

$$= \frac{\zeta_{m-1}^{2a} - 1}{1 - \zeta_{m-1}^{2a}} + \frac{\zeta_{m-1}^{-2a} - 1}{1 - \zeta_{m-1}^{-2a}} = -2. \quad (18.51)$$

The contribution to the sum from $p = m$ is zero; if instead we substitute -2 for $\sum_{\chi \neq \chi_0} \chi^2(m)$, our error is $O(\frac{1}{\log m})$ and hence negligible. Therefore

$$\sum_{\chi \neq \chi_0} [\chi^2(p) + \overline{\chi}^2(p)] = \begin{cases} 2(m-2) & \text{if } p \equiv \pm 1 \bmod m \\ -2 & \text{othwerwise.} \end{cases} \quad (18.52)$$

Let

$$\delta_m(p, \pm) = \begin{cases} 1 & \text{if } p \equiv \pm 1 \bmod m \\ 0 & \text{otherwise.} \end{cases} \quad (18.53)$$

Up to $O\left(\frac{1}{\log m}\right)$ we find that

$$
\begin{aligned}
S_2 &= \frac{1}{m-2} \sum_{\chi \neq \chi_0} \sum_p \frac{\log p}{\log(m/\pi)} \widehat{h}\left(2\frac{\log p}{\log(m/\pi)}\right) [\chi^2(p) + \overline{\chi}^2(p)] p^{-1} \\
&= \frac{1}{m-2} \sum_p \frac{\log p}{\log(m/\pi)} \widehat{h}\left(2\frac{\log p}{\log(m/\pi)}\right) \sum_{\chi \neq \chi_0} [\chi^2(p) + \overline{\chi}^2(p)] p^{-1} \\
&= \frac{1}{m-2} \sum_p^{m^{\sigma/2}} \frac{\log p}{\log(m/\pi)} \widehat{h}\left(2\frac{\log p}{\log(m/\pi)}\right) p^{-1}[-2 + (2m-2)\delta_m(p, \pm)] \\
&\ll \frac{1}{m-2} \sum_p^{m^{\sigma/2}} p^{-1} + \frac{2m-2}{m-2} \sum_{p \equiv 1(m)}^{m^{\sigma/2}} p^{-1} \\
&\ll \frac{1}{m-2} \sum_k^{m^{\sigma/2}} k^{-1} + \sum_{\substack{k \equiv 1(m) \\ k \geq m+1}}^{m^{\sigma/2}} k^{-1} + \sum_{\substack{k \equiv -1(m) \\ k \geq m-1}}^{m^{\sigma/2}} k^{-1} \\
&\ll \frac{1}{m-2} \log(m^{\sigma/2}) + \frac{1}{m} \sum_k^{m^{\sigma/2}} k^{-1} + \frac{1}{m} \sum_k^{m^{\sigma/2}} k^{-1} + O(\frac{1}{m}) \\
&\ll \frac{\log m}{m} + \frac{\log m}{m} + \frac{\log m}{m}. \tag{18.54}
\end{aligned}
$$

Therefore $S_2 = O(\frac{\log m}{m})$, so for all σ there is no contribution.

Exercise[h] 18.2.4. *Justify the bounds in (18.54) from the sums of $k \equiv \pm 1$ mod m.*

18.2.5 1-Level Density for Prime Conductors

Using our bounds on the sums S_1 and S_2 in the Explicit Formula (Theorem 18.1.15), we immediately obtain

Theorem 18.2.5 (1-Level Density for Prime Conductors). *Let $\widehat{h}$ be an even Schwartz function with $\mathrm{supp}(\widehat{h}) \subset (-2, 2)$, m a prime and $\mathcal{F}_m = \{\chi : \chi$ is non-trivial mod $m\}$; note $|\mathcal{F}_m| = m - 2$. Write the zeros of $L(s, \chi)$ as $\frac{1}{2}_i \gamma_\chi$. Assuming GRH we have $\gamma_\chi \in \mathbb{R}$, and*

$$
\frac{1}{|\mathcal{F}_m|} \sum_{\chi \in \mathcal{F}_m} \sum_{\gamma_\chi} h\left(\gamma_\chi \frac{\log(m/\pi)}{2\pi}\right) = \int_{-\infty}^\infty h(y)dy + O\left(\frac{1}{\log m}\right). \tag{18.55}
$$

As $\int_{-\infty}^\infty h(y)dy = \widehat{h}(0)$, we may rewrite the above as

$$
\frac{1}{|\mathcal{F}_m|} \sum_{\chi \in \mathcal{F}_m} \sum_{\gamma_\chi} h\left(\gamma_\chi \frac{\log(m/\pi)}{2\pi}\right) = \widehat{h}(0) + O\left(\frac{1}{\log m}\right). \tag{18.56}
$$

The above result agrees with random matrix theory. Instead of considering all non-trivial Dirichlet characters with prime conductor m, one could consider m

square-free (see [Mil2]), and again the density is seen to agree with unitary matrices. One could also consider quadratic Dirichlet characters (see [Rub2]). In this case, the density agrees not with unitary matrices, but symplectic. Here S_1 still does not contribute for small support, but S_2 *does* have a non-zero contribution. This is not surprising as for quadratic characters $\chi(p)^2 = 1$ whenever $p \neq m$. Thus the second sum will be different. This is reminiscent of the behavior of eigenvalues from d-regular random graphs. There we also had a thin subset, and saw different behavior. See §15.5.1.

An enormous advantage of studying all non-trivial Dirichlet characters with prime conductor is that we have terrific averaging formulas (compare Lemma 3.3.15 to Remark 15.3.15). In fact, more is true: we also have simple, tractable formulas for the characters. Such detail is not available in general and more sophisticated tools are needed to obtain more complicated and weaker averaging formulas.

Remark 18.2.6. GRH is assumed only for interpretation purposes in Theorem 18.2.5. If we do not assume GRH, the sum on the left side of (18.55) is over all zeros in the critical strip. If GRH fails, we can no longer write the zeros in increasing order on a given line. If the zeros are all on a line then there is a powerful analogy with Random Matrix Theory, as the eigenvalues of real symmetric, complex Hermitian and unitary matrices can be written in terms of a real parameter.

An important application of the 1-level density is in bounding the number of characters that have zeros at the central point $s = \frac{1}{2}$. For a character χ let m_χ denote the multiplicity of the zero at $s = \frac{1}{2}$; if there are no zeros then $m_\chi = 0$. Let

$$z_\chi = \begin{cases} 1 & \text{if } m_\chi \geq 1 \text{ (i.e., there is a zero at } s = \frac{1}{2}), \\ 0 & \text{if } m_\chi = 0 \text{ (i.e., there is no zero at } s = \frac{1}{2}). \end{cases} \qquad (18.57)$$

Theorem 18.2.7. *Assuming GRH, as $m \to \infty$ through prime values a positive percent of non-trivial χ have $L(\frac{1}{2}, \chi) \neq 0$; in other words, a positive percent of $L(s, \chi)$ do not vanish at the central point.*

Proof. Let $h(x)$ be a non-negative test function such that $h(0) > 0$ and $\operatorname{supp}(\widehat{h}) \subset (-2, 2)$. As h is non-negative we have

$$\frac{1}{|\mathcal{F}_m|} \sum_{\chi \in \mathcal{F}_m} m_\chi h(0) \leq \frac{1}{|\mathcal{F}_m|} \sum_{\chi \in \mathcal{F}_m} \sum_{\gamma_\chi} h\left(\gamma_\chi \frac{\log(m/\pi)}{2\pi}\right); \qquad (18.58)$$

this follows immediately because the sum over the zeros not at $s = \frac{1}{2}$ is non-negative, so removing their contribution gives a smaller answer. Thus by (18.56) we have

$$\frac{1}{|\mathcal{F}_m|} \sum_{\chi \in \mathcal{F}_m} m_\chi h(0) \leq \widehat{h}(0) + O\left(\frac{1}{\log m}\right). \qquad (18.59)$$

As $z_\chi \leq m_\chi$ and $h(0) > 0$ we immediately find that

$$\frac{1}{|\mathcal{F}_m|} \sum_{\chi \in \mathcal{F}_m} z_\chi \leq \frac{\widehat{h}(0)}{h(0)} + O\left(\frac{1}{\log m}\right). \qquad (18.60)$$

The left hand side is just the percent of non-trivial characters χ with prime conductor m with a zero at the central point $s = \frac{1}{2}$. Thus if we can find an h (satisfying the above conditions) such that $\frac{\widehat{h}(0)}{h(0)} < 1$ then as $m \to \infty$ through prime values, assuming GRH we will have shown a positive percent of such $L(s, \chi)$ are non-zero at the central point! See Exercise 18.2.8 for such an h. $\qquad\qquad\square$

See Appendix A of [ILS] for a discussion of good choices of test functions h satisfying conditions like those above and [IS1, IS2] for applications and other results concerning non-vanishing of L-functions at the central point. The above argument is another example of a common technique, positivity. If h was not a non-negative function then we would not be able to deduce non-vanishing bounds. See §2.2 for more on the interplay between positivity, average order and non-vanishing.

Exercise 18.2.8. *We give a satisfactory choice for the test function. Let* $h_\sigma(x) = \left(\frac{\sin(2\pi \frac{\sigma}{2} x)}{2\pi x}\right)^2$ *for any* $\sigma < 2$. *Show* $\widehat{h_\sigma}(u) = \frac{1}{4}(\sigma - |u|)$ *for* $|u| \leq \sigma$ *and* 0 *otherwise. Thus* $\mathrm{supp}(\widehat{h_\sigma}) \subset (-2, 2)$ *if* $\sigma < 2$, *and* h_σ *satisfies the conditions of Theorem 18.2.5. Show this implies that the percent of* $\chi \in \mathcal{F}_m$ *such that* $L(\frac{1}{2}, \chi) = 0$ *is at most* $\frac{1}{\sigma} + O\left(\frac{1}{\log m}\right)$. *Thus we can show that, in the limit, about half of the* $L(\frac{1}{2}, \chi)$ *do not vanish.*

Exercise 18.2.9. *It is conjectured that Theorem 18.2.5 holds for any test function* h *of compact support. Assume as* $m \to \infty$ *through prime values that we may take* $\mathrm{supp}(\widehat{h})$ *to be arbitrarily large; i.e., there is a monotone increasing function* $g(m)$ *such that* $\lim_{m\to\infty} g(m) = \infty$ *and for any* m *we may take* $\mathrm{supp}(\widehat{h}) \subset (-g(m), g(m))$. *Prove this implies that for any* $\epsilon > 0$ *that as* $m \to \infty$ *through prime values, we have at least* $(100 - \epsilon)\%$ *of the* $L(s, \chi)$ *are non-zero at* $s = \frac{1}{2}$ *for non-trivial characters* χ *of conductor* m.

Remark 18.2.10. It is worthwhile commenting on our assumptions. While assuming GRH does give us a lot of information about the location of the zeros, it is not inconsistent with GRH for our L-functions to vanish at the central point (just unexpected), and hence our non-vanishing results are non-trivial observations. Further, if GRH were false and there were zeros off the critical line $\Re s = \frac{1}{2}$, then $h\left(\gamma_\chi \frac{\log(m/\pi)}{2\pi}\right)$ is no longer necessarily non-negative (or even real!). Thus even though we can still show that sums over zeros are behaving like sums over eigenvalues if GRH is false, we cannot deduce certain consequences of this correspondence that we would like.

Exercise 18.2.11. *Generalize these arguments to study* $\mathcal{F}_{N,\mathrm{quad}}$, *the family of all Dirichlet L-functions of a quadratic character with modulus m, $m \in [N, 2N]$ prime. A* **quadratic character** *χ of modulus m is such that $\chi(n)^2 = 1$ for $(n, m) = 1$. For suitably restricted test functions, the answer is*

$$\int_{-\infty}^{\infty} \phi(y)\,dy - \frac{1}{2}\phi(0) + O\left(\frac{1}{\log m}\right). \qquad (18.61)$$

Note this is different from the family of all Dirichlet L-functions with modulus m. Those had unitary symmetry; these have symplectic symmetry. See [Rub2] for complete proofs.

Exercise 18.2.12. *Generalize these arguments to handle $\mathcal{F}_{N,\text{sq-free}}$, the family of all Dirichlet L-functions with characters of modulus m, $m \in [N, 2N]$ square-free. See [Mil2] for complete details. Hint: First study m is a product of r primes, with r fixed and $m \to \infty$.*

Research Project 18.2.13. Generalize these arguments to handle the case of m not square-free. The functional equation and explicit formula are more complicated in this case. See [Da2].

18.3 SUMMARY OF CALCULATIONS

The Explicit Formula (see §18.1) is the Number Theory analogue of the Eigenvalue Trace Formula (Theorem 15.2.1). We wanted to study the eigenvalues of matrices; however, it was the matrix elements that we chose at random, and it was the matrix elements where we had good averaging formulas. The Eigenvalue Trace Formula allowed us to pass from knowledge of the matrix entries to knowledge of the eigenvalues. The Explicit Formula plays a similar role. We want to understand the zeros of an L-function, but our summation formulas are for the coefficients of the L-functions. As the Explicit Formula connects sums over zeros to sums over the coefficients, this allows us to pass from our knowledge of sums of the coefficients to properties of the zeros.

Like the Eigenvalue Trace Formula, the Explicit Formula is *the* starting point of all such investigations; to obtain useful results we then need to input summation results on the coefficients. For Dirichlet L-functions it is Lemma 3.3.15. This is the analogue of integrating the matrix coefficients against the probability density; unfortunately, the summation formulas for Number Theory are not as tractable as those from Random Matrix Theory, and this leads to restrictions on the results for L-functions. To go beyond these natural boundaries requires better summation formulations, which will involve deeper arithmetic relations (see for example Remark 3.3.1).

Finally, it is important to determine what the correct or natural scale is for each problem. When studying $x_n = n^k \alpha \mod 1$ in Chapter 12 we saw that if we looked at $x_1, \ldots, x_N$, then the distance between typical adjacent x_n's is about $\frac{1}{N}$; letting $y_1, \ldots, y_N$ be the x_n's in increasing order, we see the natural quantity to study is $\frac{y_{i+1} - y_i}{1/N}$. Similarly we have seen in §8.2.3 and §15.2.2 that for primes we should study $\frac{\log p_{i+1} - \log p_i}{\log p_i}$. In Random Matrix Theory we saw in §15.2.2 that the Central Limit Theorem and the Eigenvalue Trace Formula imply that the eigenvalues from our ensembles of real symmetric matrices are of size $\sqrt{N}$; thus we should study $\frac{\lambda_i(A)}{\sqrt{N}}$. Thus in order to study the zeros of Dirichlet L-functions of conductor m as $m \to \infty$, we need to figure out what the appropriate scale is. In the third version of the Explicit Formula, Theorem 18.1.15, we see it is reasonable to study zeros near the central point by scaling by $\frac{\log(m/\pi)}{2\pi}$; thus we study not the imaginary part of the zeros, γ, but rather $\gamma \frac{\log(m/\pi)}{2\pi}$. Again we see how the Explicit Formula is playing a similar role as the Eigenvalue Trace Lemma as both suggest the appropriate scaling.

Appendix A

Analysis Review

A.1 PROOFS BY INDUCTION

Assume for each positive integer n we have a statement $P(n)$ which we desire to show is true. $P(n)$ is true for all positive integers n if the following two statements hold:

- **Basis Step:** $P(1)$ is true;

- **Inductive Step**: whenever $P(n)$ is true, $P(n+1)$ is true.

This technique is called **Proof by Induction**, and is a very useful method for proving results; we shall see many instances of this in this appendix and Chapter 1 (indeed, throughout much of the book). The reason the method works follows from basic logic. We assume the following two sentences are true:

$$P(1) \text{ is true}$$
$$\forall n \geq 1, P(n) \text{ is true implies } P(n+1) \text{ is true.} \qquad (A.1)$$

Set $n = 1$ in the second statement. As $P(1)$ is true, and $P(1)$ implies $P(2)$, $P(2)$ must be true. Now set $n = 2$ in the second statement. As $P(2)$ is true, and $P(2)$ implies $P(3)$, $P(3)$ must be true. And so on, completing the proof. Verifying the first statement the **basis step** and the second the **inductive step**. In verifying the inductive step, note we assume $P(n)$ is true; this is called the **inductive assumption**. Sometimes instead of starting at $n = 1$ we start at $n = 0$, although in general we could start at any n_0 and then prove for all $n \geq n_0$, $P(n)$ is true.

We give three of the more standard examples of proofs by induction, and one false example; the first example is the most typical.

A.1.1 Sums of Integers

Let $P(n)$ be the statement

$$\sum_{k=1}^{n} k = \frac{n(n+1)}{2}. \qquad (A.2)$$

Basis Step: $P(1)$ is true, as both sides equal 1.
Inductive Step: Assuming $P(n)$ is true, we must show $P(n+1)$ is true. By the

inductive assumption, $\sum_{k=1}^{n} k = \frac{n(n+1)}{2}$. Thus

$$\sum_{k=1}^{n+1} k = (n+1) + \sum_{k=1}^{n} k$$

$$= (n+1) + \frac{n(n+1)}{2}$$

$$= \frac{(n+1)(n+1+1)}{2}. \tag{A.3}$$

Thus, given $P(n)$ is true, then $P(n+1)$ is true.

Exercise A.1.1. *Prove*

$$\sum_{k=1}^{n} k^2 = \frac{n(n+1)(2n+1)}{6}. \tag{A.4}$$

Find a similar formula for the sum of k^3. See also Exercise 1.2.4.

Exercise A.1.2. *Show the sum of the first n odd numbers is n^2, i.e.,*

$$\sum_{k=1}^{n} (2k-1) = n^2. \tag{A.5}$$

Remark A.1.3. We define the empty sum to be 0, and the empty product to be 1. For example, $\sum_{n\in\mathbb{N}, n<0} 1 = 0$.

See [Mil4] for an alternate derivation of sums of powers that does not use induction.

A.1.2 Divisibility

Let $P(n)$ be the statement 133 divides $11^{n+1} + 12^{2n-1}$.

Basis Step: A straightforward calculation shows $P(1)$ is true: $11^{1+1} + 12^{2-1} = 121 + 12 = 133$.
Inductive Step: Assume $P(n)$ is true, i.e., 133 divides $11^{n+1} + 12^{2n-1}$. We must show $P(n+1)$ is true, or that 133 divides $11^{(n+1)+1} + 12^{2(n+1)-1}$. But

$$11^{(n+1)+1} + 12^{2(n+1)-1} = 11^{n+1+1} + 12^{2n-1+2}$$

$$= 11 \cdot 11^{n+1} + 12^2 \cdot 12^{2n-1}$$

$$= 11 \cdot 11^{n+1} + (133 + 11)12^{2n-1}$$

$$= 11 \left(11^{n+1} + 12^{2n-1}\right) + 133 \cdot 12^{2n-1}. \tag{A.6}$$

By the inductive assumption 133 divides $11^{n+1} + 12^{2n-1}$; therefore, 133 divides $11^{(n+1)+1} + 12^{2(n+1)-1}$, completing the proof.

Exercise A.1.4. *Prove 4 divides $1 + 3^{2n+1}$.*

A.1.3 The Binomial Theorem

We prove the Binomial Theorem. First, recall that

Definition A.1.5 (Binomial Coefficients). *Let n and k be integers with $0 \le k \le n$. We set*

$$\binom{n}{k} = \frac{n!}{k!(n-k)!}. \tag{A.7}$$

Note that $0! = 1$ and $\binom{n}{k}$ is the number of ways to choose k objects from n (with order not counting).

Lemma A.1.6. *We have*

$$\binom{n}{k} = \binom{n}{n-k}, \quad \binom{n}{k} + \binom{n}{k-1} = \binom{n+1}{k}. \tag{A.8}$$

Exercise A.1.7. *Prove Lemma A.1.6.*

Theorem A.1.8 (The Binomial Theorem). *For all positive integers n we have*

$$(x+y)^n = \sum_{k=0}^{n} \binom{n}{k} x^{n-k} y^k. \tag{A.9}$$

Proof. We proceed by induction.
Basis Step: For $n = 1$ we have

$$\sum_{k=0}^{1} \binom{1}{k} x^{1-k} y^k = \binom{1}{0} x + \binom{1}{1} y = (x+y)^1. \tag{A.10}$$

Inductive Step: Suppose

$$(x+y)^n = \sum_{k=0}^{n} \binom{n}{k} x^{n-k} y^k. \tag{A.11}$$

Then using Lemma A.1.6 we find that

$$
\begin{aligned}
(x+y)^{n+1} &= (x+y)(x+y)^n \\
&= (x+y) \sum_{k=0}^{n} \binom{n}{k} x^{n-k} y^k \\
&= \sum_{k=0}^{n} \binom{n}{k} x^{n+1-k} y^k + \binom{n}{k} x^{n-k} y^{k+1} \\
&= x^{n+1} + \sum_{k=1}^{n} \left\{ \binom{n}{k} + \binom{n}{k-1} \right\} x^{n+1-k} y^k + y^{n+1} \\
&= \sum_{k=0}^{n+1} \binom{n+1}{k} x^{n+1-k} y^k. \tag{A.12}
\end{aligned}
$$

This establishes the induction step, and hence the theorem. $\qquad \square$

A.1.4 False Proofs by Induction

Consider the following: let $P(n)$ be the statement that in any group of n people, everyone has the same name. We give a (false!) proof by induction that $P(n)$ is true for all n!

Basis Step: Clearly, in any group with just 1 person, every person in the group has the same name.

Inductive Step: Assume $P(n)$ is true, namely, in any group of n people, everyone has the same name. We now prove $P(n+1)$. Consider a group of $n+1$ people:

$$\{1, 2, 3, \ldots, n-1, n, n+1\}. \tag{A.13}$$

The first n people form a group of n people; by the inductive assumption, they all have the same name. So, the name of 1 is the same as the name of 2 is the same as the name of 3 $\ldots$ is the same as the name of n.

Similarly, the last n people form a group of n people; by the inductive assumption they all have the same name. So, the name of 2 is the same as the name of 3 $\ldots$ is the same as the name of n is the same as the name of $n+1$. Combining yields everyone has the same name! Where is the error?

If $n = 4$, we would have the set $\{1, 2, 3, 4, 5\}$, and the two sets of 4 people would be $\{1, 2, 3, 4\}$ and $\{2, 3, 4, 5\}$. We see that persons 2, 3 and 4 are in both sets, providing the necessary link.

What about smaller n? What if $n = 1$? Then our set would be $\{1, 2\}$, and the two sets of 1 person would be $\{1\}$ and $\{2\}$; there is no overlap! The error was that we assumed n was "large" in our proof of $P(n) \Rightarrow P(n+1)$.

Exercise A.1.9. *Show the above proof that $P(n)$ implies $P(n+1)$ is correct for $n \geq 2$, but fails for $n = 1$.*

Exercise A.1.10. *Similar to the above, give a false proof that any sum of k integer squares is an integer square, i.e., $x_1^2 + \cdots + x_n^2 = x^2$. In particular, this would prove all positive integers are squares as $m = 1^2 + \cdots + 1^2$.*

Remark A.1.11. There is no such thing as *Proof By Example*. While it is often useful to check a special case and build intuition on how to tackle the general case, checking a few examples is not a proof. For example, because $\frac{16}{64} = \frac{1}{4}$ and $\frac{19}{95} = \frac{1}{5}$, one might think that in dividing two digit numbers if two numbers on a diagonal are the same one just cancels them. If that were true, then $\frac{12}{24}$ should be $\frac{1}{4}$. Of course this is *not* how one divides two digit numbers!

A.2 CALCULUS REVIEW

We briefly review some of the results from Differential and Integral Calculus. We recall some notation: $[a, b] = \{x : a \leq x \leq b\}$ is the set of all x between a and b, including a and b; $(a, b) = \{x : a < x < b\}$ is the set of all x between a and b, not including the endpoints a and b. For a review of continuity see §A.3.

A.2.1 Intermediate Value Theorem

Theorem A.2.1 (Intermediate Value Theorem (IVT)). *Let f be a continuous function on $[a, b]$. For all C between $f(a)$ and $f(b)$ there exists a $c \in [a, b]$ such that $f(c) = C$. In other words, all intermediate values of a continuous function are obtained.*

Sketch of the proof. We proceed by **Divide and Conquer**. Without loss of generality, assume $f(a) < C < f(b)$. Let x_1 be the midpoint of $[a, b]$. If $f(x_1) = C$ we are done. If $f(x_1) < C$, we look at the interval $[x_1, b]$. If $f(x_1) > C$ we look at the interval $[a, x_1]$.

In either case, we have a new interval, call it $[a_1, b_1]$, such that $f(a_1) < C < f(b_1)$ and the interval has half the size of $[a, b]$. We continue in this manner, repeatedly taking the midpoint and looking at the appropriate half-interval.

If any of the midpoints satisfy $f(x_n) = C$, we are done. If no midpoint works, we divide infinitely often and obtain a sequence of points x_n in intervals $[a_n, b_n]$. This is where rigorous mathematical analysis is required (see §A.3 for a brief review, and [Rud] for complete details) to show x_n converges to an $x \in (a, b)$.

For each n we have $f(a_n) < C < f(b_n)$, and $\lim_{n\to\infty} |b_n - a_n| = 0$. As f is continuous, this implies $\lim_{n\to\infty} f(a_n) = \lim_{n\to\infty} f(b_n) = f(x) = C$. $\square$

A.2.2 Mean Value Theorem

Theorem A.2.2 (Mean Value Theorem (MVT)). *Let $f(x)$ be differentiable on $[a, b]$. Then there exists a $c \in (a, b)$ such that*

$$f(b) - f(a) = f'(c) \cdot (b - a). \tag{A.14}$$

We give an interpretation of the Mean Value Theorem. Let $f(x)$ represent the distance from the starting point at time x. The average speed from a to b is the distance traveled, $f(b) - f(a)$, divided by the elapsed time, $b - a$. As $f'(x)$ represents the speed at time x, the Mean Value Theorem says that there is some intermediate time at which we are traveling at the average speed.

To prove the Mean Value Theorem, it suffices to consider the special case when $f(a) = f(b) = 0$; this case is known as Rolle's Theorem:

Theorem A.2.3 (Rolle's Theorem). *Let f be differentiable on $[a, b]$, and assume $f(a) = f(b) = 0$. Then there exists a $c \in (a, b)$ such that $f'(c) = 0$.*

Exercise A.2.4. *Show the Mean Value Theorem follows from Rolle's Theorem.* Hint: *Consider*

$$h(x) = f(x) - \frac{f(b) - f(a)}{b - a}(x - a) - f(a). \tag{A.15}$$

Note $h(a) = f(a) - f(a) = 0$ and $h(b) = f(b) - (f(b) - f(a)) - f(a) = 0$. The conditions of Rolle's Theorem are satisfied for $h(x)$, and

$$h'(c) = f'(c) - \frac{f(b) - f(a)}{b - a}. \tag{A.16}$$

Proof of Rolle's Theorem. Without loss of generality, assume $f'(a)$ and $f'(b)$ are non-zero. If either were zero we would be done. Multiplying $f(x)$ by -1 if needed, we may assume $f'(a) > 0$. *For convenience, we assume $f'(x)$ is continuous.* This assumption simplifies the proof, but is not necessary. In all applications in this book this assumption will be met.

Case 1: $f'(b) < 0$: As $f'(a) > 0$ and $f'(b) < 0$, the Intermediate Value Theorem applied to $f'(x)$ asserts that all intermediate values are attained. As $f'(b) < 0 < f'(a)$, this implies the existence of a $c \in (a, b)$ such that $f'(c) = 0$.

Case 2: $f'(b) > 0$: $f(a) = f(b) = 0$, and the function f is increasing at a and b. If x is real close to a then $f(x) > 0$ if $x > a$. This follows from the fact that

$$f'(a) = \lim_{x \to a} \frac{f(x) - f(a)}{x - a}. \tag{A.17}$$

As $f'(a) > 0$, the limit is positive. As the denominator is positive for $x > a$, the numerator must be positive. Thus $f(x)$ must be greater than $f(a)$ for such x. Similarly $f'(b) > 0$ implies $f(x) < f(b) = 0$ for x slightly less than b.

Therefore the function $f(x)$ is positive for x slightly greater than a and negative for x slightly less than b. If the first derivative were always positive then $f(x)$ could never be negative as it starts at 0 at a. This can be seen by again using the limit definition of the first derivative to show that if $f'(x) > 0$ then the function is increasing near x. Thus the first derivative cannot always be positive. Either there must be some point $y \in (a, b)$ such that $f'(y) = 0$ (and we are then done) or $f'(y) < 0$. By the Intermediate Value Theorem, as 0 is between $f'(a)$ (which is positive) and $f'(y)$ (which is negative), there is some $c \in (a, y) \subset [a, b]$ such that $f'(c) = 0$. $\square$

A.2.3 Taylor Series

Using the Mean Value Theorem we prove a version of the nth **Taylor series** Approximation: if f is differentiable at least $n+1$ times on $[a, b]$, then for all $x \in [a, b]$, $f(x) = \sum_{k=0}^{n} \frac{f^{(k)}(a)}{k!}(x - a)^k$ plus an error that is at most $\max_{a \le c \le x} |f^{(n+1)}(c)| \cdot |x - a|^{n+1}$.

Assuming f is differentiable $n + 1$ times on $[a, b]$, we apply the Mean Value Theorem multiple times to bound the error between $f(x)$ and its Taylor Approximations. Let

$$f_n(x) = \sum_{k=0}^{n} \frac{f^{(k)}(a)}{k!}(x - a)^k$$
$$h(x) = f(x) - f_n(x). \tag{A.18}$$

$f_n(x)$ is the nth Taylor series Approximation to $f(x)$. Note $f_n(x)$ is a polynomial of degree n and its first n derivatives agree with the derivatives of $f(x)$ at $x = 0$. We want to bound $|h(x)|$ for $x \in [a, b]$. Without loss of generality (basically, for notational convenience), we may assume $a = 0$. Thus $h(0) = 0$. Applying the Mean Value Theorem to h yields

$$\begin{aligned}
h(x) &= h(x) - h(0) \\
&= h'(c_1) \cdot (x - 0) \quad \text{with } c_1 \in [0, x] \\
&= (f'(c_1) - f_n'(c_1))\, x \\
&= \left(f'(c_1) - \sum_{k=1}^{n} \frac{f^{(k)}(0)}{k!} \cdot k(c_1 - 0)^{k-1} \right) x \\
&= \left(f'(c_1) - \sum_{k=1}^{n} \frac{f^{(k)}(0)}{(k-1)!} c_1^{k-1} \right) x \\
&= h_1(c_1)x.
\end{aligned}$$
(A.19)

We now apply the Mean Value Theorem to $h_1(u)$. Note that $h_1(0) = 0$. Therefore

$$\begin{aligned}
h_1(c_1) &= h_1(c_1) - h_1(0) \\
&= h_1'(c_2) \cdot (c_1 - 0) \quad \text{with } c_2 \in [0, c_1] \subset [0, x] \\
&= (f''(c_2) - f_n''(c_2))\, c_1 \\
&= \left(f''(c_2) - \sum_{k=2}^{n} \frac{f^{(k)}(0)}{(k-1)!} \cdot (k-1)(c_2 - 0)^{k-2} \right) c_1 \\
&= \left(f''(c_2) - \sum_{k=2}^{n} \frac{f^{(k)}(0)}{(k-2)!} c_2^{k-2} \right) c_1 \\
&= h_2(c_2)c_1.
\end{aligned}$$
(A.20)

Therefore,

$$h(x) = f(x) - f_n(x) = h_2(c_2)c_1 x, \quad c_1, c_2 \in [0, x].$$
(A.21)

Proceeding in this way a total of n times yields

$$h(x) = \left(f^{(n)}(c_n) - f^{(n)}(0) \right) c_{n-1} c_{n-2} \cdots c_2 c_1 x.$$
(A.22)

Applying the Mean Value Theorem to $f^{(n)}(c_n) - f^{(n)}(0)$ gives $f^{(n+1)}(c_{n+1}) \cdot (c_n - 0)$. Thus

$$h(x) = f(x) - f_n(x) = f^{(n+1)}(c_{n+1})c_n \cdots c_1 x, \quad c_i \in [0, x].$$
(A.23)

Therefore

$$|h(x)| = |f(x) - f_n(x)| \le M_{n+1}|x|^{n+1}$$
(A.24)

where

$$M_{n+1} = \max_{c \in [0,x]} |f^{(n+1)}(c)|.$$
(A.25)

Thus if f is differentiable $n + 1$ times then the nth Taylor series approximation to $f(x)$ is correct within a multiple of $|x|^{n+1}$; further, the multiple is bounded by the maximum value of $f^{(n+1)}$ on $[0, x]$.

Exercise A.2.5. *Prove (A.22) by induction.*

Exercise A.2.6. *Calculate the first few terms of the Taylor series expansions at 0 of* $\cos(x), \sin(x), e^x$, *and* $2x^3 - x + 3$. *Calculate the Taylor series expansions of the above functions at* $x = a$ *Hint: There is a fast way to do this.*

Exercise A.2.7 (Advanced). *Show* all *the Taylor coefficients for*

$$f(x) = \begin{cases} e^{-1/x^2} & \text{if } x \neq 0 \\ 0 & \text{if } x = 0 \end{cases} \tag{A.26}$$

expanded about the origin vanish. What does this imply about the uniqueness of a Taylor series expansion? Warning: be careful differentiating at zero. More is strangely true. Borel showed that if $\{a_n\}$ *is any sequence of real numbers then there exists an infinitely differentiable* f *such that* $\forall n \geq 0$, $f^{(n)}(0) = a_n$ *(for a constructive proof see [GG]). Ponder the Taylor series from* $a_n = (n!)^2$.

A.2.4 Advanced Calculus Theorems

For the convenience of the reader we record exact statements of several standard results from advanced calculus that are used at various points of the text.

Theorem A.2.8 (Fubini). *Assume* f *is continuous and*

$$\int_a^b \int_c^d |f(x,y)| dx dy < \infty. \tag{A.27}$$

Then

$$\int_a^b \left[\int_c^d f(x,y) dy \right] dx = \int_c^d \left[\int_a^b f(x,y) dx \right] dy. \tag{A.28}$$

Similar statements hold if we instead have

$$\sum_{n=N_0}^{N_1} \int_c^d f(x_n, y) dy, \qquad \sum_{n=N_0}^{N_1} \sum_{m=M_0}^{M_1} f(x_n, y_m). \tag{A.29}$$

For a proof in special cases, see [BL, VG]; an advanced, complete proof is given in [Fol]. See Exercise 11.4.12 for an example where the orders of integration cannot be changed.

Theorem A.2.9 (Green's Theorem). *Let* C *be a simply closed, piecewise-smooth curve in the plane, oriented clockwise, bounding a region* D. *If* $P(x,y)$ *and* $Q(x,y)$ *have continuous partial derivatives on some open set containing* D, *then*

$$\int_C P(x,y) dx + Q(x,y) dy = \int\int_D \left(\frac{\partial Q}{\partial x} - \frac{\partial P}{\partial y} \right) dx dy. \tag{A.30}$$

For a proof, see [Rud], Theorem 9.50 as well as [BL, La5, VG].

Exercise A.2.10. *Prove Green's Theorem. Show it is enough to prove the theorem for* D *a rectangle, which is readily checked.*

Theorem A.2.11 (Change of Variables). *Let V and W be bounded open sets in $\mathbb{R}^n$. Let $h : V \to W$ be a 1-1 and onto map, given by*

$$h(u_1,\ldots,u_n) = (h_1(u_1,\ldots,u_n),\ldots,h_n(u_1,\ldots,u_n)).\tag{A.31}$$

Let $f : W \to \mathbb{R}$ be a continuous, bounded function. Then

$$\int \cdots \int_W f(x_1,\ldots,x_n)dx_1\cdots dx_n$$
$$= \int \cdots \int_V f\left(h(u_1,\ldots,u_n)\right) J(u_1,\ldots,u_v)du_1\cdots du_n.\tag{A.32}$$

*where J is the **Jacobian***

$$J = \begin{vmatrix} \frac{\partial h_1}{\partial u_1} & \cdots & \frac{\partial h_1}{\partial u_n} \\ \vdots & \ddots & \vdots \\ \frac{\partial h_n}{\partial u_1} & \cdots & \frac{\partial h_n}{\partial u_n} \end{vmatrix}.\tag{A.33}$$

For a proof, see [La5, Rud].

A.3 CONVERGENCE AND CONTINUITY

We recall some needed definitions and results from real analysis. See [Rud] for more details.

Definition A.3.1 (Convergence). *A sequence $\{x_n\}_{n=1}^\infty$ converges to x if given any $\epsilon > 0$ there exists an N (possibly depending on ϵ) such that for all $n > N$, $|x_n - x| < \epsilon$. We often write $x_n \to x$.*

Exercise A.3.2. *If $x_n = \frac{3n^2}{n^2+1}$, prove $x_n \to 3$.*

Exercise A.3.3. *If $\{x_n\}$ converges, show it converges to a unique number.*

Exercise A.3.4. *Let $\alpha > 0$ and set $x_{n+1} = \frac{1}{2}\left(x_n + \frac{\alpha}{x_n}\right)$. If $x_0 = \alpha$, prove x_n converges to $\sqrt{\alpha}$. Can you generalize this to find p^{th} roots? This formula can be derived by Newton's Method (see §1.2.4).*

Definition A.3.5 (Continuity). *A function f is continuous at a point x_0 if given an $\epsilon > 0$ there exists a $\delta > 0$ (possibly depending on ϵ) such that if $|x - x_0| < \delta$ then $|f(x) - f(x_0)| < \epsilon$.*

Definition A.3.6 (Uniform Continuity). *A continuous function is uniformly continuous if given an $\epsilon > 0$ there exists a $\delta > 0$ such that $|x - y| < \delta$ implies $|f(x) - f(y)| < \epsilon$. Note that the same δ works for all x.*

Usually we will work with functions that are uniformly continuous on some fixed, finite interval.

Theorem A.3.7. *Any continuous function on a closed, finite interval is uniformly continuous.*

Exercise A.3.8. *Show x^2 is uniformly continuous on $[a, b]$ for $-\infty < a < b < \infty$. Show $\frac{1}{x}$ is not uniformly continuous on $(0, 1)$, even though it is continuous. Show x^2 is not uniformly continuous on $[0, \infty)$.*

Exercise A.3.9. *Show the sum or product of two uniformly continuous functions is uniformly continuous. In particular, show any finite polynomial is uniformly continuous on $[a, b]$.*

We sketch a proof of Theorem A.3.7. We first prove

Theorem A.3.10 (Bolzano-Weierstrass). *Let $\{x_n\}_{n=1}^{\infty}$ be a sequence in a finite closed interval. Then there is a subsequence $\{x_{n_k}\}_{k=1}^{\infty}$ such that x_{n_k} converges.*

Sketch the proof. Without loss of generality, assume the finite closed interval is $[0, 1]$. We proceed by divide and conquer. Consider the two intervals $I_1 = [0, \frac{1}{2}]$ and $I_2 = [\frac{1}{2}, 1]$. At least one of these (possibly both) must have infinitely many points of the original sequence as otherwise there would only be finitely many x_n's in the original sequence. Choose a subinterval (say I_a) with infinitely many x_n's, and choose any element of the sequence in that interval to be x_{n_1}.

Consider all x_n with $n > n_1$. Divide I_a into two subintervals I_{a1} and I_{a2} as before (each will be half the length of I_a). Again, at least one subinterval must contain infinitely many terms of the original sequence. Choose such a subinterval, say I_{ab}, and choose any element of the sequence in that interval to be x_{n_2} (note $n_2 > n_1$). We continue in this manner, obtaining a sequence $\{x_{n_k}\}$. For $k \geq K$, x_{n_k} is in an interval of size $\frac{1}{2^K}$. We we leave it as an exercise to the reader to show how this implies there is an x such that $x_{n_k} \to x$. □

Proof of Theorem A.3.7. If $f(x)$ is not uniformly continuous, given $\epsilon > 0$ for each $\delta = \frac{1}{2^n}$ there exist points x_n and y_n with $|x_n - y_n| < \frac{1}{2^n}$ and $|f(x_n) - f(y_n)| > \epsilon$. By the Bolzano-Weierstrass Theorem, we construct sequences $x_{n_k} \to x$ and $y_{n_{k_j}} \to y$. One can show $x = y$, and $|f(x_{n_{k_j}}) - f(y_{n_{k_j}})| > \epsilon$ violates the continuity of f at x. □

Exercise A.3.11. *Fill in the details of the above proof.*

Definition A.3.12 (Bounded). *We say $f(x)$ is bounded (by B) if for all x in the domain of f, $|f(x)| \leq B$.*

Theorem A.3.13. *Let $f(x)$ be uniformly continuous on $[a, b]$. Then $f(x)$ is bounded.*

Exercise A.3.14. *Prove the above theorem. Hint: Given $\epsilon > 0$, divide $[a, b]$ into intervals of length δ.*

A.4 DIRICHLET'S PIGEON-HOLE PRINCIPLE

Theorem A.4.1 (Dirichlet's Pigeon-Hole Principle). *Let $A_1, A_2, \ldots, A_n$ be a collection of sets with the property that $A_1 \cup \cdots \cup A_n$ has at least $n + 1$ elements. Then at least one of the sets A_i has at least two elements.*

This is called the Pigeon-Hole Principle for the following reason: if $n+1$ pigeons go to n holes, at least one of the holes must be occupied by at least two pigeons. Equivalently, if we distribute k objects in n boxes and $k > n$, one of the boxes contains at least two objects. The Pigeon-Hole Principle is also known as the Box Principle. One application of the Pigeon-Hole Principle is to find good rational approximations to irrational numbers (see Theorem 5.5.4). We give some examples to illustrate the method.

Example A.4.2. *If we choose a subset S from the set $\{1, 2, \ldots, 2n\}$ with $|S| = n + 1$, then S contains at least two elements a, b with $a|b$.*

Write each element $s \in S$ as $s = 2^\sigma s_0$ with s_0 odd. There are n odd numbers in the set $\{1, 2, \ldots, 2n\}$, and as the set S has $n + 1$ elements, the Pigeon-Hole Principle implies that there are at least two elements a, b with the same odd part; the result is now immediate.

Exercise A.4.3. *If we choose 55 numbers from $\{1, 2, 3, \ldots, 100\}$ then among the chosen numbers there are two whose difference is ten (from [Ma]).*

Exercise A.4.4. *Let $a_1, \ldots, a_{n+1}$ be distinct integers in $\{1, \ldots, 2n\}$. Prove two of them add to a number divisible by $2n$.*

Exercise A.4.5. *Let $a_1, \ldots, a_n$ be integers. Prove that there is a subset whose sum is divisible by n.*

Example A.4.6. *Let $\{a_1, a_2, a_3, a_4, a_5\}$ be distinct real numbers. There are indices i, j with $0 < a_i - a_j < 1 + a_i a_j$.*

As the function $\tan : (-\frac{\pi}{2}, \frac{\pi}{2}) \to \mathbb{R}$ is surjective, there are angles $\theta_i \in (-\frac{\pi}{2}, \frac{\pi}{2})$ with $a_i = \tan \theta_i$, $1 \le i \le 5$. Divide the interval $(-\frac{\pi}{2}, \frac{\pi}{2})$ into four equal pieces, each of length $\frac{\pi}{4}$. As we have five angles, at least two of them must lie in the same small interval, implying that there are i, j with $0 < \theta_i - \theta_j < \frac{\pi}{4}$. Applying $\tan$ to the last inequality and using the identity

$$\tan(x - y) = \frac{\tan x - \tan y}{1 + \tan x \tan y} \tag{A.34}$$

gives the result.

Exercise A.4.7. *Let $\phi_1, \phi_2, \ldots, \phi_K$ be angles. Then for any $\epsilon > 0$ there are infinitely many $n \in \mathbb{N}$ such that*

$$\left| K - \sum_{j=1}^{K} \cos(n\phi_k) \right| < \epsilon. \tag{A.35}$$

Exercise[(h)] A.4.8. *The Pigeon-Hole Principle ensures that, if there are N boxes and $N + 1$ objects, then at least one box has two objects. What if we lower our sites and ask only that there is a high probability of having a box with two elements; see for example the birthday problem (Exercise 8.1.29). Specifically, let us assume that each object is equally likely to be in any of the N boxes. For each fixed k, show there is a positive probability of having at least k objects in a box if there are $N^{\frac{k-1}{k}}$ objects.*

A.5 MEASURES AND LENGTH

We discuss sizes of subsets of $[0, 1]$. It is natural to define the length of an interval $I = [a, b]$ (or $[a, b)$ and so on) as $b - a$. We denote this by $|I|$, and refer to this as the **length** or **measure** of I. Our definition implies a point a has zero length. What about more exotic sets, such as the rationals and the irrationals? What are the measures of these sets? A proper explanation is given by measure theory (see [La5, Rud]); we introduce enough for our purposes. We assume the reader is familiar with countable sets (see Chapter 5).

Let I be a countable union of disjoint intervals $I_n \subset [0, 1)$; thus $I_n \cap I_m$ is empty if $n \neq m$. *It is natural* (but see §5.1 as a warning for how *natural* statements are often wrong) to say

$$|I| = \sum_n |I_n|. \tag{A.36}$$

It is important to take a countable union. Consider an uncountable union with $I_x = \{x\}$ for $x \in [0, 1]$. As each singleton $\{x\}$ has length zero, we expect their union to also have length zero; however, their union is $[0, 1]$, which has length 1. If $A \subset B$, it is natural to say $|A|$ (the length of A) is at most $|B|$ (the length of B). Note our definition implies $[a, b)$ and $[a, b]$ have the same length.

A.5.1 Measure of the Rationals

Our assumptions imply that the rationals in $[0, 1]$ have zero length (hence the irrationals in $[0, 1]$ have length 1).

Theorem A.5.1. *The rationals $\mathbb{Q}$ have zero measure.*

Sketch of the proof. We claim it suffices to show $Q = \mathbb{Q} \cap [0, 1]$ has measure zero. To prove $|Q| = 0$ we show that given any $\epsilon > 0$ we can find a countable set of intervals I_n such that

1. $|Q| \subset \cup_n I_n$;

2. $\sum_n |I_n| < \epsilon$.

As the rationals are countable, we can enumerate Q, say $Q = \{x_n\}_{n=0}^\infty$. For each n let

$$I_n = \left[x_n - \frac{\epsilon}{4 \cdot 2^n}, x_n + \frac{\epsilon}{4 \cdot 2^n}\right], \quad |I_n| = \frac{\epsilon}{2 \cdot 2^n}. \tag{A.37}$$

Clearly $Q \subset \cup_n I_n$. The intervals I_n are not necessarily disjoint, but

$$|\cup_n I_n| \leq \sum_n |I_n| = \epsilon, \tag{A.38}$$

which completes the proof.   □

Exercise A.5.2. *Show that if $Q = \mathbb{Q} \cap [0, 1]$ has measure zero, then $\mathbb{Q}$ has measure zero.*

Exercise A.5.3. *Show any countable set has measure zero; in particular, the algebraic numbers have length zero.*

Definition A.5.4 (Almost all). *Let A^c be the compliment of $A \subset \mathbb{R}$: $A^c = \{x : x \notin A\}$. If A^c is of measure zero, we say almost all x are in A.*

Thus the above theorem shows that not only are almost all real numbers are irrational but almost all real numbers are transcendental.

A.5.2 Measure of the Cantor Set

The Cantor set is a fascinating subset of $[0, 1]$. We construct it in stages. Let $C_0 = [0, 1]$. We remove the middle third of C_0 and obtain $C_1 = [0, \frac{1}{3}] \cup [\frac{2}{3}, 1]$. Note C_1 is a union of two closed intervals (we keep all endpoints). To construct C_2 we remove the middle third of all remaining intervals and obtain

$$C_2 = \left[0, \frac{1}{9}\right] \cup \left[\frac{2}{9}, \frac{3}{9}\right] \cup \left[\frac{6}{9}, \frac{7}{9}\right] \cup \left[\frac{8}{9}, 1\right]. \qquad (A.39)$$

We continue this process. Note C_n is the union of 2^n closed intervals, each of size 3^{-n}, and

$$C_0 \supset C_1 \supset C_2 \supset \cdots . \qquad (A.40)$$

Definition A.5.5 (Cantor Set). *The Cantor set C is defined by*

$$C = \bigcap_{n=1}^{\infty} C_n = \{x \in \mathbb{R} : \forall n, x \in C_n\}. \qquad (A.41)$$

Exercise A.5.6. *Show the length of the Cantor set is zero.*

If x is an endpoint of C_n for some n, then $x \in C$. At first, one might expect that these are the only points, especially as the Cantor set has length zero.

Exercise A.5.7. *Show $\frac{1}{4}$ and $\frac{3}{4}$ are in C, but neither is an endpoint. Hint: Proceed by induction. To construct C_{n+1} from C_n, we removed the middle third of intervals. For each sub-interval, what is left looks like the union of two pieces, each one-third the length of the previous. Thus, we have shrinking maps fixing the left and right parts $L, R : \mathbb{R} \to \mathbb{R}$ given by $L(x) = \frac{x}{3}$ and $R(x) = \frac{x+2}{3}$, and $C_{n+1} = R(C_n) + L(C_n)$.*

Exercise A.5.8. *Show the Cantor set is also the set of all numbers $x \in [0, 1]$ which have no 1's in their base three expansion. For rationals such as $\frac{1}{3}$, we may write these by using repeating 2's: $\frac{1}{3} = .02222\ldots$ in base three. By considering base two expansions, show there is a one-to-one and onto map from $[0, 1]$ to the Cantor set.*

Exercise A.5.9 (From the *American Mathematical Monthly*). *Use the previous exercise to show that every $x \in [0, 2]$ can be written as a sum $y + z$ with $y, z \in C$.*

Remark A.5.10. The above exercises show the Cantor set is uncountable and is in a simple correspondence to all of $[0,1]$, *but* it has length zero! Thus, the notion of "length" is different from the notion of "cardinality": two sets can have the same cardinality but very different lengths.

Exercise A.5.11 (Fat Cantor Sets)**.** *Instead of removing the middle third in each step, remove the middle $\frac{1}{m}$. Is there a choice of m which yields a set of positive length? What if at stage n we remove the middle $\frac{1}{a_n}$. For what sequences a_n are we left with a set of positive length? If the a_n are digits of a simple continued fraction, what do you expect to be true for "most" such numbers?*

For more on the Cantor set, including dynamical interpretations, see [Dev, Edg, Fal, SS3].

A.6 INEQUALITIES

The first inequality we mention here is the Arithmetic Mean and Geometrically Mean Inequality (AM–GM); see [Mil3] for some proofs. For positive numbers $a_1, \ldots, a_n$, the arithmetic mean is $\frac{a_1 + \cdots + a_n}{n}$ and the geometric mean is $\sqrt[n]{a_1 \cdots a_n}$.

Theorem A.6.1 (AM-GM)**.** *Let $a_1, \ldots, a_n$ be positive real numbers. Then*

$$\sqrt[n]{a_1 \cdots a_n} \leq \frac{a_1 + \cdots + a_n}{n}, \tag{A.42}$$

with equality if and only if $a_1 = \cdots = a_n$.

Exercise A.6.2. *Prove the AM-GM when $n = 2$. Hint: For $x \in \mathbb{R}$, $x^2 \geq 0$; this is one of the most useful inequalities in mathematics. We will see it again when we prove the Cauchy-Schwartz inequality.*

Exercise A.6.3. *Prove the AM-GM using mathematical induction.*

There is an interesting generalization of the AM-GM; AM-GM is the case $p_1 = \cdots = p_n = \frac{1}{n}$ of the following theorem.

Theorem A.6.4. *Let $a_1, \ldots, a_n$ be as above, and let $p_1, \ldots, p_n$ be positive real numbers. Set $P = p_1 + \cdots + p_n$. Then*

$$a_1^{p_1} \cdots a_n^{p_n} \leq \left(\frac{p_1 a_1 + \cdots + p_n a_n}{P} \right)^P, \tag{A.43}$$

and equality holds if and only if $a_1 = \cdots = a_n$.

This inequality is in turn a special case of the following important theorem:

Theorem A.6.5 (Jensen's Inequality)**.** *Let f be a real continuous function on $[a, b]$ with continuous second derivative on (a, b). Suppose that $f''(x) \leq 0$ for all $x \in (a, b)$. Then for $a_1, \ldots, a_n \in [a, b]$ and $p_1, \ldots, p_n$ positive real numbers, we have*

$$f \left(\frac{p_1 a_1 + \cdots + p_n a_n}{p_1 + \cdots + p_n} \right) \leq \frac{p_1 f(a_1) + \cdots + p_n f(a_n)}{p_1 + \cdots + p_n}. \tag{A.44}$$

Exercise A.6.6. *Prove Jensen's inequality. Hint: Draw a picture; carefully examine the case $n = 2$, $p_1 = p_2 = \frac{1}{2}$. What does $f''(x) \leq 0$ mean in geometric terms?*

Exercise A.6.7. *Investigate the cases where Jensen's inequality is an equality.*

Exercise A.6.8. *Show that Jensen's inequality implies the AM-GM and its generalization Theorem A.6.4. Hint: Examine the function $f(x) = -\log x$, $x > 0$.*

Our final inequality is the **Cauchy-Schwarz inequality**. There are a number of inequalities that are referred to as the Cauchy-Schwarz inequality. A useful version is the following:

Lemma A.6.9 (Cauchy-Schwarz). *For complex-valued functions f and g,*

$$\int_0^1 |f(x)g(x)|dx \leq \left(\int_0^1 |f(x)|^2 dx\right)^{\frac{1}{2}} \cdot \left(\int_0^1 |g(x)|^2 dx\right)^{\frac{1}{2}}. \qquad (A.45)$$

Proof. For notational simplicity, assume f and g are non-negative functions. Working with $|f|$ and $|g|$ we see there is no harm in the above assumption. As the proof is immediate if either of the integrals on the right hand side of (A.45) is zero or infinity, we assume both integrals are non-zero and finite. Let

$$h(x) = f(x) - \lambda g(x), \quad \lambda = \frac{\int_0^1 f(x)g(x)dx}{\int_0^1 g(x)^2 dx}. \qquad (A.46)$$

As $\int_0^1 h(x)^2 dx \geq 0$ we have

$$0 \leq \int_0^1 (f(x) - \lambda g(x))^2\, dx$$

$$= \int_0^1 f(x)^2 dx - 2\lambda \int_0^1 f(x)g(x)dx + \lambda^2 \int_0^1 g(x)^2 dx$$

$$= \int_0^1 f(x)^2 dx - 2\frac{\left(\int_0^1 f(x)g(x)dx\right)^2}{\int_0^1 g(x)^2 dx} + \frac{\left(\int_0^1 f(x)g(x)dx\right)^2}{\int_0^1 g(x)^2 dx}$$

$$= \int_0^1 f(x)^2 dx - \frac{\left(\int_0^1 f(x)g(x)dx\right)^2}{\int_0^1 g(x)^2 dx}. \qquad (A.47)$$

This implies

$$\frac{\left(\int_0^1 f(x)g(x)dx\right)^2}{\int_0^1 g(x)^2 dx} \leq \int_0^1 f(x)^2 dx, \qquad (A.48)$$

or equivalently

$$\left(\int_0^1 f(x)g(x)dx\right)^2 \leq \int_0^1 f(x)^2 dx \cdot \int_0^1 g(x)^2 dx. \qquad (A.49)$$

Taking square roots completes the proof. $\qquad\square$

Again, note that both the AG-GM and the Cauchy-Schwartz inequalities are clever applications of $x^2 \geq 0$ for $x \in \mathbb{R}$.

Exercise A.6.10. *For what f and g is the Cauchy-Schwarz Inequality an equality?*

Exercise A.6.11. *One can also prove the Cauchy-Schwartz inequality as follows: consider $h(x) = af(x) + bg(x)$ where $a = \sqrt{\int_0^1 |f(x)|^2 dx}$, $b = \sqrt{\int_0^1 |g(x)|^2 dx}$, and then integrate $h(x)^2$.*

Remark A.6.12. The Cauchy-Schwarz Inequality is often useful when $g(x) = 1$. In this special case, it is important that we integrate over a finite interval.

Exercise A.6.13. *Suppose $a_1, \ldots, a_n$ and $b_1, \ldots, b_n$ are two sequences of real numbers. Prove the following Cauchy-Schwarz inequality:*

$$|a_1 b_1 + a_2 b_2 + \cdots + a_n b_n| \leq (a_1^2 + \ldots a_n^2)^{\frac{1}{2}} (b_1^2 + \cdots + b_n^2)^{\frac{1}{2}}. \qquad \text{(A.50)}$$

Exercise A.6.14. *Let $f, g : \mathbb{R} \to \mathbb{C}$ be such that $\int_{\mathbb{R}} |f(x)|^2 dx, \int_{\mathbb{R}} |g(x)|^2 dx < \infty$. Prove the following Cauchy-Schwarz inequality:*

$$\left| \int_{-\infty}^{\infty} f(x)g(x)dx \right|^2 \leq \int_{-\infty}^{\infty} |f(x)|^2 dx \cdot \int_{-\infty}^{\infty} |g(x)|^2 dx. \qquad \text{(A.51)}$$

Appendix B

Linear Algebra Review

We give a brief review of the definitions and results of Linear Algebra that we need for Part 5, leaving many of the proofs to the reader. For more details the reader should consult a textbook in Linear Algebra, for example [St].

B.1 DEFINITIONS

Definition B.1.1 (Transpose, Complex Conjugate Transpose). *Given an $n \times m$ matrix A (where n is the number of rows and m is the number of columns), the transpose of A, denoted A^T, is the $m \times n$ matrix where the rows of A^T are the columns of A. The complex conjugate transpose, A^*, is the complex conjugate of the transpose of A.*

Exercise B.1.2. *Prove $(AB)^T = B^T A^T$ and $(A^T)^T = A$.*

Definition B.1.3 (Real Symmetric, Complex Hermitian). *If an $n \times n$ real matrix A satisfies $A^T = A$ then we say A is real symmetric; if an $n \times n$ complex matrix A satisfies $A^* = A$ then we say A is complex Hermitian.*

Definition B.1.4 (Dot or Inner Product). *The dot (or inner) product of two real vectors v and w is defined as $v^T w$; if the vectors are complex, we instead use $v^* w$. If v and w have n components, $v^T w = v_1 w_1 + \cdots + v_n w_n$ and $v^* w = \overline{v}_1 w_1 + \cdots + \overline{v}_n w_n$.*

Definition B.1.5 (Orthogonality). *Two real vectors are orthogonal (or perpendicular) if $v^T w = 0$; for complex vectors, the equivalent condition is $v^* w = 0$.*

Definition B.1.6 (Length of a vector). *The length of a real vector v is $|v| = \sqrt{v^T v}$; for a complex vector, we have $|v| = \sqrt{v^* v}$.*

Definition B.1.7 (Linearly Dependent, Independent). *Vectors $v_1, \ldots, v_n$ are linearly independent over a field $\mathbb{F}$ if*

$$a_1 v_1 + \cdots + a_n v_n = 0 \tag{B.1}$$

implies $a_1 = \cdots = a_n = 0$. If the vectors are not linearly independent we say they are linearly dependent. We often take $\mathbb{F}$ to be $\mathbb{Q}, \mathbb{R}$ or $\mathbb{C}$.

A vector v has a length and direction. Usually applying a matrix A to v gives a new vector Av with different length and direction; the vectors v such that Av is in the same direction, and the amount the length changes by, are very useful in understanding the matrix A. Specifically, we say

Definition B.1.8 (Eigenvalue, Eigenvector). *λ is an eigenvalue and v is an eigenvector if $Av = \lambda v$ and v is not the zero vector.*

Exercise B.1.9. *If v is an eigenvector of A with eigenvalue λ, show $w = av$, $a \in \mathbb{C}$, is also an eigenvector of A with eigenvalue λ. Therefore, given an eigenvalue λ and an eigenvector v, one can always find an eigenvector w with the same eigenvalue but $|w| = 1$.*

If v is an eigenvector (and hence non-zero), then $Av = \lambda v$ is equivalent to $(\lambda I - A)v = 0$. If $\lambda I - A$ were invertible then v would be zero; hence $\lambda I - A$ is not invertible which implies its determinant is zero. Thus to find the eigenvalues we solve the equation $\det(\lambda I - A) = 0$. This gives a polynomial $p(\lambda) = 0$. We call $p(\lambda)$ the **characteristic polynomial**.

Definition B.1.10 (Degrees of Freedom). *The number of degrees of freedom in a matrix is the number of elements needed to completely specify it; a general $n \times m$ real matrix has nm degrees of freedom.*

Exercise B.1.11. *Show an $n \times n$ real symmetric matrix has $\frac{n(n+1)}{2}$ degrees of freedom, and determine the number of degrees of freedom of an $n \times n$ complex hermitian matrix.*

Exercise B.1.12. *If A and B are symmetric, show AB is symmetric.*

Exercise B.1.13 (Matrix Multiplication). *Show that it costs $2n^3 - n^2$ additions and multiplications to multiply two $n \times n$ matrices in the "obvious" way (by taking dot products). If one can store some multiplications, this can be improved. Show, using memory, one can square a 2×2 matrix with just 11 additions and multiplications. Use this to show that for a $2^k \times 2^k$ matrix, using memory one needs at most $\frac{15}{8}n^3 + O(n^2)$ operations to square it. While this is an improvement, it is still an order n^3 algorithm. In practice, multiplications are more expensive than additions (there are more digit operations in multiplying two k digit numbers than in adding them; see Exercise 1.2.3). See page 202 of [BB] or [Wilf] for an algorithm requiring only $O(n^{\log_2 7}) = O(n^{2.81\cdots})$ multiplications to multiply any two $n \times n$ matrices!*

B.2 CHANGE OF BASIS

Given a matrix A with n rows and m columns we denote the element in the ith row and j^{th} column by a_{ij}. We represent a vector v as a column of elements with the ith being v_i. A nice way to see matrix-vector multiplication is that the v_i give the *coefficients* by which the columns of A are linearly mixed together. For the product $w = Av$ to make sense, the dimension of v must equal m, and the dimension of w will be n. A is therefore a linear transform map m-dimensional to n-dimensional space.

Multiple transformations appear written backwards: if we apply A then B then C to a vector, we write $w = CBAv$. Note that taking the product of two $n \times n$ matrices requires $O(n^3)$ effort.

Definition B.2.1 (Invertible Matrices). *A is invertible if a matrix B can be found such that $BA = AB = I$. The inverse is then written $B = A^{-1}$.*

Exercise B.2.2. *Prove if A is invertible then A must be a square matrix (same number of rows as columns).*

A matrix A is a linear transformation; to write it in matrix form requires us to choose a coordinate system (basis), and the transformation will look different in different bases. Consider the scalar quantity $x = w^T A v$, where A, v and w are written relative to a given basis, say $u_1, \ldots, u_n$. If M is an invertible matrix, we can write these quantities in a new basis, $M u_1, \ldots, M u_n$. We find $v' = M v$ and $w' = M w$.

How does the matrix A look in the new basis? We require x to remain unchanged for any change of basis M for all choices of v and w (as a scalar must). This requires that A become $A' = (M^T)^{-1} A M^{-1}$ in the new basis:

$$x' = w'^T A' v' = (M w)^T ((M^T)^{-1} A M^{-1})(M v) = w^T I A I v = w^T A v = x.$$
(B.2)

This is a **similarity transformation** and represents A in the new basis; our primary use of this is in Theorem B.5.1.

B.3 ORTHOGONAL AND UNITARY MATRICES

Definition B.3.1 (Orthogonal Matrices). *Q is an orthogonal $n \times n$ matrix if it has real entries and $Q^T Q = Q Q^T = I$.*

Note Q is invertible, with inverse Q^T.

Exercise B.3.2. *Show the product of two orthogonal matrices is an orthogonal matrix. Show that the dot product is invariant under orthogonal transformation. That is, show that given two vectors v and w and an orthogonal Q, $(Qw)^T (Qv) = v^T w$.*

Exercise B.3.3. *Let A be a real matrix such that $(Av)^T (Av) = v^T v$ for all vectors v. Must A be an orthogonal matrix?*

Exercise B.3.4. *Show the number of degrees of freedom in an orthogonal matrix is $\frac{n(n-1)}{2}$.*

The set of orthogonal matrices of order n forms a **continuous** or **topological group**, which we call $O(n)$. For the group properties:

- Associativity follows from that of matrix multiplication.

- The identity matrix acts as an identity element, since it is in the group.

- Inverse is the transpose: $Q^{-1} = Q^T$.

- Closure is satisfied because any product of orthogonal matrices is itself orthogonal.

Exercise B.3.5. *Prove the last assertion.*

For $n = 2$, a general orthogonal matrix can be written as

$$\begin{pmatrix} \cos\theta & -\sin\theta \\ \sin\theta & \cos\theta \end{pmatrix} \quad \text{or} \quad \begin{pmatrix} \cos\theta & -\sin\theta \\ -\sin\theta & -\cos\theta \end{pmatrix}, \tag{B.3}$$

where $0 \le \theta < 2\pi$ is a real angle. The determinant of the first is $+1$ and defines the **special** (i.e., unit determinant) **orthogonal group** $SO(2)$; this is a subgroup of $O(2)$ and has identity I. The second has determinant -1 and corresponds to rotations with a reflection; this subgroup is disjoint from $SO(2)$.

Note that $SO(2)$, alternatively written as the family of planar rotations $R(\theta)$, is **isomorphic** to the unit length complex numbers under the multiplication operation:

$$R(\theta) \longleftrightarrow e^{i\theta}; \tag{B.4}$$

there is a bijection between the two sets. Therefore we have $R(\theta_1)R(\theta_2) = R(\theta_1 + \theta_2)$. This commutativity relation does *not* hold in higher $n > 2$.

Exercise B.3.6. *In three dimensions a general rotation involves three angles (for example, azimuth, elevation, and roll). How many angles are needed in four dimensions? In three dimensions one rotates about a line (the set of points which do not move under rotation); what object do you rotate about in four dimensions? See also Project 16.4.7.*

If an orthogonal matrix Q is used for conjugation of a general square matrix A, then the rule for transformation (B.2) becomes $A' = QAQ^T$.

Exercise B.3.7. *Recall the Taylor expansions of* sin *and* cos:

$$\cos(x) = \sum_{n=0}^{\infty} \frac{(-x)^{2n}}{(2n)!}, \quad \sin(x) = \sum_{n=0}^{\infty} \frac{-(-x)^{2n+1}}{(2n+1)!}. \tag{B.5}$$

For small x, we have

$$\cos(x) = 1 + O(x^2), \quad \sin(x) = x + O(x^3). \tag{B.6}$$

For small ϵ, show a rotation by ϵ is

$$\begin{pmatrix} \cos(\epsilon) & -\sin(\epsilon) \\ \sin(\epsilon) & \cos(\epsilon) \end{pmatrix} = \begin{pmatrix} 1 + O(\epsilon^2) & \epsilon + O(\epsilon^3) \\ -\epsilon + O(\epsilon^3) & 1 + O(\epsilon^2) \end{pmatrix}. \tag{B.7}$$

B.4 TRACE

The **trace** of a matrix A, denoted $\text{Trace}(A)$, is the sum of the diagonal entries of A:

$$\text{Trace}(A) = \sum_{i=1}^{n} a_{ii}. \tag{B.8}$$

Theorem B.4.1 (Eigenvalue Trace Formula). $\text{Trace}(A) = \sum_{i=1}^{n} \lambda_i$, *where the λ_i's are the eigenvalues of A.*

Proof. The proof relies on writing out the characteristic equation and comparing powers of λ with the factorized version. As the polynomial has roots λ_i, we can write

$$\det(\lambda I - A) = p(\lambda) = \prod_{i=1}^{n}(\lambda - \lambda_i). \tag{B.9}$$

Note the coefficient of λ^n is 1, thus we have $\prod_i(\lambda - \lambda_i)$ and not $c\prod_i(\lambda - \lambda_i)$ for some constant c. By expanding out the right hand side, the coefficient of λ^{n-1} is $-\sum_{i=1}^{n} \lambda_i$. Expanding the left hand side, the lemma follows by showing the coefficient of λ^{n-1} in $\det(\lambda I - A)$ is $-\mathrm{Trace}(A)$. $\square$

Exercise B.4.2. *Prove* $\det(A) = \prod_{i=1}^{n} \lambda_i$, *where* λ_i *are the eigenvalues of A.*

Lemma B.4.3. *We have* $\det(A_1 \cdots A_m) = \det(A_1)\cdots\det(A_m)$. *Note* $\det(I) = 1$.

Corollary B.4.4. $\mathrm{Trace}(A)$ *is invariant under a rotation of basis:* $\mathrm{Trace}(Q^T A Q) = \mathrm{Trace}(A)$.

Exercise B.4.5. *Prove the above corollary.*

Definition B.4.6 (Unitary Matrices). *U is an unitary $n \times n$ matrix if it has complex entries and $U^*U = UU^* = I$.*

Note unitary matrices are the complex analogue of orthogonal matrices.

B.5 SPECTRAL THEOREM FOR REAL SYMMETRIC MATRICES

Theorem B.5.1 (Spectral Theorem). *Let A be a real symmetric $n \times n$ matrix. Then there exists an orthogonal $n \times n$ matrix Q and a real diagonal matrix Λ such that $Q^T A Q = \Lambda$, and the n eigenvalues of A are the diagonal entries of Λ.*

This result is remarkable: any real symmetric matrix is diagonal when rotated into an appropriate basis; because of this, this theorem is often called the Diagonalization Theorem. In other words, the operation of a matrix A on a vector v can be broken down into three steps:

$$Av = Q\Lambda Q^T v = (\text{undo the rotation})(\text{stretch along the coordinate axes})(\text{rotate})v. \tag{B.10}$$

We shall only prove the theorem in the case when the eigenvalues are distinct (note a generic matrix has n distinct eigenvalues, so this is not a particularly restrictive assumption). A similar theorem holds for complex hermitian matrices; the eigenvalues are again real, except instead of conjugating by an orthogonal matrix Q we must now conjugate by a unitary matrix U.

Remark B.5.2. The spectral theorem allows us to calculate the effect of high powers of a matrix quickly. Given a vector v, write v as a linear combination of the eigenvectors v_i: $v = \sum_i c_i v_i$. Then $A^m v = \sum c_i \lambda_i^m v_i$; this is significantly faster than calculating the entries of A^m.

B.5.1 Preliminary Lemmas

For the Spectral Theorem we prove a sequence of needed lemmas:

Lemma B.5.3. *The eigenvalues of a real symmetric matrix are real.*

Proof. Let A be a real symmetric matrix with eigenvalue λ and eigenvector v. Therefore $Av = \lambda v$. Note that we do not yet know that v has only real coordinates. Take the dot product of both sides with the vector v^*, the complex conjugate transpose of v:

$$v^* A v = \lambda v^* v. \tag{B.11}$$

The two sides are clearly complex numbers. As A is real symmetric, $A^* = A$. Taking the complex conjugate transpose of the left hand side of (B.11) gives $v^* A v$. Therefore, both sides of (B.11) are real. As $v^* v$ is real and non-zero, we obtain λ is real. □

Exercise B.5.4. *Prove the eigenvalues of an orthogonal matrix have absolute value 1; thus to each eigenvalue λ there is a $\theta \in \mathbb{R}$ such that $\lambda = e^{i\theta}$. Show the eigenvalues occur in complex conjugate pairs.*

Exercise B.5.5. *Calculate the eigenvalues and eigenvectors of $\left(\begin{smallmatrix} 0 & -1 \\ 1 & 0 \end{smallmatrix}\right)$. Show if v is an eigenvector then $v^T v = 0$ but $v^* v > 0$. It is because of examples like this that we had to use complex conjugates in the proof of Lemma B.5.3.*

Lemma B.5.6. *The eigenvectors of a real symmetric matrix are real.*

Proof. The eigenvectors solve the equation $(\lambda I - A)v = 0$. As $\lambda I - A$ is real, Gaussian elimination shows v is real. □

Lemma B.5.7. *If λ_1 and λ_2 are two distinct eigenvalues of a real symmetric matrix A, then their corresponding eigenvectors are perpendicular.*

Proof. We study $v_1^T A v_2$. Now

$$v_1^T A v_2 = v_1^T(Av_2) = v_1^T(\lambda_2 v_2) = \lambda_2 v_1^T v_2. \tag{B.12}$$

Also

$$v_1^T A v_2 = v_1^T A^T v_2 = (v_1^T A^T)v_2 = (Av_1)^T v_2 = (\lambda_1 v_1)^T v_2 = \lambda_1 v_1^T v_2. \tag{B.13}$$

Therefore

$$\lambda_2 v_1^T v_2 = \lambda_1 v_1^T v_2 \text{ or } (\lambda_1 - \lambda_2)v_1^T v_2 = 0. \tag{B.14}$$

As $\lambda_1 \neq \lambda_2$, $v_1^T v_2 = 0$. Thus the eigenvectors v_1 and v_2 are perpendicular. □

B.5.2 Proof of the Spectral Theorem

We prove the Spectral Theorem for real symmetric matrices **If there are n distinct eigenvectors**. Let λ_1 to λ_n be the n distinct eigenvectors and let v_1 to v_n be the corresponding eigenvectors chosen so that each v_i has length 1. Consider the matrix

Q where the first column of Q is v_1, the second column of Q is v_2, all the way to the last column of Q which is v_n:

$$Q = \begin{pmatrix} \uparrow & \uparrow & & \uparrow \\ v_1 & v_2 & \cdots & v_n \\ \downarrow & \downarrow & & \downarrow \end{pmatrix}. \tag{B.15}$$

The transpose of Q is

$$Q^T = \begin{pmatrix} \leftarrow & v_1 & \rightarrow \\ & \vdots & \\ \leftarrow & v_n & \rightarrow \end{pmatrix}. \tag{B.16}$$

Exercise B.5.8. *Show that Q is an orthogonal matrix. Use the fact that the v_i all have length one and are perpendicular to each other.*

Consider $B = Q^T A Q$. To find its entry in the ith row and j^{th} column, we look at

$$e_i^T B e_j \tag{B.17}$$

where the e_k are column vectors which are 1 in the kth position and 0 elsewhere. Thus we need only show that $e_i^T B e_j = 0$ if $i \neq j$ and equals λ_j if $i = j$.

Exercise B.5.9. *Show $Q e_i = v_i$ and $Q^T v_i = e_i$.*

We calculate

$$\begin{aligned} e_i^T B e_j &= e_i^T Q^T A Q e_j \\ &= (e_i^T Q^T) A (Q e_j) \\ &= (Q e_i)^T A (Q e_j) \\ &= v_i^T A v_j \\ &= v_i^T (A v_j) \\ &= v_i^T \lambda_j v_j = \lambda_j v_i^T v_j. \end{aligned} \tag{B.18}$$

As $v_i^T v_j$ equals 0 if $i \neq j$ and 1 if $i = j$, this proves the claim.

Thus the off-diagonal entries of $Q^T A Q$ are zero and the diagonal entries are the eigenvalues λ_j. This shows that $Q^T A Q$ is a diagonal matrix whose entries are the n eigenvalues of A.

Note that in the case of n distinct eigenvalues, not only can we write down the diagonal matrix but we can easily write down what Q should be. Further, by reordering the columns of Q, we see we can reorder the positioning of the eigenvalues on the diagonal.

Exercise B.5.10. *Prove similar results for a complex Hermitian matrix A. In particular, show the eigenvalues are real, and if the eigenvalues of A are distinct, then $A = U^* \Lambda U$ for a unitary U.*

Exercise B.5.11. *Prove if A has n linearly independent eigenvectors then there exists an invertible matrix S such that $S^{-1} A S$ is diagonal.*

Lemma B.5.12 (Triangularization Lemma). *Show for any square matrix A there is a unitary matrix U such that $U^{-1}AU$ is an upper triangular matrix; if A is real symmetric, one may take U to be an orthogonal matrix.*

Exercise B.5.13. *Prove the Triangularization Lemma.* Hint: *Every matrix must have at least one eigenvector v; use Gram-Schmidt to extend v to a unitary matrix, and proceed by induction.*

Exercise B.5.14. *Using the Triangularization Lemma, prove any real symmetric or complex Hermitian matrix can be diagonalized. More is true. A matrix N is* **normal** *if $N^*N = NN^*$. Prove all normal matrices can be diagonalized.*

Exercise B.5.15. *Prove that if a matrix A does not have n linearly independent eigenvectors than A cannot be diagonalized. Prove $\left(\begin{smallmatrix} 0 & 1 \\ 0 & 0 \end{smallmatrix}\right)$ cannot be diagonalized. The closest property to diagonalization that holds for all matrices is Jordan Canonical Form; see [St]. For a nice instance of a matrix that cannot be diagonalized, consider the* see and say *sequence (sometimes called the* look and say *sequence), whose terms are given by 1, 11 (one 1), 21 (two 1's), 1211 (one 2, one 1), 111221 (one 1, one 2, two 1's), 312211 (three 1's, two 2's, one 1) and so on. For more see [Conw, CG] and the entries in Mathworld [We].*

Appendix C

Hints and Remarks on the Exercises

Chapter 1: Mod p Arithmetic, Group Theory and Cryptography

Exercise 1.1.4: *Hint:* Give each person a different point on the curve. Given $n + 1$ points (x_i, y_i) in the plane such that no two points have the same x-coordinate, show there is a unique polynomial $f(x)$ with real coefficients of degree n such that $y_i = f(x_i)$. As an application, given any sequence of n numbers we can justify any number as the next term in the sequence. Thus, while it is natural to assume that if our sequence is $2, 4, 8, 16$ that the next term is 32, show there exists a polynomial $f(x)$ of degree 4 such that $f(i) = 2^i$ for $i \in \{1, 2, 3, 4\}$ but $f(5) = 11988$. The point of all this is that one should always be careful in extrapolating patterns from small data sets. See [Sl] for a website where you can input integers and see what interesting sequences have those as the first few terms.

Exercise 1.2.25: *Hint:* Let x_0 be any point in $[a, b]$. Set $x_1 = f(x_0)$, $x_2 = f(x_1)$, $\ldots$, $x_{n+1} = f(x_n)$. Use the Mean Value Theorem (Theorem A.2.2) to show

$$|x_{n+1} - x_n| = |f(x_n) - f(x_{n-1})| < C|x_n - x_{n-1}|. \tag{C.1}$$

By Induction show $|x_n - x_{n-1}| < C^{n-1}|x_1 - x_0|$. Thus $\{x_n\}_{n=0}^{\infty}$ converges, and it converges to the fixed point.

Exercise 1.2.29: *Hint:* Both sides are the number of ways of partitioning subsets of $\mathcal{O}$ of size ℓ into two sets, where all the elements of the first (resp., second) are in $\mathcal{M}$ (resp., $\mathcal{N}$). Can you find a generalization for a sum of three binomial coefficients?

Exercise 1.2.33: *Hint:* The answer is $\binom{d+n-1}{n-1}$.

Exercise 1.2.34: *Hint:* Use the binomial theorem to write

$$n^k - (n-1)^k = ((n-1) - 1)^k - (n-1)^k = \sum_{k_1=0}^{k-1} \binom{k}{k_1}(n-1)^{k_1}. \tag{C.2}$$

Keep applying the binomial theorem. *General advice:* for problems like this, it is often best to try doing special cases first to detect the pattern. For example, first study $k = 2$ and $k = 3$ and then try to generalize.

Exercise 1.2.35: *Hint:* This problem uses many common techniques: interchange the orders of summation, recognize that we are one term shy of being able to use the binomial theorem and getting $(\ell + 1)^d$ (and this can be remedied by cleverly adding zero), and then note that we have a telescoping series.

Exercise 1.3.4: *Hint:* Use the Euclidean algorithm to find the inverses.

Exercise 1.4.11: *Hint:* Use Dirichlet's Pigeon-Hole Principle (see §A.4).

Exercise 1.4.16: *Hint:* See Exercise 1.4.9.

Exercise 1.4.20: *Hint:* Fix a prime p and proceed by induction on a. Use the bi-nomial theorem to expand $(a + 1)^p - (a + 1)$. Note that for p prime the binomial coefficients are divisible by p except for the first and last. This is a very important property and occurs in many proofs of congruence results.

Exercise 1.4.26: *Hint:* Write a number $a_n a_{n-1} \ldots a_1 a_0$ as

$$a_n a_{n-1} \ldots a_1 a_0 = a_n 10^n + a_{n-1} 10^{n-1} + \cdots + a_1 10 + a_0. \qquad \text{(C.3)}$$

Note $10 = 3 \cdot 3 + 1 \equiv 1 \bmod 3$, $100 = 33 \cdot 3 + 1 \equiv 1 \bmod 3$, and so on.

Exercise 1.4.27: *Hint:* By Fermat's Little Theorem, $10^{p-1} \equiv 10^0 \bmod p$; thus $10^{m(p-1)+k} \equiv 10^k \bmod p$ and we need only determine the remainders of $10^k \bmod p$ for $k \in \{0, 1, \ldots, p-2\}$.

Exercise 1.4.28: *Hint:* Use $10^{\phi(x)} \equiv 1 \bmod x$ to construct a number z whose dig-its are all zeros and ones and $x | z$.

Exercise 1.4.31: *Hint:* Use Exercise 1.4.30 to reduce to the case when m and n are relatively prime by changing x. Look at appropriate powers of xy, using $(xy)^r \equiv x^r y^r \bmod p$.

Exercise 1.4.33: *Hint:* Show any polynomial (that is not identically zero) of de-gree d has at most d roots modulo p by long division. Namely, if a is a root of $f(x) \equiv 0 \bmod p$, then the remainder of $\frac{f(x)}{x-a}$ must be zero. We then have $f(x) = (x - a)g(z)$, with $g(x)$ of smaller degree than $f(x)$.

Exercise 1.5.6: *Hint:* Use the Euclidean algorithm. The ability to quickly find d given e and $(p - 1)(q - 1)$ is crucial in implementing RSA.

Exercise 1.6.4: *Hint:* If $(a, p) = 1$, $a^{\frac{p-1}{2}}$ squared is $a^{p-1} \equiv 1$, so $a^{\frac{p-1}{2}} \equiv \pm 1 \bmod p$.

Chapter 2: Arithmetic Functions

Exercise 2.1.3: *Hint:* For the ϕ-function, apply the inclusion-exclusion principle (see Exercise 2.3.19) to $n - \phi(n)$.

Exercise 2.1.4: *Hints:* For the first part, we need to subtract from p^α those numbers that are *not* coprime to p^α, i.e., multiples of p. The second part follows from multiplicativity and the first part. The third part is straightforward. For the fourth part see Theorem 1.4.22 (this is the raison d'être of the ϕ-function); however, the first three parts allow us to prove this independent of group theory.

Exercise 2.1.15: *Hint:* Write $\phi = \mu * I$, where $I(n) = n$. Now use the fact that I is completely multiplicative.

Exercise 2.2.27: *Hint:* Show that $I_p(s) - p^{-s}\zeta(s)$ has a series expansion that is absolutely convergent at $s = \frac{1}{2}$. We thank Akshay Venkatesh for suggesting this problem.

Exercise 2.3.6: *Hint:* Determining if q is a square modulo p is equivalent to seeing which residue class $q \bmod p$ is in; by Dirichlet's Theorem for Primes in Arithmetic Progressions, to first order for each a relatively prime to p we have the same number of primes congruent to $a \bmod p$.

Exercise 2.3.11: *Hint:* Write x as $a \cdot 2n + b$ for some $b \in \{0, \ldots, 2n - 1\}$.

Exercise 2.3.16: *Hint:* Use Chebyshev's Theorem and a dyadic decomposition. Write the interval $[1, x]$ as a union of intervals of the form $[2^k, 2^{k+1}]$ and estimate the contribution from the primes in each interval.

Chapter 3: Zeta and L-Functions

Exercise 3.1.1: *Hint:* Use the integral test (note this is commonly referred to as a p-series in calculus, where here $p = s$).

Exercise 3.1.2: *Hint:* If a prime p divides ab, prove $p|a$ or $p|b$.

Exercise 3.1.4: *Hint:* Let $S(m, n)$ denote the sum and consider $S(m, n) - rS(m, n)$; this is a telescoping sum. If $|r| < 1$ one may take the limit as $n \to \infty$. This is also an example where a multiple of the original sum is easy to evaluate.

Exercise 3.1.5: *Hint:* To avoid having to deal with infinite sums and products, truncate the sum at N and the product at P, and show the difference tends to 0 as $N, P \to \infty$.

Exercise 3.1.6: *Hint:* An alternate proof of the divergence of the harmonic series is to sum terms 2^k to $2^{k+1} - 1$. There are 2^k such terms, each term at least $\frac{1}{2^k}$. For a proof by differentiating identities, see Appendix D of [Mil4].

Exercise 3.1.11: *Hint:* Take logarithms to convert the infinite product to an infinite sum; this is a common technique to analyze such products.

Exercise 3.1.18: *Hint:* Use the functional equation of the Γ-function: $\Gamma(s + 1) = s\Gamma(s)$.

Exercise 3.1.23: *Hint:* Let $[x]$ denote the greatest integer at most x; thus $x = [x] + \{x\}$. Show

$$\sum_{n=1}^{\infty} \frac{1}{n^s} = \sum_{n=1}^{\infty} n \left[n^{-s} - (n+1)^{-s} \right]. \qquad (C.4)$$

Replace the nth summand with an integral over $[n, n+1]$ times $n = [x]$, and then replace $[x]$ with $x - \{x\}$. See Remark 3.2.22 for complete details.

Exercise 3.1.25: *Hint:* Take the logarithmic derivative of

$$\sin(z) = z \prod_{n=1}^{\infty} \left(1 - \frac{z^2}{\pi^2 n^2} \right) \qquad (C.5)$$

and show

$$z \cot(z) = 1 + \sum_{k=1}^{\infty} \frac{(-4)^k B_{2k}}{(2k)!} z^{2k}$$

$$= 1 + 2 \sum_{n=1}^{\infty} \frac{z^2}{z^2 - n^2 \pi^2} = 1 - 2 \sum_{k=1}^{\infty} \zeta(2k) \frac{z^{2k}}{\pi^{2k}}. \qquad (C.6)$$

Exercise 3.1.26: *Hint:* Use the functional equation.

Exercise 3.2.15: *Hint:* $\log \zeta(s) = -\sum_p \log(1 - p^{-s})$. For $\Re s > 1$ we have $|p^{-s}| < 1$. We take derivatives and expand by using the geometric series formula (3.4).

Exercise 3.2.23: *Hint:* Use the functional equation of the zeta function.

Exercise 3.2.29: *Hint:* The number of changes of sign for $f(t)$ is the same as the number of changes of sign for $f(a - t)$.

Exercise 3.2.40: *Hint:* Use mathematical induction. Write

$$\left[1 - \left(1 - \frac{1}{\rho} \right)^{n+1} \right] = \left[1 - \left(1 - \frac{1}{\rho} \right)^n \right] + \frac{1}{\rho} \left(1 - \frac{1}{\rho} \right)^n.$$

Exercise 3.2.42: *Hint:* Note that the multiset $\mathcal{R}_\zeta$ satisfies $\mathcal{R}_\zeta = 1 - \mathcal{R}_\zeta$, where

$$1 - \mathcal{R}_\zeta = \{r : r = 1 - \rho, \rho \in \mathcal{R}_\zeta\}.$$

For a multiset $\mathcal{R}$ set

$$S(\mathcal{R}) = \sum_{r \in \mathcal{R}} \left[1 - \left(1 - \frac{1}{r} \right)^{-n} \right].$$

Then

$$S(1 - \mathcal{R}_\varsigma) = \sum_{r \in 1 - \mathcal{R}_\varsigma} \left[1 - \left(1 - \frac{1}{r} \right)^{-n} \right]$$

$$= \sum_{\rho \in \mathcal{R}_\varsigma} \left[1 - \left(1 - \frac{1}{1 - \rho} \right)^{-n} \right]$$

$$= \sum_{\rho \in \mathcal{R}_\varsigma} \left[1 - \left(\frac{-\rho}{1 - \rho} \right)^{-n} \right]$$

$$= \sum_{\rho \in \mathcal{R}_\varsigma} \left[1 - \left(\frac{1 - \rho}{-\rho} \right)^{n} \right]$$

$$= \sum_{\rho \in \mathcal{R}_\varsigma} \left[1 - \left(1 - \frac{1}{\rho} \right)^{n} \right] = S(\mathcal{R}_\varsigma).$$

Exercise 3.2.43: *Hint:* Use

$$\left(\frac{\rho}{\rho - 1} \right)^n = e^{n \sum_{k=1}^{\infty} \frac{1}{k\rho^k}} . \tag{C.7}$$

Exercise 3.3.12: *Hint:* Show $rn \equiv y \bmod m$ has a unique solution in $n \in \mathbb{Z}/m\mathbb{Z}$ for all $y \in \mathbb{Z}/m\mathbb{Z}$.

Exercise 3.3.14: *Hint:* Use Exercise 3.3.13.

Exercise 3.3.16: *Hint:* Use Exercise 3.3.14.

Exercise 3.3.20: *Hint:* Use Exercise 3.3.4.

Exercise 3.3.21: *Hint:* Use unique factorization, the complete multiplicativity of χ, and mimic the proof of Theorem 3.1.3.

Exercise 3.3.25: *Hint:* As the only character with a pole is χ_0 (and it is a simple pole of order 1), if a complex character has $L(1, \chi) = 0$ then $L(1, \overline{\chi}) = 0$, and this contradicts $\prod_\chi L(1, \chi) \geq 1$. See Chapters 4 and 6 of [Da2] for complete details.

Exercise 3.3.29: *Hint:* Show

$$\int_0^1 \frac{dx}{1 + x^2} = \frac{\pi}{4} \tag{C.8}$$

by trigonometric substitution ($x = \tan \theta$). Expand $\frac{1}{1+x^2}$ by using the geometric series formula, interchange the integration and summation, and then integrate term by term. Note you must justify both using the geometric series and interchanging the order of integration and summation. One way to do this is to consider the interval $[0, 1 - \epsilon]$ and the first N terms in the geometric series expansion, and show the error from this tends to zero as $\epsilon \to 0$ and $N \to \infty$.

Exercise 3.3.30: *Hint:* Write $\sin^{2m+1}(x)dx$ as $\sin^{2m}(x) \cdot \sin(x)dx$ and integrate by parts, noting that $\cos^2(x) = 1 - \sin^2(x)$.

Exercise 3.3.34: *Hint:* Consider $|c(m,\chi)|^2 = c(m,\chi)\overline{c(m,\chi)}$. Bounding a quantity S by studying $S\overline{S}$ is a very powerful technique; we see this again in Chapters 4, 13 and 14. Studying the absolute value square leads to a sum in two variables, say x and y modulo p. It is best to consider a change of variables $y \to xy \bmod p$; see §4.4.4 for a complete proof.

Chapter 4: Solutions to Diophantine Equations

Exercise 4.2.7: *Hint:* Try factoring $x^3 + ax + b$.

Exercise 4.3.4: *Hint:* See §A.6.

Exercise 4.3.6: *Hint:* Consider squares of length 1 centered at rational points. How many squares does the boundary of the circle with radius $\sqrt{B}$ centered at the origin intersect?

Exercise 4.4.1: *Hint:* If p, q are any two distinct primes, the claim is true for $(x^2 - p)(x^2 - q)(x^2 - pq)$. It is sometimes easier to prove statements in general rather than for specific values. For example, here one could think there is something "special" about 3 and 5, and such thoughts often color our thinking and attempts.

Exercise 4.4.4: *Hint:* $\left(\frac{-1}{p}\right) = (-1)^{\frac{p-1}{2}} = -1$.

Exercise 4.4.5: *Hint:* Show $x \to x^3$ is an automorphism of $(\mathbb{Z}/p\mathbb{Z})^*$.

Chapter 5: Algebraic and Transcendental Numbers

Exercise 5.3.8: *Hint:* For $\cup_{i=0}^{\infty} A_i$, mimic the proof used to show $A \times B$ is countable.

Exercise 5.3.14: *Hint:* Consider four such triangles. Assuming rotation and translation does not change areas (a reasonable assumption), use these four triangles to build a square with sides of length $a + b$ which has a square with sides of length c inside.

Exercise 5.5.2: *Hint:* For a given $e > \tau(\xi)$ choose a small $\delta > 0$ such that $e - \delta >$

$\tau(\xi)$. Then (5.34) becomes

$$\left| \xi - \frac{p}{q} \right| < \frac{C}{q^{\delta}} \frac{1}{q^{e-\delta}}.$$ (C.9)

Exercise 5.5.5: *Hint:* $\left| \xi - \frac{p}{q} \right| < \frac{1}{Q}$ and we may take Q arbitrarily large.

Exercise 5.5.6: *Hint:* Since π is irrational there are infinitely many relatively prime integers p, q such that $\left| \frac{p}{q} - \pi \right| < q^{-2}$. It suffices to show that for infinitely many n, $|\cos n|^{n}$ does not tend to zero. Write

$$\cos n = \cos(n - \pi m_n + \pi m_n) = (-1)^{m_n} \cos(n - \pi m_n).$$ (C.10)

If

$$\left| \frac{n}{m_n} - \pi \right| = \frac{\theta_n}{m_n^2} \le \frac{1}{m_n^2},$$ (C.11)

then $\frac{n}{4} \le m_n \le n$, $\theta_n \le 1$, and Taylor expanding gives

$$\cos(n - \pi m_n) = \cos\left(\frac{\theta_n}{m_n} \right) = 1 - \frac{\theta_n^2}{m_n^2} + O\left(\frac{1}{m_n^4} \right).$$ (C.12)

The proof is completed by raising to the nth power and using $e^{-\theta_n^2 \cdot n / m_n^2}$ tends to 1 as $n \to \infty$.

Exercise 5.6.2: *Hint:* If $f(x) = g(x)h(x)$ with $g(x)$ and $h(x)$ polynomials with rational coefficient, clear denominators.

Chapter 7: Introduction to Continued Fractions

Exercise 7.1.1: *Hint:* It suffices to prove this for $x = \frac{p}{q}$ with $p < q$ relatively prime. Assume $(q, 10) = 1$ and use Theorem 1.4.22 to find a k such that $10^k \equiv 1 \bmod q$. Then $10^k x - x$ is an integer, and the proof of this case is completed by noting that zero's decimal expansion is unique. Generalize the proof to any q; note the only difficulty will be when there are unequal powers of 2 and 5 in q's factorization, but these can be handled easily. Alternatively, this can be shown by using Dirichlet's Pigeon-Hole Principle: as we divide p by q, eventually we get the same remainder twice in the divisions; use this to bound the size of the denominator.

Exercise 7.2.9: *Hint:* Often one must write a remainder r as $\frac{1}{1/r}$ and then "rationalize" the denominator. For example, if $r = 3 + 2\sqrt{5}$ then $\frac{1}{r} = \frac{-3 + 2\sqrt{5}}{-3^2 + 2^2 5}$.

Exercise 7.6.22: *Hint:* Use the method of §4.1.1.

Exercise 7.7.7: *Hint:* It suffices to verify this for the matrices S, T (and their inverses) from Lemma 7.7.5).

Chapter 8: Introduction to Probability

Exercise 8.1.18: *Hint:* Let a_n be the probability that there are at least 3 consecutive heads in n tosses. Show a_n satisfies the recurrence relation

$$a_n = \frac{1}{2}a_{n-1} + \frac{1}{4}a_{n-2} + \frac{1}{8}a_{n-3} + \frac{1}{8}. \tag{C.13}$$

The presence of the final term, $\frac{1}{8}$, greatly complicates matters; we cannot use the methods of Exercise 7.3.9 or §9.3 to solve the recurrence relation. It is much easier to study b_n, the probability that there are not 3 consecutive heads in n tosses; note $a_n = 1 - b_n$. Show b_n satisfies

$$b_n = \frac{1}{2}b_{n-1} + \frac{1}{4}b_{n-2} + \frac{1}{8}b_{n-3}. \tag{C.14}$$

More generally, determine the probability of observing at least k heads in n tosses of a coin with probability p of heads. If $p = \frac{1}{2}$ show that the roots of the characteristic polynomial of the recurrence relation are at most $\left(1 - 2^{-k}\right)^{1/k}$. One application of this is to roulette, where the probability of getting red (or black) is $16/38$ because there are two green spaces. This shows there is a large enough probability of consecutive losses so that the strategy of double plus one (bet \$1 on the first spin; if you lose bet \$2 on the second, if you lose again bet \$4 on the third, if you lose again bet \$8 on the fourth, and so on; it does not matter when your color finally comes up – you always win \$1) will fail in general, as too quickly you reach the house limit (maximum allowable bet) and lose a lot.

Exercise 8.1.31: *Hint:* Let $X_{[m]}$ denote the largest of player one's rolls, and $Y_{[n]}$ the largest of player two's rolls. For $a \in \{1, \ldots, k\}$,

$$\text{Prob}(X_{[m]} = a) = \frac{a^m - (a-1)^m}{k^m}; \tag{C.15}$$

this follows from

$$\text{Prob}(X_{[m]} = a) = \sum_{\ell=1}^{m} \binom{m}{\ell} \frac{1}{k^\ell} \left(\frac{a-1}{k}\right)^{m-\ell}, \tag{C.16}$$

the binomial theorem and noticing we have a telescoping sum. The proof is completed by noting that

$$\text{Prob(Player one wins)} = \sum_{a=2}^{k} \text{Prob}(X_{[m]} = a) \cdot \text{Prob}(Y_{[n]} \leq a - 1). \tag{C.17}$$

$X_{[m]}$ and $Y_{[n]}$ are examples of order statistics; see also Exercise 12.7.7.

Exercise 8.2.7: *Hint:* Let

$$f(\lambda) = \sum_{k=0}^{\infty} \frac{\lambda^k}{k!} - e^\lambda \tag{C.18}$$

Differentiate once to determine the mean, twice to determine the variance.

Chapter 9: Applications of Probability: Benford's Law and Hypothesis Testing

Exercise 9.3.3: *Hint:* Consider $a_0 = a_1 = a_2 = 1$. This recurrence relation was constructed by starting with the characteristic polynomial $(r-2)^2(r-1)$ and then finding initial conditions so that the coefficients of the $\lambda_1 = \lambda_2 = 2$ eigenvalues vanish. In searching for counter-examples, it is significantly easier here to specify the roots of the characteristic polynomial first, and find the actual recurrence relation second.

Exercise 9.3.4: *Hint:* Consider a recurrence relation of length k with k distinct roots. By specifying k terms (say $a_0, \ldots, a_{k-1}$), the coefficients of the roots λ_i are determined. We must solve

$$u_1 \lambda_1^n + \cdots + u_k \lambda_k^n = a_n, \quad n \in \{0, \ldots, k-1\}. \tag{C.19}$$

We may write this in matrix form as

$$
\begin{pmatrix}
1 & 1 & \cdots & 1 \\
\lambda_1 & \lambda_2 & \cdots & \lambda_k \\
\lambda_1^2 & \lambda_2^2 & \cdots & \lambda_k^2 \\
\vdots & \vdots & \ddots & \vdots \\
\lambda_1^{k-1} & \lambda_2^{k-1} & \cdots & \lambda_k^{k-1}
\end{pmatrix}
\begin{pmatrix}
u_1 \\
u_2 \\
u_3 \\
\vdots \\
u_k
\end{pmatrix}
=
\begin{pmatrix}
a_0 \\
a_1 \\
a_2 \\
\vdots \\
a_{k-1}
\end{pmatrix}. \tag{C.20}
$$

The matrix of eigenvalues is a Vandermonde matrix; by Exercise 17.3.9 its determinant is non-zero when $\lambda_i \neq \lambda_j$. Thus its inverse exists, and the initial conditions which lead to $u_1 = 0$ are a hyperplane in $\mathbb{C}^k$, which shows that almost all initial conditions lead to $u_1 \neq 0$.

Chapter 12: $\{n^k \alpha\}$ and Poissonian Behavior

Exercise 12.3.12: *Hint:* If $z_n = n\alpha$, then $z_{n+h} - z_n = h\alpha$.

Exercise 12.3.13: *Hint:* If $z_n = n^2\alpha$, then $z_{n+h} - z_n = n(2h\alpha) + h^2\alpha$.

Exercise 12.3.14: *Hint:* Proceed by induction.

Exercise 12.3.22: *Hint:* Use Weyl's criterion. In order to handle the summation, use Euler's summation formula (Lemma 2.2.9).

Exercise 12.7.4: *Hint:* Assume the x_i's are drawn from a continuous distribution $f(x)$ such that, when $f(x) \neq 0$, $f(x)$ has at least a second order Taylor series. For each N, consider intervals $[a_N, b_N]$ such that $\int_{a_N}^{b_N} f_N(x)dx = N^\delta/N$; thus the proportional of the total mass in such intervals is $N^{\delta-1}$. By the Central Limit Theorem, the number of x_i in this interval is $N^\delta + O(N^{\delta/2})$. The average spacing between adjacent x_i is of size $1/Nf(a_N)$ plus lower order terms. The analysis is completed by arguing as in the proof of Theorem 12.7.3.

Exercise 12.7.8: *Hint:* Use Stirling's formula to estimate the factorials. For t slightly smaller than $\widetilde{\mu}$,

$$\int_{-\infty}^{t} p(x)dx = \frac{1}{2} - p(\widetilde{\mu})(\widetilde{\mu} - t) - \frac{p'(\widetilde{\mu}) \cdot (\widetilde{\mu} - t)^2}{2!} + O\left((\widetilde{\mu} - t)^3\right); \quad \text{(C.21)}$$

note unless t is near $\widetilde{\mu}$ the integral is negligible; t greater than $\widetilde{\mu}$ is handled similarly.

Exercise 12.7.12: *Hint:* See Exercise 12.6.3.

Chapter 13: Introduction to the Circle Method

Exercise 13.1.3: *Hint:* The combinatorial bounds from §1.2.4 might be a useful starting point.

Exercise 13.1.6: *Hint:*

$$(1 + x)(1 + x^2)(1 + x^3) \cdots = \frac{1 - x^2}{1 - x} \frac{1 - x^4}{1 - x^2} \frac{1 - x^6}{1 - x^3} \cdots$$

$$= \frac{1}{(1 - x)(1 - x^3)(1 - x^5) \cdots}. \quad \text{(C.22)}$$

Exercise 13.3.8: *Hint:* Rewrite the sum as two sums by using the Legendre symbol (see §1.6.1).

Chapter 14: Circle Method: Heuristics for Germain Primes

Exercise 14.2.2: *Hint:* For $F_N(x)$, use the Cauchy-Schwartz inequality.

Exercise 14.6.19: *Hint:* Take the logarithm, and Taylor expand.

Chapter 15: From Nuclear Physics to L-Functions

Exercise 15.4.7: *Hint:* Fix s and a large integer m and let $I_j = [\frac{j-1}{m}s, \frac{j}{m}s]$ for $1 \leq j \leq m$. The first assumption gives an extremely simple (approximate) formula for the probability of observing an eigenvalue in I_j (in terms of an unknown proportionality constant a). Now the probability of a gap of size about s is equal to the probability that $I_1, \ldots, I_{m-1}$ contain no eigenvalues but I_m contains an eigenvalue. Let $m \to \infty$ to find a formula for the eigenvalue spacing in terms of a and use the unit mean spacing condition to determine the value of a.

Exercise 15.6.5: *Hint:* Let $\vec{v}$ be an eigenvector of U and consider $(U\vec{v})^*(U\vec{v})$.

Chapter 16: Random Matrix Theory: Eigenvalue Densities

Exercise 16.1.13: *Hint:* Write the tuple as $a_{i_1 i_2} a_{i_2 i_3} \cdots a_{i_k i_1} = a_{x_1 y_1}^{r_1} \cdots a_{x_\ell y_\ell}^{r_\ell}$. If any pair (x, y) occurs only once, show $\mathbb{E}[a_{x_1 y_1}^{r_1} \cdots a_{x_\ell y_\ell}^{r_\ell}] = 0$. Thus each $r_j \geq 2$, and $\sum_j r_j = k$. Show there are $o(N^{1+\frac{k}{2}})$ such tuples, implying these terms do not contribute.

Exercise 16.1.15: *Hint:* Any tuple with an $r_j = 1$ contributes 0. Show there are at most $o(N^{1+m})$ tuples where all $r_j \geq 2$ and at least one $r_j \geq 3$.

Exercise 16.1.19: *Hint:* Prove a recursion relation.

Chapter 17: Random Matrix Theory: Spacings between Adjacent Eigenvalues

Exercise 17.2.2: *Hint:* Use trigonometric identities to simplify the resulting form. Solve $(A - \lambda_1 I)\vec{v}_1 = \vec{0}$.

Exercise 17.2.6: *Hint:* Express the polynomial in terms of the eigenvalues.

Exercise 17.3.9: *Hint:* Prove the recursion relation.

Chapter 18: The Explicit Formula and Density Conjectures

Exercise 18.2.1: *Hint:* Consider comparisons involving 0 and i, and remember that multiplication by -1 reverses inequalities.

Exercise 18.2.4: *Hint:* Argue along the lines of Remark 18.2.3.

Appendix A: Analysis Review

Exercise A.4.8: *Hint:* Use inclusion-exclusion to obtain upper and lower bounds. For example, if there are m balls, let $E_{k;N,m}$ denote the event that at least one of the N boxes has at least k of the m balls, $E_{k,i;N,m}$ the event that there are at least k balls in box i, and $F_{n,i;N,m}$ the event that there are exactly n balls in box i. Then

$$E_{k;N,m} = \bigcup_{i=1}^{N} E_{k,i;N,m}, \quad E_{k,i;N,m} = \bigcup_{n=k}^{m} F_{n,i;N,m}.$$

If $|E|$ is the probability of E, by symmetry $|E_{k;N,m}| \leq N \cdot |E_{k,1;N,m}|$, and

$$F_{n,1;N,m} = \sum_{n=k}^{m} |F_{n,1;N,m}|, \quad F_{n,1;N,m} = \binom{m}{n} \frac{1}{N^k} \binom{m-n}{m-n} \left(1 - \frac{1}{N}\right)^{m-n}.$$

For a lower bound one must analyze events such as $F_{n_1,i_1,n_2,i_2;N,m}$, the event of exactly n_1 balls in box i_1 and n_2 objects in box i_2.

Appendix D

Concluding Remarks

This book is meant as an introduction to a vast, active subject. It is our hope that the reader will pursue these topics further through the various projects and references mentioned in the introduction and chapters above. We also hope that we have shown how similar tools, techniques and concepts arise in different parts of mathematics. We briefly summarize some of what we have seen.

The first is the Philosophy of Square Root Cancellation. As a general principle, many "nice" sums of N terms of absolute value 1 are approximately of size $\sqrt{N}$. Examples range from the Gauss sums of §3.3.5 (which were then used in our investigations of the number of solutions to Diophantine equations in §4.4.3) to the average value of generating functions encountered in the Circle Method in Chapters 13 and 14 to the Central Limit Theorem of §8.4 (which shows that for a wide class of populations, the distribution of the mean of a large sample is independent of the fine properties of the underlying distribution).

Similar to the universality of the Central Limit Theorem, many different systems after normalization follow the same spacing laws. We have seen numerical and theoretical evidence showing that spacings between primes, the fractional parts of $n^k\alpha$ (for certain k and α) and numbers uniformly chosen in $[0,1]$ are the same (see Chapter 12), while in Chapters 15 to 18 we see similar behavior in energy levels of heavy nuclei, eigenvalues of matrices (of random matrix ensembles as well as adjacency matrices attached to d-regular graphs) and zeros of L-functions.

Throughout our investigations, certain viewpoints have consistently proven useful. Among the most important are Fourier Analysis (Chapter 11) and the structure of numbers (Chapters 2 and 7). From Fourier Analysis we obtain Poisson Summation and the Fourier Transform (which are useful for investigating problems as varied as the first digits of sequences (§9.4.2), the functional equation of $\zeta(s)$ (§3.1.2) and in Chapter 18 the zeros of L-functions). Other applications range from Weyl's Theorem (Chapter 12.3) on the equidistribution of sequences to the Circle Method and representing numbers as the sum of primes or integer powers (§13 and 14). We have used the structure of numbers in finding good rational approximations (§7.9), Roth's Theorem (Chapter 6), and studying the properties of $n^k\alpha$ mod 1 (Chapter 12).

Finally, we have tried to emphasize in the text which techniques appear throughout mathematics. Some of the most common are adding zero or multiplying by one, divide and conquer, dyadic decomposition, no integers are in $(0,1)$, the Pigeon-Hole Principle, positivity, and splitting integrals or sums; see the *techniques* entry in the index for more details.

Bibliography

Links to many of the references below are available online at
http://www.math.princeton.edu/mathlab/book/index.html

[Acz] A. Aczel, *Fermat's Last Theorem: Unlocking the Secret of an Ancient Mathematical Problem*, Four Walls Eight Windows, New York, 1996.

[AKS] R. Adler, M. Keane, and M. Smorodinsky, *A construction of a normal number for the continued fraction transformation*, J. Number Theory **13** (1981), no. 1, 95–105.

[AgKaSa] M. Agrawal, N. Kayal and N. Saxena, *PRIMES is in P*, Ann. of Math. (2) **160** (2004), no. 2, 781–793.

[Al] L. Ahlfors, *Complex Analysis*, 3rd edition, McGraw-Hill, New York, 1979.

[AZ] M. Aigner and G. M. Ziegler, *Proofs from THE BOOK*, Springer-Verlag, Berlin, 1998.

[AGP] W. R. Alford, A. Granville, and C. Pomerance, *There are infinitely many Carmichael numbers*, Ann. Math. **139** (1994), 703–722.

[AMS] AMS MathSciNet, http://www.ams.org/msnmain?screen=Review

[AB] U. Andrews IV and J. Blatz, *Distribution of digits in the continued fraction representations of seventh degree algebraic irrationals*, Junior Thesis, Princeton University, Fall 2002.

[Ap] R. Apéry, *Irrationalité de $\zeta(2)$ et $\zeta(3)$*, Astérisque **61** (1979) 11–13.

[Apo] T. Apostol, *Introduction to Analytic Number Theory*, Springer-Verlag, New York, 1998.

[ALM] S. Arms, A. Lozano-Robledo and S. J. Miller, *Constructing One-Parameter Families of Elliptic Curves over $\mathbb{Q}(T)$ with Moderate Rank*, preprint.

[Art] M. Artin, *Algebra*, Prentice-Hall, Englewood Cliffs, NJ, 1991.

[Ay] R. Ayoub, *Introduction to the Analytic Theory of Numbers*, AMS, Providence, RI, 1963.

[Bai] Z. Bai, *Methodologies in spectral analysis of large-dimensional random matrices, a review*, Statist. Sinica **9** (1999), no. 3, 611–677.

[B] A. Baker, *Transcendental Number Theory*, Cambridge University Press, Cambridge, 1990.

[BM] R. Balasubramanian and C. J. Mozzochi, *Siegel zeros and the Goldbach problem*, J. Number Theory **16** (1983), no. 3, 311–332.

[BR] K. Ball and T. Rivoal, *Irrationalité d'une infinité valeurs de la fonction zeta aux entiers impairs*, Invent. Math. **146** (2001), 193–207.

[BT] V. V. Batyrev and Yu. Tschinkel, *Tamagawa numbers of polarized algebraic varieties*, Nombre et répartition de points de hauteur bornée (Paris, 1996), Astérisque (1998), No. 251, 299–340.

[BL] P. Baxandall and H. Liebeck, *Vector Calculus*, Clarendon Press, Oxford, 1986.

[Be] R. Beals, *Notes on Fourier series*, Lecture Notes, Yale University, 1994.

[Bec] M. Beceanu, *Period of the continued fraction of $\sqrt{n}$*, Junior Thesis, Princeton University, 2003.

[Ben] F. Benford, *The law of anomalous numbers*, Proceedings of the American Philosophical Society **78** (1938) 551–572.

[BBH] A. Berger, Leonid A. Bunimovich, and T. Hill, *One-dimensional dynamical systems and Benford's Law*, Trans. Amer. Math. Soc. **357** (2005), no. 1, 197–219.

[BEW] B. Berndt, R. Evans, and K. Williams, *Gauss and Jacobi Sums*, Canadian Mathematical Society Series of Monographs and Advanced Texts, Vol. 21, Wiley-Interscience Publications, John Wiley & Sons, New York, 1998.

[Ber] M. Bernstein, *Games, hats, and codes*, lecture at the SUMS 2005 Conference.

[BD] P. Bickel and K. Doksum, *Mathematical Statistics: Basic Ideas and Selected Topics*, Holden-Day, San Francisco, 1977.

[Bi] P. Billingsley, *Probability and Measure*, 3rd edition, Wiley, New York, 1995.

[Bl1] P. Bleher, *The energy level spacing for two harmonic oscillators with golden mean ratio of frequencies*, J. Stat. Phys. **61** (1990) 869–876.

[Bl2] P. Bleher, *The energy level spacing for two harmonic oscillators with generic ratio of frequencies*, J. Stat. Phys. **63** (1991), 261–283.

[Bob] J. Bober, *On the randomness of modular inverse mappings*, Undergraduate Mathematics Laboratory report, Courant Institute, NYU, 2002.

[Bol] B. Bollobás, *Random Graphs*, Cambridge Studies in Advanced Mathematics, Cambridge University Press, Cambridge, 2001.

[BoLa] E. Bombieri and J. Lagarias, *Complements to Li's criterion for the Riemann ypothesis*, J. Number Theory **77** (1999), no. 2, 274–287.

[BP] E. Bombieri and A. van der Poorten, *Continued fractions of algebraic numbers*. Pages 137–152 in *Computational Algebra and Number Theory (Sydney, 1992)*, Mathematical Applications, Vol. 325, Kluwer Academic, Dordrecht, 1995.

[Bon] D. Boneh, *Twenty years of attacks on the RSA cryptosystem*, Notices of the American Mathematical Society, **46** (1999), no. 2, 203–213.

[BS] Z. Borevich and I. Shafarevich, *Number Theory*, Academic Press, New York, 1968.

[BB] J. Borwein and P. Borwein, *Pi and the AGM: A Study in Analytic Number Theory and Computational Complexity*, John Wiley and Sons, New York, 1987.

[BK] A. Boutet de Monvel and A. Khorunzhy, *Some elementary results around the Wigner semicircle law*, lecture notes.

[BoDi] W. Boyce and R. DiPrima, *Elementary differential equations and boundary value problems*, 7th edition, John Wiley & Sons, New York, 2000.

[Bre1] R. Brent, *The distribution of small gaps between successive primes*, Math. Comp. **28** (1974), 315–324.

[Bre2] R. Brent, *Irregularities in the distribution of primes and twin primes*, Collection of articles dedicated to Derrick Henry Lehmer on the occasion of his seventieth birthday, Math. Comp. **29** (1975), 43–56.

[BPR] R. Brent, A. van der Poorten, and H. te Riele, *A comparative study of algorithms for computing continued fractions of algebraic numbers*. Pages 35–47 in *Algorithmic number theory (Talence, 1996)*, Lecture Notes in Computer Science, Vol. 1122, Springer, Berlin, 1996.

[deBr] R. de la Bretèche, *Sur le nombre de points de hauteur bornée d'une certaine surface cubique singulière*. Pages 51–77 in *Nombre et répartition de points de hauteur bornée (Paris, 1996)*, Astérisque, (1998) no. 251, 51–77.

[BBD] R. de la Bretèche, T. D. Browning, and U. Derenthal, *On Manin's conjecture for a certain singular cubic surface*, preprint.

[BPPW] B. Brindza, A. Pintér, A. van der Poorten, and M. Waldschmidt, *On the distribution of solutions of Thue's equation*. Pages 35–46 in *Number theory in progress (Zakopane-Koscielisko, 1997)*, Vol. 1, de Gruyter, Berlin, 1999.

[BFFMPW] T. Brody, J. Flores, J. French, P. Mello, A. Pandey, and S. Wong, *Random-matrix physics: spectrum and strength fluctuations*, Rev. Mod. Phys. **53** (1981), no. 3, 385–479.

[BrDu] J. Brown and R. Duncan, *Modulo one uniform distribution of the sequence of logarithms of certain recursive sequences*, Fibonacci Quarterly **8** (1970) 482–486.

[Bro] T. Browning, *The density of rational points on a certain singular cubic surface*, preprint.

[BDJ] W. Bryc, A. Dembo, T. Jiang, *Spectral measure of large random Hankel, Markov and Toeplitz matrices*, preprint.

[Bry] A. Bryuno, *Continued frations of some algebraic numbers*, U.S.S.R. Comput. Math. & Math. Phys. **4** (1972), 1–15.

[Bur] E. Burger, *Exploring the Number Jungle: A Journey into Diophantine Analysis*, AMS, Providence, RI, 2000.

[BuP] E. Burger and A. van der Poorten, *On periods of elements from real quadratic number fields*. Pages 35–43 in *Constructive, Experimental, and Nonlinear Analysis (Limoges, 1999)*, CMS Conf. Proc., **27**, AMS, Providence, RI, 2000.

[CaBe] G. Casella and R. Berger, *Statistical Inference*, 2nd edition, Duxbury Advanced Series, Pacific Grove, CA, 2002.

[CGI] G. Casati, I. Guarneri, and F. M. Izrailev, *Statistical properties of the quasienergy spectrum of a simple integrable system*, Phys. Lett. A **124** (1987), 263–266.

[Car] L. Carleson, *On the convergence and growth of partial sums of Fourier series*, Acta Math. **116** (1966), 135–157.

[Ca] J. W. S. Cassels, *An Introduction to Diophantine Approximation*, Cambridge University Press, London 1957.

[Ch] D. Champernowne, *The construction of decimals normal in the scale of ten*, J. London Math. Soc. 8 (1933), 254–260.

[Cha] K. Chang, *An experimental approach to understanding Ramanujan graphs*, Junior Thesis, Princeton University, Spring 2001.

[ChWa] J. R. Chen and T. Z. Wang, *On the Goldbach problem*, Acta Math. Sinica **32** (1989), 702–718.

[Chr] J. Christiansen, *An introduction to the moment problem*, lecture notes.

[Ci] J. Cisneros, *Waring's problem*, Junior Thesis, Princeton University, Spring 2001.

[CW] J. Coates and A. Wiles, *On the conjecture of Birch and Swinnterton-Dyer*, Invent. Math. **39** (1977), 43–67.

[CB] S. Chatterjee and A. Bose, *A new method for bounding rates of convergence of empirical spectral distributions*, J. Theoret. Probab. **17** (2004), no. 4, 1003–1019.

[CL1] H. Cohen and H. W. Lenstra, Jr., *Heuristics on class groups of number fields*. Pages 33–62 in *Number Theory*, Lecture Notes in Mathematics, Vol. 1068, Springer-Verlag, Berlin, 33–62.

[CL2] H. Cohen and H. W. Lenstra, Jr., *Heuristics on class groups*, in *Number Theory*, Lecture Notes in Mathematics, Vol. 1052, Springer-Verlag, Berlin, 26–36.

[Coh] P. Cohen, *The independence of the continuum hypothesis*, Proc. Nat. Acad. Sci. U.S.A, **50** (1963), 1143–1148; **51** (1964), 105–110.

[Cohn] J. Cohn, *The length of the period of simple continued fractions*, Pacific Journal of Mathematics, **71** (1977), no. 1, 21–32.

[Con1] J. B. Conrey, *L-Functions and random matrices*. Pages 331–352 in *Mathematics unlimited − 2001 and Beyond*, Springer-Verlag, Berlin, 2001.

[Con2] J. B. Conrey, *The Riemann hypothesis*, Notices of the AMS, **50** (2003), no. 3, 341–353.

[CFKRS] B. Conrey, D. Farmer, P. Keating, M. Rubinstein and N. Snaith, *Integral moments of L-functions*, preprint.

[Conw] J. H. Conway, *The weird and wonderful chemistry of audioactive decay*. Pages 173–178 in *Open Problems in Communications and Computation*, ed. T. M. Cover and B. Gopinath, Springer-Verlag, New York, 1987.

[CG] J. H. Conway and R. Guy, *The Book of Numbers*, Springer-Verlag, Berlin, 1996.

[CS] J. H. Conway and N. J. A. Sloane, *Lexicographic Codes: Error-Correcting Codes from Game Theory*, IEEE Trans. Inform. Theory, **32** (1986), no. 3, 219–235.

[Corl] R. M. Corless,*Continued fractions and chaos*. Amer. Math. Monthly **99** (1992), no. 3, 203–215.

[Cor1] Cornell University, *arXiv*, http://arxiv.org

[Cor2] Cornell University, *Project Euclid*, http://projecteuclid.org/

[CFS] I. P. Cornfeld, S. V. Fomin, and I. G. Sinai, *Ergodic Theory*, Grundlehren Der Mathematischen Wissenschaften, Springer-Verlag, Berlin, 1982.

[Da1] H. Davenport, *The Higher Arithmetic: An Introduction to the Theory of Numbers*, 7th edition, Cambridge University Press, Cambridge, 1999.

[Da2] H. Davenport, *Multiplicative Number Theory*, 2nd edition, revised by H. Montgomery, Graduate Texts in Mathematics, Vol. 74, Springer-Verlag, New York, 1980.

[Da3] H. Davenport, *On the distribution of quadratic residues (mod p)*, London Math. Soc. **6** (1931), 49–54.

[Da4] H. Davenport, *On character sums in finite fields*, Acta Math. **71** (1939), 99–121.

[DN] H. A. David and H. N. Nagaraja, *Order Statistics*, 3rd edition, Wiley Interscience, Hoboken, NJ, 2003.

[DSV] G. Davidoff, P. Sarnak, and A. Valette, *Elementary Number Theory, Group Theory, and Ramanujan Graphs*, London Mathematical Society, Student Texts, Vol. 55, Cambridge University Press, Cambridge 2003.

[Dev] R. Devaney, *An Introduction to Chaotic Dynamical Systems*, 2nd edition, Westview Press, Cambridge, MA, 2003.

[Dia] P. Diaconis, *Patterns in eigenvalues: the 70th Josiah Williard Gibbs lecture*, Bulletin of the American Mathematical Society, **40** (2003), no. 2, 155–178.

[Di] T. Dimofte, *Rational shifts of linearly periodic continued fractions*, Junior Thesis, Princeton University, 2003.

[DM] E. Dueñez and S. J. Miller, *The Low Lying Zeros of a* $GL(4)$ *and a* $GL(6)$ *family of L-functions*, preprint.

[Du] R. Durrett, *Probability: Theory and Examples*, 2nd edition, Duxbury Press, 1996.

[Dy1] F. Dyson, *Statistical theory of the energy levels of complex systems: I, II, III*, J. Mathematical Phys., **3** (1962) 140–156, 157–165, 166–175.

[Dy2] F. Dyson, *The threefold way. Algebraic structure of symmetry groups and ensembles in quantum mechanics*, J. Mathematical Phys., **3** (1962) 1199–1215.

[Edg] G. Edgar, *Measure, Topology, and Fractal Geometry*, 2nd edition, Springer-Verlag, 1990.

[Ed] H. M. Edwards, *Riemann's Zeta Function*, Academic Press, New York, 1974.

[EST] B. Elias, L. Silberman and R. Takloo-Bighash, *On Cayley's theorem*, preprint.

[EE] W. J. Ellison and F. Ellison, *Prime Numbers*, John Wiley & Sons, New York, 1985.

[Est1] T. Estermann, *On Goldbach's problem: Proof that almost all even positive integers are sums of two primes*, Proc. London Math. Soc. Ser. 2 **44** (1938) 307–314.

[Est2] T. Estermann, *Introduction to Modern Prime Number Theory*, Cambridge University Press, Cambridge, 1961.

[Fal] K. Falconer, *Fractal Geometry: Mathematical Foundations and Applications*, 2nd edition, John Wiley & Sons, New York, 2003.

[Fef] C. Fefferman, *Pointwise convergence of Fourier series*, Ann. of Math. Ser. 2 **98** (1973), 551–571.

[Fe] W. Feller, *An Introduction to Probability Theory and Its Applications*, 2nd edition, Vol. II, John Wiley & Sons, New York, 1971.

[Fi] D. Fishman, *Closed form continued fraction expansions of special quadratic irrationals*, Junior Thesis, Princeton University, 2003.

[Fol] G. Folland, *Real Analysis: Modern Techniques and Their Applications*, 2nd edition, Pure and Applied Mathematics, Wiley-Interscience, New York, 1999.

[For] P. Forrester, *Log-gases and random matrices*, book in progress.

[Fou] E. Fouvry, *Sur la hauteur des points d'une certaine surface cubique singulière*. In *Nombre et répartition de points de hauteur bornée (Paris, 1996)*, Astérisque, (1999) no. 251, 31–49.

[FSV] P. J. Forrester, N. C. Snaith, and J. J. M. Verbaarschot, *Developments in Random Matrix Theory*. In *Random matrix theory*, J. Phys. A **36** (2003), no. 12, R1–R10.

[Fr] J. Franklin, *Mathematical Methods of Economics: Linear and Nonlinear Programming, Fixed-Point Theorem*, Springer-Verlag, New York, 1980.

[Ga] P. Garrett, *Making, Breaking Codes: An Introduction to Cryptography*, Prentice-Hall, Englewood Cliffs, NJ, 2000.

[Gau] M. Gaudin, *Sur la loi limite de l'espacement des valeurs propres d'une matrice aléatoire*, Nucl. Phys. **25** (1961) 447–458.

[Gel] A. O. Gelfond, *Transcendental and Algebraic Numbers*, Dover, New York, 1960.

[Gl] A. Gliga, *On continued fractions of the square root of prime numbers*, Junior Thesis, Princeton University, 2003.

[Gö] K. Gödel, *On Formally Undecidable Propositions of Principia Mathematica and Related Systems*, Dover, New York, 1992.

[Gol1] D. Goldfeld, *The class number of quadratic fields and the conjectures of Birch and Swinnerton-Dyer*, Ann. Scuola Norm. Sup. Pisa Cl. Sci. 3, **4** (1976), 624–663.

[Gol2] D. Goldfeld, *The Elementary proof of the Prime Number Theorem, An Historical Perspective*. Pages 179–192 in *Number Theory, New York Seminar 2003*, eds. D. and G. Chudnovsky, M. Nathanson, Springer-Verlag, New York, 2004.

[Gold] L. Goldmakher, *On the limiting distribution of eigenvalues of large random regular graphs with weighted edges*, American Institute of Mathematics Summer REU, 2003.

[GV] D. A. Goldston and R. C. Vaughan, *On the Montgomery-Hooley asymptotic formula*. Pages 117–142 in *Sieve Methods, Exponential Sums and their Applications in Number Theory*, ed. G. R. H. Greaves, G. Harman, and M. N. Huxley, Cambridge University Press, Cambridge, 1996.

[GG] M. Golubitsky and V. Guillemin, *Stable Mappings and Their Singularities*, Graduate Texts in Mathematics, Vol. 14, Springer-Verlag, New York, 1973.

[GKP] R. L. Graham, D. E. Knuth, and O. Patashnik, *Concrete Mathematics: A Foundation for Computer Science*, Addison-Wesley, Reading, MA, 1988.

[GK] A. Granville and P. Kurlberg, *Poisson statistics via the Chinese remainder theorem*, preprint.

[GT] A. Granville and T. Tucker, *It's as easy as abc*, Notices of the AMS, **49** (2002), no. 10, 224–1231.

[GZ] B. Gross and D. Zagier, *Heegner points and derivatives of L-series*, Invent. Math. **84** (1986), no. 2, 225–320.

[Guy] R. Guy, *Unsolved Problems in Number Theory (Problem Books in Mathematics)*, 2nd edition, Springer-Verlag, New York, 1994.

[HM] C. Hammond and S. J. Miller, *Eigenvalue spacing distribution for the ensemble of real symmetric Toeplitz matrices*, Journal of Theoretical Probability **18** (2005), no. 3, 537–566.

[HL1] G. H. Hardy and J. E. Littlewood, *A new solution of Waring's problem*, Q. J. Math. **48** (1919), 272–293.

[HL2] G. H. Hardy and J. E. Littlewood, *Some problems of "Partitio Numerorum." A new solution of Waring's problem*, Göttingen Nach. (1920), 33–54.

[HL3] G. H. Hardy and J. E. Littlewood, *Some problems of "Partitio Numerorum." III. On the expression of a number as a sum of primes*, Acta Math. **44** (1923), 1–70.

[HL4]   G. H. Hardy and J. E. Littlewood, *Some problems of "Partitio Numerorum."* *IV. Further researches in Waring's problem*, Math. Z., **23** (1925) 1–37.

[HW]   G. H. Hardy and E. Wright, *An Introduction to the Theory of Numbers*, 5th edition, Oxford Science Publications, Clarendon Press, Oxford, 1995.

[HR]   G. H. Hardy and S. Ramanujan, *Asymptotic formulae in combinatorial analysis*, Proc. London Math. Soc. **17** (1918), 75–115.

[Hata]   R. Hata, *Improvement in the irrationality measures of π and π^2*, Proc. Japan. Acad. Ser. A Math. Sci. **68** (1992), 283–286.

[Ha]   B. Hayes, *The spectrum of Riemannium*, American Scientist, **91** (2003), no. 4, 296–300.

[He]   R. Heath-Brown, *The density of rational points on Cayley's cubic surface*, preprint.

[Hei]   H. Heillbronn, *On the average length of a class of finite continued fractions.* In *Number Theory and Analysis (A collection of papers in honor of E. Landau)*, VEB Deutscher Verlag, Berlin, 1968.

[Hej]   D. Hejhal, *On the triple correlation of zeros of the zeta function*, Internat. Math. Res. Notices (1994), no. 7, 294–302.

[Hil]   D. Hilbert, *Beweis für die Darstellbarkeit der ganzen zahlen durch eine feste Anzahl n^{ter} Potenzen (Waringsches Problem)*, Mat. Annalen **67** (1909), 281–300.

[Hi1]   T. Hill, *The first-digit phenomenon*, American Scientist **86** (1996), 358–363.

[Hi2]   T. Hill, *A statistical derivation of the significant-digit law*, Statistical Science **10** (1996), 354–363.

[HS]   M. Hindry and J. Silverman, *Diophantine Geometry: An Introduction*, Graduate Texts in Mathematics, Vol. 201, Springer-Verlag, New York, 2000.

[HJ]   K. Hrbacek and T. Jech, *Introduction to Set Theory*, Pure and Applied Mathematics, Marcel Dekker, New York, 1984.

[Hua]   Hua Loo Keng, *Introduction to Number Theory*, Springer-Verlag, New York, 1982.

[HuRu]   C. Hughes and Z. Rudnick, *Mock Gaussian behaviour for linear statistics of classical compact groups*, J. Phys. A **36** (2003) 2919–2932.

[Hu]   J. Hull, *Options, Futures, and Other Derivatives*, 5th edition, Prentice-Hall, Englewood Cliffs, NJ, 2002.

[IR]   K. Ireland and M. Rosen, *A Classical Introduction to Modern Number Theory*, Graduate Texts in Mathematics, Vol. 84, Springer-Verlag, New York, 1990.

[Iw] H. Iwaniec, *Topics in Classical Automorphic Forms*, Graduate Studies in Mathematics, Vol. 17, AMS, Providence, RI, 1997.

[IK] H. Iwaniec and E. Kowalski, *Analytic Number Theory*, AMS Colloquium Publications, Vol. 53, AMS, Providence, RI, 2004.

[ILS] H. Iwaniec, W. Luo, and P. Sarnak, *Low lying zeros of families of L-functions*, Inst. Hautes Études Sci. Publ. Math. **91** (2000), 55–131.

[IS1] H. Iwaniec and P. Sarnak, *Dirichlet L-functions at the central point*. Pages 941–952 in *Number Theory in Progress, (Zakopane-Kościelisko, 1997)*, Vol. 2, de Gruyter, Berlin, 1999.

[IS2] H. Iwaniec and P. Sarnak, *The non-vanishing of central values of automorphic L-functions and Landau-Siegel zeros*, Israel J. Math. **120** (2000), 155–177.

[JMRR] D. Jakobson, S. D. Miller, I. Rivin, and Z. Rudnick, *Eigenvalue spacings for regular graphs*. Pages 317–327 in *Emerging Applications of Number Theory (Minneapolis, 1996)*, The IMA Volumes in Mathematics and its Applications, Vol. 109, Springer, New York, 1999.

[J] N. Jacobson, *Basic Algebra I*, 2nd edition, W H Freeman & Co, San Francisco, 1985.

[Je] R. Jeffrey, *Formal Logic: Its Scope and Limits*, McGraw-Hill, New York, 1989.

[Ka] S. Kapnick, *Continued fraction of cubed roots of primes*, Junior Thesis, Princeton University, Fall 2002.

[KS1] N. Katz and P. Sarnak, *Random Matrices, Frobenius Eigenvalues and Monodromy*, AMS Colloquium Publications, Vol. 45, AMS, Providence, RI, 1999.

[KS2] N. Katz and P. Sarnak, *Zeros of zeta functions and symmetries*, Bull. AMS **36** (1999), 1–26.

[KeSn] J. P. Keating and N. C. Snaith, *Random matrices and L-functions*. In *Random Matrix Theory*, J. Phys. A **36** (2003), no. 12, 2859–2881.

[Kel] D. Kelley, *Introduction to Probability*, Macmillan Publishing Company, London, 1994.

[Kh] A. Y. Khinchin, *Continued Fractions*, 3rd edition, University of Chicago Press, Chicago, 1964.

[KSS] D. Kleinbock, N. Shah, and A. Starkov, *Dynamics of subgroup actions on homogeneous spaces of Lie groups and applications to number theory*. Pages 813–930 in *Handbook of Dynamical Systems*, Vol. 1A, North-Holland, Amsterdam, 2002.

[Kn] A. Knapp, *Elliptic Curves*, Princeton University Press, Princeton, NJ, 1992.

[Knu] D. Knuth, *The Art of Computer Programming, Volume 2: Seminumerical Algorithms*, 3rd edition, Addison-Wesley, MA, 1997.

[Kob1] N. Koblitz, *Why study equations over finitess fields?*, Math. Mag. **55** (1982), no. 3, 144–149.

[Kob2] N. Koblitz, *Elliptic curve cryptosystems*, Math. Comp. **48** (1987), no. 177, 203–209.

[Kob3] N. Koblitz, *A survey of number theory and cryptography*. Pages 217-239 in *Number Theory*, Trends in Mathematics, Birkhäuser, Basel, 2000.

[Ko] V. Kolyvagin, *On the Mordell-Weil group and the Shafarevich-Tate group of modular elliptic curves*. Pages 429-436 in *Proceedings of the International Congress of Mathematicians (Kyoto, 1990)*, vols. I and II, Math. Soc. Japan, Tokyo, 1991.

[KonMi] A. Kontorovich and S. J. Miller, *Benford's law, values of L-functions and the $3x + 1$ problem*, preprint.

[KonSi] A. Kontorovich and Ya. G. Sinai, *Structure theorem for (d, g, h)-maps*, Bull. Braz. Math. Soc. (N.S.) 33 (2002), no. 2, 213–224.

[Kor] A. Korselt, *Probléme chinois*, L'intermédiaire math. **6** (1899), 143–143.

[Kos] T. Koshy, *Fibonacci and Lucas Numbers with Applications*, Wiley-Interscience, New York, 2001

[Kua] F. Kuan, *Digit distribution in the continued fraction of $\zeta(n)$*, Junior Thesis, Princeton University, Fall 2002.

[KN] L. Kuipers and H. Niederreiter, *Uniform Distribution of Sequences*, John Wiley & Sons, New York, 1974.

[KR] P. Kurlberg and Z. Rudnick, *The distribution of spacings between quadratic residues*, Duke Math. J. **100** (1999), no. 2, 211–242.

[Ku] R. Kuzmin, *Ob odnoi zadache Gaussa*, Doklady Akad. Nauk, Ser. A (1928), 375–380.

[Lag1] J. Lagarias, *The $3x + 1$ problem and its generalizations*. Pages 305-334 in *Organic mathematics (Burnaby, BC, 1995)*, CMS Conf. Proc., vol. 20, AMS, Providence, RI, 1997.

[Lag2] J. Lagarias, *The 3x+1 problem: An annotated bibliography*, preprint.

[LaSo] J. Lagarias and K. Soundararajan, *Benford's Law for the $3x + 1$ function*, preprint.

[La1] S. Lang, *Diophantine Geometry*, Interscience Publishers, New York, 1962.

[La2] S. Lang, *Introduction to Diophantine Approximations*, Addison-Wesley, Reading, MA, 1966.

[La3] S. Lang, *Undergraduate Algebra*, 2nd edition, Springer-Verlag, New York, 1986.

[La4] S. Lang, *Calculus of Several Variables*, Springer-Verlag, New York, 1987.

[La5] S. Lang, *Undergraduate Analysis*, 2nd edition, Springer-Verlag, New York, 1997.

[La6] S. Lang, *Complex Analysis*, Graduate Texts in Mathematics, Vol. 103, Springer-Verlag, New York, 1999.

[LT] S. Lang and H. Trotter, *Continued fractions for some algebraic numbers*, J. Reine Angew. Math. **255** (1972), 112–134.

[LF] R. Larson and B. Farber, *Elementary Statistics: Picturing the World*, Prentice-Hall, Englewood Cliffs, NJ, 2003.

[LP] R. Laubenbacher and D. Pengelley, *Gauss, Eisenstein, and the "third" proof of the quadratic reciprocity theorem: Ein kleines Schauspiel*, Math. Intelligencer 16 (1994), no. 2, 67–72.

[Law1] J. Law, *Kuzmin's theorem on algebraic numbers*, Junior Thesis, Princeton University, Fall 2002.

[Law2] J. Law, *The circle method on the binary goldbach conjecture*, Junior Thesis, Princeton University, Spring 2003.

[Leh] R. Lehman, *First order spacings of random matrix eigenvalues*, Junior Thesis, Princeton University, Spring 2000.

[LS] H. Lenstra and G. Seroussi, *On hats and other covers*, 2002, preprint.

[Le] P. Lévy, *Sur les lois de probabilité dont dependent les quotients complets et incomplets d'une fraction continue*, Bull. Soc. Math. **57** (1929), 178–194.

[Lidl] R. Lidl, *Mathematical aspects of cryptanalysis*. Pages 86–97 in *Number Theory and Cryptography (Sydney, 1989)*, London Mathematical Society Lecture Note Series, vol. 154, Cambridge University Press, Cambridge, 1990.

[Li] R. Lipshitz, *Numerical results concerning the distribution of* $\{n^2\alpha\}$, Junior Thesis, Princeton University, Spring 2000.

[Liu] Y. Liu, *Statistical behavior of the eigenvalues of random matrices*, Junior Thesis, Princeton University, Spring 2000.

[Mah] K. Mahler, *Arithmetische Eigenschaften einer Klasse von Dezimalbrüchen*, Amsterdam Proc. Konin. Neder. Akad. Wet. **40** (1937), 421–428.

[Ma] E. S. Mahmoodian, *Mathematical Olympiads in Iran*, Vol. I, Sharif University Press, Tehran, Iran, 2002.

[Man] B. Mandelbrot, *The Fractal Geometry of Nature*, W. H. Freeman, New York, 1982.

[Mar] J. Marklof, *Almost modular functions and the distribution of n^2x modulo one*, Int. Math. Res. Not. (2003), no. 39, 2131–2151.

[MaMc] R. Martin and W. McMillen, *An elliptic curve over $\mathbb{Q}$ with rank at least 24*, Number Theory Listserver, May 2000.

[MMS] A. Massey, S. J. Miller, and J. Sinsheimer, *Eigenvalue spacing distribution for the ensemble of real symmetric palindromic Toeplitz matrices*, preprint.

[Maz1] B. Mazur, *Modular curves and the Eisenstein ideal*, IHES Publ. Math. **47** (1977), 33–186.

[Maz2] B. Mazur, *Rational isogenies of prime degree (with an appendix by D. Goldfeld)*, Invent. Math. **44** (1978), no. 2, 129–162.

[Maz3] B. Mazur, *Number Theory as Gadfly*, Amer. Math. Monthly, **98** (1991), 593–610.

[McK] B. McKay, *The expected eigenvalue distribution of a large regular graph*, Linear Algebra Appl. **40** (1981), 203–216.

[MW] B. McKay and N. Wormald, *The degree sequence of a random graph. I. The models*, Random Structures Algorithms **11** (1997), no. 2, 97–117.

[Meh1] M. Mehta, *On the statistical properties of level spacings in nuclear spectra*, Nucl. Phys. **18** (1960), 395–419.

[Meh2] M. Mehta, *Random Matrices*, 2nd edition, Academic Press, Boston, 1991.

[Met] N. Metropolis, *The beginning of the Monte Carlo method*, Los Alamos Science, No. 15, Special Issue (1987), 125–130.

[MU] N. Metropolis and S. Ulam, *The Monte Carlo method*, J. Amer. Statist. Assoc. **44** (1949), 335–341.

[Mic1] M. Michelini, *Independence of the digits of continued fractions*, Junior Thesis, Princeton University, Fall 2002.

[Mic2] M. Michelini, *Kuzmin's extraordinaty zero measure set*, Senior Thesis, Princeton University, Spring 2004.

[Mi1] N. Miller, *Various tendencies of non-Poissonian distributions along subsequences of certain transcendental numbers*, Junior Thesis, Princeton University, Fall 2002.

[Mi2] N. Miller, *Distribution of eigenvalue spacings for band-diagonal matrices*, Junior Thesis, Princeton University, Spring 2003.

[Mill] S. D. Miller, *A simpler way to show $\zeta(3)$ is irrational*, preprint.

[Mil1] S. J. Miller, *1- and 2-level densities for families of elliptic curves: Evidence for the underlying group symmetries*, Compositio Mathematica **140** (2004), no. 4, 952–992.

[Mil2] S. J. Miller, *Density functions for families of Dirichlet characters*, preprint.

[Mil3] S. J. Miller, *The arithmetic mean and geometric inequality*, Class Notes from Math 187/487, The Ohio State University, Fall 2003.

[Mil4] S. J. Miller, *Differentiating identities*, Class Notes from Math 162: Statistics, Brown University, Spring 2005.

[Mil5] S. J. Miller, *The Pythagorean won-loss formula in baseball*, preprint.

[Mil6] S. J. Miller, *Investigations of zeros near the central point of elliptic curve L-functions*, to appear in Experimental Mathematics.

[Mil7] S. J. Miller, *Die battles and order statistics*, Class Notes from Math 162: Statistics, Brown University, Spring 2006.

[MN] S. J. Miller and M. Nigrini, *Differences of independent variables and almost Benford behavior*, preprint.

[M] V. Miller, *Use of elliptic curves in cryptography*. Pages 417–426 in *Advances in cryptology – CRYPTO '85 (Santa Barbara, CA, 1985)*, Lecture Notes in Computer Science, Vol. 218, Springer-Verlag, Berlin, 1986.

[Milne] J. S. Milne, *Elliptic Curves*, course notes.

[Min] S. Minteer, *Analysis of Benford's law applied to the $3x+1$ problem*, Number Theory Working Group, The Ohio State University, 2004.

[Mon1] H. Montgomery, *Primes in arithmetic progression*, Michigan Math. J. **17** (1970), 33–39.

[Mon2] H. Montgomery, *The pair correlation of zeros of the zeta function*. Pages 181–193 in *Analytic Number Theory*, Proceedings of Symposia in Pure Mathematics, vol. 24, AMS, Providence, RI, 1973.

[MoMc] D. Moore and G. McCabe, *Introduction to the Practice of Statistics*, W. H. Freeman and Co., London, 2003.

[MS] H. Montgomery and K. Soundararajan, *Beyond pair correlation*. Pages 507–514 in *Paul Erdös and His Mathematics, I (Budapest, 1999)*, Bolyai Society Mathematical Studies, Vol. 11, János Bolyai Math. Soc., Budapest, 2002.

[Moz1] C. J. Mozzochi, *An analytic sufficiency condition for Goldbach's conjecture with minimal redundancy*, Kyungpook Math. J. **20** (1980), no. 1, 1–9.

[Moz2] C. J. Mozzochi, *The Fermat Diary*, AMS, Providence, RI, 2000.

[Moz3] C. J. Mozzochi, *The Fermat Proof*, Trafford Publishing, Victoria, 2004.

[Mu1] R. Murty, *Primes in certain arithmetic progressions*, Journal of the Madras University, (1988), 161–169.

[Mu2] R. Murty, *Problems in Analytic Number Theory*, Springer-Verlag, New York, 2001.

[MM] M. R. Murty and V. K. Murty, *Non-Vanishing of L-Functions and Applications*, Progress in Mathematics, vol. 157, Birkhäuser, Basel, 1997.

[NS] K. Nagasaka and J. S. Shiue, *Benford's law for linear recurrence sequences*, Tsukuba J. Math. **11** (1987), 341–351.

[Nar] W. Narkiewicz, *The Development of Prime Number Theory*, Springer Monographs in Mathematics, Springer-Verlag, New York, 2000.

[Na] M. Nathanson, *Additive Number Theory: The Classical Bases*, Graduate Texts in Mathematics, Springer-Verlag, New York, 1996.

[NT] J. von Neumann and B. Tuckerman, *Continued fraction expansion of $2^{1/3}$*, Math. Tables Aids Comput. **9** (1955), 23–24.

[Ni1] T. Nicely, *The pentium bug*, http://www.trnicely.net/pentbug/pentbug.html

[Ni2] T. Nicely, *Enumeration to 10^{14} of the Twin Primes and Brun's Constant*, Virginia J. Sci. **46** (1996), 195–204.

[Nig1] M. Nigrini, *Digital Analysis and the Reduction of Auditor Litigation Risk.* Pages 69–81 in *Proceedings of the 1996 Deloitte & Touche / University of Kansas Symposium on Auditing Problems*, ed. M. Ettredge, University of Kansas, Lawrence, KS, 1996.

[Nig2] M. Nigrini, *The Use of Benford's Law as an Aid in Analytical Procedures*, Auditing: A Journal of Practice & Theory, **16** (1997), no. 2, 52–67.

[NZM] I. Niven, H. Zuckerman, and H. Montgomery, *An Introduction to the Theory of Numbers*, 5th edition, John Wiley & Sons, New York, 1991.

[Nov] T. Novikoff, *Asymptotic behavior of the random 3-regular bipartite graph*, Undergraduate Mathematics Laboratory report, Courant Institute, NYU, 2002.

[Od1] A. Odlyzko, *On the distribution of spacings between zeros of the zeta function*, Math. Comp. **48** (1987), no. 177, 273–308.

[Od2] A. Odlyzko, *The 10^{22}-nd zero of the Riemann zeta function.* Pages 139–144 in *Procreedings of the Conference on Dynamical, Spectral and Arithmetic Zeta Functions*, ed. M. van Frankenhuysen and M. L. Lapidus, Contemporary Mathematics Series, AMS, Providence, RI, 2001.

[Ok] T. Okano, *A note on the transcendental continued fractions*, Tokyo J. Math, **10** (1987), no. 1, 151–156.

[Ol] T. Oliveira e Silva, *Verification of the Goldbach conjecture up to* $6 \cdot 10^{16}$, NMBRTHRY@listserv.nodak.edu mailing list, Oct. 3, 2003, http://listserv.nodak.edu/scripts/wa.exe?A2=ind0310&L=nmbrthry&P=168 and http://www.ieeta.pt/~tos/goldbach.html

[Ols] L. Olsen, *Extremely non-normal continued fractions*, Acta Arith. **108** (2003), no. 2, 191–202.

[vdP1] A. van der Poorten, *An introduction to continued fractions.* Pages 99-138 in *Diophantine Analysis (Kensington, 1985)*, London Mathematical Society Lecture Note Series, Vol. 109, Cambridge University Press, Cambridge, 1986.

[vdP2] A. van der Poorten, *Notes on continued fractions and recurrence sequences.* Pages 86–97 in *Number theory and cryptography (Sydney, 1989)*, London Mathematical Society Lecture Note Series, Vol. 154, Cambridge University Press, Cambridge, 1990.

[vdP3] A. van der Poorten, *Continued fractions of formal power series.* Pages 453–466 in *Advances in Number Theory (Kingston, ON, 1991)*, Oxford Science Publications, Oxford University Press, New York, 1993.

[vdP4] A. van der Poorten, *Fractions of the period of the continued fraction expansion of quadratic integers*, Bull. Austral. Math. Soc. **44** (1991), no. 1, 155–169.

[vdP5] A. van der Poorten, *Continued fraction expansions of values of the exponential function and related fun with continued fractions*, Nieuw Arch. Wisk. (4) **14** (1996), no. 2, 221–230.

[vdP6] A. van der Poorten, *Notes on Fermat's Last Theorem*, Canadian Mathematical Society Series of Monographs and Advanced Texts, Wiley-Interscience, New York, 1996.

[PS1] A. van der Poorten and J. Shallit, *Folded continued fractions*, J. Number Theory **40** (1992), no. 2, 237–250.

[PS2] A. van der Poorten and J. Shallit, *A specialised continued fraction*, Canad. J. Math. **45** (1993), no. 5, 1067–1079.

[Po] C. Porter (editor), *Statistical Theories of Spectra: Fluctuations*, Academic Press, New York, 1965.

[Py] R. Pyke, *Spacings*, J. Roy. Statist. Soc. Ser. B **27** (1965), 395–449.

[QS1] R. Qian and D. Steinhauer, *Rational relation conjectures*, Junior Thesis, Princeton University, Fall 2003.

[QS2] R. Qian and D. Steinhauer, *Eigenvalues of weighted random graphs*, Junior Thesis, Princeton University, Spring 2003.

[Rai] R. A. Raimi, *The first digit problem*, Amer. Math. Monthly **83** (1976), no. 7, 521–538.

[Ra] K. Ramachandra, *Lectures on Transcendental Numbers*, Ramanujan Institute, Madras, 1969.

[Re] F. Reif, *Fundamentals of Statistical and Thermal Physics*, McGraw-Hill, New York, 1965.

[Ric] P. Richter, *An investigation of expanders and ramanujan graphs along random walks of cubic bipartite graphs*, Junior Thesis, Princeton University, Spring 2001.

[RDM] R. D. Richtmyer, M. Devaney, and N. Metropolis, *Continued fraction of algebraic numbers*, Numer. Math. **4** (1962), 68–84.

[Ri] G. F. B. Riemann, *Über die Anzahl der Primzahlen unter einer gegebenen Grösse*, Monatsber. Königl. Preuss. Akad. Wiss. Berlin, Nov. 1859, 671–680 (see [Ed] for an English translation).

[RSA] R. Rivest, A. Shamir, and L. Adleman, *A method for obtaining digital signatures and public key cryptosystems*, Comm. ACM **21** (1978), 120–126.

[Roc] D. Rockmore, *Stalking the Riemann Hypothesis: The Quest to Find the Hidden Law of Prime Numbers*, Pantheon, New York, 2005.

[Ro] K. Roth, *Rational approximations to algebraic numbers*, Mathematika **2** (1955), 1–20.

[Rub1] M. Rubinstein, *A simple heuristic proof of Hardy and Littlewood's conjecture B*, Amer. Math. Monthly **100** (1993), no. 5, 456–460.

[Rub2] M. Rubinstein, *Low-lying zeros of L-functions and random matrix theory*, Duke Math. J. **109** (2001), no. 1, 147–181.

[RubSa] M. Rubinstein and P. Sarnak, *Chebyshev's bias*, Experiment. Math. **3** (1994), no. 3, 173–197.

[Rud] W. Rudin, *Principles of Mathematical Analysis*, 3rd edition, International Series in Pure and Applied Mathematics, McGraw-Hill, New York, 1976.

[RS] Z. Rudnick and P. Sarnak, *Zeros of principal L-functions and random matrix theory*, Duke J. of Math. **81** (1996), 269–322.

[RS2] Z. Rudnick and P. Sarnak, *The pair correlation function of fractional parts of polynomials*, Comm. Math. Phys. **194** (1998), no. 1, 61–70.

[RSZ] Z. Rudnick, P. Sarnak, and A. Zaharescu, *The distribution of spacings between the fractional parts of* $n^2\alpha$, Invent. Math. **145** (2001), no. 1, 37–57.

[RZ1] Z. Rudnick and A. Zaharescu, *A metric result on the pair correlation of fractional parts of sequences*, Acta Arith. **89** (1999), no. 3, 283–293.

[RZ2] Z. Rudnick and A. Zaharescu, *The distribution of spacings between fractional parts of lacunary sequences*, Forum Math. **14** (2002), no. 5, 691–712.

[Sar] P. Sarnak *Some applications of modular forms*, Cambridge Trusts in Mathemetics, Vol. 99, Cambridge University Press, Cambridge, 1990.

[Sch] D. Schmidt, *Prime Spacing and the Hardy-Littlewood Conjecture B*, Junior Thesis, Princeton University, Spring 2001.

[Sc] P. Schumer, *Mathematical Journeys*, Wiley-Interscience, John Wiley & Sons, New York, 2004.

[Se] J. P. Serre, *A Course in Arithmetic*, Springer-Verlag, New York, 1996.

[Sh] A. Shidlovskii, *Transcendental Numbers*, Walter de Gruyter & Co., New York, 1989.

[ShTa] J. A. Shohat and J. D. Tamarkin, *The Problem of Moments*, AMS, Providence, RI, 1943.

[Sil1] J. Silverman, *The Arithmetic of Elliptic Curves*, Graduate Texts in Mathematics, Vol. 106, Springer-Verlag, New York, 1986.

[Sil2] J. Silverman, *A Friendly Introduction to Number Theory*, 2nd edition, Prentice-Hall, Englewood Cliffs, NJ, 2001.

[ST] J. Silverman and J. Tate, *Rational Points on Elliptic Curves*, Springer-Verlag, New York, 1992.

[Si] B. Simon, *The classical moment problem as a self-adjoint finite difference operator*, Adv. Math. **137** (1998), no. 1, 82–203.

[SM] S. Simon and A. Moustakas, *Eigenvalue density of correlated complex random Wishart matrices*, Bell Labs Technical Memo, 2004.

[Sk] S. Skewes, *On the difference* $\pi(x) - \mathrm{Li}(x)$, J. London Math. Soc. **8** (1933), 277–283.

[Sl] N. Sloane, *On-Line Encyclopedia of Integer Sequences*, http://www.research.att.com/~njas/sequences/Seis.html

[Sn] N. Snaith, *Derivatives of random matrix characteristic polynomials with applications to elliptic curves*, preprint.

[SS1] E. Stein and R. Shakarchi, *Fourier Analysis: An Introduction*, Princeton University Press, Princeton, NJ, 2003.

[SS2] E. Stein and R. Shakarchi, *Complex Analysis*, Princeton University Press, Princeton, NJ, 2003.

[SS3] E. Stein and R. Shakarchi, *Real Analysis: Measure Theory, Integration, and Hilbert Spaces*, Princeton University Press, Princeton, NJ, 2005.

[StTa] I. Stewart and D. Tall, *Algebraic Number Theory*, 2nd edition, Chapman & Hall, London, 1987.

[St] Strang, *Linear Algebra and Its Applications*, 3rd edition, Wellesley-Cambridge Press, Wellesley, MA 1998.

[Str] K. Stromberg, *The Banach-Tarski paradox*, Amer. Math. Monthly **86** (1979), no. 3, 151–161.

[Sz] P. Szüsz, *On the length of continued fractions representing a rational number with given denominator*, Acta Arithmetica **37** (1980), 55–59.

[Ta] C. Taylor, *The Gamma function and Kuzmin's theorem*, Junior Thesis, Princeton University, Fall 2002.

[TW] R. Taylor and A. Wiles, *Ring-theoretic properties of certain Hecke algebras*, Ann. Math. **141** (1995), 553–572.

[TrWi] C. Tracy and H. Widom, *Correlation functions, cluster functions, and spacing distributions for random matrices*, J. Statist. Phys., **92** (1998), no. 5–6, 809–835.

[Te] G. Tenenbaum, *Introduction to Analytic and Probabilistic Number Theory*, Cambridge University Press, Cambridge, 1995.

[Ti] E. C. Titchmarsh, *The Theory of the Riemann Zeta-function*, revised by D. R. Heath-Brown, Oxford University Press, Oxford, 1986.

[Va] R. C. Vaughan, *On a variance associated with the distribution of primes in arithmetic progression*, Proc. London Math. Soc. (3) **82** (2001), 533–553.

[VW] R. C. Vaughan and T. D. Wooley, *Waring's problem: a survey*. Pages 301–340 in *Number Theory for the Millennium, III (Urbana, IL, 2000)*, A. K. Peters, Natick, MA, 2002.

[Vin1] I. Vinogradov, *Representation of an odd number as the sum of three primes*, Doklady Akad. Nauk SSSR, **15** (1937), no. 6–7, 291–294.

[Vin2] I. Vinogradov, *Some theorems concerning the theory of primes*, Mat. Sbornik, **2** (1937), no. 44, 179–195.

[Vo] A. Voros, *A sharpening of Li's criterion for the Riemann hypothesis*, preprint.

[VG] W. Voxman and R. Goetschel, Jr., *Advanced Calculus*, Mercer Dekker, New York, 1981.

[Wa] L. Washington, *Elliptic Curves: Number Theory and Cryptography*, Chapman & Hall / CRC, New York, 2003.

[Wed] S. Wedeniwski, *ZetaGrid*, http://www.zetagrid.net

[Wei1] A. Weil, *Numbers of Solutions of Equations in Finite Fields*, Bull. Amer. Math. Soc. **14** (1949), 497–508.

[Wei2] A. Weil, *Prehistory of the zeta-function*. Pages 1–9 in *Number Theory, Trace Formulas and Discrete Groups (Oslo, 1987)*, Academic Press, Boston, 1989.

[Weir] B. Weir, *The local behavior of Germain primes*, Undergraduate Mathematics Laboratory report, Courant Institute, NYU, 2002.

[We] E. Weisstein, *MathWorld – A Wolfram Web Resource*, http://mathworld.wolfram.com

[Weyl] H. Weyl, *The Classical Groups: Their Invariants and Representations*, Princeton University Press, Princeton, NJ, 1946.

[Wh] E. Whittaker, *A Treatise on the Analytical Dynamics of Particles and Rigid Bodies: With an Introduction to the Problem of Three Bodies*, Dover, New York, 1944.

[WW] E. Whittaker and G. Watson, *A Course of Modern Analysis*, 4th edition, Cambridge University Press, Cambridge, 1996.

[Wig1] E. Wigner, *On the statistical distribution of the widths and spacings of nuclear resonance levels*, Proc. Cambridge Philo. Soc. **47** (1951), 790–798.

[Wig2] E. Wigner, *Characteristic vectors of bordered matrices with infinite dimensions*, Ann. of Math. **2** (1955), no. 62, 548–564.

[Wig3] E. Wigner, *Statistical Properties of real symmetric matrices*. Pages 174–184 in *Canadian Mathematical Congress Proceedings*, University of Toronto Press, Toronto, 1957.

[Wig4] E. Wigner, *Characteristic vectors of bordered matrices with infinite dimensions. II*, Ann. of Math. Ser. 2 **65** (1957), 203–207.

[Wig5] E. Wigner, *On the distribution of the roots of certain symmetric matrices*, Ann. of Math. Ser. 2 **67** (1958), 325–327.

[Wi] A. Wiles, *Modular elliptic curves and Fermat's last theorem*, Ann. Math. **141** (1995), 443–551.

[Wilf] H. Wilf, *Algorithms and Complexity*, 2nd edition, A. K. Peters, Natick, MA, 2002.

[Wir] E. Wirsing, *On the theorem of Gauss-Kuzmin-Lévy and a Frobenius-type theorem for function spaces*, Acta Arith. **24** (1974) 507–528.

[Wis] J. Wishart, *The generalized product moment distribution in samples from a normal multivariate population*, Biometrika **20 A** (1928), 32–52.

[Wor] N. C. Wormald, *Models of random regular graphs*. Pages 239–298 in *Surveys in combinatorics, 1999 (Canterbury)* London Mathematical Society Lecture Note Series, vol. 267, Cambridge University Press, Cambridge, 1999.

[Wo] T. Wooley, *Large improvements in Waring's problem*, Ann. of Math. (2), **135** (1992), no. 1, 131–164.

[Za] I. Zakharevich, *A generalization of Wigner's law*, preprint.

[Zu] W. Zudilin, *One of the numbers $\zeta(5), \zeta(7), \zeta(9), \zeta(11)$ is irrational*, Uspekhi Mat. Nauk **56** (2001), 149-150.

[Zy] A. Zygmund, *Trigonometrical Series*, vols. I and II, Cambridge University Press, Cambridge, 1968.

Index